Advances in Delays and Dynamics

Volume 4

Series editor

Silviu-Iulian Niculescu, Laboratory of Signals and Systems, Gif-sur-Yvette, France
e-mail: Silviu.Niculescu@lss.supelec.fr

About this Series

Delay systems are largely encountered in modeling propagation and transportation phenomena, population dynamics, and representing interactions between interconnected dynamics through material, energy, and communication flows. Thought as an open library on delays and dynamics, this series is devoted to publish basic and advanced textbooks, explorative research monographs as well as proceedings volumes focusing on delays from modeling to analysis, optimization, control with a particular emphasis on applications spanning biology, ecology, economy, and engineering. Topics covering interactions between delays and modeling (from engineering to biology and economic sciences), control strategies (including also control structure and robustness issues), optimization, and computation (including also numerical approaches and related algorithms) by creating links and bridges between fields and areas in a delay setting are particularly encouraged.

More information about this series at http://www.springer.com/series/11914

Iasson Karafyllis · Michael Malisoff
Frederic Mazenc · Pierdomenico Pepe
Editors

Recent Results on Nonlinear Delay Control Systems

In honor of Miroslav Krstic

Springer

Editors
Iasson Karafyllis
Department of Mathematics
National Technical University of Athens
Athens
Greece

Michael Malisoff
Department of Mathematics
Louisiana State University (LSU)
Baton Rouge, LA
USA

Frederic Mazenc
EPI Inria DISCO, Laboratory of Signals and Systems (L2S, UMR CNRS 8506)
CNRS-CentraleSupélec-Université Paris Sud
Gif-sur-Yvette
France

Pierdomenico Pepe
Department of Information Engineering, Computer Science and Mathematics
University of L'Aquila
L'Aquila
Italy

ISSN 2197-117X ISSN 2197-1161 (electronic)
Advances in Delays and Dynamics
ISBN 978-3-319-36760-6 ISBN 978-3-319-18072-4 (eBook)
DOI 10.1007/978-3-319-18072-4

Springer Cham Heidelberg New York Dordrecht London

Softcover reprint of the hardcover 1st edition 2016

Printed on acid-free paper

Springer International Publishing AG Switzerland is part of Springer Science+Business Media (www.springer.com)

We dedicate this book to Prof. Miroslav Krstic on his 50th birthday, to help honor him for his many important contributions to control theory and its applications.

The Editors

Foreword

The past 5 years have seen many significant advances in the area of control theory for time-delay systems, including constructive methods for transforming Lyapunov functions into the Lyapunov-Krasovskii functionals that are convenient for establishing stability of nonlinear time-delay systems, controllers for hybrid systems that involve continuous time subsystems and rules for switching between the subsystems, predictive results that can compensate arbitrarily long input delays, and much more. Much of this theory is now being applied in key engineering applications, including fluid dynamics, oil production, and rehabilitation methods for patients with mobility disorders. As the editor for the *Advances in Delays and Dynamics* series, it is a pleasure to present this new volume, which covers many of these advances. The volume includes papers by many leaders in the controls field, including two papers by Miroslav Krstic to whom this volume is dedicated. The collection was meticulously edited by four prominent specialists in the field of delay systems whose research is also included in this volume. I hope you enjoy reading this volume as much as I appreciate the opportunity to present this excellent volume to you.

Gif-sur-Yvette, France
June 2015

Silviu-Iulian Niculescu

Preface

Time-delay systems occur in many important engineering applications. When the delays are small, they can sometimes be safely ignored, which reduces control problems to more standard problems that can be covered by traditional frequency-domain or well-known Lyapunov function techniques for systems free of delays. However, many engineering systems contain large time delays. For instance, large input delays often arise from communication delays between sensors and actuators, or from time-consuming information gathering. Using standard controls on nonstandard systems such as delay systems can yield poor control performance, and uncompensated input delays can produce hazards when used in engineering applications. This puts time-delay systems outside of the scope of standard methods that are commonly used for systems that have no delays, and it necessitates developing more reliable methods that can avoid the pitfalls of using traditional controllers on more complicated time-delay systems.

This volume collects some recent advances in the area of time-delay systems, with special emphasis on constructive generalized Lyapunov approaches that can certify key stability properties such as uniform global asymptotic stability, and predictive methods that can compensate arbitrarily long input delays. In this preface, we summarize the main contributions of the chapters to put the contributions in context. In the first chapter, we present an overview of recent advances in delay compensation for nonlinear time-delay systems, with emphasis on Lyapunov-Krasovskii functionals, and on tests for robustness of nonlinear controllers with respect to delays in the input. The second chapter is by Ahmed-Ali, Karafyllis, Krstic, and Lamnabhi-Lagarrigue and covers robustness of time delays systems with respect to measurement and modeling uncertainties, including systems with outputs and zero-order hold.

Backstepping is an important technique for generating controllers for nonlinear systems by building up from controllers for their subsystems. In the third chapter, Bekiaris-Liberis, Jankovic, and Krstic present a generalization of backstepping for systems in strict feedback form, where the delays may be nonconstant. Although linear systems are generally much easier to control than nonlinear systems, there are still significant challenges for linear systems with time-varying coefficients. The fourth chapter by Bresch-Pietri and Petit addresses some of these remaining challenges for linear systems, using a predictive approach. The next chapter is by Cacace, Germani, and Manes and it deals with state estimation under measurement delay, through the use of a chain-type high gain observer. Then Califano and Moog provide a chapter on new results on normal forms for time-delay systems, including novel necessary and sufficient conditions for a time-delay system to be decomposable into observable and nonobservable subsystems. The next chapter is by Downey, Kamalapurkar, Fischer, and Dixon and is a change of pace, focusing on an application of predictor-based delay compensation to neuromuscular electrical stimulation, which is a developing technology that can help restore movement in certain patients with mobility disorders. Homogeneity is an important property of certain nonlinear systems that can significantly facilitate the design and analysis of controllers, and the chapter by Efimov, Perruquetti, and Richard in this volume proposes a novel extension of homogeneity methods to time-delay systems based on a Lyapunov-Razumikhin approach.

In the following chapter, authors Han, Fridman, and Spurgeon present a sliding mode observer that is based on linear matrix inequalities and provides ultimate boundedness of the observation error. Neutral systems with delay play an important role in distributed networks, heat exchangers, and population models, and are characterized by having time delays entering the time derivatives of the states. In the following chapter, authors Mazenc and Ito present a novel Lyapunov-Krasovskii functional method for proving stability properties for large-scale neutral systems, where the delays occur in both the interconnection channels and the subsystems themselves. Small gain theory is another powerful method for establishing stability of large-scale systems, based on studying growth conditions on the compositions of the comparison functions in the stability estimates for the individual subsystems. In their chapter on input-to-state stability and switched systems, authors Jiang, Lin and Wang combine small gain arguments with Razumikhin function methods to establish conditions under which a switched system consisting of input-to-state stable subsystems is itself input-to-state stable.

Returning to the important themes of Lyapunov-Krasovskii methods and neutral systems, the next chapter by Karafyllis and Pepe presents converse Lyapunov functional methods for neutral systems that generalize existing results by allowing uncertainties, delays, and more general difference operators that were not already covered in the literature. Although converse Lyapunov theory is largely nonconstructive, it has had significant theoretical ramifications for systems free of delays, so time delay converse theory has the potential for wide use across control theory. The next chapter by Liu and Teel discusses several recent efforts to extend hybrid systems theory to time-delay systems with uncertainties, by covering foundational issues such as well-posedness of solutions.

The next chapter is by Novella-Rodriguez, Witrant, and Sename, and it extends some time-delay systems notions to hyperbolic systems of PDEs, by considering transport in fluid pipes. Returning to the predictor approach, the chapter by Oguchi provides an analog of finite spectrum assignment for time-delay systems, which leads to a predictive approach that does not require numerical integration. Input constraints are another key challenge for delay compensating controllers, because saturating the control values in a stabilizing controller can lead to a failure of the control to ensure the required stability properties. In their chapter, authors Seuret, Gouaisbaut, Tarbouriech, and Gomes da Silva address input saturation problems in the context of delayed sampled data systems, by studying a convex optimization problem. The following chapter by Verriest studies a state reconstruction problem, whose goal is to find values of the state using current or delayed values of an output, including cases where the delay is not required to be constant.

Continuing with the theme of nonconstant delays but returning to the theme of sliding mode controls, the next chapter by Yan, Spurgeon, and Orlov uses Lyapunov-Razumikhin methods to provide a sliding mode observer for nonlinear systems with time-varying delays, assuming a time invariant output matrix and general conditions on the nonlinear parts of the systems. Maintaining the theme of time-varying delays, the following chapter by Yoon, Anantachaisilp, and Lin uses linear matrix inequalities to stabilize linear systems under input saturations and time-varying delays, including an application to an experimental test rig for active magnetic bearings. In the final chapter, Zheng and Richard address the problem of identifying input delays, using the theory of noncommutative rings, and they apply their methods to establish causal and noncausal observability for nonlinear time-delay systems with unknown inputs.

We dedicate this volume to Prof. Miroslav Krstic, on the occasion of his 50th birthday. Through his many excellent presentations and publications including 10 books and his helpful discussions, Miroslav has inspired each of us to work on challenging control problems that can impact several parts of engineering. Infinite thanks go to Dr. Silviu-Iulian Niculescu, our longstanding mentor, for his continuous advice and precious guidance. This volume would not have been possible without the help of the external referees who reviewed each chapter, and the technical support we received from Thomas Ditzinger and his colleagues at Springer, so we thank the reviewers and everyone at Springer for their help and for the opportunity to publish this work. We hope you enjoy reading this volume as much as we appreciated the opportunity to edit it.

Athens, Greece
Baton Rouge, LA, USA
Gif-sur-Yvette, France
L'Aquila, Italy
January 2015

Iasson Karafyllis
Michael Malisoff
Frederic Mazenc
Pierdomenico Pepe

Contents

Stabilization of Nonlinear Delay Systems: A Tutorial on Recent Results

Iasson Karafyllis, Michael Malisoff, Frederic Mazenc and Pierdomenico Pepe

Abstract Stabilization of nonlinear systems under delays is a central and challenging problem in control theory. It is also of considerable interest in engineering, because delay systems are prevalent in aerospace, biological, marine robotic, network control, and many other applications. Input delays naturally arise due to transport phenomena, time consuming information processing, and sensor designs, and they can produce complicated systems that are beyond the scope of standard frequency-domain or Lyapunov function methods. This has led to large control theoretic and engineering literatures on stabilization problems, spanning more than 40 years, based on backstepping, Lyapunov-Krasovskii functionals, prediction, and sampling controllers. In addition to input delays, there may also be state delays in the vector fields that define the system. This tutorial summarizes some recent work on stabilization under input or state delays and suggests future research directions.

Michael Malisoff: Supported by NSF grant 1102348.

I. Karafyllis (✉)
Department of Mathematics, National Technical University of Athens,
Heroon Polytechneiou 9, 15780 Athens, Greece
e-mail: iasonkar@central.ntua.gr

M. Malisoff
Department of Mathematics, 303 Lockett Hall, Louisiana State University,
Baton Rouge, LA 70803-4918, USA
e-mail: malisoff@lsu.edu

F. Mazenc
EPI DISCO INRIA-Saclay, L2S, CNRS-Supelec,
3 Rue Joliot Curie, 91192 Gif-sur-Yvette, France
e-mail: Frederic.Mazenc@lss.supelec.fr

P. Pepe
Department of Information Engineering, Computer Science, and Mathematics,
University of L'Aquila, Via Gronchi 18, 67100 L'aquila, Italy
e-mail: pierdomenico.pepe@univaq.it

I. Karafyllis et al. (eds.), *Recent Results on Nonlinear Delay Control Systems*,
Advances in Delays and Dynamics 4, DOI 10.1007/978-3-319-18072-4_1

1 General Introduction

Stabilization and robustness analysis for nonlinear control systems with input or state delays is a challenging problem that has been addressed by many authors in a large number of works. In some applications, when the delays are small enough that one can ignore them and still be assured of satisfactory performance of the controllers. However, there are many applications where the delays are too large to disregard. For instance, in bioreactors, there is often a long time lag between the time the organisms in the bioreactor are fed and the time the organisms grow. One can often model bioreactors by control systems, where the food entering the bioreactor is the control input and the numbers of organisms in the reactor are the states of the system, and then we have a control system with long input delays.

There are two important streams of research on stabilization under input delays. The first is sometimes called the *emulation approach*, and is based on a two step process. In the first step, one disregards the delays in the feedback design problem to produce a closed loop system that satisfies the desired stabilization goal when the input delays are set to zero. In the next step, one studies the potential effects of input delays on the closed loop system. This produces a feedback design problem that does not require time delay control design theory, but in general it can only compensate small input delays. The second stream involves augmenting the dynamics by a new system whose states generate the control. This is usually called *model reduction* or *prediction*, and can compensate arbitrarily large delay. However, using more complex predictive controls can make it harder to quantify the effects of uncertainty, and it may be difficult to find the required predictor map in closed form.

Studying systems with state delays produces special challenges that often do not arise when the delays only occur in the inputs. When the system has a state delay, the state entering the right side of the open loop system takes its values in an infinite dimensional set of functions, and then the usual notions of control-Lyapunov functions do not apply. For systems with either input or state delays, it is also natural to search for generalizations of backstepping, where the controller is found in a recursive way by finding controls for lower dimensional subsystems and then incorporating the other subsystems in a sequential way to build the control for the overall system. Finally, one would hope to find generalizations of the universal control formulas for undelayed control affine systems, where the feedback control is expressed in terms of Lie derivatives of a control-Lyapunov function for the system in the directions of the vector fields that define the system. For systems with state delays, such generalizations should use analogs of Lie derivatives that apply in infinite dimensional spaces, and should also ensure robustness to additive uncertainties on the controls in terms of a suitable state delayed analogs of input-to-state stability.

This tutorial addresses many of the preceding issues, by discussing several of the authors' papers and their relationship to alternative approaches such as techniques based on Razumikhin functions. Although this tutorial focuses on theoretical aspects and only provides sketches of the proofs of the theoretical results, we refer the reader to our extensive bibliography at the end of this survey, which includes many papers involving control applications and complete proofs of all results that follow.

2 Basic Definitions and Notation

The following notation and definitions will be used throughout this chapter. In all of what follows, the Euclidean space dimensions are arbitrary, unless otherwise noted. We let $\mathbb{R}^\star$ denote the extended real line $[-\infty, \infty]$. A nonnegative valued function is called positive definite provided it takes the value 0 at the origin 0 in its domain but is positive at all other points in its domain. Let $\mathcal{K}$ denote the set of all continuous and strictly increasing functions $\rho : [0, \infty) \to [0, \infty)$ that satisfy $\rho(0) = 0$. We use $\mathcal{K}_\infty$ to denote the class of all unbounded functions of class $\mathcal{K}$. Let $\mathcal{KL}$ denote the set of all continuous functions $\beta : [0, \infty) \times [0, \infty) \to [0, \infty)$ satisfying (1) $\beta(\cdot, t) \in \mathcal{K}_\infty$ for each $t \geq 0$, (2) $\beta(s, \cdot)$ is non-increasing for each $s \geq 0$, and (3) $\beta(s, t) \to 0$ as $t \to \infty$ for each $s \geq 0$, where we use ∞ to mean $+\infty$. A function $v : [0, \infty) \to \mathbb{R}^m$ is called essentially bounded provided $\text{essup}_{t \geq 0}|v(t)| < \infty$, where essup denotes the essential supremum. For each measurable function $\varphi : \mathcal{I} \to \mathbb{R}^m$, defined on an interval $\mathcal{I}$, we let $|\varphi|_{\mathcal{I}}$ denote its (essential) supremum. Then for values T_1 and T_2 such that $0 \leq T_1 < T_2$ and $[T_1, T_2] \subseteq \mathcal{I}$, we let $\varphi_{[T_1,T_2)} : [0, \infty) \to \mathbb{R}^m$ denote the function given by $\varphi_{[T_1,T_2)}(t) = v(t)$ for all $t \in [T_1, T_2)$ and $\varphi_{[T_1,T_2)}(t) = 0$ elsewhere. Let $|\cdot|$ denote the usual Euclidean norm (or the induced matrix norm), and $\mathcal{C}_n(\mathcal{I})$ denote the set of all continuous functions defined on a given interval $\mathcal{I}$ and valued in $\mathbb{R}^n$. We use a.e. to mean almost everywhere in terms of Lebesgue measure. We use $\mathcal{B}_\eta$ to denote the open ball in Euclidean space centered at 0 of radius η, where the dimension will be clear from the context.

We use $\mathcal{C}$ to denote the space of the continuous functions mapping $[-\Delta, 0]$ into $\mathbb{R}^n$, when the choice of the constant $\Delta > 0$ will be clear from the context. Also, $\mathcal{R}$ denotes the space of all bounded right-continuous functions mapping $[-\Delta, 0)$ into $\mathbb{R}^n$ that are continuous except at a finite number of points; and $||u||$ denotes the sup norm of any function u on its entire domain. A function $f : \mathcal{C} \to \mathbb{R}^{n\times m}$ is called completely continuous if it is continuous and maps closed bounded subsets of $\mathcal{C}$ into bounded subsets of $\mathbb{R}^{n\times m}$; it is called completely locally Lipschitz provided for each $\varphi \in \mathcal{C}$, there exist positive real numbers δ and η such that, for any φ_1 and φ_2 in $I_\delta(\varphi)$, the inequality $|f(\varphi_1) - f(\varphi_2)| \leq \eta\|\varphi_1 - \varphi_2\|$ holds. Here $I_\delta(\varphi) = \{\psi \in \mathcal{C} : \|\psi - \varphi\| \leq \delta\}$. We also use the standard little o notation, where $o(s)$ denotes any function $o : [0, \infty) \to [0, \infty)$ such that $\lim_{s\to\infty}(o(s)/s) = 0$.

By radial unboundedness of a function $V : \mathbb{R}^n \to [0, \infty)$, we mean that $V(x) \to \infty$ as $|x| \to \infty$. Also, $L^\infty_{loc}(A; B)$ (resp., $C^k(A; B)$) is the set of all locally essentially bounded measurable functions (resp., all functions whose partial derivatives up through order k are continuous, where C^0 corresponds to continuity) whose domains are some set A and whose ranges are some set B. Similarly, $L^\infty(A; B)$ is the set of essentially bounded measurable functions $\varphi : A \to B$.

We let $\mathbb{Z}_{\geq 0}$ denote the set of all nonnegative integers, and $[a]$ denote the least integer $j \geq a$ for each $a \in \mathbb{R}$.

For a given $t \geq 0$, we set $x_t(\theta) = x(t + \theta)$ for all $\theta \in [-\Delta, 0]$, where the maximal allowable delay Δ will again be clear from the context. By uniform global asymptotic stability (UGAS) of a system $\dot{x} = \mathscr{F}(t, x_t)$ whose state space is all of $\mathbb{R}^n$, we mean that there is a function $\beta \in \mathscr{KL}$ such that for each initial function x_{t_0} for each initial time $t_0 \geq 0$, the unique solution $x(t)$ of the system satisfies $|x(t)| \leq \beta(||x_{t_0}||, t - t_0)$ for all $t \geq t_0$; if, in addition, there are positive constants c_1, c_2, and r such that $\beta(s, t) \leq c_1 e^{-c_2 t} s$ for all $t \geq 0$ and $s \in [0, r]$, then we also say that $\dot{x} = \mathscr{F}(t, x_t)$ is uniformly locally exponentially stable (ULES). When the system is time invariant, we use the abbreviation GAS instead of UGAS. More generally, a system of the form $\dot{x} = \mathscr{F}(t, x_t, d)$ is called input-to-state stable (see [52, 96]) provided that there are functions $\beta \in \mathscr{KL}$ and $\gamma \in \mathscr{K}_\infty$ such that for each condition $x_{t_0} = x_0$, each measurable locally essentially bounded function $d : [t_0, \infty) \to \mathbb{R}^m$, and each initial time $t_0 \geq 0$, the unique solution $x(t)$ of the system satisfies $|x(t)| \leq \beta(||x_{t_0}||, t - t_0) + \gamma(|d|_{[t_0,t]})$ for all $t \geq t_0$, in which case we call γ the gain for the disturbance d. We use the convention that the initial time is always $t_0 = 0$ when the system is autonomous. A partition $\{T_i\}_{i=0}^{\infty}$ of $[0, \infty)$ is any increasing sequence of times such that $T_0 = 0$ and $T_i \to \infty$.

3 Delay in the Input Channel

3.1 Introduction

This section discusses the emulation and dynamic extension approaches to handling input delays. For simplicity, we only deal with constant input delays, but analogous considerations apply under time-varying delays, under certain growth conditions on the delay functions; see [7]. Also, we will not cover systems with outputs, which can model cases where the current state is not available for measurement. Instead, see for example [69] for a discussion of output feedbacks with an application to bioreactors. Also, see [66] for a discussion on observers, including a new class of interval observers that provide framing intervals containing the corresponding components of the unmeasured states. The reader can refer to [1, 26, 51] for observers of nonlinear systems with constant and arbitrarily long delays in the measurements, based on a chain algorithm that was introduced in [26]. Observers for nonlinear systems with time-varying delays in the measurements are considered in [10, 11], the former dealing with suitably small delays, and the latter with arbitrarily long delays, in the chain formulation. For observer-based control laws for nonlinear systems with delayed measurements, see [12].

3.2 Preserving Stability Under Bounds on the Delays

We begin by discussing the first stream of input delay research, where the stabilization is first done with the delays set to 0, and then the effects of the delays on the closed loop system performance are studied. Given a feedback stabilized nonlinear system $\dot{x}(t) = f(t, x(t), u(t, x(t)))$ where the controller u has been designed to render the closed loop system uniformly globally (or locally) asymptotically stable (UGAS or ULAS) to 0, it is natural to inquire how big an upper bound $\bar{\Delta} > 0$ we can find such that the input delayed system

$$\dot{x}(t) = f\big(t, x(t), u(t, x(t - \Delta))\big) \tag{1}$$

is still UGAS or ULAS to 0 for all constants delays $\Delta \in [0, \bar{\Delta}]$. Our ULAS condition is the same as UGAS, except the corresponding convergence condition is only required for initial functions valued in a neighborhood of the equilibrium 0. For the special case of time invariant systems, we use GAS and LAS to mean UGAS and ULAS, respectively.

The rationale for only having a delay in the state argument of the control in (1) is that the current time is normally easy to measure, but the current state may be difficult to measure. Finding an upper bound $\bar{\Delta}$ for the admissible delays is appealing, for two reasons. First, it is generally much easier to design controllers for systems without input delays. Second, if in the specific application of interest, the delays are unknown but are known to stay below $\bar{\Delta}$, then it suffices to design the control for the system free of delays. While many systems have long input delays (especially bioreactors, where there is usually a substantial time lag between the time the organisms are fed and the time they actually grow), many mechanical systems have relatively short delays, which is another motivation for searching for the upper bound $\bar{\Delta}$.

However, one can find nonlinear UGAS systems that become unstable when we introduce any positive input delay. Here is an example from [67] where this occurs:

Example 1 Consider the input delayed system

$$\dot{x}_1(t) = -x_1(t) + x_1^4(t)x_2(t), \quad \dot{x}_2(t) = u(x(t - \Delta)) \tag{2}$$

with state space $\mathbb{R}^2$ and a constant delay $\Delta \geq 0$. When $\Delta = 0$, this system is rendered GAS by the feedback $u(x) = -x_2 - x_1^5$, and $V(x) = |x|^2$ is a Lyapunov function for the corresponding closed-loop system, because $\dot{V} \leq -|x|^2$ along all of the closed loop trajectories. However, this system is not globally asymptotically stabilizable by any continuous $u_s(x(t - \Delta), x(t - 2\Delta))$ for any constant $\Delta > 0$. To see why, suppose that such u_s and Δ existed. Then the closed loop system must be forward complete for all initial functions $x_0 : [-2\Delta, 0] \to \mathbb{R}^2$. Let $\delta > 0$ be any constant such that (i) $-(\Delta/3)|u_s(0, 0)| + \delta \geq \delta/2$ and (ii) $\delta \geq (6e^{\Delta}/\Delta)^{1/4}$. Pick any initial function x_0 that is zero on $[-2\Delta, -2\Delta/3]$, nondecreasing in each component, and such that $x_0(t) = (\delta, \delta)^\top$ for all $t \in [-\Delta/3, 0]$.

Then for all $t \in [0, \Delta/3)$, we get $x_2(t) = u_s(0,0)t + x_2(0) = u_s(0,0)t + \delta$. Our condition (i) on δ gives $\dot{x}_1(t) \geq -x_1(t) + \delta x_1^4(t)/2$ for all $t \in [0, \Delta/3)$ for which the solution is defined. Setting $X_1(t) = e^t x_1(t)$, we then have $\dot{X}_1(t) \geq \delta e^{-3t} X_1^4(t)/2 \geq \delta e^{-\Delta} X_1^4(t)/2$ for all such $t \in [0, \Delta/3)$. Dividing through both sides of the inequality $\dot{X}_1(t) \geq \delta e^{-\Delta} X_1^4(t)/2$ by $X_1^4(t) > 0$ and integrating, we get $X_1^3(t) \geq \delta^3/(1 - \delta^4 e^{-\Delta} t/2)$ for all such $t \in [0, \Delta/3)$. Our condition (ii) on δ gives $\Delta \delta^4 e^{-\Delta}/6 \geq 1$, so X_1 cannot be bounded on the interval $[0, \Delta/3)$. This means the solution could not have been defined on $[0, \infty)$. Hence, no such control u_s can exist for any delay $\Delta > 0$.

In the preceding example, the instability under any positive delay Δ followed because the closed loop input delayed system had a finite escape time. This suggests the possibility that when the vector fields of the system satisfy suitable growth conditions, then any controller that renders a closed loop system UGAS when there are no input delays should also render the closed loop system UGAS under a small enough positive delay. This motivated the work [72], which gives growth conditions under which a controller that provides UGAS under zero input delays also ensures UGAS when a sufficiently small delay is introduced into the state component of the control. The paper [72] applies to control affine systems, i.e., systems of the form

$$\dot{x}(t) = f(t, x(t)) + g(t, x(t))[u_s(t, \xi_\Delta(t)) + d(t)] \tag{3}$$

where $\xi_\Delta(t) = (x_1(t - \Delta_1), x_2(t - \Delta_2), \ldots, x_n(t - \Delta_n))$ and the constant delays Δ_i satisfy $0 \leq \Delta_i \leq \bar{\Delta}$ for all i for some bound $\bar{\Delta} \geq 0$. Here the actuator error d is valued in the set $\mathscr{L}_m^\infty([0, \infty))$ of all measurable essentially bounded functions $[0, \infty) \to \mathbb{R}^m$. The function d represents the uncertainty that would typically occur when implementing the control u_s in actual engineering systems. The use of different delays in each component of the state is motivated by networked control applications [78].

The strategy in [72] is to use Lyapunov-Krasovskii functionals to prove that (3) is input-to-state stable (ISS). To explain the contributions from [72] precisely, we need several additional definitions, which we will use throughout this chapter. We say that (3) has *globally well defined solutions* provided for all choices of $t_o \geq 0$, Δ, $x_o \in \mathscr{C}_n([t_o - \bar{\Delta}, t_o])$, and $d \in \mathscr{L}_m^\infty([0, \infty))$, the initial value problem

$$\begin{aligned} &\dot{x}(t) = f(t, x(t)) + g(t, x(t))[u(t, \xi_\Delta(t)) + d(t)], \quad t \geq t_0 \\ &x(r) = x_o(r) \ \ \forall r \in [t_o - \bar{\Delta}, t_o]. \end{aligned} \tag{IP}$$

has a unique solution $t \mapsto x(t; t_o, x_o, d, \Delta)$ defined on $[t_o - \bar{\Delta}, \infty)$. Solutions of differential equations must be understood to satisfy the equation for (Lebesgue) almost all $t \geq t_0$. In terms of this unique solution, our ISS condition from the previous section can therefore be written as follows:

Definition 1 Assume that (3) has globally well defined solutions. Given a vector delay $\Delta \in (0, \bar{\Delta}]^n$, we call (3) *(uniformly) input-to-state stable (ISS)* provided that there exist $\beta \in \mathcal{KL}$ and $\gamma \in \mathcal{K}_\infty$ such that for all $d \in \mathcal{L}_m^\infty([0,\infty))$, we have

$$|x(t; t_o, x_o, d, \Delta)| \leq \beta(|x_o|_{[t_o-\bar{\Delta}, t_o]}, t - t_o) + \gamma(|d|_{[t_o,t]}) \tag{4}$$

for all $t_o \geq 0$, $x_o \in \mathcal{C}_n([t_o - \bar{\Delta}, t_o])$, and $t \geq t_o$.

The special case of the preceding definition where d is identically equal to zero is therefore UGAS. Let $\bar{\Delta}$ be a positive real number and κ be a nonnegative integer.

Definition 2 A continuous functional $U : \mathcal{C}_n(\mathbb{R}) \times [0,\infty) \rightarrow [0,\infty)$ is called an *ISS Lyapunov-Krasovskii functional (ISS-LKF)* for (3) provided for all constant vectors $\Delta \in [0, \bar{\Delta}]^n$ and all trajectories $x(t) = x(t; t_o, x_o, d, \Delta)$ of (3), the function $t \mapsto U(t, x_t)$ is locally absolutely continuous and there exist functions $\alpha_i \in \mathcal{K}_\infty$ for $i = 1, 2, 3, 4$ and $\kappa \in \mathbb{N}$ such that for all $\varphi \in \mathcal{C}_n([-\kappa\bar{\Delta}, 0])$, all trajectories $x(t)$ of (3), and all $t \geq t_o + \kappa\bar{\Delta}$, we have (i) $\alpha_1(|\varphi(0)|) \leq U(\varphi, t) \leq \alpha_2(|\varphi|_{[-\kappa\bar{\Delta},0]})$ and (ii) $(d/dt)U(t, x_t) \leq -\alpha_3(U(t, x_t)) + \alpha_4(|d|_{[t_o,t]})$ a.e.

The special case of Definition 2 where the perturbations d are identically equal to zero and $\kappa = 1$ is the definition of a Lyapunov-Krasovskii functional for the corresponding unperturbed system. Functions satisfying condition (i) in the preceding definitions for some $\mathcal{K}_\infty$ functions α_1 and α_2 are called uniformly proper and positive definite. As in the case of system free of delays, the existence of an ISS-LKF for (3) will imply that the system is ISS [72]. An additional advantage of having an ISS-LKF is that it can be used to find the comparison functions in the ISS estimate [72].

We next discuss the growth conditions that are required in [72]. Set

$$F(t, x, u_s) = f(t, x) + g(t, x)u_s(t, x). \tag{5}$$

Consider (3) with $d \in \mathcal{L}_m^\infty([0,\infty))$ being an unknown disturbance, a constant vector $\Delta \in [0, \bar{\Delta}]^n$, and $f : [0,\infty) \times \mathbb{R}^n \rightarrow \mathbb{R}^n$, $g : [0,\infty) \times \mathbb{R}^n \rightarrow \mathbb{R}^{n\times m}$ and $u_s : \mathbb{R} \times \mathbb{R}^n \rightarrow \mathbb{R}^m$ satisfying:

Assumption 1 The function u_s is C^1 and $u_s(t, 0) = 0$ for all $t \geq 0$, and f and g are locally Lipschitz. Also, there exist a $\sigma \in \mathcal{K}_\infty$ for which $\sigma(r) \leq r$ for all $r \geq 0$; a C^1 uniformly proper and positive definite function $V : [0,\infty) \times \mathbb{R}^n \rightarrow [0,\infty)$; positive constants $\bar{L}$ and K_1; and constants $K_i \geq 0$ ($i = 2, 3, 4$) such that

H1 $V_t(l, x) + V_x(l, x)F(l, x, u_s) \leq -\sigma^2(\sqrt{n}|x|)$,

H2 $|V_x(l, x)g(l, x)| \leq K_1\sigma(|x|)$, $|(\partial u_s/\partial x)(l, x)| \leq \bar{L}$,

H3 $|f(l, x)|^2 \leq K_2\sigma^2(|x|)$ and $|g(l, x)|^2 \leq K_3(\sigma(|x|) + 1)$, and

H4 $[|g(l, x)||u_s(l, q)|]^2 \leq K_4[\sigma^2(|x|) + \sigma^2(|q|)]$

hold for all $x \in \mathbb{R}^n$, $q \in \mathbb{R}^n$, and $l \geq 0$.

Assumption 1 allows many cases where (3) is a stable linear system with bounded g (using a quadratic Lyapunov function), and cases where the closed loop system is not exponentially stable or g is unbounded [72]. In the following theorem, we set

$$\bar{\Delta} = \frac{0.49}{\bar{L} K_1 \sqrt{2K_2 + 8K_4 + 0.25}}, \tag{6}$$

but see the remarks below for larger delay bounds.

Theorem 1 *Let Assumption* 1 *hold, and define* $\bar{\Delta}$ *by* (6). *Then, for any constant vector* $\Delta \in (0, \bar{\Delta}]^n$, *the feedback delay systems* (3), *in closed loop with the feedback* u_s, *admits the ISS-LKF*

$$U(t, x_t) = V(t, x(t)) + \frac{1}{4\bar{\Delta}} \int_{t-2\bar{\Delta}}^{t} \left(\int_{r}^{t} \sigma^2(\sqrt{n}|x(l)|)\mathrm{d}l \right) \mathrm{d}r \tag{7}$$

and therefore is ISS.

The complete proof of Theorem 1 is in [72], so we do not prove the theorem here. Instead, we remark that the construction (7) is based on a delayed system analog of the strictification approach in [62]. The strictification framework in [62] involves transforming nonstrict Lyapunov functions for nondelayed systems into strict Lyapunov functions for the same systems; see [99] for a recent application of strictification to aerospace models. Nonstrict Lyapunov functions V are defined in exactly the same way as strict ones, except instead of requiring the time derivative $\dot{V}$ to be negative for all states outside the equilibrium, we only require that $\dot{V} \leq 0$ holds everywhere. In conjunction with LaSalle invariance, it is often possible to use nonstrict Lyapunov functions to prove UGAS. However, it is usually necessary to have strict Lyapunov functions to prove ISS and other key robustness properties [62]. The general idea of the strictification process for undelayed systems in [62] is to add terms to the nonstrict Lyapunov function that produce a strict decay condition at those state values outside the equilibrium where $\dot{V}$ is zero. The formulas for the additional terms often involve higher order Lie derivatives of V in the directions of the vector fields defining the system, or in the directions of an auxiliary vector field [62]. The construction (7) is analogous, insofar that the double integral term being added to the original V from Assumption 1 is chosen to compensate for the extra terms produced by the delay, and thereby obtain the strict decay condition from the ISS-LKF definition.

In the special case of our control affine system (3) where the drift term f is identically zero, it is natural to conjecture that scaling the controller u_s by suitable small positive constants should make it possible to prove UGAS of the input delayed system under arbitrarily long input delays. The work [72] shows that this conjecture is in fact true for a large class of drift free systems. To see why, we consider the case where all of the delays Δ_i are equal to some common value Δ. This produces the drift free system

$$\dot{x}(t) = g(t, x(t))[u_s(t, x(t - \Delta)) + \delta(t)]. \tag{8}$$

As shown in [72], it is then possible to prove ISS under the following relaxed version of Assumption 1:

Assumption 2 The function $u_s \in C^1$. Also, there exist a $\sigma \in \mathscr{K}_\infty$ for which $\sigma(r) \leq r$ for all $r \geq 0$, a C^1 uniformly proper and positive definite function $V : [0, \infty) \times \mathbb{R}^n \to [0, \infty)$, and constants $K_i > 0$ (for $i = 1, 2, 3$) such that for all $x \in \mathbb{R}^n, q \in \mathbb{R}^n, l \geq 0$, and $t \geq 0$, the following four conditions hold: (H1′) $V_t(t, x) + V_x(t, x)g(t, x)u_s(t, x) \leq -\sigma^2(|x|)$; (H2′) $|V_x(t, x)g(t, x)| \leq K_1\sigma(|x|)$; (H3′) $|\partial u_s/\partial x(t, x)g(x, l)|^2 \leq K_2(\sigma(|x|)+1)$; and (H4′) $[|\partial u_s/\partial x(t, x)g(x, l)|\, |u_s(q, l)|]^2 \leq K_3[\sigma^2(|x|) + \sigma^2(|q|)]$.

Next note that if (8) satisfies Assumption 2 with $V_t = 0$ and if $\eta \in (0, 1)$ is any fixed constant, then Assumption 2 also holds with u_s, σ, K_1, and K_3 replaced by u_s^η, $\eta^{1/2}\sigma$, $K_1/\eta^{1/2}$, and $\eta^3 K_3$ respectively, with the feedback $u_s^\eta(x, t) = \eta u_s(x, t)$. Also, a slight variant of the proof of Theorem 1 in [72] gives ISS of (8) when Assumption 2 holds and

$$0 < \Delta \leq \bar{\Delta}_c = \frac{1}{4K_1\sqrt{4K_3}}. \tag{9}$$

Combining the preceding observations, we conclude that if (8) satisfies Assumption 2 with $V_t = 0$, then $u = u_s^\eta = \eta u_s$ renders the system (8) ISS as long as

$$0 < \Delta \leq \frac{1}{4\frac{K_1}{\eta^{1/2}}\sqrt{4\eta^3 K_3}} = \frac{1}{8\eta K_1\sqrt{K_3}} =: \bar{\Delta}_\eta.$$

Since $\bar{\Delta}_\eta \to \infty$ as $\eta \to 0$, we get ISS with arbitrarily large delay bounds $\bar{\Delta} = \bar{\Delta}_\eta$, if u_s is properly selected. Moreover, we can find explicit ISS-LKFs for all $\Delta > 0$.

On the other hand, there are interesting classes of systems with drift that violate our Assumption 1. For example, we have the following result, which is [71, Lemma 1]:

Lemma 1 *Let $f : \mathbb{R}^n \to \mathbb{R}^n$ be C^1 and bounded. If a system $\dot{x} = f(x) + u$ is rendered GAS on $\mathbb{R}^n$ by a bounded feedback $u_s(x)$, then for each Lyapunov function $V(t, x)$ of the closed loop system, the requirements of Assumption 1 on the delay systems $\dot{x}(t) = f(x(t)) + u_s(x(t - \tau))$ fail to hold.*

The preceding lemma was one motivation for the main theorem from [71], which proves UGAS for (3) when the perturbation d is identically zero, under a suitable bound on the delays Δ_i and under less restrictive conditions on the corresponding undelayed system than Assumption 1. Another contribution of [71] is that it establishes UGAS under sampling in the controller, under an upper bound on the maximal sampling intervals. We will discuss sampling in detail in Sect. 3.5 in the alternative setting of predictive controllers.

3.3 Backstepping

Backstepping is an important technique for constructing stabilizing controls for certain classes of nonlinear dynamical systems. It has been developed in many contributions and described in many research monographs, e.g. [20, 52, 57, 62, 94]. Backstepping applies to wide families of systems including those of the form

$$\begin{cases} \dot{x}(t) = f(x(t)) + g(x(t))z(t) \\ \dot{z}(t) = u(t) + h(x(t), z(t)), \end{cases} \tag{10}$$

where x is valued in $\mathbb{R}^n$, z is valued in $\mathbb{R}$, and the input u is valued in $\mathbb{R}$. The design of the control laws and the stability analysis frequently rely on the design of a Lyapunov function. The technique can be applied repeatedly, which makes it possible to stabilize systems with a lower triangular structure.

Backstepping has been adapted to the several families of systems with delay in several contributions, notably in [16, 43, 67, 68, 74, Chap. 6]. The papers [16, 67, 74] are devoted to the case where the delay is in the input. The difficulty of this particular case is in the fact that (10) is nonlinear and from the fact that, in general, the control laws and the Lyapunov functions deduced from the backstepping approach in the absence of delay do not allow us to apply the results from [71, 72].

The work [67] adapts backstepping to the case where there is a pointwise known constant delay in the input. Its key ideas are the following. Instead of the system (10), the work [67] studies systems of the form

$$\begin{cases} \dot{x}(t) = f(x(t)) + g(x(t))z(t) \\ \dot{z}(t) = u(t-\tau) + h(x(t-\tau), z(t-\tau)) \end{cases} \tag{11}$$

where $\tau \geq 0$. It is assumed that there exists a control law $z_s(x)$ of class C^1 such that the system:

$$\dot{x}(t) = f(x(t)) + g(x(t))z_s(x(t-\tau)) \tag{12}$$

is GAS. Extra technical assumptions are imposed. Essentially, they are growth conditions on f and g, which ensure that the finite escape time phenomenon does not occur and make it possible to construct a Lyapunov-Krasovskii functional for the system (12). A condition ensuring forward completeness for systems (11) is needed to obtain a global stabilizability result.

The control design begins with the introduction of the operator

$$Z(t) = z(t) - z_s(x(t-\tau)), \tag{13}$$

which plays the role of a change of coordinates. It gives

$$\begin{cases} \dot{x}(t) = f(x(t)) + g(x(t))[Z(t) + z_s(x(t-\tau))], \\ \dot{Z}(t) = u(t-\tau) + h(x(t-\tau), z(t-\tau)) \\ \qquad -\frac{\partial z_s}{\partial x}(x(t-\tau))[f(x(t-\tau)) + g(x(t-\tau))z(t-\tau)]. \end{cases} \tag{14}$$

Consequently, when $\varepsilon > 0$ is any constant and the control law

$$\begin{aligned} u(t) = & -\varepsilon[z(t) - z_s(x(t-\tau))] - h(x(t), z(t)) \\ & + \frac{\partial z_s}{\partial x}(x(t))[f(x(t)) + g(x(t))z(t)] \end{aligned} \tag{15}$$

is applied, then for all $t \geq \tau$, we get

$$\begin{cases} \dot{x}(t) = f(x(t)) + g(x(t))[Z(t) + z_s(x(t-\tau))], \\ \dot{Z}(t) = -\varepsilon Z(t-\tau). \end{cases} \tag{16}$$

Then we can find an explicit formula for a constant $\bar{\varepsilon} > 0$ such that the following is true: We can build a Lyapunov-Krasovskii functional for (16) and a positive definite function $\alpha : [0, \infty) \to [0, \infty)$ such that for each constant $\varepsilon \in (0, \bar{\varepsilon}]$, we have

$$(d/dt)V(t, x_t, z_t) \leq -\alpha(|(x(t), z(t))|) \tag{17}$$

along all trajectories of (16) for all $t \geq \tau$.

From this property and the fact that the finite escape time phenomenon cannot occur, global asymptotic stability for the closed loop system can be deduced. In general, the assumptions are only satisfied when the delay is smaller than a certain bound. However, in some cases, an arbitrarily large delay can be considered by selecting an appropriate function z_s according to the size of the delay.

The contribution [16] proposes another extension of the backstepping approach for systems with a known constant pointwise input delay. The main result of [16] applies to systems of the form

$$\begin{cases} \dot{x}_1(t) &= x_2(t) \\ &\vdots \\ \dot{x}_{n-1}(t) &= x_n(t) \\ \dot{x}_n(t) &= z(t) \\ \dot{z}(t) &= u(t-\tau) + h_1(x(t), z(t)), \end{cases} \tag{18}$$

with $x = (x_1, \ldots, x_n)$ valued in $\mathbb{R}^n$ and u valued in $\mathbb{R}$, under various assumptions on the growth properties of the term h_1. In contrast to the feedbacks proposed in [67] and [74], the control laws of [16] depend on the past values of the controls, and they have the form

$$u(t) = K\left(x_1(t), \ldots, x_{n-1}(t), x_n(t) + \int_{t-\tau}^{t} u(m)\mathrm{d}m\right)^{\top}. \tag{19}$$

The work [16] presents local and global results, depending on the properties of h_1, and in some cases, the results hold in the presence of arbitrarily large delays. The stability analyses are carried out by applying Razumikhin's theorem with a positive definite quadratic function. The recent work [70] based on the reduction model approach applies to the system (18) as well, but no comparison between the performances of the control laws provided by [16, 70] has been made.

The recent paper [74] improves and extends [67]. Its main stabilization result applies to a family of systems that are more general than (11), namely, systems of the form

$$\begin{cases} \dot{x}(t) = f(x(t)) + g(x(t))z(t) \,, \\ \dot{z}(t) = u(t-\tau) + h_1(x(t), z(t)) + h_2(x(t-\tau), z(t-\tau)) \,. \end{cases} \tag{20}$$

The difference between (11) and (20) is that the system (20) has the undelayed term $h_1(x(t), z(t))$ in the z dynamics, which is not present in (11). This difference is important, because the family (20) encompasses systems of the form (18), which may result from an attempt to linearize a single-input single-output system including a single delay in the input. This linearization cannot be completed when $\tau > 0$, since the term $h_1(x(t), z(t))$ cannot be removed through a change of feedback.

In contrast with the feedbacks in [67], the feedbacks in [74] only depend on $x(t)$ and not on both $x(t)$ and $x(t-\tau)$. Moreover, the stability analysis in [74] relies on the construction of a classical Lyapunov-Krasovskii functional V that admits a positive definite function α such that (17) holds along all trajectories of the closed loop system for all $t \geq 0$. The main result of [74] can be summarized as follows.

First, a classical assumption is imposed, namely, the existence of a C^1 positive definite, radially unbounded function V and a C^2 function $z_s(x)$ such that $z_s(0) = 0$ and such that

$$W(x) = -\frac{\partial V}{\partial x}(x)[f(x) + g(x)z_s(x)] \,, \tag{21}$$

is positive definite. Second, the preceding classical assumption is complemented by some growth assumptions on the functions f, g, h_1, V and W and a restriction on the size of the delay. However, as in [67], there are particular cases where an arbitrarily long delay can be handled by selecting an appropriate function z_s. These assumptions ensure that the origin of the system (20) is globally asymptotically stabilized by a feedback of the family

$$\begin{aligned} &u(x,z) = \\ &Le^{\tau L}(z - z_s(x)) + e^{\tau L}\left[\frac{\partial z_s}{\partial x}(x)(f(x) + g(x)z) - h_1(x,z)\right] - h_2(x,z) \end{aligned} \tag{22}$$

provided that L is a negative constant with a sufficiently small absolute value. The preceding result applies under assumptions that do not imply that (20) admits an exponentially stabilizable linear approximation at the origin.

The Lyapunov-Krasovskii functional for the closed loop system uses an operator that is reminiscent of the one introduced in [73]. The operator in [74] is

$$\mathscr{G}(x_t, Z_t) = Z(t) + \int_{t-\tau}^{t} e^{L(t-m-\tau)}\big[u(x(m), z(m)) + h_2(x(m), z(m))\big]\mathrm{d}m\ , \tag{23}$$

where $Z = z - z_s(x)$, and where L is the tuning parameter in the expression for u. Along all closed-loop system trajectories, its time derivative is

$$\dot{\mathscr{G}}(t) = L\mathscr{G}(x_t, Z_t). \tag{24}$$

This can be used to a obtain a strict Lyapunov-Krasovskii functional of the form

$$\begin{aligned} U(x_t, Z_t) = V(x(t)) &+ k_1 \int_{t-\tau}^{t}\int_{\ell}^{t} Z^2(m)\mathrm{d}m\mathrm{d}\ell + k_2 \int_{t-\tau}^{t} W(x(m))\mathrm{d}m \\ &+k_3 \int_{t-\tau}^{t}\int_{\ell}^{t} W(x(m))\mathrm{d}m\mathrm{d}\ell + k_4\mathscr{G}^2(x_t, Z_t), \end{aligned} \tag{25}$$

where k_i for $i = 1$ to 4 are positive constants. Using this Lyapunov-Krasovskii functional, one can prove that for a family of systems with additive disturbances, the proposed control laws ensure ISS with respect to additive disturbances.

3.4 Stabilization Under Large Input Delays Using Predictors

Predictor feedback for linear systems with long input delays has been used for a long time; see, e.g., the pioneering works [3, 58, 63], the more recent works [75, 85, 104], and the historical information in [55, 56]. However, the use of predictor feedback for nonlinear systems is relatively recent. The first researcher who proposed the use of predictor feedback for nonlinear systems with large input delays seems to have been Krstic; see [56]. In his first works on the topic, Krstic was inspired by representing the input delay by a system of first order hyperbolic partial differential equations.

We next review the main results from [56]. Consider the system

$$\dot{x}(t) = f(x(t), u(t - \tau)) \tag{26}$$

where x and u are valued in $\mathbb{R}^n$ and $\mathbb{R}^m$ respectively, $f : \mathbb{R}^n \times \mathbb{R}^m \to \mathbb{R}^n$ is a locally Lipschitz vector field satisfying $f(0, 0) = 0$, and $\tau > 0$ is any constant. Global stabilization of system (26) by predictor feedback is usually studied under two crucial assumptions.

Assumption 3 The delay-free version

$$\dot{x}(t) = f(x(t), u(t)) \tag{27}$$

of system (26) admits a locally Lipschitz mapping $k : \mathbb{R}^n \to \mathbb{R}^m$ satisfying $k(0) = 0$ such that (27) in closed loop with

$$u(t) = k(x(t)) \tag{28}$$

is GAS.

Assumption 4 For every initial condition $x(0) \in \mathbb{R}^n$ and for every input $u \in L^{\infty}_{loc}([0, \infty); \mathbb{R}^m)$ the corresponding solution of (27) exists for all $t \geq 0$.

Assumption 4 means that system (27) is forward complete, and Assumption 3 says that the corresponding system free of delays can be globally asymptotically stabilized by a locally Lipschitz controller. The feedback law $k : \mathbb{R}^n \to \mathbb{R}^m$ in Assumption 3 is called the nominal feedback and is designed for the corresponding delay-free system (27). Assumption 4 is a necessary assumption, since the control action applied to the system (26) for the first $\tau > 0$ units is arbitrary.

To make the prediction method precise, let Φ be the mapping such that $\Phi(x_0, v)$ is the solution of (27) at time $t = \tau$ for the initial state $x_0 \in \mathbb{R}^n$ and corresponding to input $u(t) = v(t)$ for $t \in [0, \tau)$. The mapping $\Phi : \mathbb{R}^n \times L^{\infty}([0, \tau); \mathbb{R}^m) \to \mathbb{R}^n$ is well-defined by Assumption 4 and the solution of (26) satisfies

$$x(t + \tau) = \Phi(x(t), \delta_\tau \tilde{u}_t) \tag{29}$$

for all $t \geq 0$, where $(\tilde{u}_t)(s) = u(t+s)$ for $s \in [-\tau, 0)$ and δ_τ is defined by the formula $(\delta_\tau u)(s) = u(s-\tau)$ for all $s \in [0, \tau)$. In (29), the mapping $\Phi(x(t), \delta_\tau \tilde{u}_t)$ predicts the value of the state vector and this is the reason why $\Phi : \mathbb{R}^n \times L^{\infty}([0, \tau); \mathbb{R}^m) \to \mathbb{R}^n$ is called the predictor mapping.

The following result is proved in [56]:

Theorem 2 *Let* (27) *satisfy Assumptions* 3–4. *Then* (26), *in closed loop with*

$$u(t) = k(\Phi(x(t), \delta_\tau \tilde{u}_t)) \tag{30}$$

for all $t \geq 0$, *admits a function* $\beta \in \mathcal{KL}$ *such that for every* $x_0 \in \mathbb{R}^n$ *and* $u_0 \in L^{\infty}([-\tau, 0); \mathbb{R}^m)$, *the solution of* (26) *in closed loop with* (30) *for the initial conditions* $x(0) = x_0$ *and* $u(s) = u_0(s)$ *for* $s \in [-\tau, 0)$ *exists for all* $t \geq 0$ *and satisfies*

$$|x(t)| + \|\tilde{u}_t\| \leq \beta(|x_0| + \|u_0\|, t) \tag{31}$$

for all $t \geq 0$.

Theorem 2 led to new directions in the research on nonlinear time delay systems, because it implies that predictor feedback can be applied to nonlinear systems in a straightforward way. Other results in [56] provided Lyapunov-Krasovskii functionals for the closed-loop system. Krstic extended Theorem 2 to other important cases that are not covered by this tutorial, including systems with state and input delays [6], state-dependent input delays [8], and adaptive estimation of the input delay [9]. See also the book [7] for a detailed study of nonconstant input delays.

Other research directions were suggested in [46]. In [46], it was shown that predictor feedback can also handle cases with both input delays and measurement delays. In [46], the predictor mapping $\Phi : \mathbb{R}^n \times L^\infty([-r-\tau, 0); \mathbb{R}^m) \to \mathbb{R}^n$ was redefined for all $r \geq 0$ and $\tau \geq 0$ satisfying $r + \tau > 0$ in the following way:

"For every $x_0 \in \mathbb{R}^n$ and $u \in L^\infty([-r-\tau, 0); \mathbb{R}^m)$, the solution $x(t)$ of the dynamics (26) with the initial condition $x(-r) = x_0$ corresponding to the initial input $u \in L^\infty([-r-\tau, 0); \mathbb{R}^m)$ satisfies $x(\tau) = \Phi(x_0, u)$."

The following result is [46, Theorem 2.1]:

Theorem 3 *Let the constants $T > 0$, $r \geq 0$, and $\tau \geq 0$ be given, and assume that $r + \tau > 0$, and that Assumptions 3–4 hold for the system (27). Then the closed-loop system (26) with*

$$u(t) = k(z(t)) \tag{32}$$

for all $t > 0$, where

$$\dot{z}(t) = f(z(t), u(t)) \text{ for all } t \in [iT, (i+1)T) \text{ and } i \in \mathbb{Z}_{\geq 0}, \tag{33}$$

$$z((i+1)T) = \Phi\left(y((i+1)T), \tilde{u}_{(i+1)T}\right) \text{ for all } i \in \mathbb{Z}_{\geq 0}, \text{ and} \tag{34}$$

$$y(iT) = x(iT - r) \text{ for all } i \in \mathbb{Z}_{\geq 0} \tag{35}$$

is GAS, in the following sense: There is a function $\tilde{\beta} \in \mathcal{KL}$ such that for every $(x_0, z_0, u_0) \in C^0([-r, 0]; \mathbb{R}^n) \times \mathbb{R}^n \times L^\infty([-r-\tau, 0]; \mathbb{R}^m)$, the solution $(x(t), z(t), u(t))$ of the closed-loop system given by (26) and (32)–(35) with the initial conditions $z(0) = z_0 \in \mathbb{R}^n$, $u(s) = (u_0)(s)$ for all $s \in [-r-\tau, 0]$, and $x(s) = (x_0)(s)$ for all $s \in [-r, 0]$, satisfies

$$|z(t)| + \|x_t\| + \|u_t\| \leq \tilde{\beta}\left(|z_0| + \|x_0\| + \|u_0\|, t\right) \tag{36}$$

for all $t \geq 0$.

Theorem 3 differs from Theorem 2 in various ways. First, Theorem 3 handles the measurement delay by increasing the prediction horizon of the predictor mapping. Two other differences are:

1. The predictor feedback is implemented in a hybrid way in Theorem 3, which requires sampled measurements, instead of continuous measurements.
2. The predictor mapping in Theorem 3 is only used at discrete times, and between two consecutive sampling times, the prediction is generated by the system (33).

Those differences can be exploited for practical purposes (if needed). Formula (30) is directly implementable for many nonlinear systems for which the predictor mapping is known. More specifically, the predictor mapping can be explicitly constructed in the following cases:

1. Linear systems $\dot{x} = Ax + Bu$, where A and B are constant matrices with x and u valued in $\mathbb{R}^n$ and $\mathbb{R}^m$, respectively. In this case, the predictor mapping $\Phi : \mathbb{R}^n \times L^\infty([0,\tau);\mathbb{R}^m) \to \mathbb{R}^n$ is given by $\Phi(x,u) = \exp(A\tau)x + \int_0^\tau \exp(A(\tau - w))Bu(w)dw$.
2. Bilinear systems $\dot{x} = Ax + Bu + uCx$, with x and u valued in $\mathbb{R}^n$ and $\mathbb{R}$, respectively, and A and C such that $AC = CA$. Here, A, B, and C are constant matrices. In this case, the predictor mapping $\Phi : \mathbb{R}^n \times L^\infty([0,\tau);\mathbb{R}) \to \mathbb{R}^n$ is

$$\begin{aligned}\Phi(x,u) = \exp(A\tau)\exp\left(C\int_0^\tau u(s)\mathrm{d}s\right)x \\ + \int_0^\tau \exp(A(\tau - w))\exp\left(C\int_w^\tau u(s)ds\right)Bu(w)\mathrm{d}w\,.\end{aligned} \tag{37}$$

3. Nonlinear systems of form

$$\begin{aligned}\dot{x}_1 &= a_1(u)x_1 + f_1(u)\\ \dot{x}_2 &= a_2(u,x_1)x_2 + f_2(u,x_1)\\ &\vdots\\ \dot{x}_n &= a_n(u,x_1,\ldots,x_{n-1})x_n + f_n(u,x_1,\ldots,x_{n-1}),\end{aligned} \tag{38}$$

where $x = (x_1,\ldots,x_n)^\top$ and u are valued in $\mathbb{R}^n$ and $\mathbb{R}^m$, respectively, and all of the mappings a_i and f_i (for $i = 1,\ldots,n$) are locally Lipschitz. In this case, the predictor mapping $\Phi : \mathbb{R}^n \times L^\infty([0,\tau);\mathbb{R}^m) \to \mathbb{R}^n$ can be constructed inductively. The predictor mapping is

$$\begin{aligned}\Phi(x,u) = \exp\left(\int_0^\tau a_1(u(s))\mathrm{d}s\right)x \\ + \int_0^\tau \exp\left(\int_w^\tau a_1(u(s))\mathrm{d}s\right)f_1(u(w))\mathrm{d}w\end{aligned} \tag{39}$$

when $n = 1$.
4. Nonlinear systems $\dot{x} = f(x,u)$ that admit a global diffeomorphism $\Theta : \mathbb{R}^n \to \mathbb{R}^n$ such that the change of coordinates $z = \Theta(x)$ transforms the system into one of the above cases.

However, for nonlinear systems where the predictor mapping is not available, the predictor feedback law (30) is not directly implementable. In such cases, we can approximate the predictor mapping in three different ways:

1. Using successive approximations. This idea is used in [41, 47] for globally Lipschitz systems.
2. Using approximations generated by appropriate dynamical systems. This idea is used in [2] for globally Lipschitz systems and [48] for systems with a compact absorbing set.
3. Using numerical approximations. This idea is used in [49].

Here, we only present a theorem that uses the third way of approximating the predictor map. It will use the following assumption:

Assumption 5 There is a continuously differentiable function $k \in C^1(\mathbb{R}^n; \mathbb{R}^m)$ such that $k(0) = 0$ and such that (27) in closed loop with $u(t) = k(x(t))$ is UGAS and ULES.

We then have the following result, which is [49, Theorem 1.1]:

Theorem 4 *Consider the delay-free system (27), where $f : \mathbb{R}^n \times \mathbb{R}^m \to \mathbb{R}^n$ is a C^1 mapping satisfying $f(0,0) = 0$. Assume that Assumptions 3–5 hold. Then for all constants $\tau > 0$ and $r > 0$, there exist a locally bounded mapping $N : [0, \infty) \to \{1, 2, 3, \ldots\}$, a constant $\omega > 0$, and a locally Lipschitz, non-decreasing function $C : [0, \infty) \to [0, \infty)$ satisfying $C(0) = 0$, such that the following is true: For each partition $\{T_i\}_{i=0}^{\infty}$ of $[0, \infty)$ satisfying $\sup_{i \geq 0}(T_{i+1} - T_i) \leq r$, every $x_0 \in \mathbb{R}^n$, and every $u_0 \in L^{\infty}([-\tau, 0); \mathbb{R}^m)$, the unique solution $x(t)$ of (26) in closed loop with $u(t) = z(t)$, where $z(t)$ is defined by*

$$\dot{z}(t) = f(z(t), k(z(t))) \text{ for all } t \in [T_i, T_{i+1}) \text{ and } i \geq 0\,, \tag{40}$$

$$z(T_i) = z_N,\ N = N(|x(T_i)| + \sup_{T_i - \tau \leq s < T_i} |u(s)|),\ h = \frac{\tau}{N},\ z_0 = x(T_i), \text{ and} \tag{41}$$

$$z_{j+1} = z_j + \int_{jh}^{(j+1)h} f(z_j, u(T_i - \tau + s))ds \text{ for } j = 0, \ldots, N-1 \tag{42}$$

with the initial condition $x(0) = x_0$ and $u(s) = u_0(s)$ for $s \in [-\tau, 0)$ satisfies

$$|x(t)| + \sup_{t-\tau \leq s < t} |u(s)| \leq \exp(-\omega t) C\left(|x_0| + ||u_0||\right) \tag{43}$$

for all $t \geq 0$.

Notice that (42) is the application of the explicit Euler numerical scheme to the control system (26) with step size $h = \tau/N$. Since the number of grid points $N = N(|x(T_i)| + \sup_{T_i - \tau \leq s < T_i} |u(s)|)$ is a function of the state and the input, different time steps are used each time that a new measurement arrives. Theorem 4 is proved using combined Lyapunov and small-gain methodology and its proof is constructive. Its proof in [49] provides formulas for

$$N = N\left(|x(T_i)| + \sup_{T_i - \tau \leq s < T_i} |u(s)|\right), \tag{44}$$

which are found using an appropriate Lyapunov function for the closed-loop system (27) with $u(t) = k(x(t))$. Since the function $C : [0, \infty) \to [0, \infty)$ in (43) is locally Lipschitz and $C(0) = 0$, the preceding theorem provides a constant $\Omega > 0$ such that the estimate

$$|x(t)| + \sup_{t-\tau \leq s < t} |u(s)| \leq \Omega \exp(-\omega t)\left(|x_0| + ||u_0||\right) \tag{45}$$

holds for all $t \geq 0$ and all initial conditions such that $|x_0| + \sup_{-\tau \leq s < 0} |u_0(s)|$ is sufficiently small. This is a local exponential stability result for the complicated closed-loop system (26) with $u(t) = k(z(t))$, where z is given by (40) and (42), which is a delayed system with hybrid features. Therefore, both global asymptotic stability and local exponential stability are preserved, despite the delay, the sampled measurements, and the numerical approximation.

The work [50] extends the preceding numerical approximation so that it also applies to time-varying systems and time-varying feedback laws. Also, [50] applies the time-varying generalization to a tracking problem for neuromuscular electrical stimulation. The works [2, 41, 47, 48] give further results on predictor feedback under partial measurement of the state and under various disturbances, including modeling errors, measurement errors and actuator errors. In summary, predictor feedback can be directly implemented to globally feedback stabilizable forward complete nonlinear systems for which either the predictor mapping is available or which can be locally exponentially stabilized [2, 41, 47, 48].

3.5 Stabilization by Sampled-Data Feedback Under Input Delays

The design and existence of sampled-data feedback controllers for nonlinear systems is a recent problem. Sampling arises simultaneously with input and output delays in many control problems, most notably in control over networks. The simultaneous appearance of the input delay and the application of the control action with zero-order hold leads to a high level of complexity for the underlying stability analysis. By zero-order hold, we mean that the controller value is held constant throughout the half open intervals $[T_i, T_{i+1})$ between the sample times T_i in the partition of $[0, \infty)$. Sampling and delays are considered simultaneously in the literature on control of linear and nonlinear systems over networks [23, 31, 76, 77, 79, 97, 98]. These works cover cases where the delays are either physical or sampling-induced, but almost all available results rely on delay-dependent conditions for the existence of stabilizing feedback.

There is another connection between control systems under sampled-data feedback and time delay systems. An interesting point of view that was recently explored in linear systems theory (in [21, 22]) involves considering the closed-loop system under zero-order hold as a time delay system (with varying state delay). The approach was extended to nonlinear systems in [44].

In this section, we focus on global results for nonlinear systems. As in the case where the input can be continuously adjusted, sampled-data feedback controllers can be designed by using either predictor feedback or the emulation approach. Emulation is the use of controllers designed for the delay-free case with continuously adjusted input. One recent result based on the emulation approach was [71, Theorem 1], which we present next in the special case where the system is autonomous and control affine with a single input and a time-invariant feedback. Consider the system

$$\dot{x}(t) = f(x(t)) + g(x(t))u(t - \tau) \tag{46}$$

where the state x and input u are valued in $\mathbb{R}^n$ and $\mathbb{R}$, respectively, $f \in C^0(\mathbb{R}^n; \mathbb{R}^n)$ and $g \in C^0(\mathbb{R}^n; \mathbb{R}^n)$ are locally Lipschitz vector fields with $f(0) = 0$, and the constant $\tau > 0$ represents the delay. We assume that:

Assumption 6 There exist a function $k \in C^1(\mathbb{R}^n; \mathbb{R})$ satisfying $k(0) = 0$, a positive definite, radially unbounded function $V \in C^1(\mathbb{R}^n; [0, \infty))$, a positive definite function $W \in C^0(\mathbb{R}^n; [0, \infty))$, and constants $c_i > 0$ (for $i = 1, 2, 3, 4$) such that

$$\nabla V(x) f(x) + \nabla V(x) g(x) k(x) \leq -W(x), \tag{47}$$

$$|\nabla k(x) g(x)|^2 \leq c_1, \tag{48}$$

$$|\nabla V(x) g(x)|^2 \leq c_2 W(x), \tag{49}$$

$$|\nabla k(x)\,(f(x) + g(x)k(x))|^2 \leq c_3 W(x), \tag{50}$$

and

$$|\nabla V(x) g(x) k(x)| \leq c_4 (V(x) + 1) \tag{51}$$

hold for all $x \in \mathbb{R}^n$.

The following result follows from the special case of [46, Theorem 2.1] where the control and the system are time invariant.

Theorem 5 *If the system* (46) *satisfies Assumption* 6, *then there exists* $\beta \in \mathcal{KL}$ *such that for every* $x_0 \in C^0([-\tau, 0]; \mathbb{R}^n)$ *and every partition* $\{T_i\}_{i=0}^{\infty}$ *of* $[0, \infty)$ *such that*

$$\tau + \sup_{i \geq 0} (T_{i+1} - T_i) \leq \frac{1}{2\sqrt{c_1 + 2c_2c_3}}, \tag{52}$$

the solution of the system

$$\dot{x} = f(x(t)) + g(x(t))k\,(x(T_i - \tau)) \text{ for all } t \in [T_i, T_{i+1}) \text{ and } i \in \mathbb{Z}_{\geq 0} \tag{53}$$

with initial condition $x(s) = (x_0)(s)$ *for* $s \in [-\tau, 0]$ *satisfies*

$$\|x_t\| \leq \beta\,(\|x_0\|\,, t) \tag{54}$$

for all $t \geq 0$.

Theorem 5 as stated above differs from Theorem 1 in [71] because no lower diameter of the sampling partition is assumed and because it can be applied under arbitrary continuous initial conditions. However, the reader may notice that the proof of Theorem 1 in [71] can be carried out without assuming a lower diameter of the sampling partition and that by using Lemma 2.18 on page 96 in [43] the result of Theorem 1 in [71] can be applied to arbitrary continuous initial conditions. Theorem 5 allows the stabilization of systems that cannot be locally exponentially stabilized by smooth feedback; see the discussion in [71].

One of the main characteristics of the emulation approach is the restriction of the magnitude of the input delay. This restriction can be removed by the use of predictor feedback. Prediction-based control methods were used in [15, 60]. However, to our knowledge, the first work that proposed such a systematic design method for nonlinear systems was [46]. We next present [46, Theorem 3.2].

Consider the system

$$\dot{x}(t) = f(x(t), u(t-\tau)) \tag{55}$$

where the state x and input u are valued in $\mathbb{R}^n$ and $\mathbb{R}^m$, respectively, $f \in C^0(\mathbb{R}^n \times \mathbb{R}^m; \mathbb{R}^n)$ is a locally Lipschitz vector field satisfying $f(0,0) = 0$, and $\tau > 0$ is a constant. We assume the following two conditions involving an input applied with zero-order hold:

Assumption 7 There exist a C^1 function $k : \mathbb{R}^n \to \mathbb{R}^m$, a function $g \in \mathcal{K}_\infty$, and a constant $T > 0$ such that

$$|k(x)| \leq g(|x|) \quad \text{for all} \quad x \in \mathbb{R}^n \tag{56}$$

and such that the sampled-data system

$$\dot{x}(t) = f(x(t), k(x(iT))) \ , \ t \in [iT, (i+1)T) \text{ and } i \in \mathbb{Z}_{\geq 0} \tag{57}$$

is UGAS.

Assumption 8 System (55) with $\tau = 0$ is forward complete in the following sense: For every initial condition $x(0) \in \mathbb{R}^n$ and for every input $u \in L^\infty_{loc}([0,\infty); \mathbb{R}^m)$, the corresponding solution of (47) exists and is uniquely defined for all $t \geq 0$.

Assumption 7 implies that system (55) with $\tau = 0$ is globally stabilizable by a feedback applied with zero order hold. This is very different from Assumption 3 of the previous section. Consider system (55) under Assumptions 7–8. In this case, the following feedback law, which is a composition of the feedback stabilizer and the delay compensator, is the natural extension of the predictor feedback methodology presented in the previous section:

$$u(t) = k\left(\Phi\left(x(iT-r), \tilde{u}_{iT}\right)\right), \ t \in [iT, (i+1)T) \text{ and } i \in \mathbb{Z}_{\geq 0} \tag{58}$$

where we refer to $T > 0$ as the sampling/holding period, the constant $r \geq 0$ is the measurement delay, and $\Phi : \mathbb{R}^n \times L^\infty([-r-\tau, 0); \mathbb{R}^m) \to \mathbb{R}^n$ is the so called predictor mapping defined for all constants $r \geq 0$ and $\tau \geq 0$ with $r + \tau > 0$ by:

"For every $x_0 \in \mathbb{R}^n$ and $u \in L^\infty([-r-\tau,0);\mathbb{R}^m)$, the solution $x(t)$ of (55) with the initial condition $x(-r) = x_0$ and initial input $u \in L^\infty([-r-\tau,0);\mathbb{R}^m)$ satisfies $x(\tau) = \Phi(x_0, u)$."

The control action (58) is applied with zero order hold, i.e., it is constant on $[iT, (i+1)T)$; however the control action affecting system (55) remains constant on the interval $[iT+\tau, (i+1)T+\tau)$. For what follows, we use the following definition: Given a positive integer j, we say that the system (57) has the dead-beat property of order jT provided that for all $x_0 \in \mathbb{R}^n$, the solution $x(t)$ of (57) with initial condition $x(0) = x_0 \in \mathbb{R}^n$ satisfies $x(t) = 0$ for all $t \geq jT$. Theorem 3.2 in [46] is stated next.

Theorem 6 *Let $T > 0$, $r \geq 0$, and $\tau \geq 0$ be any constants, and assume that $r+\tau > 0$. Let Assumptions 7–8 hold for system* (55). *Then the closed-loop system* (55) *with* (58) *is globally asymptotically stable, in the following sense: There is a function $\tilde{\beta} \in \mathcal{KL}$ such that for every $(x_0, u_0) \in C^0([-r,0];\mathbb{R}^n) \times L^\infty([-r-\tau,0);\mathbb{R}^m)$, the solution $(x(t), u(t)) \in \mathbb{R}^n \times \mathbb{R}^m$ of system* (55) *with* (58) *with the initial conditions given by $u(s) = u_0(s)$ for $s \in [-r-\tau, 0)$ and $x(s) = x_0(s)$ for $s \in [-r, 0]$ satisfies*

$$\|x_t\| + \|\tilde{u}_t\| \leq \tilde{\beta}\left(\|x_0\| + \|u_0\|, t\right) \tag{59}$$

for all $t \geq 0$. If, in addition, the system (57) *satisfies the dead-beat property of order jT for some positive integer j, then the system* (55) *with the control* (58) *satisfies the dead-beat property of order $(j+p)T$, where $p = [r/T] + [\tau/T] + 2$.*

The conclusion of the preceding theorem on the dead-beat property means that for every $(x_0, u_0) \in C^0([-r,0];\mathbb{R}^n) \times L^\infty([-r-\tau,0);\mathbb{R}^m)$, the solution $(x(t), u(t))$ of system (55) with (58) with initial conditions $u(s) = (u_0)(s)$ for $s \in [-r-\tau, 0)$ and $x(s) = x_0(s)$ for $s \in [-r, 0]$ satisfies $x(t) = 0$ for all $t \geq (j+p)T$. Implementing (58) requires knowledge of the predictor mapping. In addition to linear time-invariant systems, another class of nonlinear systems that allow the explicit construction of the predictor mapping and satisfy Assumptions 7–8 is the class of feedforward systems of the form

$$\begin{aligned}
\dot{x}_1 &= u \\
\dot{x}_2 &= x_1 + g_2(x_1, u) \\
&\vdots \\
\dot{x}_{n-1} &= x_{n-2} + g_{n-1}(x_1, \ldots, x_{n-2}, u) \\
\dot{x}_n &= x_{n-1} + g_n(x_1, \ldots, x_{n-1}, u)
\end{aligned} \tag{60}$$

where $x = (x_1, \ldots, x_n)^\top$ is valued in $\mathbb{R}^n$ and the input u is real valued, and all mappings $g_i : \mathbb{R}^{i-1} \times \mathbb{R} \to \mathbb{R}$ for $i = 2, \ldots, n$ are locally Lipschitz and where we assume that there is a smooth non-decreasing function $L \in C^0([0,\infty);[0,\infty))$ with the following property:

$$\left|g_i(\underline{x}_{i-1}, u)\right| \leq L\left(\left|(\underline{x}_{i-1}, u)\right|\right)\left|\underline{x}_{i-1}\right|^2 + L\left(\left|(\underline{x}_{i-1}, u)\right|\right)\left|\underline{x}_{i-1}\right| |u| \tag{61}$$

for all $(x, u) \in \mathbb{R}^n \times \mathbb{R}$ and $i = 2, \ldots, n$, where $\underline{x}_{i-1} = (x_1, \ldots, x_{i-1})^\top \in \mathbb{R}^{i-1}$.

See [45] for the explicit construction of discontinuous feedback laws that satisfy Assumption 7. There are also important examples of bilinear systems that satisfy Assumption 7 and allow the explicit construction of the predictor mapping, such as the non-holonomic integrator [46]. If the predictor mapping is not available, then we can approximate it using one of the ways described in the previous section. This research program was carried out in [47] and [2] for globally Lipschitz systems. The results suggested a control scheme that can be used in such cases, called the ISP-O-P-DFC control scheme. The ISP-O-P-DFC feedback design consists of:

1. An inter-sample predictor (ISP), which uses sampled, delayed and corrupted measurements of the output and provides an estimate of the (unavailable) delayed continuous output signal;
2. a global observer (O), which uses the estimate of the delayed continuous output signal and provides an estimate of the delayed state vector;
3. an approximate or exact predictor (P), which uses the estimate of the delayed state vector to provide an estimate of the future state vector; and
4. a delay-free controller (DFC) that provides global asymptotic stability for the delay-free version of the system, which in the presence of delay uses an estimate of the future state vector to provide the control action.

The ISP-O-P-DFC control scheme is considered in [47] for the case of constant delays $\tau \geq 0$ and $r \geq 0$, where the approach is applied to globally Lipschitz systems of the form

$$\begin{aligned} \dot{x}_i(t) &= f_i(x_1(t), \ldots, x_i(t)) + x_{i+1}(t) \\ &\quad + g_i(x(t), u(t))d_i(t), \quad i = 1, \ldots, n-1 \\ \dot{x}_n(t) &= f_n(x(t)) + g_n(x(t), u(t))d_n(t) + u(t-\tau) \end{aligned} \tag{62}$$

where $x(t) = (x_1(t), \ldots, x_n(t))^\top$ is valued in $\mathbb{R}^n$, the control u is scalar valued, the disturbance $d(t) = (d_1(t), \ldots, d_n(t))^\top$ is valued in $\mathbb{R}^n$, the functions $f_i : \mathbb{R}^i \to \mathbb{R}$ are globally Lipschitz and satisfy $f_i(0) = 0$ (for $i = 1, \ldots, n$), the measurement is $y = x_1 + e$, where e is the measurement error, and the functions $g_i : \mathbb{R}^i \to \mathbb{R}$ (for $i = 1, \ldots, n$) are locally Lipschitz, bounded functions. The main result in [47] provides a control to render (62) input-to-state stable with respect to the combined disturbance (d, e) (with linear gain functions for all external inputs e and d_i for $i = 1, \ldots, n$) even under perturbations of the sampling schedule, by combining the sampled-data observer design in [44] with the approximate predictor control proposed in [41]. The feedback design was based on the corresponding delay free system

$$\begin{cases} \dot{x}_i(t) = f_i(x_1(t), \ldots, x_i(t)) + x_{i+1}(t), \quad i = 1, \ldots, n-1 \\ \dot{x}_n(t) = f_n(x(t)) + u(t) \end{cases} \tag{63}$$

The results of [46] were generalized in [2] to cover general globally Lipschitz systems. The results in [48] suggest the possibility of carrying out the same research program for nonlinear systems with a compact absorbing set. More work remains to be done for other classes of nonlinear systems.

4 Delay in the State

We next consider systems described by the functional differential equations of retarded type

$$\dot{x}(t) = f(x_t) + g(x_t)(u(t) + d(t)), \tag{64}$$

where $x(t)$ is valued in $\mathbb{R}^n$, the maps $f : \mathscr{C} \to \mathbb{R}^n$ and $g : \mathscr{C} \to \mathbb{R}^{n\times m}$ are completely continuous and locally Lipschitz, $f(0) = 0$, u is the $\mathbb{R}^m$ valued control input signal, and d is the $\mathbb{R}^m$ valued unknown disturbance signal. As before, $\mathscr{C}$ is the space of all continuous functions mapping $[-\Delta, 0]$ into $\mathbb{R}^n$, where $\Delta > 0$ denotes the maximum involved time delay. We assume that all disturbance functions are Lebesgue measurable and locally essentially bounded. Many researchers have studied control problems for systems with delays in the state. For contributions concerning state feedback stabilization and the input-output state feedback linearization of nonlinear time delay systems with systems with delays in the state, see [5, 6, 8, 24, 27, 30, 33, 37, 38, 46, 59, 64, 65, 83, 86, 102].

For the case of discrete time delays, differential geometric methods have been extensively applied to time delay systems, with the goal of input-output linearization. The reader can refer to [13, 14, 24, 27, 64, 65, 81–83, 86, 100] for the application of differential geometric tools to time delay systems. Here, we will not consider observer-based stabilizers. The reader can refer to [25, 29] for observers of nonlinear systems with state delays, and [28] for an observer based stabilizing control law. Applications of observer based control laws to nonlinear systems with delays in the state can be found in [18, 84]. One feedback stabilization approach for time delay system is to develop time delayed versions of Artstein's pioneering work [4], based on control Lyapunov functions. This analog topic entails a reformulation, in terms of control Lyapunov-Krasovskii functionals. We next provide some very recent results in this direction.

4.1 Stabilization Using Lyapunov-Krasovskii Functionals

Control Lyapunov functions have been used to help stabilize a large class of time-invariant time delay systems in affine form in [37], using Lyapunov-Razumikhin functions. To the best of our knowledge, the first results concerning the use of control Lyapunov-Krasovskii functionals (CLKFs) (instead of control Lyapunov-Razumikhin functions) for building stabilizing control laws for time delay systems can be found in [36, 42, 61, 80]. The work [36] used control Lyapunov-Krasovskii functionals that can be written as the sum of three terms. The first term only depends on the value of the state at 0, and the other terms use integrals to deal with the discrete and distributed time delays. This type of control Lyapunov-Krasovskii functional can help address several interesting problems. For instance, it is shown that Sontag's formulas can be used for global stabilization purposes; see [95]. Under additional technical conditions (such as the small control property, as explained in [36, 95]), the proposed feedback control laws are at least locally Lipschitz outside the origin and continuous at the origin.

In [61] the authors propose a predictive control scheme with guaranteed closed-loop stability for non-linear time delay systems, using the type of control Lyapunov-Krasovskii functionals that were presented in [36]. In [42], the authors prove the equivalence between (a) the existence of a completely locally Lipschitz control Lyapunov-Krasovskii functional satisfying the small control property and (b) stabilizability using completely locally Lipschitz control laws, for a large class of non-linear delayed systems.

In the following sections, we summarize the main results from [42, 91] for nonlinear time delay systems in affine form. The results from [42] concern the relationship between stabilizability and existence of a CLKF, and the application of the control Lyapunov-Krasovskii method to triangular time delay nonlinear systems. The results in [91] concern the application of a version of Sontag's universal formula to practical input-to-state stabilization issues. To make our work self-contained, we also include the definitions of CLKF and the small control property.

4.2 Equivalence of Stabilization and Existence of a CLKF

Consider the time delayed system

$$\dot{x}(t) = f(d(t), x_t, u(t)), \tag{65}$$

where the state x is valued in $\mathbb{R}^n$, the disturbance $d(t)$ is valued in a compact set $D \subset \mathbb{R}^l$, the control input $u(t)$ is valued in a closed convex set $U \subset \mathbb{R}^m$, the mapping $f : D \times \mathscr{C} \times U \to \mathbb{R}^n$ is continuous and there exists a non-decreasing function $L : [0, \infty) \to [0, \infty)$ such that

$$|f(d, x, u) - f(d, y, v)| \leq L(\|x\| + \|y\| + |u| + |v|)(\|x - y\| + |u - v|) \tag{66}$$

for all $d \in D$, all x and y in $\mathscr{C}$, and all u and v in U. Moreover, we assume that $f(d, 0, 0) = 0$, for all $d \in D$. We continue to use $\Delta > 0$ to denote the maximum involved time delay.

Given a continuous functional $V : \mathscr{C} \to [0, \infty)$ that is Lipschitz on bounded sets, we define the upper right-hand derivative $D^+V : \mathscr{C} \times \mathbb{R}^m \to \mathbb{R}^\star$ of the functional V along the direction $w \in \mathbb{R}^n$ by

$$D^+V(x,w) = \limsup_{h\to 0^+} \frac{1}{h}\left(V(x_h) - V(x)\right), \tag{67}$$

where for all $h \in [0, \Delta)$, $x_h \in \mathscr{C}$ is given by

$$x_h(\theta) = \begin{cases} x(\theta + h), & \theta \in [-\Delta, -h), \\ x(0) + w(\theta + h), & \theta \in [-h, 0] \end{cases} \tag{68}$$

The preceding upper derivative is called the Driver's form [19]; see [40, 87–89] for results using Driver's form. We need the following definitions:

Definition 3 A functional $V : \mathscr{C} \to [0, \infty)$ that is Lipschitz on bounded sets is called a control Lyapunov-Krasovskii functional (CLKF) for (65) if there exist functions γ_1 and γ_2 of class $\mathscr{K}_\infty$; a mapping $\Psi : \mathbb{R}^p \times U \to \mathbb{R}$ that satisfies $\Psi(0,0) = 0$ and is such that $\Psi(\cdot, u)$ is upper semi-continuous for each $u \in U$; a completely locally Lipschitz mapping $\Phi : \mathscr{C} \to \mathbb{R}^p$; and a continuous positive definite function $\gamma_3 : [0, \infty) \to [0, \infty)$ of class $\mathscr{K}$ such that the following properties hold:

(i) $\gamma_1(|x(0)|) \le V(x) \le \gamma_2(\|x\|)$ for all $x \in \mathscr{C}$.
(ii) For every $x \in \mathscr{C}$, for every finite set $\{u_1, u_2, \ldots, u_N\} \subset U$ and for all $\lambda_i \in [0, 1]$ $(i = 1, \ldots, N)$ such that $\sum_{i=1}^N \lambda_i = 1$, it holds that

$$\sup_{d\in D} D^+V\left(x, f\left(d, x, \sum_{i=1}^N \lambda_i u_i\right)\right) \le -\gamma_3(V(x)) + \max_{i=1,\ldots,N}\{\Psi(\Phi(x), u_i)\}.$$

(iii) $\inf_{u\in U} \Psi(\varphi, u) \le 0$ for all $\varphi \in \mathbb{R}^p$.

Definition 4 A CLKF $V : \mathscr{C} \to [0, \infty)$ for (65) is said to satisfy the small control property provided that for each positive real number ε, there exist a positive constant δ (depending on ε) such that for any $\varphi \in \mathscr{C}$ that satisfies $\|\varphi\| \le \delta$ and $\varphi(0) \ne 0$, the inequality

$$\inf_{u\in\mathbb{R}^m, |u|\le\varepsilon} D^+V(\varphi, u) < 0 \tag{69}$$

holds. A CLKF $V : \mathscr{C} \to [0, \infty)$ for (65) is said to satisfy the local stabilization property provided there exist a positive constant $\eta > 0$ and a locally Lipschitz mapping $K : \mathscr{B}_{4\eta} \to U$ such that $K(0) = 0$ and such that $\Psi(\Phi(x), K(\Phi(x))) \le 0$ for all $x \in \mathscr{C}$ satisfying $|\Phi(x)| \le 2\eta$.

The small control property is related to the possibility of finding a state feedback law that is at least continuous; see [95] for systems described by ordinary differential equations. The following theorem is a reduced version of a much more general result from [42]. See also [43, Theorems 6.4 and 6.5], where CLKF methods are discussed in detail. For what follows, we say that (65) with some control $u(t) = k(x_t)$ is robustly globally asymptotically stable or RGAS provided there is a function $\beta \in \mathscr{KL}$ such that $|x(t)| \leq \beta(||x_0||, t)$ holds for all of the closed loop trajectories of (65), all initial functions x_0, and all choices of the uncertainties d.

Theorem 7 *There exists a locally Lipschitz CLKF $V : \mathscr{C} \to [0, \infty)$ for (65) satisfying the local stabilization property if and only if there exists a completely locally Lipschitz mapping $k : \mathscr{C} \to U$ such that the closed-loop system (65) with $u(t) = k(x_t)$ is RGAS.*

Example 2 Consider the uncertain control system:

$$\begin{aligned} \dot{x}_1(t) &= d(t) \max_{t-\Delta \leq s \leq t} x_1^2(s) + x_2(t) \\ \dot{x}_2(t) &= u(t) \end{aligned} \tag{70}$$

where $x(t) = (x_1(t), x_2(t)) \in \mathbb{R}^2$, $u(t) \in \mathbb{R}$ and $d(t) \in [-1, 1]$. Let $\sigma > 0$ be a given constant and define

$$Q(\xi) = \xi_1^2 + \left(\xi_2 + c\xi_1 + c\xi_1^3\right)^2 \text{ for all } (\xi_1, \xi_2) \in \mathbb{R}^2 \text{ and} \tag{71}$$

$$V(x) = \max_{-\Delta \leq s \leq 0} \exp(2\sigma s) Q(x(s)) \text{ for all } x \in C^0\left([-\Delta, 0]; \mathbb{R}^2\right) \tag{72}$$

where $c > 0$ is a constant that satisfies

$$c \geq \exp(2\sigma\Delta) + \sigma. \tag{73}$$

Since $Q : \mathbb{R}^2 \to [0, \infty)$ is a continuous, positive definite and radially unbounded function, there exist a_1 and a_2 in $\mathscr{K}_\infty$ such that

$$a_1(|\xi|) \leq Q(\xi) \leq a_2(|\xi|) \tag{74}$$

for all $\xi \in \mathbb{R}^2$. Therefore, inequalities (74) in conjunction with definition (72) imply that

$$\exp(-2\sigma\Delta) a_1(\|x\|) \leq V(x) \leq a_2(\|x\|) \tag{75}$$

hold for all $x \in C^0\left([-\Delta, 0]; \mathbb{R}^2\right)$. Since $Q : \mathbb{R}^2 \to [0, \infty)$ is a smooth function, it then follows from Lemma 6.7 on page 322 in [43] that the functional $V : C^0\left([-\Delta, 0]; \mathbb{R}^2\right) \to [0, \infty)$ defined by (72) is Lipschitz on bounded sets and satisfies the following differential inequalities:

$$\begin{aligned} &D^+V(x, w) \leq -2\sigma V(x) \text{ for all } x \in C^0\left([-\Delta, 0]; \mathbb{R}^2\right) \text{ and } w \in \mathbb{R}^2 \\ &\text{such that } Q(x(0)) < V(x) \end{aligned} \tag{76}$$

and

$$D^+V(x,w) \le 2\max\{-\sigma V(x), x_1(0)w_1 + \left(x_2(0)+cx_1(0)+cx_1^3(0)\right)\left(w_2+cw_1+3cx_1^2(0)w_1\right)\} \tag{77}$$

for all $x \in C^0\left([-\Delta,0];\mathbb{R}^2\right)$ and $w=(w_1,w_2)\in\mathbb{R}^2$

satisfying $Q(x(0)) = V(x)$.

Defining

$$f(d,x,u) = \left(d \max_{-\Delta\le s\le 0} x_1^2(s) + x_2(0), u\right) \tag{78}$$

for all $(d,x,u) \in [-1,1]\times C^0\left([-\Delta,0];\mathbb{R}^2\right)\times\mathbb{R}$, we obtain the following from (77):

$$\begin{aligned}&D^+V(x,f(d,x,u)) \le 2\max\{-\sigma V(x), W(d,x,u)\},\\ &\text{for all } x\in C^0\left([-\Delta,0];\mathbb{R}^2\right)\\ &\text{such that } Q(x(0))=V(x),\end{aligned} \tag{79}$$

where

$$\begin{aligned}W(d,x,u) &= x_1(0)\left(d\max\nolimits_{-\Delta\le s\le 0}\left(x_1^2(s)\right) - cx_1(0) - cx_1^3(0)\right)\\ &+\left(x_2(0)+cx_1(0)+cx_1^3(0)\right)\left(u+cx_2(0)+x_1(0)+3cx_1^2(0)x_2(0)\right)\\ &+c\left(x_2(0)+cx_1(0)+cx_1^3(0)\right)\left(1+3x_1^2(0)\right)d\max_{-\Delta\le s\le 0}x_1^2(s)\end{aligned} \tag{80}$$

When $Q(x(0)) = V(x)$, we can use (71) and (72) to get

$$\begin{aligned}\max_{-\Delta\le s\le 0} x_1^2(s) &\le \exp(2\sigma\Delta)\,x_1^2(0)\\ &+\exp(2\sigma\Delta)\left(x_2(0)+cx_1(0)+cx_1^3(0)\right)^2.\end{aligned} \tag{81}$$

Combining the previous inequality with (80), and recalling that $|d|\le 1$, we get the following for the case $Q(x(0)) = V(x)$:

$$\begin{aligned}W(d,x,u) \le\ &\exp(2\sigma\Delta)\,|x_1(0)|\,x_1^2(0) - cx_1^2(0) - cx_1^4(0)\\ &+\left(x_2(0)+cx_1(0)+cx_1^3(0)\right)\left(u+cx_2(0)+x_1(0)+3cx_1^2(0)x_2(0)\right)\\ &+\exp(2\sigma\Delta)\,|x_1(0)|\left(x_2(0)+cx_1(0)+cx_1^3(0)\right)^2\\ &+c\exp(2\sigma\Delta)\left(x_2(0)+cx_1(0)+cx_1^3(0)\right)\left(1+3x_1^2(0)\right)x_1^2(0)\\ &+c\exp(2\sigma\Delta)\left(x_2(0)+cx_1(0)+cx_1^3(0)\right)^3\left(1+3x_1^2(0)\right)\end{aligned}$$

Since $|x_1(0)| \leq 1 + x_1^2(0)$ and $c \geq \exp(2\sigma\Delta) + \sigma$, the above inequality implies that for the case $Q(x(0)) = V(x)$, we have

$$\begin{aligned} W(d,x,u) \leq & -\sigma\, x_1^2(0) - \sigma\left(x_2(0) + cx_1(0) + cx_1^3(0)\right)^2 + \\ & \left(x_2(0) + cx_1(0) + cx_1^3(0)\right)(u - k(x(0))) \end{aligned} \tag{82}$$

where

$$\begin{aligned} k(\xi) = & -c\xi_2 - \xi_1 - 3c\xi_1^2\xi_2 \\ & - \left(\sigma + \exp(2\sigma\Delta)\left(1+\xi_1^2\right)\right)\left(\xi_2 + c\xi_1 + c\xi_1^3\right) \\ & - c\exp(2\sigma\Delta)\left(1 + 3\xi_1^2\right)\left(\xi_1^2 + \left(\xi_2 + c\xi_1 + c\xi_1^3\right)^2\right)\,. \end{aligned} \tag{83}$$

Combining (76), (79), and (82), it follows that

$$\begin{aligned} \sup_{d \in [-1,1]} \; & D^+ V(x, f(d,x,u)) \leq -2\sigma\, V(x) + \\ & 2\max\left\{0\,,\,\left(x_2(0) + cx_1(0) + cx_1^3(0)\right)(u - k(x(0)))\right\} \end{aligned} \tag{84}$$

holds for all $(x,u) \in C^0\left([-\Delta,0];\mathbb{R}^2\right) \times \mathbb{R}$. Inequalities (75) and (84) imply that the functional $V : C^0\left([-\Delta,0];\mathbb{R}^2\right) \to [0,\infty)$ defined by (72) is a CLKF for system (70). More specifically, all inequalities in the definition of the CLKF hold with

$$\gamma_1(s) = \exp(-2\sigma\Delta)a_1(s),\;\; \gamma_2(s) = a_2(s),\, \gamma_3(s) = 2\sigma\, s,$$

$$\Phi(x) = \begin{bmatrix} \varphi_1(x) \\ \varphi_2(x) \end{bmatrix} = \begin{bmatrix} x_2(0) + cx_1(0) + cx_1^3(0) \\ k(x(0)) \end{bmatrix},\;\text{and}$$

$$\Psi(\varphi_1, \varphi_2, u) = 2\max\{0\,,\,\varphi_1\,(u - \varphi_2)\}\;.$$

Finally, notice that the local stabilization property holds for arbitrary $\eta > 0$ with $K(\varphi_1,\varphi_2) = \varphi_2$. A smooth feedback stabilizer that robustly, globally stabilizes system (70) is given by the delay-free feedback law $u(t) = k(x(t))$; see Example 3. ◁

4.3 Backstepping Under State Delays

In Sect. 3.3, we explained how to use backstepping to design globally asymptotically stabilizing controllers under input delays, but there is no obvious analog of the results from Sect. 3.3 for systems with state delays. The literature on backstepping for time delay systems with state delays has focused on the class of triangular time delay nonlinear systems of the form

$$\begin{aligned} \dot{x}_i &= f_i(d, x_{1,t}, \ldots, x_{i,t}) + g_i(d, x_{1,t}, \ldots, x_{i,t})x_{i+1}\;,\; i = 1,\ldots,n-1 \\ \dot{x}_n &= f_n(d, x_t) + g_n(d, x_t)u, \end{aligned} \tag{85}$$

where $x = (x_1, \dots, x_n)^\top$ is valued in $\mathbb{R}^n$, the disturbance d is valued in a compact set $D \subseteq \mathbb{R}^l$, the functions $f_i : D \times C^0([-\Delta, 0]; \mathbb{R}^i) \to \mathbb{R}$ and $g_i : D \times C^0([-\Delta, 0]; \mathbb{R}^i) \to \mathbb{R}$ (for $i = 1, \dots, n$) are completely locally Lipschitz with respect to $x \in C^0([-\Delta, 0]; \mathbb{R}^i)$, and $f_i(d, 0) = 0$ for all $d \in D$. The complete local Lipschitzness condition implies that

$$\sup \left\{ \frac{|f_i(d,x) - f_i(d,y)| + |g_i(d,x) - g_i(d,y)|}{\|x - y\|} : d \in D, x \in S, y \in S, x \neq y \right\} < \infty$$

holds for every bounded set $S \subseteq C^0([-\Delta, 0]; \mathbb{R}^i)$.

Lyapunov-based feedback design for various special cases of systems described by delayed functional differential equations was first used in [32, 37, 38, 67, 80, 105], including the input-delayed case. More specifically, the stabilization problem for autonomous and disturbance-free systems of the form (85) was studied in [37, 38, 80, 105]. The results in [37, 38] used the Razumikhin theorem to design a smooth delay-free globally stabilizing feedback law of the form $u(t) = k(x(t))$. On the other hand, the work [42] proposed a recursive procedure for the design of a smooth delay-free globally stabilizing feedback law of the form $u(t) = k(x(t))$ based on Lyapunov-Krasovskii functionals. We next describe the procedure from [42]. We state the main result in [42] on a class of triangular time delay control systems of the form (85).

Theorem 8 *Assume that there is a non-decreasing function $\varphi \in C^\infty([0, \infty); (0, \infty))$ such that*

$$\frac{1}{\varphi(\|x\|)} \leq g_i(d, x) \leq \varphi(\|x\|) \tag{86}$$

holds for all $(x, d) \in C^0([-\Delta, 0]; \mathbb{R}^i) \times D$ and $i = 1, \dots, n$. Then for each constant $\sigma > 0$, there are functions $\mu_i \in C^\infty(\mathbb{R}^i; (0, \infty))$ for $i = 1, \dots, n$ such that with the choices

$$k_1(\xi_1) = -\mu_1(\xi_1)\xi_1 \tag{87}$$

and

$$\begin{aligned} &k_j(\xi_1, \dots, \xi_j) \\ &= -\mu_j(\xi_1, \dots, \xi_j)\left(\xi_j - k_{j-1}(\xi_1, \dots, \xi_{j-1})\right), j = 2, \dots, n,) \end{aligned} \tag{88}$$

the functional

$$V(x) = \max_{\theta \in [-\Delta, 0]} e^{2\sigma\theta} \left(x_1^2(\theta) + \sum_{j=2}^{n} \left| x_j(\theta) - k_{j-1}(x_1(\theta), \dots, x_{j-1}(\theta)) \right|^2 \right) \tag{89}$$

admits class $\mathcal{K}_\infty$ *functions* a_1 *and* a_2 *such that*

$$a_1(\|x\|) \le V(x) \le a_2(\|x\|) \tag{90}$$

and

$$D^+V(x;v) \le -2\sigma V(x) \tag{91}$$

hold for all $(x,d) \in C^0([-\Delta,0];\mathbb{R}^n) \times D$ *with the choice*

$$v = (f_1(d,x_1) + g_1(d,x_1)x_2, \ldots, f_n(d,x) + g_n(d,x)k_n(x))^\top \in \mathbb{R}^n. \tag{92}$$

Moreover, the closed-loop system (85) *with the choice* $u(t) = k_n(x(t))$ *is RGAS.*

It is worth comparing Theorem 8 with [20, Theorem 5.1], which studies the triangular finite-dimensional case. The functional $V(x)$ defined by (89) is a CLKF for system (85) with the feedback $u(t) = k_n(x(t))$. Notice that the feedback law $u(t) = k_n(x(t))$ is delay-independent. As in the finite-dimensional case, the feedback design and the construction of the state robust control Lyapunov functional are done in parallel.

There are important differences between Theorem 8 and the results [37, 38] by Jankovic. In [37, 38], Razumikhin functions are constructed in parallel with feedback design while Theorem 8 provides Lyapunov functionals. Furthermore, Theorem 8 covers uncertain systems that were not covered by [37, 38]. The results in [103, 105] also provide Lyapunov functionals, but are restricted to the disturbance free case (85) with a single discrete delay and where the g_i's (for $i = 1, \ldots, n$) are free of delay, i.e., $g_i(d, x_{1,t}, \ldots, x_{i,t}) = g_i(x_1(t), \ldots, x_i(t))$ for $i = 1, \ldots, n$.

The construction of the CLKF can be automated by the following algorithm. Let $\sigma > 0$ be a given constant. Let $L : [0,\infty) \to [0,\infty)$ be a C^∞, non-decreasing function that satisfies the inequalities $|f_i(d,x)| \le L(\|x\|)\,\|x\|$, for all $(d,x) \in D \times C^0([-\Delta,0];\mathbb{R}^i)$ and $i = 1, \ldots, n$. We define functions

$$\mu_i \in C^\infty(\mathbb{R}^i;(0,\infty)),\ \ \gamma_i \in C^\infty([0,\infty);(0,\infty)),\ \text{ and } b_i \in C^\infty([0,\infty);(0,\infty))$$

for $i = 1, \ldots, n$ using the following algorithm.

ALGORITHM:

Step $i = 1$: We define

$$\mu_1(\xi_1) = \frac{\gamma_1(1+\xi_1^2) + n\sigma}{b_1(1+\xi_1^2)}, \tag{93}$$

where

$$\begin{aligned} \gamma_1(s) &= \exp(\sigma\Delta)L\,(s\exp(\sigma\Delta)) + \varphi\,(s\exp(\sigma\Delta)) \quad \text{and} \\ b_1(s) &= \frac{1}{\varphi\,(s\exp(\sigma\Delta))} \end{aligned} \tag{94}$$

Step $i \geq 2$: Using the functions $\mu_j \in C^\infty(\mathbb{R}^j; (0, \infty))$ for $j = 1, \ldots, i-1$ from the previous steps, we define the function $\mu_i \in C^\infty(\mathbb{R}^i; (0, \infty))$, in terms of several other functions that we must define first. First define

$$k_0 = 0, \quad k_1(\xi_1) = -\mu_1(\xi_1)\xi_1, \tag{95}$$

$$k_j(\xi_1, \ldots, \xi_j) = -\mu_j(\xi_1, \ldots, \xi_j)\left(\xi_j - k_{j-1}(\xi_1, \ldots, \xi_{j-1})\right) \quad \text{if } 2 \leq j \leq i-1, \tag{96}$$

$$\begin{aligned}\gamma_j(s) = {} & \exp(\sigma\Delta)L\left(s\exp(\sigma\Delta)B_j\left(s\exp(\sigma\Delta)\right)\right)B_j\left(s\exp(\sigma\Delta)\right) \\ & +\varphi\left(s\exp(\sigma\Delta)B_j\left(s\exp(\sigma\Delta)\right)\right) \quad \text{if } 1 \leq j \leq i,\end{aligned} \tag{97}$$

and

$$b_j(s) = \frac{1}{\varphi\left(s\exp(\sigma\Delta)B_j\left(s\exp(\sigma\Delta)\right)\right)} \quad \text{if } 1 \leq j \leq i, \tag{98}$$

where $B_1(s) = 1$ and $B_j \in C^\infty([0, \infty); (0, \infty))$ for $j = 2, \ldots, i$ are any non-decreasing functions that satisfy

$$B_j(s) \geq \max\left\{1 + \sum_{l=1}^{j-1}\mu_l(\xi_1, \ldots, \xi_l) : \max_{l=1,\ldots,j}|\xi_l - k_{l-1}(\xi_1, \ldots, \xi_{l-1})| \leq s\right\} \tag{99}$$

for all $s \geq 0$ and $j \geq 2$. Let $\rho_j \in C^\infty([0, \infty); (0, \infty))$ (for $j = 1, \ldots, i-1$) and $\delta_j \in C^\infty(\mathbb{R}^j; (0, \infty))$ (for $j = 0, \ldots, i-1$) be any functions such that

$$b_j(s') - b_j(s) + s\gamma_j(s) - s'\gamma_j(s') \leq (s - s')\rho_j(s) \quad \text{and} \tag{100}$$

$$\delta_j(\xi_1, \ldots, \xi_j) \geq \left|\nabla k_j(\xi_1, \ldots, \xi_j)\right|\left(1 + \mu_1(\xi_1) + \cdots + \mu_j(\xi_1, \ldots, \xi_j)\right) \tag{101}$$

hold for all $s \geq 0$ and $s' \geq 0$ that satisfy $s \geq s' \geq 0$ and all $(\xi_1, \ldots, \xi_j) \in \mathbb{R}^j$.

We next define

$$\begin{aligned}\mu_i(\xi_1, \ldots, \xi_i) = {} & \tfrac{1}{b_i(p)}\left[(n+1-i)\sigma + \tfrac{i-1}{4\sigma}a^2(p, \xi_1, \ldots, \xi_{i-1}) + \gamma_i(p)\right. \\ & \left.+c_{i-1}(p)\delta_{i-1}(\xi_1, \ldots, \xi_{i-1})\right]\end{aligned} \tag{102}$$

where

$$p = \frac{i}{2} + \frac{1}{2}\sum_{j=1}^{i} \left|\xi_j - k_{j-1}(\xi_1, \ldots, \xi_{j-1})\right|^2,$$
$$c_0 = 0, \quad c_j(s) = \sum_{k=1}^{j} \gamma_k(s) \text{ for } j = 1, \ldots, i, \tag{103}$$

and

$$a(s, \xi_1, \ldots, \xi_{i-1}) = c_{i-1}(s)\delta_{i-1}(\xi_1, \ldots, \xi_{i-1}) + c_i(s) + \left(1 + \sum_{j=1}^{i-1} \left(s\mu_j(\xi_1, \ldots, \xi_j) + \delta_{j-1}(\xi_1, \ldots, \xi_{j-1})\right)\right) \sum_{k=1}^{i-1} \rho_k(s) . \tag{104}$$

At every step $i \geq 2$ of the above algorithm, we only need to compute the functions $\gamma_i(s)$, $b_i(s)$, $B_i(s)$, $\rho_{i-1}(s)$, $\delta_{i-1}(\xi_1, \ldots, \xi_{i-1})$ and $\mu_i(\xi_1, \ldots, \xi_i)$.

Example 3 Consider again the uncertain control system (70). In Example 2, we showed that the functional $V : C^0\left([-\varDelta, 0]; \mathbb{R}^2\right) \to [0, \infty)$ defined by (72) is a CLKF for system (70). The reader can verify that the functional $V : C^0\left([-\varDelta, 0]; \mathbb{R}^2\right) \to [0, \infty)$ defined by (72) was constructed by means of the above algorithm. Notice that system (70) has the form of system (85) with $n = 2$ and satisfies (86) with $\varphi(s) \equiv 1$, and $L(s) = 1 + s$. More specifically, the application of the above algorithm will give us $k_1(\xi_1) = -c_1\xi_1 - c_2\xi_1^3$, for positive constants c_1, c_2. Appropriate values of the constants c_1, c_2 can be found by direct substitution (in Example 2, we showed that the selection $c_1 = c_2 = c$, where c satisfies (73), is an appropriate selection so that the functional V given by (72) is a CLKF for (70)). Hence, by Theorem 8, the closed loop system in Example 2 with $u(t) = k(x(t))$ is RGAS. ◁

A different backstepping procedure was used in [39] to get finite-time global stabilization by time-varying distributed delay feedback for these classes of systems:

1. The case of triangular control systems

$$\begin{cases} \dot{x}_i(t) = f_i(t, x_1(t), \ldots, x_i(t)) + x_{i+1}(t), \quad i = 1, \ldots, n-1 \\ \dot{x}_n(t) = f_n(t, x(t)) + u(t) \end{cases} \tag{105}$$

where $x \in \mathbb{R}^n$ and the input u is scalar valued, with no delays.

2. Chains of delayed integrators

$$\begin{cases} \dot{x}_i(t) = x_{i+1}(t - \tau_i) , \ i = 1, \ldots, n-1 \\ \dot{x}_n(t) = u(t - \tau_n) \end{cases} \tag{106}$$

with no limitation on the size of the delays where x and u are valued in $\mathbb{R}^n$ and $\mathbb{R}$, respectively, and $\tau_i \geq 0$ $i = 1, \ldots, n$ are the delays

3. Triangular control systems

$$\begin{cases} \dot{x}_i(t) = f_i(t, x_1(t - \tau_{i,1}), \ldots, x_i(t - \tau_{i,i})) + x_{i+1}(t), \ i = 1, \ldots, n-1 \\ \dot{x}_n(t) = f_n(t, x_1(t - \tau_{n,1}), \ldots, x_n(t - \tau_{n,n})) + u(t) \end{cases}$$

with delayed drift terms, where $x \in \mathbb{R}^n$ and the input u is scalar valued, where $\min_{i=1,\ldots,n-1} \min_{j=1,\ldots,i} \ \tau_{i,j} > 0$.

The recent work [103] gave backstepping methods that ensure global stabilization of specific classes of nonlinear delay systems by dynamic gain-based backstepping. The approach in [103] does not require an upper bound on the delay. This contrasts with Theorem 8 above, which requires the upper bound $\Delta > 0$ for the delay. Finally, note that the theory of feedback linearization for time delay systems has been developed in [24, 27, 64, 65, 83]. Feedback linearization methods are directly connected with backstepping methodologies (as in the finite-dimensional case) and can provide smooth nonlinear feedback stabilizers for systems with state delays.

4.4 Practical Input-to-State Stabilization Under State Delays

We next present results on input-to-state practical stabilization of nonlinear systems with state delays that are affine in the control, meaning, they have the form

$$\dot{x}(t) = f(x_t) + g(x_t)u(t), \tag{107}$$

where u is the control as before. Input-to-state practical stability is defined exactly as we defined ISS, except one adds a positive constant to the right side of the usual ISS estimate. In particular, this practical stability property means that there may be a positive asymptotic bound on the norm of the state, even if the disturbance is the zero function. Invariantly differentiable functionals (which we borrow from [53, 54]) play an important role in this topic; see their definition below. Invariant differentiability is helpful for finding a robustifying control law, yielding input-to-state stability with respect to actuator disturbances, as shown in [90, 93]. The same definition is useful for applying Sontag's well known universal formula from [95] to control-affine nonlinear systems with state delays; see [91]. Given a CLKF V, it may well happen that the induced Sontag's formula provides a discontinuous control law [36]. While in systems described by ordinary differential equations, the discontinuity may only occur at the origin, the discontinuity may occur throughout significant subsets of the infinite dimensional state space in the state delayed case.

Aside from theoretical issues of existence and uniqueness of the solution, non Lipschitz feedback controls may cause implementation difficulties; see, e.g., [20, Sect. 5.3.1]. In [91], an extension of Sontag's formula to delayed systems is modified in the critical subsets of the infinite dimensional state space where the Lipschitz property of the related feedback control law may be lost. Then, a Sontag input-to-state stabilizing term (as explained in [90, 96]) is added to the control law, to achieve attenuation of the actuator disturbance and attenuation of the bounded error due to the modification of Sontag's formula. The resulting control law is locally Lipschitz. We next sketch the main results from [91], omitting the technical details.

Consider the functional differential equations of retarded type

$$\begin{cases} \dot{x}(t) = f(x_t) + g(x_t)u(t), & t \geq 0 \\ x(t) = \xi_0(t), & t \in [-\Delta, 0] \end{cases} \tag{108}$$

with initial functions $\xi_0 \in \mathscr{C}$, where $x(t)$ is valued in $\mathbb{R}^n$, $\Delta > 0$ is the maximum time delay involved, the functions $f : \mathscr{C} \to \mathbb{R}^n$ and $g : \mathscr{C} \to \mathbb{R}^{n \times m}$ are completely continuous and locally Lipschitz in $\mathscr{C}$ and f satisfies $f(0) = 0$, and the input signal u is valued in $\mathbb{R}^m$ and is Lebesgue measurable and locally essentially bounded.

The following definition of invariant differentiable functionals is from Definitions 2.2.1 and 2.5.2 in [54, Chap. 2], although we modify the formalism from [54] slightly to make the notation more uniform in this chapter. For any $x \in \mathbb{R}^n$, $\varphi \in \mathscr{R}$, continuous function $\mathscr{Y} : [0, \Delta] \to \mathbb{R}^n$ satisfying $\mathscr{Y}(0) = x$, and $h \in [0, \Delta)$, we define $\psi_h^{(x,\varphi,\mathscr{Y})} \in \mathscr{R}$ by

$$\psi_0^{(x,\varphi,\mathscr{Y})} = \varphi \ \text{ and } \ \psi_h^{(x,\varphi,\mathscr{Y})}(s) = \begin{cases} \varphi(s+h), & s \in [-\Delta, -h) \\ \mathscr{Y}(s+h), & s \in [-h, 0) \end{cases} \tag{109}$$

for all $h > 0$. For all $\varphi \in \mathscr{C}$ and $h \in [0, \Delta)$, we define $\varphi^h \in \mathscr{C}$ by

$$\varphi^h(s) = \begin{cases} \varphi(s+h), & s \in [-\Delta, -h) \\ \varphi(0), & s \in [-h, 0] \end{cases} \tag{110}$$

We then have the following definition of invariant differentiability from [54]:

Definition 5 A functional $V : \mathbb{R}^n \times \mathscr{R} \to [0, \infty)$ is called invariantly differentiable provided at each pair $(x, \varphi) \in \mathbb{R}^n \times \mathscr{R}$, the following conditions hold:

1. For each continuous function $\mathscr{Y} : [0, \Delta] \to \mathbb{R}^n$ satisfying $\mathscr{Y}(0) = x$, the right-hand derivative

$$D_h^+ V(x, \psi_h^{(x,\varphi,\mathscr{Y})}) \tag{111}$$

   of the function

$$h \mapsto V(x, \psi_h^{(x,\varphi,\mathscr{Y})}) \tag{112}$$

   exists at $h = 0$ and is independent of the function $\mathscr{Y}$;
2. the derivative $(\partial V/\partial x)(x, \varphi)$ exists; and
3. for each continuous function $\mathscr{Y} : [0, \Delta] \to \mathbb{R}^n$ satisfying $\mathscr{Y}(0) = x$, the equality

$$V\left(x+z,\psi_h^{(x,\varphi,\mathscr{Y})}\right)-V(x,\varphi)= \\ (\partial V/\partial x)(x,\varphi)z+D_\ell^+V\left(x,\psi_\ell^{(x,\varphi,\mathscr{Y})}\right)\Big|_{\ell=0}h+o\left(\sqrt{|z|^2+h^2}\right) \tag{113}$$

holds for all $z\in\mathbb{R}^n$ and all $h\in[0,\Delta)$.

In the following, $\mathscr{V}$ denotes the class of all functionals $V:\mathbb{R}^n\times\mathscr{R}\to[0,\infty)$ that satisfy the following: (i) V is locally Lipschitz in $\mathbb{R}^n\times\mathscr{R}$ and invariantly differentiable; (ii) the maps

$$\varphi\mapsto D_h^+V\left(\varphi(0),\varphi_{[-\Delta,0)}^h\right)\Big|_{h=0}\quad\text{and}\quad\varphi\to\left.\frac{\partial V(x,\varphi_{[-\Delta,0)})}{\partial x}\right|_{x=\varphi(0)} \tag{114}$$

are completely continuous and locally Lipschitz in $\mathscr{C}$.

We now consider the perturbed functional differential equation

$$\begin{cases}\dot{x}(t)=f(x_t)+g(x_t)(u(t)+d(t)), & t\geq 0\\ x(t)=\xi_0(t), & t\in[-\Delta,0]\end{cases} \tag{115}$$

with initial functions $\xi_0\in\mathscr{C}$, where the input signal is the sum of the control signal $u(t)\in\mathbb{R}^m$ plus the actuator disturbance $d(t)\in\mathbb{R}^m$. We assume that d is Lebesgue measurable and locally essentially bounded, and that our assumptions on (108) also hold for (115). For a given functional $V:\mathbb{R}^n\times\mathscr{R}\to[0,\infty)$ in $\mathscr{V}$, let the map $a:\mathscr{C}\to[0,\infty)$ and the row vector valued function $b:\mathscr{C}\to\mathbb{R}^m$ be defined by

$$a(\varphi)=D_h^+V\left(\varphi(0),\varphi_{[-\Delta,0)}^h\right)\Big|_{h=0}+\left.\frac{\partial V(x,\varphi_{[-\Delta,0)})}{\partial x}\right|_{x=\varphi(0)}f(\varphi) \tag{116}$$

and

$$b(\varphi)=\left.\frac{\partial V(x,\varphi_{[-\Delta,0)})}{\partial x}\right|_{x=\varphi(0)}g(\varphi)\,. \tag{117}$$

For each constant $r>0$, we define $k_r:\mathscr{C}\to\mathbb{R}^m$ by

$$k_r(\varphi)=\begin{cases}-\frac{a(\varphi)+\sqrt{a^2(\varphi)+|b(\varphi)|^4}}{|b(\varphi)|^2}b^T(\varphi), & |b(\varphi)|>r\\ \\ -\frac{a(\varphi)+\sqrt{a^2(\varphi)+|b(\varphi)|^4}}{r^2}b^T(\varphi), & |b(\varphi)|\leq r\end{cases} \tag{118}$$

For ease of exposition, we make use, in the following assumption, of the well known M_2 norm [87], denoted with the symbol $\|\cdot\|_{M_2}$. Hence,

$$\|\varphi\|_{M_2}=\left(\varphi^\top(0)\varphi(0)+\int_{-\Delta}^0\varphi^\top(s)\varphi(s)\mathrm{d}s\right)^{1/2}$$

for all $\varphi \in \mathscr{R}$. The reader can refer to [91] for a weaker assumption which makes use of more general M_a functionals, including the supremum norm and suitable semi-norms as special cases; see [34, 35, 90, 101].

Assumption 9 There exist a class $\mathscr{V}$ functional $V : \mathbb{R}^n \times \mathscr{R} \to [0, \infty)$ and corresponding maps a and b; functions α_1, α_2, and α_3 of class $\mathscr{K}_\infty$; and positive real constants r and p such that

1. $\alpha_1(|\varphi(0)|) \leq V(\varphi(0), \varphi_{[-\Delta,0)}) \leq \alpha_2(\|\varphi\|_{M_2})$;
2. $a(\varphi) \leq 0$ holds for all $\varphi \in \mathscr{C}$ such that $b(\varphi) = 0$;
3. $a^2(\varphi) + |b(\varphi)|^4 \geq \alpha_3^2(\|\varphi\|_{M_2})$; and
4. $\sup\{a(\psi)/|b(\psi)| : \psi \in \mathscr{C},\ 0 < |b(\psi)| \leq r\} \leq p$

hold for all $\varphi \in \mathscr{C}$.

The following is then a special case of the main result in [91, Theorem 9]:

Theorem 9 *Let Assumption 9 hold, and r, p, and q be any positive constants. Then the map* (118) *is completely continuous and locally Lipschitz. Also, there are functions $\beta \in \mathscr{KL}$ and $\gamma \in \mathscr{K}$ (that are both independent of p, q, and r) such that for each initial function x_0, the unique solution of* (115) *in closed loop with*

$$u(t) = k_r(x_t) - qb^T(x_t), \tag{119}$$

exists for all $t \geq 0$ and satisfies

$$|x(t)| \leq \beta(\|x_0\|, t) + \gamma\left(\sqrt{\frac{2}{q}}\|d_{[0,t)}\|\right) + \gamma\left(\sqrt{\frac{2}{q}}(2p + r)\right) \tag{120}$$

for all $t \geq 0$.

If the disturbance is bounded, then an arbitrarily small neighborhood of the origin can be reached, by choosing the tuning parameter q to be large enough.

Remark 1 In general, the aim of avoiding discontinous state feedbacks may make it harder to use Lyapunov-Krasovskii functionals. Implementation problems as well as theoretical issues related to discontinuous feedbacks can be solved by the sampling-and-holding process [17]. Results in this direction can be found in [92], where it is proved that state feedbacks (that may or may not be continuous), induced by suitable control Lyapunov-Krasovskii functionals, yield stabilization in the sample-and-hold sense (which is a semi-global, practical stabilization that is introduced in [17]).

5 Conclusions

We summarized several recent control design and stabilization approaches for systems with input or state delays. One method involved emulation, where one first designs the feedback stabilizer with the delay set equal to zero, and then one analyzes how large a delay can be introduced into the controller without destroying the stability properties. A totally different approach involves dynamic predictor controllers, which can often ensure stabilization under arbitrarily long delays, even if there is sampling in the controller. Our approaches involve suitable generalizations of the standard Lyapunov functions that are commonly used for systems without delays. These generalizations include analogs of control-Lyapunov functionals, as well as Lyapunov-Krasovskii functionals. We also showed how the technical subtleties that arise when considering state delays, where the domain of the right side of the dynamics is an infinite dimensional space, can lead to complex problems in infinite dimensional analysis. Although the existing theory applies to a wide class of systems and control objectives, much work remains to be done. For instance, it would be useful to cover state constrained problems, where in addition to stabilizing an equilibrium, one must ensure that all of the closed loop solutions remain in a given subset of the state space. Control problems with state constraints naturally arise in collision avoidance problems. It would also be useful to generalize the existing work to hyperbolic or parabolic partial differential equations with input or state delays and state constraints. We leave these extensions for future papers.

References

1. Ahmed-Ali, T., Cherrier, E., Lamnabhi-Lagarrigue, F.: Cascade high gain predictor for a class of nonlinear systems. IEEE Trans. Autom. Control **57**(1), 221–226 (2012)
2. Ahmed-Ali, T., Karafyllis, I., Krstic, M., Lamnabhi-Lagarrigue, F.: Robust stabilization of nonlinear globally Lipschitz delay systems. In: Karafyllis, I., Malisoff, M., Mazenc, F., Pepe, P. (eds.) Recent Results on Nonlinear Time Delayed Systems. Series Advances in Delays and Dynamics. Springer, Berlin (2015)
3. Artstein, Z.: Linear systems with delayed controls: a reduction. IEEE Trans. Autom. Control **27**(4), 869–879 (1982)
4. Artstein, Z.: Stabilization with relaxed controls. Nonlinear Analysis: Theory, Methods and Applications **7**(1), 1163–1173 (1983)
5. Bekiaris-Liberis, N., Jankovic, M., Krstic, M.: Compensation of state-dependent state delay for nonlinear systems. Syst. Control Lett. **61**(8), 849–856 (2012)
6. Bekiaris-Liberis, N., Krstic, M.: Compensation of time-varying input and state delays for nonlinear systems. J. Dyn. Syst. Meas. Control **134**(1), Paper 011009 (2012)
7. Bekiaris-Liberis, N., Krstic, M.: Nonlinear Control Under Nonconstant Delays. Siam, Philadelphia (2013)
8. Bekiaris-Liberis, N., Krstic, M.: Compensation of state-dependent input delay for nonlinear systems. IEEE Trans. Autom. Control **58**(2), 275–289 (2013)
9. Bresch-Prietri, D., Krstic, M.: Delay-adaptive control for nonlinear systems. IEEE Trans. Autom. Control **59**(5), 1203–1218 (2014)
10. Cacace, F., Germani, A., Manes, C.: An observer for a class of nonlinear systems with time varying observation delay. Syst. Control Lett. **59**(5), 305–312 (2010)

11. Cacace, F., Germani, A., Manes, C.: A chain observer for nonlinear systems with multiple time-varying measurement delays. SIAM J. Control Optim. **52**(3), 1862–1885 (2014)
12. Cacace, F., Germani, A., Manes, C.: State estimation and control of nonlinear systems with large and variable measurement delays. In: Karafyllis, I., Malisoff, M., Mazenc, F., Pepe, P. (eds.) Recent Results on Nonlinear Time Delayed Systems. Series Advances in Delays and Dynamics. Springer, Berlin (2015)
13. Califano, C., Marquez-Martinez, L., Moog, C.: Extended Lie brackets for nonlinear time delay systems. IEEE Trans. Autom. Control **56**(9), 2213–2218 (2011)
14. Califano, C., Marquez-Martinez, L., Moog, C.: Linearization of time delay systems by input-output injection and output transformation. Automatica **49**(6), 1932–1940 (2013)
15. Castillo-Toledo, B., Di Gennaro, S., Sandoval Castro, G.: Stability analysis for a class of sampled nonlinear systems with time delay. In: Proceedings of the 49th IEEE Conference on Decision and Control, pp. 1575–1580 (2010)
16. Choi, H., Lim, J.: Asymptotic stabilization of an input-delayed chain of integrators with nonlinearity. Syst. Control Lett. **59**(6), 374–379 (2010)
17. Clarke, F., Ledyaev, Y., Sontag, E., Subbotin, A.: Asymptotic controllability implies feedback stabilization. IEEE Trans. Autom. Control **42**(10), 1394–1407 (1997)
18. Di Ciccio, M., Bottini, M., Pepe, P., Foscolo, P.: Observer-based nonlinear control law for a continuous stirred tank reactor with recycle. Chem. Eng. Sci. **66**(20), 4780–4797 (2011)
19. Driver, R.: Existence and stability of solutions of a delay-differential system. Arch. Ration. Mech. Anal. **10**(1), 401–426 (1962)
20. Freeman, R., Kokotovic, P.: Robust Nonlinear Control Design-State Space and Lyapunov Techniques. Birkhauser, Boston (1996)
21. Fridman, E., Seuret, A., Richard, J.-P.: Robust sampled-data stabilization of linear systems: an input-delay approach. Automatica **40**(8), 1441–1446 (2004)
22. Fridman, E., Shaked, U., Suplin, V.: Input/output delay approach to robust sampled-data H_∞ control. Syst. Control Lett. **54**(3), 271–282 (2005)
23. Gao, H., Chen, T., Lam, J.: A new system approach to network-based control. Automatica **44**(1), 39–52 (2008)
24. Germani, A., Manes, C., Pepe, P.: Local asymptotic stability for nonlinear state feedback delay systems. Kybernetika **36**(1), 31–42 (2000)
25. Germani, A., Manes, C., Pepe, P.: An asymptotic state observer for a class of nonlinear delay systems. Kybernetika **37**(4), 459–478 (2001)
26. Germani, A., Manes, C., Pepe, P.: A new approach to state observation of nonlinear systems with delayed output. IEEE Trans. Autom. Control **47**(1), 96–101 (2002)
27. Germani, A., Manes, C., Pepe, P.: Input-output linearization with delay cancellation for non-linear delay systems: the problem of the internal stability. Int. J. Robust Nonlinear Control **13**(9), 909–937 (2003)
28. Germani, A., Manes, C., Pepe, P.: Observer-based stabilizing control for a class of nonlinear retarded systems. In: Sipahi, R., Vyhlidal, T., Niculescu, S.-I., Pepe, P. (eds.) Time Delay Systems: Methods, Applications and New Trends, pp. 331–342. Springer, Berlin (2012)
29. Germani, A., Pepe, P.: A state observer for a class of nonlinear systems with multiple discrete and distributed time delays. Eur. J. Control **11**(3), 196–205 (2005)
30. Gomes da Silva, J., Tarbouriech, S., Garcia, G.: Anti-windup design for time delay systems subject to input saturation an LMI-based approach. Eur. J. Control **6**(12), 622–634 (2006)
31. Heemels, M., Teel, A., van de Wouw, N., Nesic, D.: Networked control systems with communication constraints: tradeoffs between transmission intervals, delays and performance. IEEE Trans. Autom. Control **55**(8), 1781–1796 (2010)
32. Hua, C., Feng, G., Guan, X.: Robust controller design of a class of nonlinear time delay systems via backstepping methods. Automatica **44**(2), 567–573 (2008)
33. Hua, C., Guan, X., Shi, P.: Robust stabilization of a class of nonlinear time delay systems. Appl. Math. Comput. **155**(3), 737–752 (2004)
34. Ito, H., Jiang, Z.-P., Pepe, P.: Construction of Lyapunov-Krasovskii functionals for networks of iISS retarded systems in small-gain formulation. Automatica **49**(11), 3246–3257 (2013)

35. Ito, H., Pepe, P., Jiang, Z.-P.: A small-gain condition for iISS of interconnected retarded systems based on Lyapunov-Krasovskii functionals. Automatica **46**(10), 1646–1656 (2010)
36. Jankovic, M.: Extension of control Lyapunov functions to time delay systemss. In: Proceedings of the IEEE Conference on Decision and Control, pp. 4403–4408 (2000)
37. Jankovic, M.: Control Lyapunov-Razumikhin functions and robust stabilization of time delay systems. IEEE Trans. Autom. Control **46**(7), 1048–1060 (2001)
38. Jankovic, M.: Stabilization of nonlinear time delay systems with delay independent feedback. In: Proceedings of the American Control Conference, pp. 4253–4258 (2005)
39. Karafyllis, I.: Finite-time global stabilization by means of time-varying distributed delay feedback. SIAM J. Control Optim. **45**(1), 320–342 (2006)
40. Karafyllis, I.: Lyapunov theorems for systems described by RFDEs. Nonlinear Anal. Theory Methods Appl. **64**(3), 590–617 (2006)
41. Karafyllis, I.: Stabilization by means of approximate predictors for systems with delayed input. SIAM J. Control Optim. **49**(3), 1100–1123 (2011)
42. Karafyllis, I., Jiang, Z.-P.: Necessary and sufficient Lyapunov-like conditions for robust nonlinear stabilization. ESAIM Control Optim. Calc. Var. **16**(4), 887–928 (2010)
43. Karafyllis, I., Jiang, Z.-P.: Stability and Stabilization of Nonlinear Systems. Springer, London (2011)
44. Karafyllis, I., Kravaris, C.: Global stability results for systems under sampled-data control. Int. J. Robust Nonlinear Control **19**(10), 1105–1128 (2009)
45. Karafyllis, I., Krstic, M.: Global stabilization of feedforward systems under perturbations in sampling schedule. SIAM J. Control Optim. **50**(3), 1389–1412 (2012)
46. Karafyllis, I., Krstic, M.: Nonlinear stabilization under sampled and delayed measurements, and with inputs subject to delay and zero-order hold. IEEE Trans. Autom. Control **57**(5), 1141–1154 (2012)
47. Karafyllis, I., Krstic, M.: Stabilization of nonlinear delay systems using approximate predictors and high-gain observers. Automatica **49**(12), 3623–3631 (2013)
48. Karafyllis, I., Krstic, M., Ahmed-Ali, T., Lamnabhi-Lagarrigue, F.: Global stabilization of nonlinear delay systems with a compact absorbing set. Int. J. Control **87**(5), 1010–1027 (2014)
49. Karafyllis, I., Krstic, M.: Numerical schemes for nonlinear predictor feedback. Math. Control Signals Systems **26**(4), 519–546 (2014)
50. Karafyllis, I., Malisoff, M., de Queiroz, M., Krstic, M., Yang, R.: Predictor-based tracking for neuromuscular electrical stimulation. Int. J. Robust Nonlinear Control (to appear). doi:10.1002/rnc.3211
51. Kazantzis, N., Wright, R.: Nonlinear observer design in the presence of delayed output measurements. Syst. Control Lett. **54**(9), 877–886 (2005)
52. Khalil, H.: Nonlinear Systems, 2nd edn. Prentice-Hall, Englewood Cliffs (1996)
53. Kim, A.: On the Lyapunov's functionals method for systems with delays. Nonlinear Anal. Theory Methods Appl. **28**(4), 673–687 (1997)
54. Kim, A.: Functional Differential Equations. Application of I-Smooth Calculus. Kluwer Academic Publishers, Dordrecht (1999)
55. Krstic, M.: Delay Compensation for Nonlinear, Adaptive, and PDE Systems. Birkhauser, Boston (2009)
56. Krstic, M.: Input delay compensation for forward complete and strict-feedforward nonlinear systems. IEEE Trans. Autom. Control **55**(2), 287–303 (2010)
57. Krstic, M., Kanellakopoulos, I., Kokotovic, P.: Nonlinear and Adaptive Control Design. Wiley, New York (1995)
58. Kwon, W., Pearson, A.: Feedback stabilization of linear systems with delayed control. IEEE Trans. Autom. Control **25**(2), 266–269 (1980)
59. Lien, C.: Global exponential stabilization for several classes of uncertain nonlinear systems with time-varying delay. Nonlinear Dyn. Syst. Theory **4**(1), 15–30 (2004)
60. Lozano, R., Castillo, P., Garcia, P., Dzul, A.: Robust prediction-based control for unstable delay systems: application to the yaw control of a mini-helicopter. Automatica **40**(4), 603–612 (2004)

61. Mahboobi Esfanjani, R., Nikravesh, S.: Stabilising predictive control of non-linear time delay systems using control Lyapunov-Krasovskii functionals. IET Control Theory Appl. **3**(10), 1395–1400 (2009)
62. Malisoff, M., Mazenc, F.: Constructions of Strict Lyapunov Functions. Springer, New York (2009)
63. Manitius, A., Olbrot, A.: Finite spectrum assignment for systems with delays. IEEE Trans. Autom. Control **24**(4), 541–553 (1979)
64. Marquez-Martinez, L., Moog, C.: Input-output feedback linearization of time delay systems. IEEE Trans. Autom. Control **49**(5), 781–785 (2004)
65. Marquez-Martinez, L., Moog, C.: New insights on the analysis of nonlinear time delay systems: application to the triangular equivalence. Syst. Control Lett. **56**(2), 133–140 (2007)
66. Mazenc, F., Bernard, O.: ISS interval observers for nonlinear systems transformed into triangular systems. Int. J. Robust Nonlinear Control **24**(7), 1241–1261 (2014)
67. Mazenc, F., Bliman, P.: Backstepping design for time delay nonlinear systems. IEEE Trans. Autom. Control **51**(1), 149–154 (2006)
68. Mazenc, F., Ito, H.: Lyapunov technique and backstepping for nonlinear neutral systems. IEEE Trans. Autom. Control **58**(2), 512–517 (2013)
69. Mazenc, F., Jiang, Z.-P.: Global output feedback stabilization of a chemostat with an arbitrary number of species. IEEE Trans. Autom. Control **55**(11), 2570–2575 (2010)
70. Mazenc, F., Malisoff, M.: Local stabilization of nonlinear systems through the reduction model approach. IEEE Trans. Autom. Control **59**(11), 3033–3039 (2014)
71. Mazenc, F., Malisoff, M., Dinh, T.: Robustness of nonlinear systems with respect to delay and sampling of the controls. Automatica **49**(6), 1925–1931 (2013)
72. Mazenc, F., Malisoff, M., Lin, Z.: Further results on input-to-state stability for nonlinear systems with delayed feedbacks. Automatica **44**(9), 2415–2421 (2008)
73. Mazenc, F., Niculescu, S.-I., Bekaik, M.: Generating positive and stable solutions through delayed state feedback. Automatica **47**(3), 525–533 (2011)
74. Mazenc, F., Niculescu, S.-I., Bekaik, M.: Backstepping for nonlinear systems with delay in the input revisited. SIAM J. Control Optim. **49**(6), 2239–2262 (2011)
75. Mondie, S., Michiels, W.: Finite spectrum assignment of unstable time delay systems with a safe implementation. IEEE Trans. Autom. Control **48**(12), 2207–2212 (2003)
76. Naghshtabrizi, P., Hespanha, J., Teel, A.: Stability of delay impulsive systems with application to networked control systems. In: Proceedings of the American Control Conference, pp. 4899–4904 (2007)
77. Nesic, D., Liberzon, D.: A unified framework for design and analysis of networked and quantized control systems. IEEE Trans. Autom. Control **54**(4), 732–747 (2009)
78. Nesic, D., Teel, A.: Input-to-state stability of networked control systems. Automatica **40**(12), 2121–2128 (2008)
79. Nesic, D., Teel, A., Carnevale, D.: Explicit computation of the sampling period in emulation of controllers for nonlinear sampled-data systems. IEEE Trans. Autom. Control **54**(3), 619–624 (2009)
80. Nguang, S.: Robust stabilization of a class of time delay nonlinear systems. IEEE Trans. Autom. Control **45**(4), 756–762 (2000)
81. Oguchi, T.: A finite spectrum assignment for retarded non-linear systems and its solvability condition. Int. J. Control **80**(6), 898–907 (2007)
82. Oguchi, T., Watanabe, A.: Input-output linearization of nonlinear systems with time delays in state variables. Int. J. Syst. Sci. **29**(6), 573–578 (1998)
83. Oguchi, T., Watanabe, A., Nakamizo, T.: Input-output linearization of retarded non-linear systems by using an extension of Lie derivative. Int. J. Control **75**(8), 582–590 (2002)
84. Palumbo, P., Pepe, P., Panunzi, S., De Gaetano, A.: Time-delay model-based control of the glucose-insulin system, by means of a state observer. Eur. J. Control **18**(6), 591–606 (2012)
85. Partington, J., Makila, P.: Rational approximation of distributed-delay controllers. Int. J. Control **78**(16), 1295–1301 (2005)

86. Pepe, P.: Adaptive output tracking for a class of nonlinear time delay systems. Int. J. Adapt. Control Signal Process. **18**(6), 489–503 (2004)
87. Pepe, P., Jiang, Z.-P.: A Lyapunov-Krasovskii methodology for ISS and iISS of time delay systems. Syst. Control Lett. **55**(12), 1006–1014 (2006)
88. Pepe, P.: On Liapunov-Krasovskii functionals under Carathéodory conditions. Automatica **43**(4), 701–706 (2007)
89. Pepe, P.: The problem of the absolute continuity for Lyapunov-Krasovskii functionals. IEEE Trans. Autom. Control **52**(5), 953–957 (2007)
90. Pepe, P.: Input-to-state stabilization of stabilizable, time delay, control affine, nonlinear systems. IEEE Trans. Autom. Control **54**(7), 1688–1693 (2009)
91. Pepe, P.: On Sontag's formula for the input-to-state practical stabilization of retarded control-affine systems. Syst. Control Lett. **62**(11), 1018–1025 (2013)
92. Pepe, P.: Stabilization in the sample-and-hold sense of nonlinear retarded systems. SIAM J. Control Optim. **52**(5), 3053–3077 (2014)
93. Pepe, P., Ito, H.: On saturation, discontinuities and delays, in iISS and ISS feedback control redesign. IEEE Trans. Autom. Control **57**(5), 1125–1140 (2012)
94. Sastry, S.: Nonlinear Systems, Analysis. Stability and Control. Springer, New York (1999)
95. Sontag, E.: A "universal" construction of Artstein's theorem on nonlinear stabilization. Syst. Control Lett. **13**(2), 117–123 (1989)
96. Sontag, E.: Smooth stabilization implies coprime factorization. IEEE Trans. Autom. Control **34**(4), 435–443 (1989)
97. Tabbara, M., Nesic, D., Teel, A.: Networked control systems: emulation based design. In: Liu, D., Wang, F.-Y. (eds.) Networked Control Systems, pp. 57–94. Springer, London (2008)
98. Tabuada, P.: Event-triggered real-time scheduling of stabilizing control tasks. IEEE Trans. Autom. Control **52**(9), 1680–1685 (2007)
99. Thakur, D., Mazenc, F., Akella, M.: Lyapunov strictification and angular velocity observers for smooth attitude tracking controllers. In: Proceedings of the AIAA Space and Astronautics Forum and Exposition: AIAA/AAS Astrodynamics Specialist Conference, 18pp. (2014)
100. Xia, X., Marquez-Martinez, L., Zagalak, P., Moog, C.: Analysis of nonlinear time delay systems using modules over non-commutative rings. Automatica **38**(9), 1549–1555 (2002)
101. Yeganefar, N., Pepe, P., Dambrine, M.: Input-to-state stability of time delay systems: a link with exponential stability. IEEE Trans. Autom. Control **53**(6), 1526–1531 (2008)
102. Zhang, X., Cheng, Z.: Global stabilization of a class of time delay nonlinear systems. Int. J. Syst. Sci. **36**(8), 461–468 (2005)
103. Zhang, X., Lin, Y.: Global stabilization of high-order nonlinear time delay systems by state feedback. Syst. Control Lett. **65**, 89–95 (2014)
104. Zhong, Q.-C.: Robust Control of Time-Delay Systems. Springer, London (2010)
105. Zhou, S., Geng, G., Nguang, S.: Comments on robust stabilization of a class of time delay nonlinear systems. IEEE Trans. Autom. Control **47**(9), 1586 (2002)

Robust Stabilization of Nonlinear Globally Lipschitz Delay Systems

Tarek Ahmed-Ali, Iasson Karafyllis, Miroslav Krstic and Francoise Lamnabhi-Lagarrigue

Abstract This paper studies the application of a recently proposed control scheme to globally Lipschitz nonlinear systems for which the input is delayed and applied with zero order hold, the measurements are sampled and delayed, and only an output is measured (i.e., the state vector is not available). The control scheme consists of an observer for the delayed state vector, an inter-sample predictor for the output signal, an approximate predictor for the future value of the state vector, and the nominal feedback law applied with zero order hold and computed for the predicted value of the future state vector. The resulting closed-loop system is robust with respect to modeling and measurement errors and robust to perturbations of the sampling schedule.

1 Introduction

Predictor feedback is used frequently in the literature for systems with large input delays. The literature on predictor feedback under non-constant input delays is reviewed in [1, 2] (where time-varying delays and state-dependent input delays

---

T. Ahmed-Ali (✉)
Laboratoire GREYC CNRS-ENSICAEN, 06 Boulevard du Marechal Juin,
14050 Caen Cedex, France
e-mail: tarek.ahmed-ali@ensicaen.fr

I. Karafyllis
Department of Mathematics, National Technical University of Athens,
Heroon Polytechneiou 9, 15780 Athens, Greece
e-mail: iasonkar@central.ntua.gr

M. Krstic
Department of Mechanical and Aerospace Engineering, University of California,
San Diego, La Jolla, CA 92093-0411, USA
e-mail: krstic@ucsd.edu

F. Lamnabhi-Lagarrigue
Centre National de la Recherche Scientifique, CNRS-EECI SUPELEC,
3 Rue Joliot Curie, 91192 Gif-Sur-Yvette, France
e-mail: lamnabhi@lss.supelec.fr

I. Karafyllis et al. (eds.), *Recent Results on Nonlinear Delay Control Systems*,
Advances in Delays and Dynamics 4, DOI 10.1007/978-3-319-18072-4_2

are studied). For nonlinear systems with constant input delays, various forms of predictors have been used:

- Exact predictors based on the knowledge of the solution mapping [3–5].
- Approximate predictors based on successive approximations [6, 7].
- Approximate predictors based on numerical schemes [8, 9].
- Approximate predictors for which the prediction is provided by the output of a properly constructed control system [10–12].

Moreover, the literature on predictor feedback for nonlinear systems with constant input delays has considered cases where the input is applied continuously (as in [1–3, 6, 8, 12], or [9]) or is applied with zero order hold or ZOH (as in [3] and [7]). There is also a wide literature of predictor feedback design and implementation for linear systems with constant input delays; see the references in [4] and [5].

This chapter considers the application of a recently proposed control scheme to globally Lipschitz nonlinear systems for which the input is applied with ZOH, the measurements are sampled and delayed, and only an output is measured (since the state vector is not available). Moreover, we also consider the effect of possible modeling errors and measurement noise. The control scheme consists of an observer for the delayed state vector, an inter-sample predictor for the output signal, an approximate predictor for the future value of the state vector, and the nominal feedback law applied with ZOH and computed for the predicted value of the future state vector. The control scheme has been applied to globally Lipschitz nonlinear systems previously (as in [7]) but in this work we have used a different prediction action, namely, we are using approximate predictors for which the prediction is provided by the output of a properly constructed control system (namely, dynamic approximate predictors) instead of predictors that are based on successive approximations. The chapter generalizes the results provided in [7] to various directions:

- We show that the convergence is independent of the lower diameter of the sampling schedule (in contrast with [7], where the estimates depended on the lower diameter of the sampling schedule).
- We provide assumptions which can be applied to general nonlinear globally Lipschitz systems (in contrast with [7], where only triangular single input systems were considered).
- We provide explicit formulae for the asymptotic gains of various inputs (in contrast with [7], where only qualitative estimates were provided).
- We provide explicit inequalities for the upper diameter of the sampling partition and the holding period, which can be used in straightforward way by the potential control practitioner.

The application of the proposed control scheme guarantees robustness with respect to modeling errors, measurement noise and perturbations of the sampling schedule.

Prior to the submission of the present chapter, we were informed of the work [13]. The results in [13] also deal with globally Lipschitz systems using the control scheme proposed in [7]. The results [13] cover various cases of transmission protocols and generalized the results of [7] to non-triangular globally Lipschitz systems.

The results of the present paper are less conservative than the results in [13] and the main difference between the present work and the results of [13] is the use of dynamic approximate predictors instead of predictors which are based on successive approximations.

2 Notation

Throughout this chapter, we adopt the following notation:

- For a vector $x \in \mathbb{R}^n$ we denote by $|x|$ its usual Euclidean norm, by x' its transpose. For a real matrix $A \in \mathbb{R}^{n\times m}$, $A' \in \mathbb{R}^{m\times n}$ denotes its transpose and $|A| = \sup\{\, |Ax|\,;\ x \in \mathbb{R}^n\ ,\ |x| = 1\,\}$ is its induced norm. $I \in \mathbb{R}^{n\times n}$ denotes the identity matrix.
- $\mathbb{R}_+$ denotes the set of non-negative real numbers. For every $t \geq 0$, $[t]$ denotes the integer part of $t \geq 0$, i.e., the largest integer being less or equal to $t \geq 0$. A partition $\pi = \{T_i\}_{i=0}^{\infty}$ of $\mathbb{R}_+$ is an increasing sequence with $T_0 = 0$ and $T_i \to +\infty$.
- Let $x : [a-r, b) \to \mathbb{R}^n$ with $b > a \geq 0$ and $r \geq 0$. By x_t we denote the "history" of x from $t-r$ to t, i.e., $(x_t)(\theta) = x(t+\theta)\ ;\ \theta \in [-r, 0]$, for $t \in [a, b)$. By $\breve{x}_t$ we denote the "open history" of x from $t-r$ to t, i.e., $(\breve{x}_t)(\theta) = x(t+\theta)\ ;\ \theta \in [-r, 0)$, for $t \in [a, b)$.
- Let $I \subseteq \mathbb{R}_+$ be an interval. By $L^\infty(I; U)$ ($L^\infty_{loc}(I; U)$) we denote the space of measurable and (locally) bounded functions $u(\cdot)$ defined on I and taking values in $U \subseteq \mathbb{R}^m$. Notice that we do not identify functions in $L^\infty(I;U)$ which differ on a measure zero set. For $L^\infty([-r, 0]; \mathbb{R}^n)$ or $x \in L^\infty([-r, 0); \mathbb{R}^n)$ we define $\|x\| = \sup_{\theta\in[-r,0]} |x(\theta)|$ or $\|x\| = \sup_{\theta\in[-r,0)} |x(\theta)|$. Notice that $\sup_{\theta\in[-r,0]} |x(\theta)|$ is not the essential supremum but the actual supremum. By $PC(I; \mathbb{R}^m)$ we denote the space of piecewise continuous functions $u(\cdot)$ defined on I and taking values in $\mathbb{R}^m$.
- By $C^0(A; \Omega)$, where $A \subseteq \mathbb{R}^n$ and $\Omega \subseteq \mathbb{R}^m$, we denote the class of continuous functions taking values in $\Omega \subseteq \mathbb{R}^m$. A continuous mapping $F : C^0([-r, 0]; \mathbb{R}^l) \times \mathbb{R}^m \to \mathbb{R}^n$ is said to be Lipschitz on bounded sets if there exists a non-decreasing function $Q : \mathbb{R}_+ \to \mathbb{R}_+$ such that $|F(x, u) - F(y, u)| \leq Q\,(\|x\| + \|y\| + |u|)\,\|x - y\|$ for all $x, y \in C^0([-r, 0]; \mathbb{R}^l)$ and for all $u \in \mathbb{R}^m$.

3 Statement of Main Results

We consider a time-invariant control system of the form

$$\begin{array}{l} \dot{x}(t) = f(x(t), u(t-\tau)) + Gv(t) \\ x \in \mathbb{R}^n\ ,\ u \in \mathbb{R}^m\ ,\ v \in \mathbb{R}^q \end{array} ,\ t \geq 0 \tag{1}$$

where $f : \mathbb{R}^n \times \mathbb{R}^m \to \mathbb{R}^n$ is a continuous vector field with $f(0,0) = 0$, $\tau > 0$ is a constant, and $G \in \mathbb{R}^{n\times q}$ is a real constant matrix. The input $v(t) \in \mathbb{R}^q$ quantifies the effect of possible modeling errors. We assume that the following assumptions hold for system (1).

(H1) *There exist a continuous mapping $k : \mathbb{R}^n \to \mathbb{R}^m$ with $k(0) = 0$, a constant $\mu > 0$ and a symmetric, positive definite matrix $P \in \mathbb{R}^{n\times n}$ such that the following inequality holds for all $x \in \mathbb{R}^n$:*

$$x' P f(x, k(x)) \leq -4\mu |x|^2 \tag{2}$$

(H2) *There exist constants $L_1, L_2, K \geq 0$ such that the following inequalities hold for all $x, y \in \mathbb{R}^n$, $u, v \in \mathbb{R}^m$:*

$$|k(x) - k(y)| \leq K |x - y| \tag{3}$$

$$|f(x,u) - f(y,u)| \leq L_1 |x - y| \tag{4}$$

$$|f(x,u) - f(x,v)| \leq L_2 |u - v| \tag{5}$$

(H3) *There exist matrices $L \in \mathbb{R}^{n\times p}$, $H \in \mathbb{R}^{p\times n}$, a constant $\omega > 0$ and a symmetric, positive definite matrix $Q \in \mathbb{R}^{n\times n}$ such that the following inequality holds for all $x, e \in \mathbb{R}^n$, $u \in \mathbb{R}^m$:*

$$e' Q \left(f(x+e, u) - f(x,u) + LHe\right) \leq -2\omega |e|^2 \tag{6}$$

Discussion of the assumptions: Assumption (H1) guarantees that the "continuously applied" feedback law $u(t) = k(x(t))$ would globally exponentially stabilize the equilibrium point $0 \in \mathbb{R}^n$ of system (1) if the input delay τ were absent, i.e., if $\tau = 0$. Assumption (H2) guarantees that both the "nominal" feedback law $k : \mathbb{R}^n \to \mathbb{R}^m$ and the mapping $f : \mathbb{R}^n \times \mathbb{R}^m \to \mathbb{R}^n$ are globally Lipschitz mappings. Assumption (H3) guarantees that the system $\dot{z}(t) = f(z(t), u(t-\tau)) + L\,(Hz(t) - y(t))$ would be a global exponential observer for system (1) provided that the output $y(t) = Hx(t)$ were available for all $t \geq 0$ and that no modeling errors were present.

System (1) under assumptions (H1), (H2), and (H3) would be globally exponentially stabilized by the dynamic output feedback law $\dot{z}(t) = f(z(t), u(t)) + L\,(Hz(t) - y(t))$ with $u(t) = k(z(t))$ if (a) the input delay τ were absent, (b) the input $u(t)$ were allowed to be continuously adjusted, (c) no modeling errors were present, and (d) the output $y(t) = Hx(t)$ were available for all $t \geq 0$. In this work, we will assume that none of the previous requirements hold. More specifically, we assume that:

- The output measurement is sampled, corrupted and delayed, i.e., there is a partition $\{\tau_i\}_{i=0}^{\infty}$ of $\mathbb{R}_+$ with $\sup_{i\geq 0}(\tau_{i+1} - \tau_i) \leq T_s$ where $T_s > 0$ is a constant, an input $\xi \in L^{\infty}_{loc}(\mathbb{R}_+; \mathbb{R}^p)$ and a constant $r \geq 0$ so that $y(t) = Hx(\tau_i) + \xi(\tau_i)$, for all

$t \in [\tau_i, \tau_{i+1})$, $i = 0, 1, 2, \ldots$. The number $T_s > 0$ is called the upper diameter of the partition and is known, while the sampling times $\{\tau_i\}_{i=0}^{\infty}$ are not known.

- The input delay τ is present and modeling errors are present as well.
- The input cannot be continuously adjusted and can only be applied with ZOH, i.e., there exists a constant $T_H > 0$ (the holding period) such that $u(t) = u_l \in \mathbb{R}^m$, for all $t \in [lT_H, (l+1)T_H)$, $l = 0, 1, 2, \ldots$.

One (or all) of the above complications are present when the control system is networked and there are communication or computation delays along the operation of the network; see the discussion in [3]. Furthermore, the existence of sampled, corrupted and delayed measurements is common for (bio)chemical processes and the inability of continuous adjustment of the input is also common for many systems. Our main result is given next.

Theorem 1 *Consider system* (1) *under assumptions (H1), (H2), and (H3). Let* $T_s > 0$, $T_H > 0$, *and* $r \geq 0$ *be real constants and* $N > 0$ *be an integer that satisfy the inequalities*

$$\begin{array}{l} \frac{\exp(L_1 T_H)(L_1+L_2K)T_H}{1-\exp(L_1 T_H)(L_1+L_2K)T_H} \left| P \right| L_2 K \leq 2\mu \text{ and} \\ \sqrt{\frac{K_4}{2K_3}} \left| QL \right| \left| H \right| L_1 T_s < \omega, L_1(r+\tau) < N \end{array} \tag{7}$$

where K_3 *and* $K_4 > 0$ *are constants that satisfy* $K_3 |x|^2 \leq x'Qx \leq K_4 |x|^2$ *for all* $x \in \mathbb{R}^n$. *Then for every* $c > 0$, *there exist constants* $\theta > 0$ *and* $\Theta > 0$ *such that for every partition* $\{\tau_i\}_{i=0}^{\infty}$ *of* $\mathbb{R}_+$ *satisfying*

$$\sup_{i \geq 0} (\tau_{i+1} - \tau_i) \leq T_s \tag{8}$$

and for all choices of $x_0 \in C^0([-r, 0]; \mathbb{R}^n)$, $z_{j,0} \in C^0([-\delta, 0]; \mathbb{R}^n)$ $(j = 0, \ldots, N)$, $\breve{u}_0 \in L^\infty([-r-\tau, 0); \mathbb{R}^m)$, $v \in L^\infty_{loc}(\mathbb{R}_+; \mathbb{R}^q)$, *and* $\xi \in L^\infty_{loc}(\mathbb{R}_+; \mathbb{R}^p)$, *the solution*

$$(x(t), z_0(t), \ldots, z_N(t), u(t), w(t)) \in \mathbb{R}^n \times \mathbb{R}^{Nn} \times \mathbb{R}^m \times \mathbb{R}^p \tag{9}$$

of the system (1) *with the choices*

$$\dot{z}_0(t) = f(z_0(t), u(t-\tau-r)) + L\,(Hz_0(t) - w(t))\,,\ t \geq 0 \tag{10}$$

$$\begin{array}{l} \dot{z}_j(t) = \dot{z}_{j-1}(t) \\ \quad + f(z_j(t), u(t+j\delta-r-\tau)) - f(z_j(t-\delta), u(t+(j-1)\delta-r-\tau)) \\ \quad - c\left(z_j(t) - z_{j-1}(t) - \int_{t-\delta}^{t} f(z_j(s), u(s+j\delta-r-\tau))ds\right), \\ \quad t \geq 0,\ j = 1, \ldots, N \end{array} \tag{11}$$

$$\dot{w}(t) = Hf(z(t), u(t-\tau-r)), t \in [\tau_i, \tau_{i+1}), i = 0, 1, 2, \ldots \tag{12}$$

$$w(\tau_i) = Hx(\tau_i - r) + \xi(\tau_i), i = 0, 1, 2, ... \tag{13}$$

$$u(t) = k\left(z_N(lT_H)\right), t \in [lT_H, (l+1)T_H), l = 0, 1, 2, ... \tag{14}$$

and $\delta = (r+\tau)/N$, *and with the initial conditions* $x(s) = (x_0)(s)$ *for* $s \in [-r, 0]$, $z_j(s) = (z_{j,0})(s)$ *for* $s \in [-\delta, 0]$, *and* $u(s) = (\breve{u}_0)(s)$ *for* $s \in [-r-\tau, 0)$ *corresponding to any inputs* $v \in L^\infty_{loc}(\mathbb{R}_+; \mathbb{R}^q)$ *and* $\xi \in L^\infty_{loc}(\mathbb{R}_+; \mathbb{R}^p)$, *is unique, exists for all* $t \geq 0$ *and satisfies the following estimates*:

$$\begin{aligned}
&|x(t)| \leq \sqrt{\tfrac{K_2}{K_1}} \exp\left(-\theta(t-\tau)\right) \sup_{0\leq s\leq\tau} (|x(s)|) \\
&+\Xi_1\Xi_2\Omega\Lambda^N \tfrac{|QL|\exp(\theta\,(\tau+T_H))}{1-g} \sup_{0\leq s\leq t} (|\xi\,(s)|) \\
&+\Xi_1\left(|PG| + |G|\,|P|\,C + \Omega\exp\left(\theta\,T_H\right)\tfrac{\Lambda^{N+1}-\Lambda}{\Lambda-1}\,|G|\right)\exp\left(\theta\,\tau\right) \sup_{0\leq s\leq t} (|v(s)|) \\
&+\Xi_1\Xi_2\Omega\Lambda^N \tfrac{\exp(\theta\,(\tau+T_H))}{1-g}\left(|QG| \times \exp(-\theta\,r) + |QL|\,|HG|\,T_s\right) \sup_{0\leq s\leq t} (|v\,(s)|) \\
&+\Xi_1\Omega\exp\left(-\theta\,(t-\tau-T_H)\right)\tfrac{\Lambda^{N+1}-\Lambda}{\Lambda-1}\left(L_2\delta\,\|\breve{u}_0\| + 3\max_{l=1,\dots,N}\left(\left\|z_{l,0}\right\|\right)\right) \\
&+\Xi_1\Omega\exp\left(-\theta\,(t-r-\tau-T_H)\right) \sup_{0\leq s\leq r} (|x\,(s+\tau) - z_N(s)|) \\
&+\Xi_1\Omega\exp\left(-\theta\,(t-r-\tau-T_s-T_H)\right)\tfrac{\Lambda^N}{1-g}\sqrt{\tfrac{K_4}{K_3}} \sup_{r\leq s\leq r+T_s} (|x\,(s-r) - z_0\,(s)|) \\
&+\Xi_1\Omega\exp\left(-\theta\,(t-r-\tau-T_H)\right)\tfrac{\Lambda^N-1}{\Lambda-1} \\
&\times \max_{j=1,\dots,N}\left(\sup_{-\delta\leq s\leq r}\left(\left|x\,(s-r+j\delta) - z_j\,(s)\right|\right)\right)
\end{aligned} \tag{15}$$

and

$$\begin{aligned}
\|x_t\| + \textstyle\sum_{j=0}^N \left\|z_{j,t}\right\| + \|\breve{u}_t\| \leq \Theta\Big(&\exp\left(-\theta\,t\right)\left(\|x_0\| + \textstyle\sum_{j=0}^N \left\|z_{j,0}\right\| + \|\breve{u}_0\|\right) + \\
&\sup\nolimits_{0\leq s\leq t} (|v(s)|) + \sup\nolimits_{0\leq s\leq t} (|\xi\,(s)|)\Big),
\end{aligned} \tag{16}$$

where

$$\begin{aligned}
&\Omega = |P|\,L_2K(1+C),\ \ \Xi_1 = \sqrt{\tfrac{K_2}{2\mu K_1(\mu-\theta K_2)}}, \\
&\Xi_2 = \sqrt{\tfrac{K_4}{2\omega K_3(\omega-\theta K_4)}},\ \ \Lambda = \tfrac{\theta}{\theta-L_1(\exp(\theta\,\delta)-1)}, \\
&g = \sqrt{\tfrac{K_4}{2\omega K_3(\omega-\theta K_4)}}\,|QL|\,|H|\,L_1\tfrac{\exp(\theta\,T_s)-1}{\theta},\ \text{ and } C = \tfrac{\exp(L_1T_H)L_2KT_H}{1-\exp(L_1T_H)(L_1+L_2K)T_H}.
\end{aligned} \tag{17}$$

Inequality (16) guarantees the input-to-state stability property (as defined in [14]) for the closed-loop system given by (1), (10)–(14) with respect to modeling errors and measurement noise. More specifically, estimate (16) shows that the gain function for the external inputs v and ξ is linear. On the other hand, inequality (15) allow us to estimate explicitly the asymptotic gains of the external inputs v and ξ to the output $Y(t) = x(t)$: the asymptotic gain of the modeling error v is guaranteed to be any number greater than

$$\Xi_1 \left(|PG| + |G||P|C + \Omega \frac{\Lambda^{N+1} - \Lambda}{\Lambda - 1} |G| + \frac{\Xi_2 \Omega \Lambda^N}{1-g} |QG| + \frac{\Xi_2 \Omega \Lambda^N}{1-g} |QL||HG| T_s \right)$$

and the asymptotic gain of the measurement error ξ is guaranteed to be any number greater than

$$\Xi_1 \Xi_2 \Omega \Lambda^N \frac{|QL|}{1-g}, \quad \text{where}$$

$$\begin{aligned} &\Omega = |P| L_2 K (1+C),\ \Xi_1 = \tfrac{1}{\mu}\sqrt{\tfrac{K_2}{2K_1}},\ \Xi_2 = \tfrac{1}{\omega}\sqrt{\tfrac{K_4}{2K_3}},\ \Lambda = \tfrac{1}{1-L_1\delta}, \\ &g = \Xi_2 |QL||H| L_1 T_s,\ \text{and } C = \tfrac{\exp(L_1 T_H) L_2 K T_H}{1-\exp(L_1 T_H)(L_1+L_2 K) T_H}. \end{aligned} \tag{18}$$

Robustness to perturbations of the sampling schedule is also guaranteed.

Estimates (15) and (16) are independent of the lower diameter of the sampling partition (i.e., of $\inf_{i \geq 0}(\tau_{i+1} - \tau_i)$). This feature is in sharp contrast with the result in [7]. This difference is explained by a different methodology in the proof; if the same methodology were followed in [7] then a similar result would be proved. A few words are needed for the explanation of the hybrid dynamic feedback given by (10)–(14).

- (10) is an observer for the delayed state vector $x(t-r)$. However, (10) does not use the continuous signal $y(t) = Hx(t-r)$, which is not available. The signal $w(t)$ replaces the output signal $y(t) = Hx(t-r)$.
- (12), (13) is an inter-sample predictor for the non-available output signal; it uses the output values at the sampling times and "tries" to predict the output signal between two consecutive sampling times.
- System (11) is an approximate predictor of the future value of the state vector $x(t+\tau)$. The approximate predictor uses the estimated value $z_0(t)$ of the delayed state vector $x(t-r)$, which is provided by the observer, and provides $z_N(t)$, which is an approximation of $x(t+\tau)$.
- Finally, (14) is the "nominal" feedback law computed at the predicted value of the future state vector $x(t+\tau)$ applied with ZOH (emulation).

Remarks:

- Contrary to the approach in [15], no tradeoff between the delays (in the input and the output) and the upper diameter of the sampling partition and the holding period is present for our control scheme.
- From (7), we see that for long delays, we can use sufficiently many predictors that ensure the robustness properties (15) and (16). The counterpart of this is that the gains corresponding to measurement and modeling errors will increase.

4 Key Lemmas

The proof of Theorem 1 is in the following section and is demanding because even the existence/uniqueness of the solution of the overall closed-loop system is not trivial, since the closed-loop system is a hybrid system with delays. For the proof of Theorem 1, we need the following two lemmas, which are stated below. Their proofs are simple and are omitted.

Lemma 1 *Let $r \geq 0$ be a constant and $F : C^0([-r, 0]; \mathbb{R}^n) \times \mathbb{R}^m \to \mathbb{R}^n$ be a continuous mapping that is Lipschitz on bounded sets and satisfies the inequality $|F(x, u)| \leq L \|x\| + M |u|$ for all $(x, u) \in C^0([-r, 0]; \mathbb{R}^n) \times \mathbb{R}^m$ for certain constants $L > 0$ and $M \geq 0$. Then for every $t_0 \geq 0$, $b \in (t_0, +\infty)$, $x_0 \in C^0([-r, 0]; \mathbb{R}^n)$, and $u \in L^\infty([t_0 - r, b); \mathbb{R}^m)$, the unique solution of $\dot{x}(t) = F(x_t, u(t))$ with initial condition $x(t_0 + s) = (x_0)(s)$ for all $s \in [-r, 0]$ exists for all $t \in [t_0, b]$ and satisfies the estimate*

$$\|x_t\| \leq \exp\left(L(t - t_0)\right)\left(\left\|x_{t_0}\right\| + M \int_{t_0}^{t} |u(s)|\, ds\right) \tag{19}$$

for all $t \in [t_0, b]$.

Lemma 2 *Let $P \in \mathbb{R}^{n\times n}$ be a symmetric, positive definite matrix and $K_1 > 0$ and $K_2 > 0$ be constants such that the inequality $K_1 |x|^2 \leq x'Px \leq K_2 |x|^2$ holds for all $x \in \mathbb{R}^n$. Let $x : [t_0, b) \to \mathbb{R}^n$ be an absolutely continuous mapping that satisfies the inequality*

$$x'(t)P\dot{x}(t) \leq -c\,|x(t)|^2 + \sum_{j=1}^{m} a_j \left|v_j(t)\right|^2 \tag{20}$$

for $t \in [t_0, b)$ a.e., where $t_0 \geq 0$, $b \in (t_0, +\infty]$, $c > 0$, and $a_j \geq 0$ $(j = 1, ..., m)$ are constants, $p_j \geq 1$ are integers and $v_j \in L^\infty_{loc}(\mathbb{R}_+; \mathbb{R}^{p_j})$ $(j = 1, ..., m)$ are measurable and locally essentially bounded functions. Then for every $\mu \in (0, c/K_2)$ and for every $t \in [t_0, b)$, the estimate

$$\begin{aligned} &\sup\nolimits_{t_0 \leq s \leq t} \left(|x(s)| \exp(\mu s)\right) \leq \\ &\sqrt{\tfrac{K_2}{K_1}}\, |x(t_0)| \exp(\mu t_0) + \sqrt{\tfrac{K_2}{K_1(c-\mu K_2)}} \sum_{j=1}^{m} \sqrt{a_j} \sup_{t_0 \leq s \leq t} \left(\left|v_j(s)\right| \exp(\mu s)\right) \end{aligned} \tag{21}$$

holds.

5 Proof of Theorem 1

We divide the proof into the following three parts:

Part I: Existence and Uniqueness of Solutions

In this part of the proof, we show that for every $x_0 \in C^0([-r,0];\mathbb{R}^n)$, $z_{j,0} \in C^0([-\delta,0];\mathbb{R}^n)$ $(j = 0,\ldots,N)$, $\breve{u}_0 \in L^\infty([-r-\tau,0);\mathbb{R}^m)$, $v \in L^\infty_{loc}(\mathbb{R}_+;\mathbb{R}^q)$, and $\xi \in L^\infty_{loc}(\mathbb{R}_+;\mathbb{R}^p)$, the closed-loop system given by (1), (10)-(14) has a unique solution with initial condition $x(s) = (x_0)(s)$ for $s \in [-r,0]$, $z_j(s) = (z_{j,0})(s)$ for $s \in [-\delta,0]$, and $u(s) = (\breve{u}_0)(s)$ for $s \in [-r-\tau,0)$ corresponding to inputs $v \in L^\infty_{loc}(\mathbb{R}_+;\mathbb{R}^q)$ and $\xi \in L^\infty_{loc}(\mathbb{R}_+;\mathbb{R}^p)$, and defined for all $t \geq 0$.

Part II: Proof of (15)

In this part of the proof, we show that estimate (15) holds for an appropriate constant $\theta > 0$.

Part III: Proof of (16)

In this part of the proof, we show that estimate (16) holds.

Part I: Existence and Uniqueness of Solutions

First we prove that for every $x_0 \in C^0([-r,0];\mathbb{R}^n)$, $z_{j,0} \in C^0([-\delta,0];\mathbb{R}^n)$ $(j = 0,\ldots,N)$, $\breve{u}_0 \in L^\infty([-r-\tau,0);\mathbb{R}^m)$, $v \in L^\infty_{loc}(\mathbb{R}_+;\mathbb{R}^q)$, and $\xi \in L^\infty_{loc}(\mathbb{R}_+;\mathbb{R}^p)$, the closed-loop system given by (1)–(14) has a unique solution with initial condition $x(s) = (x_0)(s)$ for $s \in [-r,0]$, $z_j(s) = (z_{j,0})(s)$ for $s \in [-\delta,0]$, and $u(s) = (\breve{u}_0)(s)$ for $s \in [-r-\tau,0)$ corresponding to inputs $v \in L^\infty_{loc}(\mathbb{R}_+;\mathbb{R}^q)$ and $\xi \in L^\infty_{loc}(\mathbb{R}_+;\mathbb{R}^p)$. The solution is defined for all $t \geq 0$ and is constructed step-by-step using the following claim:

Claim 1 Assume that there exist an integer $a \geq 0$ and $x \in C^0([-r,aT_H];\mathbb{R}^n)$ and $z_j \in C^0([-\delta,aT_H];\mathbb{R}^n)$ for $j = 0,\ldots,N$ that are absolutely continuous on $[0,aT_H]$, and $w \in PC([0,aT_H];\mathbb{R}^p)$ and $u \in L^\infty([-r-\tau,aT_H);\mathbb{R}^m)$ that satisfy $x(s) = (x_0)(s)$ for $s \in [-r,0]$ and $z_j(s) = (z_{j,0})(s)$ for all $s \in [-\delta,0]$ and $u(s) = (\breve{u}_0)(s)$ for all $s \in [-r-\tau,0)$, as well as Eqs. (1), (10), (11), and (12) for $t \in [0,aT_H)$ a.e., Eq. (13) for all integers $i \geq 0$ with $\tau_i \leq aT_H$, and Eq. (14) for $l = 0,\ldots,a-1$ (only when $a > 0$). Then there exist $x \in C^0([-r,(a+1)T_H];\mathbb{R}^n)$ and $z_j \in C^0([-\delta,(a+1)T_H];\mathbb{R}^n)$ $(j = 0,\ldots,N)$ that are absolutely continuous on $[0,(a+1)T_H]$, $w \in PC([0,(a+1)T_H];\mathbb{R}^p)$, and $u \in L^\infty([-r-\tau,(a+1)T_H);\mathbb{R}^m)$ that satisfy $x(s) = (x_0)(s)$ for $s \in [-r,0]$, $z_j(s) = (z_{j,0})(s)$ for $s \in [-\delta,0]$, $u(s) = (\breve{u}_0)(s)$ for $s \in [-r-\tau,0)$, Eqs. (1), (10)–(12) for $t \in [0,(a+1)T_H)$ a.e., Eq. (13) for all integers $i \geq 0$ with $\tau_i \leq (a+1)T_H$, and Eq. (14) for $l = 0,\ldots,a$.

Proof (Claim 1) Using (14) for $l = a$, we can (uniquely) define u on $[aT_H,(a+1)T_H)$. Since u is constant on $[aT_H,(a+1)T_H)$, we know that $u \in L^\infty([-r-\tau,(a+1)T_H);\mathbb{R}^m)$.

Since the right-hand side of (1) satisfies a linear growth condition and since u is defined on $[-r-\tau,(a+1)T_H)$, it follows from Lemma 1 that we can uniquely define x on $[aT_H,(a+1)T_H]$. The extended mapping $x : [-r,(a+1)T_H] \to \mathbb{R}^n$

satisfies $x \in C^0([-r,(a+1)T_H];\mathbb{R}^n)$, is absolutely continuous on $[0,(a+1)T_H]$, and satisfies (1) for $t \in [0,(a+1)T_H)$ a.e.

Since $\lim_{i\to\infty}\tau_i = +\infty$, there are only a finite number of sampling times τ_i in the interval $[aT_H,(a+1)T_H]$ (and possibly none). The right-hand sides of (10) and (12) satisfy a linear growth condition and since u is defined on $[-r-\tau,(a+1)T_H)$ and x is defined on $[-r,(a+1)T_H]$, it follows from Lemma 1 that we can uniquely define (z_0,w) on $[aT_H,(a+1)T_H]$. The extended mapping $z_0:[-\delta,(a+1)T_H]\to\mathbb{R}^n$ satisfies $z_0\in C^0([-\delta,(a+1)T_H];\mathbb{R}^n)$, is absolutely continuous on $[0,(a+1)T_H]$ and satisfies (10) and (12) for $t\in[0,(a+1)T_H)$ a.e.. Moreover, the extended mapping $w:[0,(a+1)T_H]\to\mathbb{R}^p$ satisfies $w\in PC([0,(a+1)T_H];\mathbb{R}^p)$.

Finally, using (11) and Lemma 1, we can define z_1 and next $z_2,\ldots,z_N$ on $[aT_H,(a+1)T_H]$. The extended mappings $z_j:[-\delta,(a+1)T_H]\to\mathbb{R}^n$ (for $j=1,\ldots,N$) satisfy $z_j\in C^0([-\delta,(a+1)T_H];\mathbb{R}^n)$, are absolutely continuous on $[0,(a+1)T_H]$ and satisfy (11) for $t\in[0,(a+1)T_H)$ a.e.
Therefore, the claim holds. ◁

Part II: Proof of (15)

We next present three inequalities, which are direct consequences of (1), (2), (6) and (10):

$$\begin{aligned}&(z_0(t)-x(t-r))'\,Q\tfrac{d}{dt}\,(z_0(t)-x(t-r))\le -2\omega\,|z_0(t)-x(t-r)|^2\\&\quad-(z_0(t)-x(t-r))'\,QGv(t-r)-(z_0(t)-x(t-r))'\,QL\,(w(t)-Hx(t-r))\\&\text{for a.e. } t\ge T,\end{aligned}\tag{22}$$

$$\begin{aligned}&x'(t)P\dot{x}(t)\le\\&-4\mu\,|x(t)|^2+x'(t)PGv(t)-x'(t)P\,(f(x(t),k(x(t)))-f(x(t),u(t-\tau)))\,,\\&\text{for a.e. } t\ge 0,\ \text{and}\end{aligned}\tag{23}$$

$$\begin{aligned}&z_0'(t)Q\dot{z}_0(t)\le -2\omega\,|z_0(t)|^2+z_0'(t)Qf(0,u(t-r-\tau))-z_0'(t)QLw(t),\\&\text{for a.e. } t\ge 0,\end{aligned}\tag{24}$$

where $T=\min\{\tau_i\,:\,\tau_i\ge r\,,\,i=1,2,\ldots\}$ is the smallest sampling time for which $\tau_i\ge r$ holds.

The following equations hold for all $j=1,\ldots,N$, $t\ge 0$ and are direct consequences of (11):

$$\begin{aligned}&z_j(t)=z_{j-1}(t)+\int_{t-\delta}^{t}f(z_j(s),u(s+j\delta-r-\tau))ds+\\&exp\,(-ct)\left(z_j(0)-z_{j-1}(0)-\int_{-\delta}^{0}f(z_j(s),u(s+j\delta-r-\tau))ds\right)\\&\text{for all } t\ge 0\end{aligned}\tag{25}$$

$$\begin{aligned} &x(t-r+j\delta) = \\ &x(t-r+(j-1)\delta) + \int_{t-\delta}^{t} f(x(s-r+j\delta), u(s+j\delta-r-\tau))ds \\ &+ \int_{t-r+(j-1)\delta}^{t-r+j\delta} Gv(s)ds \text{ for all } t \geq r \end{aligned} \tag{26}$$

Completing the squares in (22) and (24), and using (23) and (5), we get:

$$\begin{aligned} &(z_0(t) - x(t-r))' \, Q \tfrac{d}{dt} (z_0(t) - x(t-r)) \leq \\ &-\omega \, |z_0(t) - x(t-r)|^2 + \tfrac{1}{2\omega} \, |QG|^2 \, |v(t-r)|^2 \\ &+ \tfrac{1}{2\omega} \, |QL|^2 \, |w(t) - Hx(t-r)|^2 \text{ for a.e. } t \geq T \end{aligned} \tag{27}$$

$$\begin{aligned} &x'(t+\tau) P \dot{x}(t+\tau) \leq \\ &-4\mu \, |x(t+\tau)|^2 + |PG| \, |v(t+\tau)| \, |x(t+\tau)| \\ &+ |P| \, L_2 \, |x(t+\tau)| \, |k(x(t+\tau)) - u(t)| \text{ for a.e. } t \geq 0. \end{aligned} \tag{28}$$

$$\begin{aligned} z_0'(t) Q \dot{z}_0(t) \leq -\omega \, |z_0(t)|^2 + \tfrac{1}{2\omega} \, |Q|^2 \, L_2^2 \, |u(t-r-\tau)|^2 \\ + \tfrac{1}{2\omega} \, |QL|^2 \, |w(t)|^2 \text{, for a.e. } t \geq 0. \end{aligned} \tag{29}$$

For all $t \geq 0$, we can use (14) to get $u(t) = k\left(z_N(lT_H)\right)$, where $l = [t/T_H]$. It follows from Eq. (1) that

$$\begin{aligned} &|x(t+\tau) - x(lT_H+\tau)| \leq \\ &\int_{lT_H+\tau}^{t+\tau} |f(x(s), u(s-\tau))| \, ds + |G| \int_{lT_H+\tau}^{t+\tau} |v(s)| \, ds. \end{aligned} \tag{30}$$

Therefore, we get the following for all $t \geq 0$, where $l = [t/T_H]$:

$$\begin{aligned} &|x(t+\tau) - x(lT_H+\tau)| \leq L_1 \int_{lT_H}^{t} |x(s+\tau) - x(lT_H+\tau)| \, ds \\ &+ (L_1 + L_2 K) \, T_H \, |x(lT_H+\tau)| \\ &+ L_2 K T_H \, |z_N(lT_H) - x(lT_H+\tau)| + |G| \, T_H \sup_{lT_H \leq s \leq t} \left(|v(s+\tau)|\right) \end{aligned}$$

(using the triangle inequality, (4) and (5) and the fact that $t \in [lT_H, (l+1)T_H)$) and

$$\begin{aligned} &|x(t+\tau) - x(lT_H+\tau)| \leq \exp\left(L_1 T_H\right) (L_1 + L_2 K) \, T_H \, |x(lT_H+\tau)| \\ &+ \exp\left(L_1 T_H\right) L_2 K T_H \, |z_N(lT_H) - x(lT_H+\tau)| \\ &+ \exp\left(L_1 T_H\right) |G| \, T_H \sup_{lT_H \leq s \leq t} \left(|v(s+\tau)|\right) \end{aligned}$$

(using the Gronwall-Bellman lemma and the fact that $t \in [lT_H, (l+1)T_H)$), and

$$\begin{aligned} &|x(t+\tau) - x(lT_H+\tau)| \leq \tfrac{\exp(L_1 T_H)(L_1 + L_2 K) T_H}{1 - \exp(L_1 T_H)(L_1 + L_2 K) T_H} \, |x(t+\tau)| \\ &+ \tfrac{\exp(L_1 T_H) L_2 K T_H}{1 - \exp(L_1 T_H)(L_1 + L_2 K) T_H} \, |z_N(lT_H) - x(lT_H+\tau)| \\ &+ \tfrac{\exp(L_1 T_H) |G| T_H}{1 - \exp(L_1 T_H)(L_1 + L_2 K) T_H} \sup_{lT_H \leq s \leq t} \left(|v(s+\tau)|\right) \end{aligned}$$

(using the triangle inequality and the fact that $\exp\left(L_1 T_H\right) (L_1 + L_2 K) \, T_H < 1$).

The above inequality in conjunction with the triangle inequality, (3), (28) and (14) (which implies that $u(t) = k\,(z_N(lT_H))$, where $l = [t/T_H]$) and the inequality

$$\tfrac{\exp(L_1 T_H)(L_1+L_2K)T_H}{1-\exp(L_1 T_H)(L_1+L_2K)T_H}\,|P|\,L_2K \le 2\mu \tag{31}$$

give

$$\begin{aligned} & x'(t+\tau)P\dot{x}(t+\tau) \le -2\mu\,|x(t+\tau)|^2 \\ & +\tfrac{|P|L_2K(1-L_1T_H\exp(L_1T_H))}{1-\exp(L_1T_H)(L_1+L_2K)T_H}\,|x(t+\tau)|\,|x(lT_H+\tau)-z_N(lT_H)| \\ & +\left(\tfrac{\exp(L_1T_H)|G||P|L_2KT_H}{1-\exp(L_1T_H)(L_1+L_2K)T_H}+|PG|\right)|x(t+\tau)|\sup_{lT_H\le s\le t}\left(|v(s+\tau)|\right) \end{aligned} \tag{32}$$

for $t \ge 0$ a.e. and $l = [t/T_H]$. Completing the squares in (32), we get

$$\begin{aligned} & x'(t+\tau)P\dot{x}(t+\tau) \le \\ & -\mu\,|x(t+\tau)|^2 + \tfrac{1}{2\mu}\left(\tfrac{|P|L_2K(1-L_1T_H\exp(L_1T_H))}{1-\exp(L_1T_H)(L_1+L_2K)T_H}\right)^2|x(lT_H+\tau)-z_N(lT_H)|^2 \\ & = +\tfrac{1}{2\mu}\left(\tfrac{\exp(L_1T_H)|G||P|L_2KT_H}{1-\exp(L_1T_H)(L_1+L_2K)T_H}+|PG|\right)^2\sup_{lT_H\le s\le t}\left(|v(s+\tau)|^2\right) \end{aligned} \tag{33}$$

for $t \ge 0$ a.e. and $l = [t/T_H]$. Combining (25) and (26), we obtain the following for all $j = 1, ..., N$ and $t \ge r$:

$$\begin{aligned} & \left|z_j(t)-x(t-r+j\delta)\right| \le |G|\sup_{t-r+(j-1)\delta\le s\le t-r+j\delta}\left(|v(s)|\right)+ \\ & \left|z_{j-1}(t)-x(t-r+(j-1)\delta)\right| + L_1\int_{t-\delta}^{t}\left|z_j(s)-x(s-r+j\delta)\right|ds \\ & +\exp(-ct)\left|z_j(0)-z_{j-1}(0)-\int_{-\delta}^{0}f(z_j(s),u(s+j\delta-r-\tau))ds\right| \end{aligned} \tag{34}$$

It follows from (34) that the following inequalities hold for all $\theta \in (0, c]$, $j = 1, ..., N$, and $t \ge r$:

$$\begin{aligned} & \sup_{r\le s\le t}\left(\left|z_j(s)-x(s-r+j\delta)\right|\exp(\theta\,s)\right) \le \\ & \qquad \sup_{r\le s\le t}\left(\left|z_{j-1}(s)-x(s-r+(j-1)\delta)\right|\exp(\theta\,s)\right) \\ & \qquad +L_1\tfrac{\exp(\theta\,\delta)-1}{\theta}\sup_{r-\delta\le s\le t}\left(\left|z_j(s)-x(s-r+j\delta)\right|\exp(\theta\,s)\right) \\ & \qquad +\left|z_j(0)-z_{j-1}(0)-\int_{-\delta}^{0}f(z_j(s),u(s+j\delta-r-\tau))ds\right| \\ & \qquad +|G|\exp(\theta\,t)\sup_{0\le s\le t-r+j\delta}\left(|v(s)|\right) \end{aligned} \tag{35}$$

Since $L_1\delta < 1$, there is a small enough $\theta \in (0, c]$ such that $L_1\,(\exp(\theta\,\delta)-1) < \theta$. It follows from (35) that the following inequalities hold for all $\theta \in (0, c]$ sufficiently small, $j = 1, ..., N$ and $t \ge r$:

$$\begin{aligned}&\sup_{r\le s\le t}\left(\left|z_j(s)-x(s-r+j\delta)\right|\exp(\theta\, s)\right)\le\\&\quad\Lambda\sup_{r\le s\le t}\left(\left|z_{j-1}(s)-x(s-r+(j-1)\delta)\right|\exp(\theta\, s)\right)\\&\quad+\sup_{r-\delta\le s\le r}\left(\left|z_j(s)-x(s-r+j\delta)\right|\exp(\theta\, s)\right)\\&\quad+\Lambda\left|G\right|\exp(\theta\, t)\sup_{0\le s\le t-r+j\delta}\left(\left|v(s)\right|\right)\\&\quad+\Lambda\left|z_j(0)-z_{j-1}(0)-\int_{-\delta}^{0}f(z_j(s),u(s+j\delta-r-\tau))ds\right|\end{aligned}\tag{36}$$

where

$$\Lambda=\frac{\theta}{\theta-L_1\left(\exp(\theta\,\delta)-1\right)}.\tag{37}$$

Using (4), (5), (36), the fact that $L_1\delta<1$, and induction, we conclude that the following inequalities hold for all sufficiently small $\theta\in(0,c]$ and all $j=1,\ldots,N$ and $t\ge r$:

$$\begin{aligned}&\sup_{r\le s\le t}\left(\left|z_j(s)-x(s-r+j\delta)\right|\exp\left(\theta\, s\right)\right)\le\\&\quad\Lambda^j\sup_{r\le s\le t}\left(\left|z_0(s)-x(s-r)\right|\exp\left(\theta\, s\right)\right)\\&\quad+\tfrac{\Lambda^j-1}{\Lambda-1}\exp\left(\theta\, r\right)\max_{l=1,\ldots,j}\left(\sup_{r-\delta\le s\le r}\left(\left|z_l(s)-x(s-r+l\delta)\right|\right)\right)\\&\quad+\Lambda\tfrac{\Lambda^j-1}{\Lambda-1}\left(L_2\delta\left\|u_0\right\|+3\max_{l=1,\ldots,j}\left(\left\|z_{l,0}\right\|\right)\right)\\&\quad+\Lambda\tfrac{\Lambda^j-1}{\Lambda-1}\left|G\right|\exp(\theta\, t)\sup_{0\le s\le t-r+j\delta}\left(\left|v(s)\right|\right)\end{aligned}\tag{38}$$

Using Lemma 2 and inequalities (27) and (33) we obtain:

$$\begin{aligned}&\sup_{0\le s\le t}\left(\left|x(s+\tau)\right|\exp\left(\theta\, s\right)\right)\le\sqrt{\tfrac{K_2}{K_1}}\left|x(\tau)\right|\\&+\sqrt{\tfrac{K_2}{2\mu K_1(\mu-\theta K_2)}}\left(\left|PG\right|+\tfrac{\exp(L_1T_H)\left|G\right|\left|P\right|L_2KT_H}{1-\exp(L_1T_H)(L_1+L_2K)T_H}\right)\exp\left(\theta\, t\right)\sup_{\tau\le s\le t+\tau}\left(\left|v(s)\right|\right)\\&+\sqrt{\tfrac{K_2}{2\mu K_1(\mu-\theta K_2)}}\left(\tfrac{\left|P\right|L_2K(1-L_1T_H\exp(L_1T_H))}{1-\exp(L_1T_H)(L_1+L_2K)T_H}\right)\\&\quad\times\exp\left(\theta\, T_H\right)\sup_{0\le s\le t}\left(\left|x\left(s+\tau\right)-z_N\left(s\right)\right|\exp\left(\theta\, s\right)\right)\end{aligned}\tag{39}$$

for all $t\ge 0$ and $\theta\in(0,\mu/K_2)$, where K_1 and $K_2>0$ are constants such that the inequality $K_1\left|x\right|^2\le x'Px\le K_2\left|x\right|^2$ holds for all $x\in\mathbb{R}^n$ and

$$\begin{aligned}&\sup_{T\le s\le t}\left(\left|z_0(s)-x(s-r)\right|\exp\left(\theta\, s\right)\right)\le\sqrt{\tfrac{K_4}{K_3}}\left|z_0(T)-x(T-r)\right|\exp\left(\theta\, T\right)\\&\quad+\sqrt{\tfrac{K_4}{2\omega K_3(\omega-\theta K_4)}}\left|QG\right|\sup_{0\le s\le t-r}\left(\left|v(s)\right|\exp\left(\theta\, s\right)\right)\\&\quad+\sqrt{\tfrac{K_4}{2\omega K_3(\omega-\theta K_4)}}\left|QL\right|\sup_{T\le s\le t}\left(\left|w(s)-Hx(s-r)\right|\exp\left(\theta\, s\right)\right)\end{aligned}\tag{40}$$

for all $t \geq T$ and $\theta \in (0, \omega/K_4)$, where $T = \min\{\tau_i : \tau_i \geq r, i = 1, 2, ...\}$. Combining (39) and (38) for $j = N$, we obtain the following for all $t \geq 0$ and sufficiently small $\theta > 0$:

$$\begin{aligned}
&\sup_{0 \leq s \leq t} (|x(s+\tau)| \exp(\theta s)) \leq \sqrt{\tfrac{K_2}{K_1}}\, |x(\tau)| \\
&+\sqrt{\tfrac{K_2}{2\mu K_1(\mu-\theta K_2)}} \left(|PG| + \tfrac{\exp(L_1 T_H)|G||P|L_2 K T_H}{1-\exp(L_1 T_H)(L_1+L_2K)T_H}\right) \\
&\quad \times \exp(\theta t) \sup_{\tau \leq s \leq t+\tau} (|v(s)|) \\
&+\sqrt{\tfrac{K_2}{2\mu K_1(\mu-\theta K_2)}} \left(\tfrac{|P|L_2K(1-L_1T_H\exp(L_1T_H))}{1-\exp(L_1T_H)(L_1+L_2K)T_H}\right) \\
&\quad \times \exp(\theta (r+T_H)) \sup_{0 \leq s \leq r} (|x(s+\tau) - z_N(s)|) \\
&+\sqrt{\tfrac{K_2}{2\mu K_1(\mu-\theta K_2)}} \left(\tfrac{|P|L_2K(1-L_1T_H\exp(L_1T_H))}{1-\exp(L_1T_H)(L_1+L_2K)T_H}\right) \exp(\theta T_H) \\
&\quad \times \Lambda \tfrac{\Lambda^N - 1}{\Lambda - 1} \left(L_2 \delta \|u_0\| + 3 \max_{l=1,\dots,N} \left(\|z_{l,0}\|\right)\right) \\
&+\sqrt{\tfrac{K_2}{2\mu K_1(\mu-\theta K_2)}} \left(\tfrac{|P|L_2K(1-L_1T_H\exp(L_1T_H))}{1-\exp(L_1T_H)(L_1+L_2K)T_H}\right) \exp(\theta T_H) \\
&\quad \times \Lambda^N \sup_{r \leq s \leq t} (|x(s-r) - z_0(s)| \exp(\theta s)) \\
&+\sqrt{\tfrac{K_2}{2\mu K_1(\mu-\theta K_2)}} \left(\tfrac{|P|L_2K(1-L_1T_H\exp(L_1T_H))}{1-\exp(L_1T_H)(L_1+L_2K)T_H}\right) \exp(\theta (r+T_H)) \\
&\quad \times \tfrac{\Lambda^N - 1}{\Lambda - 1} \max_{j=1,\dots,N} \left(\sup_{-\delta \leq s \leq r} \left(|x(s - r + j\delta) - z_j(s)|\right)\right) \\
&+\sqrt{\tfrac{K_2}{2\mu K_1(\mu-\theta K_2)}} \left(\tfrac{|P|L_2K(1-L_1T_H\exp(L_1T_H))}{1-\exp(L_1T_H)(L_1+L_2K)T_H}\right) \exp(\theta (t+T_H)) \\
&\quad \times \Lambda \tfrac{\Lambda^N - 1}{\Lambda - 1} |G| \sup_{0 \leq s \leq t+\tau} (|v(s)|)
\end{aligned} \tag{41}$$

Combining (1) and (12), we obtain the following for all $t \in [\tau_i, \tau_{i+1})$ with $\tau_i \geq T$:

$$\begin{aligned}
|w(t) - Hx(t-r)| \leq |w(\tau_i) - Hx(\tau_i - r)| + |H|\, L_1 \int_{\tau_i}^t |z_0(s) - x(s-r)|\, ds \\
+ |HG| \int_{\tau_i}^t |v(s-r)|\, ds
\end{aligned} \tag{42}$$

Using (13), (42) and the facts that $t \in [\tau_i, \tau_{i+1})$, $\tau_i \geq T$, and $\sup_{i \geq 0} (\tau_{i+1} - \tau_i) \leq T_s$, we get the following for all $\theta > 0$:

$$\begin{aligned}
&|w(t) - Hx(t-r)| \exp(\theta t) \leq \\
&|\xi(\tau_i)| \exp(\theta t) + |HG|\, T_s \exp(\theta t) \sup_{\tau_i \leq s \leq t} (|v(s-r)|) \\
&+ |H|\, L_1 \sup_{\tau_i \leq s \leq t} (|z_0(s) - x(s-r)| \exp(\theta s))\, \tfrac{\exp(\theta T_s) - 1}{\theta}
\end{aligned} \tag{43}$$

Estimate (43) implies the following estimate for all $t \geq T$ and $\theta > 0$:

$$\begin{aligned}
\sup_{T \leq s \leq t} (|w(s) - Hx(s-r)| \exp(\theta s)) \leq \exp(\theta t) \sup_{T \leq s \leq t} (|\xi(s)|) \\
+ |H|\, L_1 \sup_{T \leq s \leq t} (|z_0(s) - x(s-r)| \exp(\theta s))\, \tfrac{\exp(\theta T_s) - 1}{\theta} \\
+ |HG|\, T_s \exp(\theta t) \sup_{0 \leq s \leq t-r} (|v(s)|)
\end{aligned} \tag{44}$$

Since $\sqrt{\frac{K_4}{2K_3}}\,|QL|\,|H|\,L_1 T_s < \omega$, it follows that there exists $\theta > 0$ such that

$$g = \sqrt{\frac{K_4}{2\omega K_3(\omega-\theta K_4)}}\,|QL|\,|H|\,L_1 \frac{\exp(\theta\, T_s)-1}{\theta} < 1. \tag{45}$$

Combining (40) and (44), we get this for all $t \geq T$ and for all sufficiently small $\theta > 0$:

$$\begin{aligned}&\sup_{T\leq s\leq t}\left(|z_0(s)-x(s-r)|\exp\left(\theta\, s\right)\right) \leq \\ &\quad \tfrac{1}{1-g}\sqrt{\tfrac{K_4}{K_3}}\,|z_0(T)-x(T-r)|\exp\left(\theta\, T\right) \\ &\quad +\tfrac{\exp(\theta\, t)}{1-g}\sqrt{\tfrac{K_4}{2\omega K_3(\omega-\theta K_4)}}\left(|QG|\exp\left(-\theta\, r\right)+|QL|\,|HG|\,T_s\right)\sup_{0\leq s\leq t}\left(|v(s)|\right) \\ &\quad +\tfrac{1}{1-g}\sqrt{\tfrac{K_4}{2\omega K_3(\omega-\theta K_4)}}\,|QL|\exp\left(\theta\, t\right)\sup_{T\leq s\leq t}\left(|\xi(s)|\right)\end{aligned} \tag{46}$$

Combining (41) and (46) and using the fact that $\sup_{i\geq 0}(\tau_{i+1}-\tau_i) \leq T_s$ (which implies that $T = \min\{\tau_i\ :\ \tau_i \geq r\,,\ i = 1, 2, ...\}$ satisfies $T \leq r + T_s$), we obtain estimate (15) for all $t \geq 0$ and sufficiently small $\theta > 0$.

Part III: Proof of (16)

Our strategy for the proof of estimate (16) is described next.

Using (38), (46) and (15) in conjunction with a standard causality argument, we conclude that there exists a constant $\kappa > 0$ such that the following inequality holds for all $t \geq 0$:

$$\begin{aligned}\|x_t\| + \textstyle\sum_{j=0}^{N}\|z_t\| \leq \kappa\exp\left(-\theta\, t\right)\sup_{0\leq s\leq \tau+r+T_s}\left(\|x_s\| + \|\breve{u}_s\| + \textstyle\sum_{j=0}^{N}\left\|z_{j,s}\right\|\right) \\ +\kappa\left(\sup_{0\leq s\leq t}\left(|\xi\,(s)|\right)+\sup_{0\leq s\leq t}\left(|v\,(s)|\right)\right)\end{aligned} \tag{47}$$

Using (3), (14) and the fact that $k(0) = 0$, we obtain the following for all $t \geq 0$:

$$\begin{aligned}\|\breve{u}_t\| \leq \|\breve{u}_0\|\exp\left(-\theta(t-r-\tau)\right) \\ +K\exp\left(-\theta\left(t-r-\tau-T_H\right)\right)\sup_{0\leq s\leq t}\left(|z_N(s)|\exp(\theta s)\right)\end{aligned} \tag{48}$$

Therefore, inequalities (47) and (48) allow us to conclude the existence of a constant $\bar{\kappa} > 0$ such that the following inequality holds for all $t \geq 0$:

$$\begin{aligned}&\|x_t\| + \|\breve{u}_t\| + \textstyle\sum_{j=0}^{N}\|z_t\| \leq \\ &\bar{\kappa}\exp\left(-\theta\, t\right)\sup_{0\leq s\leq \tau+r+T_s}\left(\|x_s\| + \|\breve{u}_s\| + \textstyle\sum_{j=0}^{N}\left\|z_{j,s}\right\|\right) \\ &+\bar{\kappa}\left(\sup_{0\leq s\leq t}\left(|\xi\,(s)|\right)+\sup_{0\leq s\leq t}\left(|v\,(s)|\right)\right)\end{aligned} \tag{49}$$

To show (16), we use inequality (49) and the following claim:

Claim 2 For every integer $i \geq 0$ there exists a constant $M_i > 0$ such that

$$\begin{array}{l} \|x_t\| + \|\breve{u}_t\| + \sum_{j=0}^{N} \left\|z_{j,t}\right\| \leq \\ M_i \left(\|x_0\| + \sum_{j=0}^{N} \left\|z_{j,0}\right\| + \|\breve{u}_0\| + \sup_{0\leq s\leq t} \left(|v(s)|\right) + \sup_{0\leq s\leq t} \left(|\xi(s)|\right) \right) \\ \text{for all } t \in [0, i T_H] \end{array} \quad (50)$$

Indeed, if Claim 2 holds, then inequality (16) is a direct consequence of (49). Therefore, the rest of the proof is devoted to the proof of Claim 2.

Using Lemma 2 and inequality (29) we obtain for all $t \geq 0$:

$$\begin{array}{l} \sup_{0\leq s\leq t} \left(|z_0(s)| \exp\left(\theta\, s\right)\right) \leq \sqrt{\frac{K_4}{K_3}}\, |z_0(0)| \\ +\sqrt{\frac{K_4}{2\omega K_3(\omega-\theta K_4)}}\, |Q|\, L_2 \sup_{0\leq s\leq t} \left(|u(s-r-\tau)| \exp\left(\theta\, s\right)\right) \\ +\sqrt{\frac{K_4}{2\omega K_3(\omega-\theta K_4)}}\, |QL| \sup_{0\leq s\leq t} \left(|w(s)| \exp\left(\theta\, s\right)\right) \end{array} \quad (51)$$

Using (4), (5) and (12) we obtain the following for all $t \in [\tau_i, \tau_{i+1})$ and $i = 0, 1, 2, \ldots$:

$$\begin{array}{l} |w(t)| \leq |w(\tau_i)| + |H|\, L_1 \frac{\exp(-\theta\, \tau_i)-\exp(-\theta\, t)}{\theta} \sup_{\tau_i\leq s\leq t} \left(|z_0(s)| \exp(\theta\, s)\right) \\ + |H|\, L_2 \frac{\exp(-\theta\, \tau_i)-\exp(-\theta\, t)}{\theta} \sup_{\tau_i\leq s\leq t} \left(|u(s-r-\tau)| \exp(\theta\, s)\right) \end{array} \quad (52)$$

Combining the estimate (52) with (13) and using the facts that $t \in [\tau_i, \tau_{i+1})$ and $\sup_{i\geq 0} (\tau_{i+1} - \tau_i) \leq T_s$, we obtain:

$$\begin{array}{l} \sup_{0\leq s\leq t} \left(|w(s)| \exp(\theta\, s)\right) \leq \sup_{0\leq s\leq t} \left(|\xi(s)|\right) \exp(\theta\, t) + \\ \quad |H| \exp(\theta\, (r + T_s)) \sup_{-r\leq s\leq t-r} \left(|x(s)| \exp(\theta\, s)\right) \\ \quad + |H|\, L_1 \frac{\exp(\theta\, T_s)-1}{\theta} \sup_{0\leq s\leq t} \left(|z_0(s)| \exp(\theta\, s)\right) \\ \quad + |H|\, L_2 \frac{\exp(\theta\, T_s)-1}{\theta} \sup_{0\leq s\leq t} \left(|u(s-r-\tau)| \exp(\theta\, s)\right) \end{array} \quad (53)$$

for all $t \geq 0$. Using our bound (45), the combination of (51) and (53) implies the existence of constants $\Theta_i > 0$ $(i = 1, 2, 3, 4)$ such that the following estimate holds for all $t \geq 0$:

$$\begin{array}{l} \sup_{0\leq s\leq t} \left(|z_0(s)| \exp\left(\theta\, s\right)\right) \leq \\ \Theta_1\, |z_0(0)| + \Theta_2 \sup_{0\leq s\leq t} \left(|u(s-r-\tau)| \exp\left(\theta\, s\right)\right) \\ +\Theta_3 \sup_{-r\leq s\leq t-r} \left(|x(s)| \exp\left(\theta\, s\right)\right) + \Theta_4 \exp\left(\theta\, t\right) \sup_{0\leq s\leq t} \left(|\xi(s)|\right) \end{array} \quad (54)$$

Combining (53) and (54), we obtain:

$$\begin{array}{l} \sup_{0\leq s\leq t} \left(|z_0(s)| \exp\left(\theta\, s\right)\right) + \sup_{0\leq s\leq t} \left(|w(s)| \exp\left(\theta\, s\right)\right) \leq \\ \tilde{\Theta}_1\, |z_0(0)| + \tilde{\Theta}_2 \sup_{0\leq s\leq t} \left(|u(s-r-\tau)| \exp\left(\theta\, s\right)\right) \\ +\tilde{\Theta}_3 \sup_{-r\leq s\leq t-r} \left(|x(s)| \exp\left(\theta\, s\right)\right) + \tilde{\Theta}_4 \exp\left(\theta\, t\right) \sup_{0\leq s\leq t} \left(|\xi(s)|\right) \end{array} \quad (55)$$

for all $t \geq 0$, for appropriate constants $\tilde{\Theta}_i > 0$ $(i = 1, 2, 3, 4)$.

Inequality (4) implies the following estimates for all $j = 1, ..., N$ and $t \geq 0$:

$$\begin{aligned} \left\|z_{j,t}\right\| \leq (1 + 2L_1\delta)\left\|z_{j,0}\right\| + 2\sup_{0\leq s\leq t}\left(\left|z_{j-1}(s)\right|\right) \\ + L_1 \int_0^t \left\|z_{j,s}\right\| ds + 2L_2\delta\sup_{0\leq s\leq t}\|\breve{u}_s\| \end{aligned} \tag{56}$$

Using (56) and the Gronwall lemma, we get this for all $j = 1, ..., N$ and $t \geq 0$:

$$\begin{aligned} &\left\|z_{j,t}\right\| \leq \\ &\exp(L_1 t)\left((1 + 2L_1\delta)\left\|z_{j,0}\right\| + 2\sup_{0\leq s\leq t}\left(\left|z_{j-1}(s)\right|\right) + 2L_2\delta\sup_{0\leq s\leq t}\|\breve{u}_s\|\right) \end{aligned} \tag{57}$$

Using (55) and (57) repeatedly, we obtain the following for all $t \geq 0$:

$$\begin{aligned} \sum_{j=0}^{N}\left\|z_{j,t}\right\| \leq \bar{\Theta}\exp(\sigma t)\Big(\sum_{j=0}^{N}\left\|z_{j,0}\right\| + \sup_{-r-\tau\leq s<t}(|u(s)|) \\ + \sup_{-r\leq s\leq t}(|x(s)|) + \sup_{0\leq s\leq t}(|\xi(s)|)\Big) \end{aligned} \tag{58}$$

for appropriate constants $\bar{\Theta} > 0$ and $\sigma > 0$. We are now ready to prove Claim 2.

Proof (Claim 2) We use induction. The claim holds automatically for $i = 0$. Using (3), (14), and (50) for a certain integer $i \geq 0$ and the fact that $k(0) = 0$, we get:

$$\begin{aligned} &\sup_{-r-\tau\leq s<t}(|u(s)|) \leq (1 + K)M_i\Big(\|x_0\| + \sum_{j=0}^{N}\left\|z_{j,0}\right\| + \|\breve{u}_0\| \\ &+ \sup_{0\leq s\leq t}(|v(s)|) + \sup_{0\leq s\leq t}(|\xi(s)|)\Big) \end{aligned} \tag{59}$$

for all $t \in [0, (i + 1)T_H]$. Using (4), (5), Lemma 1 and (59), we obtain

$$\begin{aligned} &\sup_{-r\leq s\leq t}(|x(s)|) \leq R_i\Big(\|x_0\| + \sum_{j=0}^{N}\left\|z_{j,0}\right\| + \|\breve{u}_0\| \\ &+ \sup_{0\leq s\leq t}(|v(s)|) + \sup_{0\leq s\leq t}(|\xi(s)|)\Big) \end{aligned} \tag{60}$$

for all $t \in [0, (i + 1)T_H]$ for an appropriate constant $R_i > 0$. Inequality (50) for $i + 1$ and an appropriate constant $M_{i+1} > 0$ is a direct consequence of (58), (59), and (60). The proof of Claim 2 is complete, so the proof of Theorem 1 is complete.

6 Concluding Remarks

In this chapter, we proposed a novel control scheme for nonlinear globally Lipschitz systems for which the input is delayed and applied with zero order hold, the measurements are sampled and delayed, and only an output is measured. The novelty of our work is in the use of a chain of approximate (dynamic) predictors to handle long delays in both input and output. Using small gain arguments, sufficient conditions on both the upper diameter of the sampling partition and the holding period, and a sufficient number of predictors in the chain, we proved that the closed loop is robust with respect to measurements and modelling errors. We also provided explicit estimates of the asymptotic gains of the external inputs v and ξ. This can be extended easily o networked control systems with uniformly globally exponentially stable scheduling protocols.

References

1. Bekiaris-Liberis, N., Krstic, M.: Robustness of nonlinear predictor feedback laws to time-and state-dependent delay perturbations. Automatica **49**(6), 1576–1590 (2013)
2. Bekiaris-Liberis, N., Krstic, M.: Nonlinear Control Under Nonconstant Delays. SIAM, Philadelphia (2013)
3. Karafyllis, I., Krstic, M.: Nonlinear stabilization under sampled and delayed measurements, and with inputs subject to delay and zero-order hold. IEEE Trans. Autom. Control **57**(5), 1141–1154 (2012)
4. Krstic, M.: Delay Compensation for Nonlinear, Adaptive, and PDE Systems. Birkhauser, Boston (2009)
5. Krstic, M.: Input delay compensation for forward complete and strict-feedforward nonlinear systems. IEEE Trans. Autom. Control **55**(2), 287–303 (2010)
6. Karafyllis, I.: Stabilization by means of approximate predictors for systems with delayed input. SIAM J. Control Optim. **49**(3), 1100–1123 (2011)
7. Karafyllis, I., Krstic, M.: Stabilization of nonlinear delay systems using approximate predictors and high-gain observers. Automatica **49**(12), 3623–3631 (2013)
8. Karafyllis, I., Krstic, M.: Numerical schemes for nonlinear predictor feedback. Math. Control Signals Systems **26**(4), 519–564 (2014)
9. Karafyllis, I., Malisoff, M., de Queiroz, M., Krstic, M., Yang, R.: Predictor-based tracking for neuromuscular electrical stimulation. Int. J. Robust Nonlinear Control doi:10.1002/rnc.3211
10. Ahmed-Ali, T., Karafyllis, I., Lamnabhi-Lagarrigue, F.: Global exponential sampled-data observers for nonlinear systems with delayed measurements. Syst. Control Lett. **62**(7), 539–549 (2013)
11. Germani, G., Manes, C., Pepe, P.: A new approach to state observation of nonlinear systems with delayed output. IEEE Trans. Autom. Control **47**(1), 96–101 (2002)
12. Karafyllis, I., Krstic, M., Ahmed-Ali, T., Lamnabhi-Lagarrigue, F.: Global stabilization of nonlinear delay systems with a compact absorbing set. Int. J. Control **87**(5), 1010–1027 (2014)
13. Sun, X.-M., Liu, K.-Z., Wen, C., Wang, W.: Predictive control of nonlinear continuous networked control systems with large time-varying transmission delays and transmission protocols. Personal Communication (2014)
14. Karafyllis, I., Jiang, Z.-P.: Stability and Stabilization of Nonlinear Systems. Springer-verlag, London (2011)
15. Heemels, W., Teel, A., van de Wouw, N., Nesic, D.: Networked control systems with communication constraints: tradeoffs between transmission intervals, delays and performance. IEEE Trans. Autom. Control **55**(8), 1781–1796 (2010)

Backstepping Designs in the Presence of Non-constant Delays on the Virtual Input

Nikolaos Bekiaris-Liberis, Mrdjan Jankovic and Miroslav Krstic

Abstract We present results on stabilization of nonlinear systems in the strict-feedback form with delays affecting the virtual inputs. We consider systems with constant and time-varying delays, as well as systems with delays that depend on the current or past states, which arise in numerous applications, such as, for example, in cooling systems. The design methodology is based on the concepts of infinite-dimensional backstepping and nonlinear predictor feedback. Several illustrative examples are provided.

1 Introduction

A plethora of methodologies exists dealing with the stabilization of nonlinear systems with input and state delays [12–15, 22, 24, 27–29, 31–33, 36–40, 42, 43, 47]. In recent years much progress has been achieved for stabilization of systems with nonconstant input and state delays [1, 4–7, 9, 11, 17, 41, 49]. The reason is that systems with nonconstant delays are ubiquitous in engineering applications. Some examples include cooling systems [10], automotive engines [11], oil drilling [18], 3D printers [21], networks [19, 20, 51], supply networks [46], and milling processes [2], to name only a few [44].

In the present chapter, we are dealing with a particular form of systems with state delays, namely systems in the strict-feedback form with delays affecting the virtual inputs which appear in applications, such as engine cooling systems [23]. We present the case of a two-block sub-class of nonlinear feedback systems. Yet, a generalization

N. Bekiaris-Liberis (✉)
University of California, Berkeley, Berkeley, CA 94720, USA
e-mail: bekiaris-liberis@berkeley.edu

M. Jankovic
Ford Motor Company, Dearborn, MI 48121, USA
e-mail: mjankov1@ford.com

M. Krstic
University of California, San Diego, La Jolla, CA 92092, USA
e-mail: krstic@ucsd.edu

I. Karafyllis et al. (eds.), *Recent Results on Nonlinear Delay Control Systems*,
Advances in Delays and Dynamics 4, DOI 10.1007/978-3-319-18072-4_3

to the n-block case, as well as to the case of simultaneous input delay, involves a recursive application of the same techniques. This fact reinforces the importance of considering this special class of systems.

Our designs are based on constructing the predictor of the state of the upper, in the integrator chain, subsystem and on a one-step backstepping procedure. Predictor feedback developed originally for delay compensation in unstable linear plants with constant input or state delays [3, 35], whereas backstepping developed originally for stabilization of finite-dimensional nonlinear systems [34].

In this chapter, we review our recently introduced predictor feedback designs for systems in the strict-feedback form with delays affecting the virtual inputs [4–6]. The degree of difficulty in our designs is proportional to the complexity of the dependence of the delays on their arguments. For this reason, we guide the reader from the simplest case of constant delay all the way through to state-dependent delays. We introduce the concept of nonlinear predictor feedback and infinite-dimensional backstepping transformation in Sect. 2 via our treatment of systems with constant delays. In Sect. 3 we consider the case of time-varying delays and highlight some of the design challenges that are imposed from the time-variation of the delay. In Sect. 4 we are dealing with state-dependent delays and we resolve the major design challenge for such systems, which is the definition of the predictor state. In Sect. 5 we extend our methodology for systems with delays that depend on current states to systems that depend on past states. We present several numerical examples in Sect. 7.

2 Constant Delay

We consider the following system

$$\dot{X}_1(t) = f_1\left(X_1(t)\right) + X_2\left(t - D\right) \tag{1}$$

$$\dot{X}_2(t) = f_2\left(X_1(t),\ X_2(t)\right) + U\left(t\right), \tag{2}$$

where, for notational simplicity, we assume $X_1,\ X_2 \in \mathbb{R}$, and $D \geq 0$. One of the challenges in designing a predictor-based control law for plant (1)–(2) is that the control signal reaches the states of the system X_1 and X_2 after different time intervals. It immediately reaches X_2, whereas it reaches X_1 through a delayed integrator. For this reason, a predictor of the state X_1 is employed in our design.

We employ a backstepping design directly on the delay system (1)–(2) which gives the following predictor-based control law

$$\begin{aligned} U(t) = &- f_2\left(X_1(t),\ X_2(t)\right) - c_2\left(X_2(t) + c_1 P_1(t) + f_1\left(P_1(t)\right)\right) \\ &- \left(c_1 + \frac{\partial f_1\left(P_1\right)}{\partial P_1}\right)\left(f_1\left(P_1(t)\right) + X_2(t)\right), \end{aligned} \tag{3}$$

where c_1, c_2 are arbitrary positive constants and

$$P_1(\theta) = X_1(t) + \int_{t-D}^{\theta} \left(f_1\left(P_1(\sigma)\right) + X_2(\sigma)\right) d\sigma, \quad t - D \leq \theta \leq t, \tag{4}$$

for $t \geq 0$. The initial predictor $P_1(\theta)$, $\theta \in [-D, 0]$ is given by (4) for $t = 0$.

The signal $P_1(t)$ represents the D time-units ahead predictor of X_1, i.e., $P_1(t) = X_1(t + D)$. In the case of linear systems, i.e., $f_1(X_1) = aX_1$, where $a \in \mathbb{R}$, the predictor $P_1(t)$ is given explicitly using the variation of constants formula, with the initial condition $P_1(t - D) = X_1(t)$, as $P_1(t) = e^{aD}X_1(t) + \int_{t-D}^{t} e^{a(t-\theta)}X_2(\theta)d\theta$. For systems that are nonlinear, $P_1(t)$ cannot be written explicitly, for the same reason as a nonlinear ordinary differential equation (ODE) cannot be solved explicitly. So we represent $P_1(t)$ implicitly using the nonlinear integral equation (4).

Our design is based on the following infinite-dimensional backstepping transformation of the state X_2

$$Z_2(\theta) = X_2(\theta) + c_1 P_1(\theta) + f_1\left(P_1(\theta)\right), \quad t - D \leq \theta \leq t, \tag{5}$$

which, together with the predictor-based control law given in relations (3), (4) transform the system (1)–(2) to the "target system" given by

$$\dot{X}_1(t) = -c_1 X_1(t) + Z_2(t - D) \tag{6}$$

$$\dot{Z}_2(t) = -c_2 Z_2(t). \tag{7}$$

Note that if X_2 was the actual input to the X_1 subsystem (1) then the term $c_1 P_1(t) + f_1\left(P_1(t)\right)$ would be the predictor feedback controller that would stabilize the X_1 subsystem in this case.

Before stating our main result, the proof of which follows from [5] for the special case of constant delay, we give the following definition.

Definition 1 The system $\dot{X} = f(X, \omega)$ is forward complete with respect to ω if for every initial condition and every measurable locally essentially bounded input signal ω the corresponding solution is defined for all $t \geq 0$.

Theorem 1 *Let $\dot{\Xi} = f_1(\Xi) + \omega$ be forward complete with respect to ω. Consider the closed-loop system consisting of the plant* (1)–(2) *and the controller* (3), (4). *There exists a class $\mathcal{KL}$ function β such that for all initial conditions $X_1 \in \mathbb{R}$, $X_2 \in C[-D, 0]$ the following holds*

$$|X_1(t)| + \sup_{t-D\leq\theta\leq t} |X_2(\theta)| \leq \beta\left(|X_1(0)| + \sup_{-D\leq\theta\leq 0} |X_2(\theta)|, t\right), \tag{8}$$

for all $t \geq 0$.

3 Time-Varying Delay

We consider the following system

$$\dot{X}_1(t) = f_1(X_1(t)) + X_2(\varphi(t)) \tag{9}$$
$$\dot{X}_2(t) = f_2(X_1(t), X_2(t)) + U(t), \tag{10}$$

where, for notational simplicity, we assume $X_1, X_2 \in \mathbb{R}$,

$$\varphi(t) = t - D(t), \tag{11}$$

and D is a positive-valued continuously differentiable functions of time t. We refer to the quantity $t - \varphi(t) = D(t)$ as the delay time. This is the time interval that indicates how long *ago* the virtual control signal that currently affects the plant was actually applied. The main goal of this section is to determine the predictor state, i.e., the quantity P_1 such that $X_1(t) = P_1(\varphi(t))$, or $X_1(\sigma(t)) = P_1(t)$, where $\sigma = \varphi^{-1}$. From now on we refer to the quantity $\sigma(t) - t$ as the prediction horizon. This is the time interval which indicates *after* how long a virtual input signal that is currently applied affects the plant. In the constant delay case, the prediction horizon is equal to the delay time, i.e., $t - \varphi(t) = D = \sigma(t) - t$.

We employ a backstepping design directly on the delay system (9)–(10) which gives the following predictor-based control law

$$\begin{aligned} U(t) =& - f_2(X_1(t), X_2(t)) - c_2(X_2(t) + c_1 P_1(t) + f_1(P_1(t))) \\ & - \left(c_1 + \frac{\partial f_1(P_1)}{\partial P_1}\right)(f_1(P_1(t)) + X_2(t))\frac{d\varphi^{-1}(t)}{dt}, \end{aligned} \tag{12}$$

where c_1, c_2 are arbitrary positive constants and

$$P_1(\theta) = X_1(t) + \int_{t-D(t)}^{\theta} \frac{(f_1(P_1(\sigma)) + X_2(\sigma))\, d\sigma}{1 - D'\left(\varphi^{-1}(\sigma)\right)}, \quad t - D(t) \le \theta \le t, \tag{13}$$

for $t \ge 0$. The initial predictor $P_1(\theta), \theta \in [-D(0), 0]$ is given by (13) for $t = 0$.

From (12) one can observe that the function

$$\frac{d\sigma(\theta)}{d\theta} = \frac{1}{\varphi'\left(\varphi^{-1}(\theta)\right)} \tag{14}$$

is employed in the control law. Therefore, one has to appropriately restrict the delay time $D(t)$ such that $\varphi'(t) \neq 0$ for all $t \ge 0$. Actually, we impose the condition $\varphi'(t) > 0$ for all $t \ge 0$. The reason is that if $\varphi'(t) > 0$ for all $t \ge 0$ then the control signal is able to reach the state X_1. Besides the condition $\varphi'(t) > 0$ for all $t \ge 0$, which can be also expressed in terms of the delay function as $\dot{D}(t) < 1$, for all $t \ge 0$, we also assume that the delay can not disappear instantaneously, i.e., φ' (or $\dot{D}$) is bounded. Also, the delay has to be positive (to guarantee the causality of the system) and bounded (such that the control signal eventually reaches the plant).

We have the following result the proof of which can be found in [6].

Theorem 2 *Let $\dot{X}_1 = f_1(X_1) + \omega$ be forward complete with respect to ω. Let the delay time $D(t) = t - \varphi(t)$ be positive and uniformly bounded from above, and its rate $\dot{D}_1(t)$ be smaller than one and uniformly bounded from below. Consider the closed-loop system consisting of the plant* (9)–(10) *and the controller* (12), (13). *There exists a class $\mathcal{KL}$ function β_v such that for all initial conditions $X_1 \in \mathbb{R}$, $X_2 \in C[-D(0), 0]$ the following holds*

$$|X_1(t)| + \sup_{t-D(t)\leq\theta\leq t} |X_2(\theta)| \leq \beta_v\left(|X_1(0)| + \sup_{-D(0)\leq\theta\leq 0} |X_2(\theta)|, t\right), \quad (15)$$

for all $t \geq 0$.

4 Delay Dependent on Current State

In this section we focus on nonlinear systems in the strict-feedback form, having a state-dependent state delay on the virtual input, i.e., we consider the system

$$\dot{X}_1(t) = f_1(X_1(t), X_2(t - D(X_1(t)))) \quad (16)$$
$$\dot{X}_2(t) = f_2(X_1(t), X_2(t)) + U(t), \quad (17)$$

where, for notational simplicity, we assume $X_2 \in \mathbb{R}$, and D is a nonnegative-valued continuously differentiable function. The main challenge in the case of systems with state-dependent delays is the determination of the predictor state. For systems with constant delays, $D = \text{const}$, the predictor of the state $X_1(t)$ is simply defined as $P_1(t) = X_1(t + D)$. For systems with state-dependent delays finding the predictor $P_1(t)$ is much trickier. The time when the virtual input affects the X_1 system depends on the value of the state X_1 at that time, namely, the following implicit relationship holds $P_1(t) = X_1(t + D(P_1(t)))$ (and $X_1(t) = P_1(t - D(X_1(t)))$).

Assuming that there exists some function κ such that the closed loop system $\dot{\Xi} = f_1(\Xi, \kappa(\Xi) + \omega)$ is input-to-state stable with respect to ω, the predictor-based control law for system (16), (17) is then given by

$$U(t) = -f_2(X_1(t), X_2(t)) - c_2(X_2(t) - \kappa(P_1(t))) + \frac{\frac{\partial\kappa(P_1)}{\partial P_1} f_1(P_1(t), X_2(t))}{1 - \nabla D(P_1(t)) f_1(P_1(t), X_2(t))}, \quad (18)$$

where c_2 is an arbitrary positive constant and the predictor P_1 of the state X_1 is defined as

$$P_1(\theta) = X_1(t) + \int_{t-D(X_1(t))}^{\theta} \frac{f_1\left(P_1(s), X_2(s)\right) ds}{1 - \nabla D\left(P_1(s)\right) f_1\left(P_1(s), X_2(s)\right)},$$
$$t - D\left(X_1(t)\right) \leq \theta \leq t, \tag{19}$$

for all $t \geq 0$. We refer to the quantity $P_1(\theta)$ given in (19) as "predictor" since $P_1(t)$ is the $D\left(P_1(t)\right)$ time units ahead predictor of $X_1(t)$, i.e., $P_1(t) = X_1\left(t + D\left(P_1(t)\right)\right)$. This fact can be seen as follows. Differentiating relation (19) with respect to θ and setting $\theta = t$ we get

$$\frac{dP_1(t)}{dt} = \frac{f_1\left(P_1(t), X_2(t)\right)}{1 - \nabla D\left(P_1(t)\right) f_1\left(P_1(t), X_2(t)\right)}. \tag{20}$$

Performing a change of variables $\tau = \sigma(t) = \varphi^{-1}(t)$ in the ODE for $X_1(\tau)$ given by $\frac{dX_1(\tau)}{d\tau} = f_1\left(X_1(\tau), X_2\left(\tau - D\left(X_1(\tau)\right)\right)\right)$, we have that

$$\frac{dX_1(\sigma(t))}{dt} = \frac{d\sigma(t)}{dt} f_1\left(X_1(\sigma(t)), X_2(t)\right). \tag{21}$$

From (21) one observes that $P_1(t)$ satisfies the same ODE in t as $X_1(\sigma(t))$ because

$$\frac{d\sigma(\theta)}{d\theta} = \frac{1}{1 - \nabla D\left(X_1(\sigma(\theta))\right) f_1\left(X_1(\sigma(\theta)), X_2(\theta)\right)}, \quad t - D\left(X_1(t)\right) \leq \theta \leq t \tag{22}$$

provided that $P_1(t) = X_1(\sigma(t))$. Since from (19) for $t = 0$ and $\theta = -D\left(X_1(0)\right)$ it follows that $P_1\left(-D\left(X_1(0)\right)\right) = X_1(0)$, we get that $P_1(0) = X_1\left(\sigma(0)\right)$. Noting that $D\left(X_1(\sigma(t))\right) = \sigma(t) - t$, differentiating this relation, we get (22). Comparing (20) with (21) we conclude with the help of (22) that $P_1(t) = X_1\left(\sigma(t)\right)$ for all $t \geq 0$.

As in the case of time-varying delays φ' and D must be positive and bounded. The positiveness of φ' (or equivalently of σ') is guaranteed by imposing the following condition on the solutions

$$\mathscr{G}_c: \quad \nabla D\left(P_1(\theta)\right) f_1\left(P_1(\theta), X_2(\theta)\right) < c, \quad \text{for all } \theta \geq -D\left(X_1(0)\right), \tag{23}$$

for $c \in (0, 1]$. We refer to $\mathscr{G}_1$ as the *feasibility condition* of the controller (18)–(19). Due to this condition, we obtain a local result. Boundness of φ' and D is then guaranteed by the boundness of the system's norm. We obtain the following result. Its proof can be found in [5].

Theorem 3 *Let $\dot{\Xi} = f_1(\Xi, \omega)$ be forward complete and $\dot{\Xi} = f_1(\Xi, \kappa(\Xi) + \omega)$ be input-to-state stable with respect to ω. Consider the plant* (16)–(17) *together with the control law given by* (18) *and* (19)*. Then there exist a class $\mathscr{K}$ function ξ_{RoA} and* *a class $\mathcal{KL}$ function β_s such that for all initial conditions $X_1 \in \mathbb{R}^n$ such that X_2 is locally Lipschitz on the interval $[-D(X_1(0)), 0]$ and which satisfy*

$$|X_1(0)| + \sup_{-D(X_1(0)) \le \theta \le 0} |X_2(\theta)| < \xi_{\mathrm{RoA}}(c), \tag{24}$$

for some $0 < c < 1$, we have

$$|X_1(t)| + \sup_{t-D(X_1(t)) \le \theta \le t} |X_2(\theta)| \le \beta_s\left(|X_1(0)| + \sup_{-D(X_1(0)) \le \theta \le 0} |X_2(\theta)|, t\right) \tag{25}$$

for all $t \ge 0$. Furthermore, there exists a class $\mathcal{K}$ function δ^*, such that for all $t \ge 0$, the following holds:

$$D(X_1(t)) \le D(0) + \delta^*(c) \tag{26}$$

$$\left|\dot{D}(X_1(t))\right| \le c. \tag{27}$$

Under the assumptions of Theorem 3, existence and uniqueness of a solution to the closed-loop system with $X_1 \in C^1[0, \infty)$, $X_2 \in C^1(0, \infty)$ it is proved in [5] (see also [9] for the case of a state-dependent input delay).

5 Delay Dependent on Past State

In the present section we consider the following plant

$$\dot{X}_1(t) = f_1(X_1(t), X_2(\varphi(t))) \tag{28}$$

$$\dot{X}_2(t) = U(t), \tag{29}$$

where $t \ge 0$, $U, X_2 \in \mathbb{R}$, $f_1 : \mathbb{R}^n \times \mathbb{R} \to \mathbb{R}^n$ is locally Lipschitz with $f_1(0, 0) = 0$ and φ satisfies

$$t = \varphi(t) + D(X_2(\varphi(t))), \tag{30}$$

where $D \in C^1(\mathbb{R}; \mathbb{R}_+)$. Note that the delay affecting X_2 in (28) is defined implicitly through relation (30) as a nonlinear function of the state X_2 at a past time which depends on the delay itself. Although this is an unusual setup such implicitly defined delays appear in networks [16, 19] and electrodynamics [48].

Remark 1 Let us highlight the importance of considering the class of nonlinear systems that satisfy (28)–(30). When one stabilizes system (28)–(30) then one can stabilize nonlinear systems with input delays that depends on the past input rather than the past state. To see this consider a nonlinear system with input delay that depends on past values of the input, i.e., consider the system

$$\dot{X}_1(t) = f_1\left(X_1(t), V\left(\varphi(t)\right)\right), \tag{31}$$

where $t = \varphi(t) + D\left(V(\varphi(t))\right)$. Then, by adding an integrator, one gets exactly equations (28)–(30) with $X_2 = V$ and $\dot{V} = U$, where U is designed in order to stabilize the system $\dot{X}_1(t) = f_1\left(X_1(t), V\left(\varphi(t)\right)\right)$, $\dot{V}(t) = U(t)$. Hence, stabilization of system (28)–(30) implies stabilization of system (31).

From plant (28), (29) one can observe that the input signal reaches the state X_2 at $t = 0$. However, it reaches X_1 through a delayed integrator. Therefore, we need to define and derive an implementable form for the predictor of the state X, i.e., the signal that satisfies $P_1(\varphi(t)) = X_1(t)$ for all $t \geq 0$. From relation (30) we get that

$$\begin{aligned}\varphi^{-1}(\theta) &= \sigma(\theta)\\ &= \theta + D(X_2(\theta)), \quad \text{for all } \varphi(t) \leq \theta \leq t.\end{aligned} \tag{32}$$

Setting $t = \sigma(\theta)$ in (28), differentiating with respect to θ and integrating the resulting expression from $\varphi(t)$ to θ, with the help of the fact that

$$P_1(\varphi(t)) = X_1(t),$$

we get

$$\begin{aligned}P_1(\theta) =& X_1(t) + \int_{\varphi(t)}^{\theta} \left(1 + D'(X_2(s))U(s)\right) f\left(P_1(s), X_2(s)\right) ds,\\ &\varphi(t) \leq \theta \leq t.\end{aligned} \tag{33}$$

We compute next φ. Differentiating (30) and recalling that

$$\varphi\left(t + D(X_2(t))\right) = \varphi(\sigma(t)) = t,$$

we get

$$\varphi(\theta) = t - \int_{\theta}^{t+D(X_2(t))} \frac{ds}{1 + D'(X_2(\varphi(s)))U(\varphi(s))}, \quad \text{for all } t \leq \theta \leq \sigma(t). \tag{34}$$

Assuming that there exists a function μ that renders $\dot{\Xi} = f_1\left(\Xi, \mu\left(\Xi\right) + \omega\right)$ input-to-state stable with respect to ω, the predictor-based control law for system (28), (29) is then given by

$$U(t) = \frac{\nabla\mu(P_1(t)) f\left(P_1(t), X_2(t)\right) - c_Z\left(X_2(t) - \mu(P_1(t))\right)}{1 - \nabla\mu(P_1(t)) f\left(P_1(t), X_2(t)\right) D'(X_2(t))}, \tag{35}$$

where $c_Z > 0$ is arbitrary. At each step in an actual implementation of the control law given by (33)–(35), one has to compute $\varphi(t)$, by numerically computing the integral in (34) and using the history of X_2 and U. Then, one computes $P_1(t)$ using $\varphi(t)$ and the history of X_2, P_1 and U. Finally, one calculates $U(t)$ from (35). However, in order to compute $\varphi(t)$ one starts the integration at $\sigma(t) = t + D(X_2(t))$. Yet, the function inside the integral evaluated at $s = \sigma(t)$, depends on $U(t)$, i.e., on the current value of the input, which is yet to be computed. Therefore, since $\sigma(t)$ is strictly increasing one can approximate $\varphi(t)$ by integrating (34) up to $s = \sigma(t - h)$, where h is the discretization step.

From (35) one can observe that besides a restriction that the denominator in (34) is positive, one has an additional condition that the denominator in (35) is also positive. Both conditions are satisfied when the following condition holds for all $\theta \geq \varphi(0)$ and for some $0 < c < 1$.

$$\mathscr{G}_c : \quad \left|D'(X_2(\theta))U(\theta)\right| + \left|\nabla\mu(P_1(\theta))f(P_1(\theta), X_2(\theta))\, D'(X_2(\theta))\right| < c. \tag{36}$$

The following is proved in [7].

Theorem 4 *Let $\dot{X}_1 = f_1(X_1, \omega)$ be forward complete and $\dot{X}_1 = f_1(X_1, \mu(X_1) + \omega)$ be input-to-state stable with respect to ω. Consider the plant* (28)–(30) *together with the control law given by* (33)–(35). *Then there exist a class $\mathscr{K}$ function ξ_{RoA} and a class $\mathscr{K}\mathscr{L}$ function $\hat{\beta}$ such that for all initial conditions for which X_2 is locally Lipschitz on the interval $[\varphi(0), 0]$, the function U is locally Lipschitz on the interval $[\varphi(0), 0)$, and* (29) *and*

$$\hat{\Omega}(0) < \xi_{\mathrm{RoA}}(c) \tag{37}$$

hold for some $c \in (0, 1)$, where

$$\hat{\Omega}(t) = |X_1(t)| + \sup_{\varphi(t)\leq\theta\leq t} |X_2(\theta)| + \sup_{\varphi(t)\leq\theta\leq t} |U(\theta)|, \tag{38}$$

and

$$\hat{\Omega}(t) \leq \hat{\beta}\left(\hat{\Omega}(0), t\right) \tag{39}$$

holds for all $t \geq 0$. Furthermore, there exists a class $\mathscr{K}$ function $\hat{\delta}^*$ such that

$$\sup_{\varphi(t)\leq\theta\leq t} D(X_2(\theta)) \leq D(0) + \hat{\delta}^*(c) \quad \text{and} \tag{40}$$

$$\sup_{\varphi(t)\leq\theta\leq t} \left|\dot{D}(X_2(\theta))\right| \leq c. \tag{41}$$

hold for all $t \geq 0$.

Under the assumptions of Theorem 4, existence and uniqueness of a solution to the closed-loop system with $X_1 \in C^1[0, \infty)$, X_2 Lipschitz on $[0, \infty)$, and U Lipschitz on $(0, \infty)$ it is proved in [7] (see also [9] for the case of a state-dependent input delay).

6 Examples

Example 1 In this example we consider the system

$$\dot{X}_1(t) = \sin(X_1(t)) + X_2(\varphi(t)) \tag{42}$$
$$\dot{X}_2(t) = U(t), \tag{43}$$

where

$$\varphi(t) = t - \frac{1+t}{1+2t}. \tag{44}$$

We choose the initial conditions of the plant as $X_1(0) = 1$ and $X_2(s) = 0$ for all $s \in [\varphi(0), 0]$. The controller for this system is

$$U(t) = -c_2\left(X_2(t) + c_1 P_1(t) + \sin(P_1(t))\right) - (c_1 + \cos(P_1(t)))\left(\sin(P_1(t)) + X_2(t)\right)\frac{d\varphi^{-1}(t)}{dt} \tag{45}$$
$$P_1(t) = X_1(t) + \int_{\varphi(t)}^{t} \frac{\left(\sin(P_1(\theta)) + X_2(\theta)\right) d\theta}{\varphi'\left(\varphi^{-1}(\theta)\right)} \tag{46}$$
$$\varphi'(t) = 1 + \frac{1}{(1+2t)^2} \tag{47}$$
$$\varphi^{-1}(t) = t + \frac{t+1}{\sqrt{(t+1)^2+1}+t}, \tag{48}$$

where we choose $c_1 = c_2 = 2$. In Fig. 1 and in Fig. 2 we show the response of the system and the control effort respectively, in comparison with the uncompensated controller, i.e., the backstepping controller (45) which assumes $\varphi(t) = t$.

Example 2 (Application to Cooling Systems) In marine transportation of materials the design of control laws for the ship's cooling system is of paramount importance due to the significant potential of the cooling system in terms of energy optimization [23]. In Fig. 3 we show a typical marine cooling circuit with one consumer, denoted by C, and a Heat Exchanger. We denote with T_{in} the input temperature towards the consumer, i.e., the output temperature of the Heat Exchanger. Due to the transportation time of the coolant (typically water) from the Heat Exchanger to the consumer C, the actual input temperature T_{in} in the consumer is delayed by H, namely $T_{\text{in}}(t - H)$. The delay time H depends on the flow rate q_2 which can be controlled through a

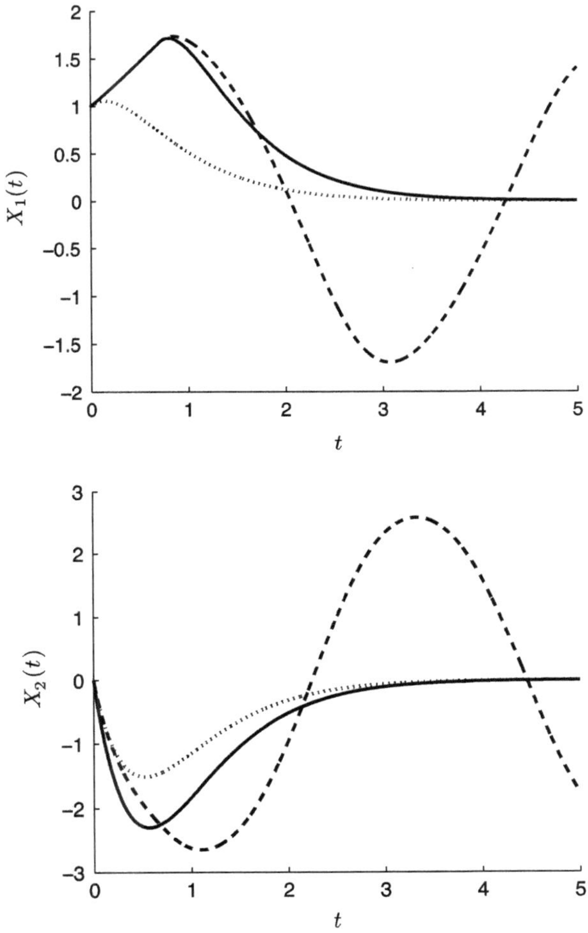

Fig. 1 System's response for Example 1. *Dotted lines* System with $\varphi(t) = t$ and the uncompensated controller. *Dashed lines* System with $\varphi(t)$ as in (44) and the uncompensated controller. *Solid lines* System with $\varphi(t)$ as in (44) and the delay-compensating controller

pump. In order to design a feedback law q_2 we take into account that the flow rate q_2 has to be proportional to the temperature at the other end of the consumer, which we denote with T_{out}. This is because it makes sense to increase the flow rate if the outer temperature of the consumer C is increasing. A simple choice is $q_2 = k_1 T_{\text{out}} + k_2$. The control objective is to regulate the temperature T_{out} to a constant set-point, say $T_{\text{eq}} > 0$. This is achieved by controlling the flow rate q_1 at the input of the Heat Exchanger through a pump.

Denoting by $T_{\text{out}} = X_1$, $T_{\text{in}} = X_2$, $q_2 = k_1 X_1 + k_2$, $H = \frac{b}{q_2} = D$ and $q_1 = U$ and by neglecting the effect of the hydraulics in the system (since the hydraulic dynamics assumed to be much faster than the heat dynamics [23]), the equations that describe the thermodynamics of the cooling circuit are

$$\dot{X}_1(t) = a\left(X_1(t) - X_2\left(t - D(X_1(t))\right)\right)\left(k_1 X_1(t) + k_2\right) \tag{49}$$

$$\dot{X}_2(t) = \left(k_1 X_1(t) + k_2\right)\left(X_1(t) - X_2(t)\right) - U(t), \tag{50}$$

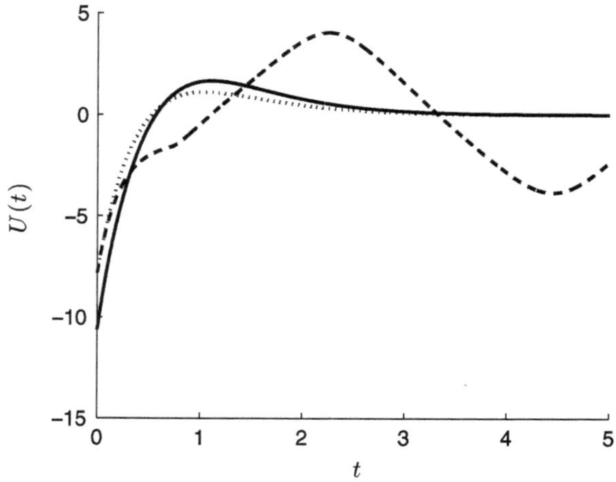

Fig. 2 Control effort for Example 1. *Dotted lines* System with $\varphi(t) = t$ and the uncompensated controller. *Dashed lines* System with $\varphi(t)$ as in (44) and the uncompensated controller. *Solid lines* System with $\varphi(t)$ as in (44) and the delay-compensating controller

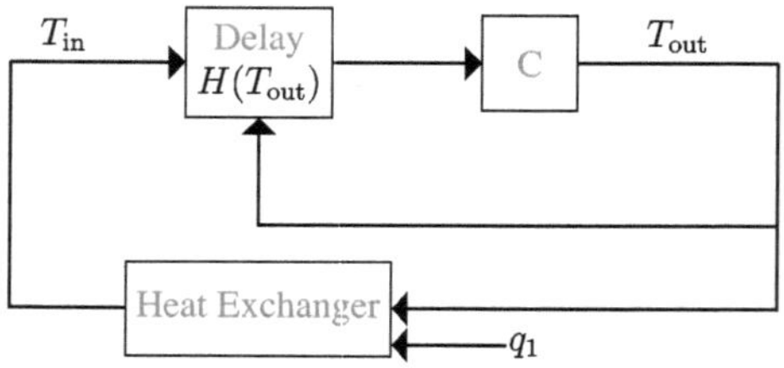

Fig. 3 A marine cooling system with one consumer from Example 2

where, $D(X_1(t)) = \frac{b}{k_1 X_1(t) + k_2}$, $a < 0$, and $b, k_1, k_2 > 0$. Since the coolant flows only in one direction $q_1 > 0$, and hence $U > 0$. Since the coolant is typically water, both T_{out} and T_{in} can not fall below zero, and hence $X_1, X_2 > 0$. In addition since the consumer always adds heat (due to its functioning), $T_{\text{out}} \geq T_{\text{in}}(t - H)$ and $T_{\text{out}} \geq T_{\text{in}}$, and hence $X_1(t) - X_2(t - D(t)) \geq 0$ and $X_1(t) - X_2(t) \geq 0$ for all $t \geq 0$. We choose the control law $\kappa(X_1) = X_1 + \frac{c_1}{a} \frac{X_1}{k_1 X_1 + k_2}$, and hence, the predictor-based control law for this system becomes

$$U(t) = (k_1 X_1(t) + k_2)(X_1(t) - X_2(t)) + c_2 \Bigg(X_2(t) - P_1(t) - \frac{c_1}{a} \frac{P_1(t) - T_{\text{eq}}}{k_1 P_1(t) + k_2}\Bigg) - \left(1 + \frac{c_1}{a k_1} \frac{T_{\text{eq}} + \frac{k_2}{k_1}}{\left(P_1(t) + \frac{k_2}{k_1}\right)^2}\right) \times \frac{(P_1(t) - X_2(t))(k_1 P_1(t) + k_2)}{R(t)}, \tag{51}$$

$$P_1(t) = \int_{t - D(X_1(t))}^{t} a \frac{(P_1(\theta) - X_2(\theta))(k_1 P_1(\theta) + k_2)\, d\theta}{R(\theta)} + X_1(t) \tag{52}$$

$$R(\theta) = 1 + \frac{b k_1 a (P_1(\theta) - X_2(\theta))}{(k_1 P_1(\theta) + k_2)^2}(k_1 P_1(\theta) + k_2). \tag{53}$$

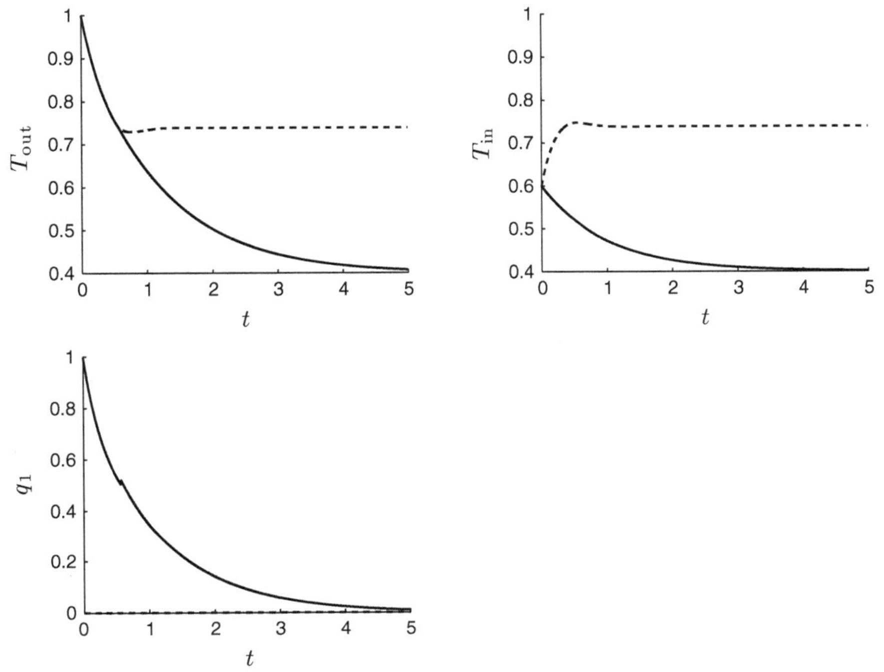

Fig. 4 Response of the cooling model (49)–(50) with the delay-compensating controller (51)–(53) (*solid line*) and in open-loop (*dashed line*)

We choose the parameters of the plant and of the controller as $a = -1$ and $c_1 = c_2 = b = k_1 = k_2 = 1$ and the initial conditions as $X_1(0) = 1$ and $X_2(\theta) = 0.2$ for all

$$-\frac{b}{k_1 X_1(0) + k_2} \leq \theta \leq 0. \tag{54}$$

In Fig. 4, we show the temperatures T_{out} and T_{in} with the input flow q_1. We compare the response of the system with the predictor-based controller and with no control. The predictor based controller regulates the temperatures T_{out} and T_{in} at the desired set point T_{eq}.

Example 3 We consider the system

$$\dot{s}(t) = v(t - r_1 \sin^2(\omega s(t))) \tag{55}$$
$$\dot{v}(t) = a(t), \tag{56}$$

where the state variables are denoted with s and v and the control variable is denoted with $a(t)$. This system resembles the model in [50] for the "soft" automatic landing. We choose the parameters of the plant as $r_1 = 0.3$ and $\omega = 15$, the initial conditions

$s(0) = 1$ and $v(\theta) = 0.1$ for all $-r_1 \sin^2(\omega s(0)) \leq \theta \leq 0$ and the parameters of the nominal controller as $c_1 = c_2 = 0.5$. The predictor-based controller is

$$a(t) = -c_1 \frac{v(t)}{1 - 2r_1\omega \sin(\omega P_1(t)) \cos(\omega P_1(t)) v(t)} - c_2 (v(t) + c_1 P_1(t)), \quad (57)$$

where for all $t - r_1 \sin^2(\omega s(t)) \leq \theta \leq t$,

$$P_1(\theta) = \int_{\varphi(t)}^{\theta} \frac{v(s)\,ds}{1 - 2r_1\omega \sin(\omega P_1(s)) \cos(\omega P_1(s)) v(s)} + s(t) \quad (58)$$

$$\varphi(t) = t - r_1 \sin^2(\omega s(t)). \quad (59)$$

The control signal reaches the state s at the time t^*, $0.3\sin^2(15(0.1t^* + 1)) = t^* = 0.0887$. In Fig. 5 we show the response of the system. Both $s(t)$, $v(t)$ converge to zero, whereas $\varphi(t)$ and $\sigma(t)$ remain increasing for all times. From Fig. 5 we observe that at the time instants where $0.3\sin^2(15s(t_1)) = 0$ we have that $\varphi(t) = t = \sigma(t)$.

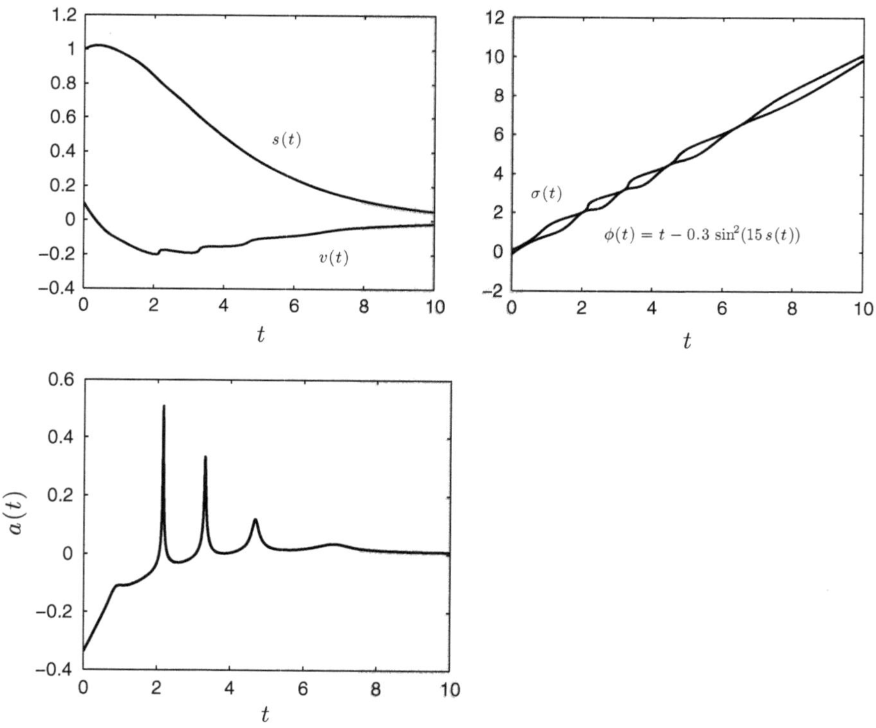

Fig. 5 Response of system (55)–(56) and the functions $\varphi(t) = t - 0.3\sin^2(15s(t))$ and $\sigma(t) = \varphi^{-1}(t) = t + 0.3\sin^2(15P_1(t))$ with the delay-compensating controller (57)–(59) and initial conditions $s(0) = 1$, $v(\theta) = 0.1$, $-r_1 \sin^2(\omega s(0)) \leq \theta \leq 0$

7 Conclusions

We presented predictor feedback designs for nonlinear systems in strict-feedback form with delays affecting the virtual inputs. We considered systems with constant delays, time-varying delays, and state-dependent delays. It would be of interest, both theoretically and practically, to consider systems in feedforward form with delays affecting various components of the vector field and to develop an infinite-dimensional version of "integrator forwarding" [45] along the lines of [26].

References

1. Ahmed, A., Verriest, E.: Estimator design for a subsonic rocket car (soft landing) based on state-dependent delay measurement. In: Proceedings of the IEEE Conference on Decision and Control, pp. 5698–5703 (2013)
2. Altintas, Y., Engin, S., Budak, E.: Analytical stability prediction and design of variable pitch cutters. ASME J. Manuf. Sci. Eng. **121**(2), 173–178 (1999)
3. Artstein, Z.: Linear systems with delayed controls: a reduction. IEEE Trans. Autom. Control **27**(4), 869–879 (1982)
4. Bekiaris-Liberis, N., Krstic, M.: Nonlinear Control Under Nonconstant Delays. SIAM, Philadelphia (2013)
5. Bekiaris-Liberis, N., Jankovic, M., Krstic, M.: Compensation of state-dependent state delay for nonlinear systems. Syst. Control Lett. **61**(8), 849–856 (2012)
6. Bekiaris-Liberis, N., Krstic, M.: Compensation of time-varying input and state delays for nonlinear systems. J. Dyn. Syst. Meas. Control **134**, 011009 (2012)
7. Bekiaris-Liberis, N., Krstic, M.: Nonlinear control under delays that depend on delayed states. Eur. J. Control **19**, 389–398 (2013)
8. Bekiaris-Liberis, N., Krstic, M.: Robustness of nonlinear predictor feedback laws to time- and state-dependent delay perturbations. Automatica **49**(6), 1576–1590 (2013)
9. Bekiaris-Liberis, N., Krstic, M.: Compensation of state-dependent input delay for nonlinear systems. IEEE Trans. Autom. Control **58**(2), 275–289 (2013)
10. Bendtsen, J., Krstic, M.: Control of thermodynamical system with input-dependent state delays. In: Proceedings of the IEEE Conference on Decision and Control, pp. 300–305 (2013)
11. Bresch-Pietri, D., Chauvin, J., Petit, N.: Prediction-based stabilization of linear systems subject to input-dependent input delay of integral-type. IEEE Trans. Autom. Control **59**(9), 2385–2399 (2014)
12. Bresch-Pietri, D., Chauvin, J., Petit, N.: Adaptive control scheme for uncertain time-delay systems. Automatica **48**(8), 1536–1552 (2012)
13. Bresch-Pietri, D., Krstic, M.: Adaptive trajectory tracking despite unknown input delay and plant parameters. Automatica **45**(9), 2075–2081 (2009)
14. Bresch-Pietri, D., Krstic, M.: Delay-adaptive predictor feedback for systems with unknown long actuator delay. IEEE Trans. Autom. Control **55**(9), 2106–2112 (2010)
15. Bresch-Pietri, D., Krstic, M.: Delay-adaptive control for nonlinear systems. IEEE Trans. Autom. Control **59**(5), 1203–1218 (2014)
16. Briat, C., Hjalmarsson, H., Johansson, K., Jonsson, U.T., Karlsson, G., Sandberg, H.: Nonlinear state-dependent delay modeling and stability analysis of internet congestion control. In: Proceedings of the IEEE Conference on Decision and Control, pp. 1484–1491 (2010)
17. Cacace, F., Germani, A., Manes, C.: An observer for a class of nonlinear systems with time varying observation delay. Syst. Control Lett. **59**(5), 305–312 (2010)

18. Cai, X., Krstic, M.: Nonlinear control under wave actuator dynamics with time- and state-dependent moving boundary. Int. J. Robust Nonlinear Control. doi:10.1002/rnc.3083
19. Choi, J., Koo, K., Kim, J., Lee, J.: Global asymptotic stability of FAST TCP in the presence of link dynamics. Int. J. Control Autom. Syst. **7**(5), 809–816 (2009)
20. Cloosterman, M., van de Wouw, N., Heemels, W., Nijmeijer, H.: Stability of networked control systems with uncertain time-varying delays. IEEE Trans. Autom. Control **54**(7), 1575–1580 (2009)
21. Diagne, M., Couenne, F., Maschke, B.: Mass transport equation with moving interface and its control as an input delay system. In: Proceedings of the 11th Workshop on Time-Delay Systems, pp. 331–336 (2013)
22. Fischer, N., Dani, A., Sharma, N., Dixon, W.: Saturated control of an uncertain nonlinear system with input delay. Automatica **49**(6), 1741–1747 (2013)
23. Hansen, M., Stoustrup, J., Bendtsen, J.: Modeling of nonlinear marine cooling systems with closed circuit flow. In: Proceedings of the IFAC World Congress, pp. 5537–5542 (2011)
24. Ito, H., Jiang, Z.-P., Pepe, P.: Construction of Lyapunov-Krasovskii functionals for networks of iISS retarded systems in small-gain formulation. Automatica **49**(11), 3246–3257 (2013)
25. Jankovic, M.: Control Lyapunov-Razumikhin functions and robust stabilization of time delay systems. IEEE Trans. Autom. Control **46**(7), 1048–1060 (2001)
26. Jankovic, M.: Cross-term forwarding for systems with time delay. IEEE Trans. Autom. Control **54**(3), 498–511 (2009)
27. Karafyllis, I.: Finite-time global stabilization by means of time-varying distributed delay feedback. SIAM J. Control Optim. **45**(1), 320–342 (2006)
28. Karafyllis, I.: Stabilization by means of approximate predictors for systems with delayed input. SIAM J. Control Optim. **49**(3), 1100–1123 (2011)
29. Karafyllis, I., Krstic, M.: Nonlinear stabilization under sampled and delayed measurements, and with inputs subject to delay and zero-order hold. IEEE Trans. Autom. Control **57**(5), 1141–1154 (2012)
30. Karafyllis, I., Krstic, M.: Stabilization of nonlinear delay systems using approximate predictors and high-gain observers. Automatica **49**(12), 3623–3631 (2013)
31. Karafyllis, I., Krstic, M., Ahmed-Ali, T., Lamnabhi-Lagarrigue, F.: Global stabilization of nonlinear delay systems with a compact absorbing set. Int. J. Control **87**(5), 1010–1027 (2014)
32. Krstic, M.: Input delay compensation for forward complete and feedforward nonlinear systems. IEEE Trans. Autom. Control **55**(2), 287–303 (2010)
33. Krstic, M.: Delay Compensation for Nonlinear, Adaptive, and PDE Systems. Birkhauser, Boston (2009)
34. Krstic, M., Kanellakopoulos, I., Kokotovic, P.: Nonlinear and Adaptive Control Design. Wiley, New York (1995)
35. Manitius, A., Olbrot, A.: Finite spectrum assignment for systems with delays. IEEE Trans. Autom. Control **24**(4), 541–553 (1979)
36. Mazenc, F., Bliman, P.: Backstepping design for time-delay nonlinear systems. IEEE Trans. Autom. Control **51**(1), 149–154 (2006)
37. Mazenc, F., Malisoff, M.: Stabilization of a chemostat model with Haldane growth functions and a delay in the measurements. Automatica **46**(9), 1428–1436 (2010)
38. Mazenc, F., Malisoff, M.: Asymptotic stabilization for feedforward systems with delayed feedbacks. Automatica **49**(3), 780–787 (2013)
39. Mazenc, F., Mondie, S., Francisco, R.: Global asymptotic stabilization of feedforward systems with delay at the input. IEEE Trans. Autom. Control **49**(5), 844–850 (2004)
40. Mazenc, F., Niculescu, S.-I.: Generating positive and stable solutions through delayed state feedback. Automatica **47**(3), 525–533 (2011)
41. Michiels, W., Verriest, E.: A systems theoretic analysis of fast varying and state dependent delays. In: Proceedings of the IEEE Conference on Decisions and Control, pp. 6313–6318 (2011)
42. Pepe, P.: On Sontag's formula for the input-to-state practical stabilization of retarded control-affine systems. Syst. Control Lett. **62**(11), 1018–1025 (2013)

43. Pepe, P., Ito, H.: On saturation, discontinuities, and delays, in iISS and ISS feedback control redesign. IEEE Trans. Autom. Control **57**(5), 1125–1140 (2012)
44. Richard, J.-P.: Time-delay systems: an overview of some recent advances and open problems. Automatica **39**(10), 1667–1694 (2003)
45. Sepulchre, R., Jankovic, M., Kokotovic, P.: Integrator forwarding: a new recursive nonlinear robust design. Automatica **33**(5), 979–984 (1997)
46. Sipahi, R., Lammer, S., Niculescu, S-I., Helbing, D.: On stability analysis and parametric design of supply networks under the presence of transportation delays. In: Proceedings of the ASME 2006 International Mechanical Engineering Congress and Exposition, pp. 135–144 (2006)
47. Teel, A.: Connections between Razumikhin-type theorems and the ISS nonlinear small gain theorem. IEEE Trans. Autom. Control **43**(7), 960–964 (1998)
48. Travis, S.: A one-dimensional two-body problem of classical electrodynamics. SIAM J. Appl. Math. **28**(3), 611–632 (1975)
49. Verriest, E.: Inconsistencies in systems with time-varying delays and their resolution. IMA J. Math. Control Inf. **28**(2), 147–162 (2011)
50. Walther, H.-O.: On a model for soft landing with state-dependent delay. J. Dyn. Syst. Differ. Equ. **19**(3), 593–622 (2007)
51. Witrant, E.: Canudas de Witt, C., Georges, D., Alamir, M.: Remote stabilization via communication networks with a distributed control law. IEEE Trans. Autom. Control **52**(8), 1480–1485 (2007)

Prediction-Based Control of Linear Systems by Compensating Input-Dependent Input Delay of Integral-Type

Delphine Bresch-Pietri and Nicolas Petit

Abstract This study addresses the problem of delay compensation via a predictor-based output feedback for a class of linear systems subject to input delay which itself depends on the input. The equation defining the delay is implicit and involves past values of the input through an integral relation, the kernel of which is a polynomial function of the input. This modeling represents systems where transport phenomena take place at the inlet of a system involving a nonlinearity, which frequently occurs in the processing industry. The conditions of asymptotic stabilization require the magnitude of the feedback gain to comply with the initial conditions. Arguments for the proof of this novel result include general Halanay inequalities for delay differential equations and take advantage of recent advances in backstepping techniques for uncertain or varying delay systems.

1 Introduction

Numerous control systems involve a physical dead-time which proves to be troublesome in the design and tuning of feedback control laws. The dead-time results from the fact that sensors and actuators are rarely co-located, for example in processes that involve the transport of materials, such as mixing processes for liquid or gaseous fluids, chemical reactors [12], automotive engine and exhaust lines [9], heat collector plants [24], networks for blending liquids and solids [8], and batch processes [22]. In all of these, and also in the crushing mill system described by [23], the lag directly depends on the control variable and is inherently input-dependent.

D. Bresch-Pietri (✉)
CNRS at GIPSA-lab, Control Department, 11 Rue des Mathématiques,
38000 Grenoble, France
e-mail: delphine.bresch-pietri@gipsa-lab.fr

N. Petit
Centre Automatique et Systèmes at MINES ParisTech, 60 Boulevard Saint-Michel,
75006 Paris, France
e-mail: nicolas.petit@mines-paristech.fr

I. Karafyllis et al. (eds.), *Recent Results on Nonlinear Delay Control Systems*,
Advances in Delays and Dynamics 4, DOI 10.1007/978-3-319-18072-4_4

It seems that the stabilization of such processes with input-dependent time-delay in the input $D(u)$ or $D(u_t)$, where u_t denotes past values over a finite horizon, has seldom been theoretically studied. Rather, a widely considered approach is to recast the delay dependence on the control, e.g., by modeling this dependence as $D(u) \approx D(t)$ (or even by a constant average value D) and expecting the controller to deal with a certain level of unstructured model variability. The advantage of such an approach is that it enables the use of a wide range of technical tools previously developed to deal with disturbances. However, recasting this input-dependency into the more general class of input-disturbance reduces performance potential. For this reason we propose to explicitly take this dependency into account and to employ a prediction-based controller, in an effort of improving transient performances.

Predictor-based control strategies [1, 17, 25], which are state-of-the-art for systems with constant input time-delays (as studied in [3, 10, 14, 18, 19] or [23] and the references therein) are much more complex to apply for time-varying delays. As shown in [20] or, more recently, in [16], to compensate a time-varying input delay, the prediction has to be calculated over a time window whose length matches the value of the future delay. In other words, future variations in the delay must be predicted. This can be done, for example, when one has access to a given delay model as in [27] or when the delay is state-dependent, by carefully predicting the future system state, as proposed in [2].

Yet, when the delay depends on the input, things become very involved. Determining the required prediction horizon is then an implicit issue, which in practice may not be resolvable or even well-posed. This implicit nature is caused by the reciprocal interactions between the (current and past) control values and the delay, yielding a closed-loop dependency.

In order to design such a prediction-based control law, we advocate a two-step method of disrupting the implicit loop, as proposed in our recent works on the topic [4, 6] and establish sufficient conditions for asymptotic stabilization.

In a first step, we considered the input-dependency as a particular form of time-variation. It enables us to then use a robust compensation result for linear systems with time-varying input delay, using the backstepping tools proposed in [15] for the analysis of input-delay system stability. Technically, this guarantees stabilization, provided the delay variations are sufficiently small. Second, a sufficient bound on the delay variations was obtained by relating them to the control tracking error, which is analyzed using the asymptotic convergence of delay differential equations (DDE) (Halanay-type inequalities [11]).

The result holds for potentially unstable linear systems of any arbitrary order $n \in \mathbb{N}$. Assuming that the kernel defining the integral is a positive polynomial function of the input, our previous result was extended to such a framework. This is the main novelty of this study. As shown, this delay model encompasses a large class of transport processes. For this class of systems, robust compensation is shown to be achievable, provided that the initial conditions are sufficiently close to equilibrium and that the feedback gain is chosen accordingly.

This chapter is organized as follows. In Sect. 2, the problem is illustrated by a rocket engine system as an motivating example. A prediction-based controller is designed in Sect. 3. Sufficient conditions for stabilization are derived through a proof of convergence, calling on Halanay-like inequalities arguments in Sect. 4.

Notation and Definitions. In the following, $|.|$ stands for the usual Euclidean norm, and $\mathscr{C}^0(S_1, S_2)$ denotes the set of all continuous functions on a set S_1 with values into a set S_2. For a given symmetric matrix, $\underline{\lambda}(.)$ and $\overline{\lambda}(.)$ stand respectively for minimum and maximum eigenvalues.

Classically, x_t refers to the function $x_t : [-\overline{D}, 0] \rightarrow x(t+s)$ defined by $x_t(s) = x(t+s)$ for a given function x and $\overline{D} > 0$ and $\max|x_t| = \max\{|x_t(s)| : s \in [-\overline{D}, 0]\}$.

For any bounded function k defined on $[-\overline{D}, 0]$ and any polynomial function π, we write

$$\pi(x_t) = \left(x(t_1), \ldots, x(t_{n-2}), \int_{t_{n-1}}^{t_n} k(t-s)x(s)ds\right) \tag{1}$$

for $(t_1, \ldots, t_n) \in [t - \overline{D}, t]^n$. We also refer to $\pi(x_t)$ as a polynomial function in x_t.

A polynomial function π in the variables $(x_1, \ldots, x_n, x_{n+1})$ is said to be *at least quadratic in* $x_1, \ldots, x_n$ if, for any given x_{n+1}, the corresponding polynomial function $\pi_{x_{n+1}}$ defined as

$$\pi_{x_{n+1}}(x_1, \ldots, x_n) = \pi(x_1, \ldots, x_n, x_{n+1})$$

has no terms of order 0 or 1, e.g., $\pi = x_1^2 + x_1x_2x_3$ and $\pi = x_2x_1 + x_3x_1^2$ are both at least quadratic in (x_1, x_2) while $\pi = x_3 + x_3x_2^2$ is not.

2 Problem Statement

Consider the following potentially unstable linear time-invariant plant sketched in Fig. 1:

$$x^{(n)}(t) + a_{n-1}x^{(n-1)}(t) + \ldots + a_1\dot{x}(t) + a_0x(t) = b_0u(t - D(t)) \tag{2}$$

driven by a delayed input, where the varying delay $D(t)$ is implicitly defined in terms of the input history by

$$\int_{t-D(t)}^{t} \varphi(u(s))ds = 1, \tag{3}$$

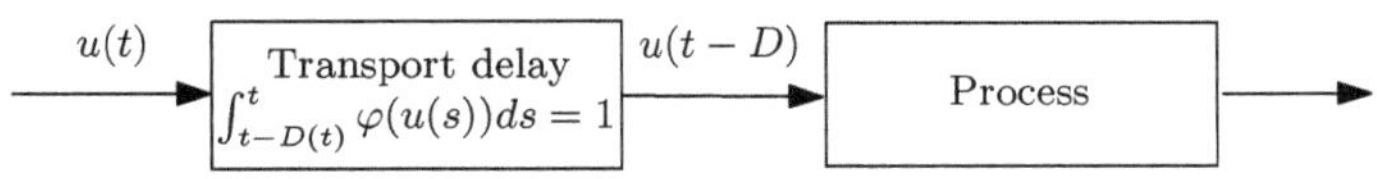

Fig. 1 The addressed problem, where the input is delayed by a transport delay which is input-varying.

in which $\varphi : \mathbb{R} \mapsto [\underline{\varphi}, \infty)$ *is a polynomial function with positive values* (i.e., $\underline{\varphi} > 0$). Physically, the integral equation (3) corresponds to a plug-flow assumption [21] in a transport phenomenon and appears in a large class of applications. The variable φ is a (normalized) flow rate, which depends polynomially on the system input. Since $\varphi \geq \underline{\varphi} > 0$, this transport delay is well defined[1] and is upper-bounded, as follows:

$$D(t) \leq \overline{D} = 1/\underline{\varphi} \text{ for all } \quad t \geq 0. \tag{4}$$

In particular, as

$$D \mapsto \int_{t-D}^{t} \varphi(u(s))ds \tag{5}$$

is strictly increasing, it is invertible and the delay can be calculated and so is assumed to be known in the following.

Motivating Example Consider a liquid non-hypergolic propellant rocket such as the one pictured on Fig. 2. The fuel and oxidizer are stored in separate tanks, blended and then fed through a system of pipes, valves, and turbopumps to a combustion chamber, where they are burned to generate thrust. As the mixture is non-hypergolic,

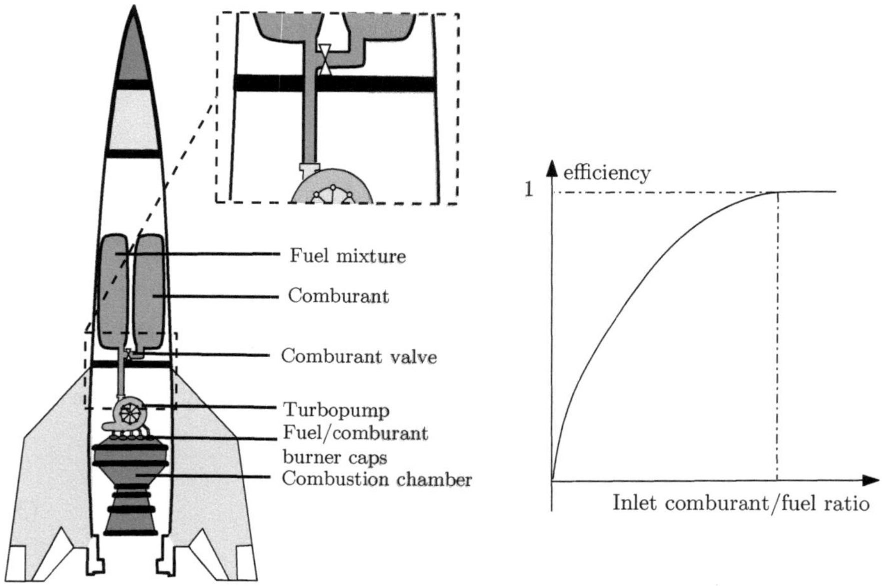

Fig. 2 *Left* Schematic view of a low cost designed (single choke) *bipropellant rocket engine*. Zoom in on the transport phenomenon in the *top* right corner. *Right* Schematic view of the evolution of the combustion efficiency with respect to the comburant/fuel ratio entering the chamber

[1]The delay is positive and, besides, its derivative can be expressed as $\dot{D}(t) = 1 - \varphi(u(t))/(\varphi(u(t-D))) < 1$ which guarantees strict causality.

combustion is started by an ignition system in the chamber. Compared to more classical setups (with dual pumps and pipes convecting the tanks directly to the combustion chamber), this architecture has been chosen for cost reasons, as it uses only one pipe, one turbopump and one choke.

After being premixed downstream of the tanks, the mixture flows to the combustion chamber, which generates a transport delay satisfying

$$\int_{t-D(t)}^{t} \frac{Q(s)}{V_P} ds = 1, \tag{6}$$

in which V_P is the pipe holdup and $Q(s) = Q_1 + Q_2(s)$ is the total volume flow rate after mixing, i.e., the sum of the (constant) outlet fuel flow rate and the comburant flow rate. Following Newton's law, the (longitudinal) dynamics of the rocket can therefore be written as

$$\ddot{x}(t) = \eta(t) F_0,$$

in which F_0 is the optimal thrust, corresponding to a stoichiometric mixture, and η is the combustion efficiency. The latter is significantly nonlinear at high flow-rate ratios, which vanishes for low ratios [26]. Hence, it can reasonably be modeled as $\eta(t) = \eta_0\sqrt{Q_2(t - D(t))}$ for example (without loss of generality), with the oxygen volume flow rate $Q_2(t - D(t))$ entering the combustion chamber. Therefore, accounting for (6) and setting

$$u(t) = \eta_0 F_0 \sqrt{Q_2(t)}, \tag{7}$$

one obtains

$$\begin{cases} \ddot{x}(t) = u(t - D(t)) \\ \displaystyle\int_{t-D(t)}^{t} \frac{1}{V_P}\left(Q_1 + \frac{u(s)^2}{\eta_0^2 F_0^2}\right) ds = 1 \end{cases}$$

which is indeed in the form (2)–(3) with

$$\varphi(u(s)) = \frac{1}{V_P}\left(Q_1 + \frac{u(s)^2}{\eta_0^2 F_0^2}\right) \tag{8}$$

being a second-order polynomial function with strictly positive values. More complex polynomial expressions can be considered to account for more detailed fit of experimental data and thrust maps.

Control Design Objective The control task is to stabilize the plant at any equilibrium point x^r such that $x^r = b_0/a_0 u^r$, where u^r is the corresponding input. For this purpose, a predictor-based feedback law will be employed here.

With this aim in view, we first formulate a state-space representation of this system as

$$\begin{cases} \dot{X}(t) = AX(t) + Bu(t - D(t)) \\ \int_{t-D(t)}^{t} \varphi(u(s))ds = 1 \end{cases}$$

where

$$A = \begin{pmatrix} 0 & 1 & & 0 \\ \vdots & & \ddots & \\ 0 & 0 & & 1 \\ -a_0 & -a_1 & \dots & -a_{n-1} \end{pmatrix} \quad \text{and} \quad B = \begin{pmatrix} 0 \\ \vdots \\ 0 \\ b_0 \end{pmatrix}. \tag{9}$$

For the sake of clarity, we assume in the following that the system state X is fully measured. Extending this to output feedback for observable systems is straightforward, following the methodology of [7]. To construct a prediction-based control law, we use the following theorem, a proof of which is given in [4] and [5].

Theorem 1 (Step 1: Preliminary result) *Consider the closed-loop single input system*

$$\dot{X}(t) = AX(t) + Bu(t - D(t)) \tag{10a}$$

$$u(t) = K\left[e^{AD(t)}X(t) + \int_{t-D(t)}^{t} e^{A(t-s)}Bu(s)ds\right], \tag{10b}$$

where $X \in \mathbb{R}^n$, $u \in \mathbb{R}$, K *is chosen such that* $A + BK$ *is Hurwitz, and* $D : \mathbb{R}_+ \to [0, \overline{D}]$ *is a time-differentiable function. Define*

$$\Upsilon_0(t) = |X(t)|^2 + \int_{t-D(t)}^{t} u(s)^2 ds + D(t)^2 \int_{t-D(t)}^{t} \dot{u}(s)^2 ds.$$

There exists $\Delta^*(K) \in (0, 1)$ *such that, if*

$$|\dot{D}(t)| < \Delta^*(K), \quad t \geq 0, \tag{11}$$

then the plant (10a) *exponentially converges to the origin, in the sense that there exist* $R > 0$ *and* $\rho > 0$ *such that* $\Upsilon_0(t) \leq R\Upsilon_0(0)e^{-\rho t}$ *for all* $t \geq 0$.

The prediction controller (10b) is a natural extension from the case of constant delay. It forecasts values of the state over a time window of varying length $D(t)$. Of course, exact compensation of the delay is not achieved with this controller. To do so, one would need to consider a time window whose length exactly matched the value of the future delay, as done in [20] and [16]. In detail, defining $\eta(t) = t - D(t)$

and assuming that its inverse exists,[2] exact delay-compensation is obtained form the feedback law $U(t) = KX(\eta^{-1}(t))$. Yet, implementing this relation requires the future variation of the delay to be predicted via $\eta^{-1}(t)$, which is not achievable in practice for an input-varying delay.

Equation (11) can be interpreted as a condition for achieving robust delay compensation.[3] The essence of this condition is that, if the delay were to vary sufficiently slowly, its current value $D(t)$ used for prediction would remain sufficiently close to its future values, and the corresponding prediction would be accurate enough to guarantee the stabilization of the plant through the feedback loop.

We now focus on an alternative sufficient condition guaranteeing (11) when the delay varies in accordance with the integral relation (3). For practical control design, this new condition involves the control gain and the initial conditions.

3 Control Design

In this section, we now establish a sufficient condition guaranteeing (11) when the delay varies in accordance with the integral relation (3). This new condition is relevant to the initial conditions (and thus, to the choice of control gain).

Theorem 2 *Consider the closed-loop system*

$$\dot{X}(t) = AX(t) + Bu(t - D(t)) \tag{12}$$

$$\int_{t-D(t)}^{t} \varphi((u(s))ds = 1 \tag{13}$$

$$u(t) = u^r + K\left[e^{AD(t)}X(t) + \int_{t-D(t)}^{t} e^{A(t-s)}Bu(s)ds - X^r\right]. \tag{14}$$

where $\varphi : \mathbb{R} \mapsto [\underline{\varphi}, \infty)$ *is a polynomial function, with* $\underline{\varphi} > 0$ *constant,* A *and* B *are defined in* (9), K *is such that* $A + BK$ *is Hurwitz,* u *is scalar,* X^r *is the state equilibrium corresponding to the original equilibrium* x^r *of the plant* (2) *and* u^r *is the corresponding (constant) reference control. Consider the functionals*

$$\Theta(t) = |X(t) - X^r| + \max_{s\in[t-\overline{D},t]} \left|[u(s) - u^r \ \dot{u}(s) \ \dots \ u^{(n-1)}(s)]^T\right| \tag{15}$$

$$\Upsilon(t) = |X(t) - X^r|^2 + \int_{t-D(t)}^{t} (u(s) - u^r)^2 ds + D(t)^2 \int_{t-D(t)}^{t} \dot{u}(s)^2 ds. \tag{16}$$

---

[2] This is the case if $\dot{D} < 1$.

[3] Interestingly, a similar condition is often stated in Linear Matrix Inequality approaches, such as [28] for example, where the delay is also assumed to be time-differentiable.

Let $u_0 \in \mathscr{C}^n([-\overline{D}, 0], \mathbb{R})$. *Then there exists* $\theta : \mathbb{R}^n \mapsto \mathbb{R}_+^\star$ *such that, if* $\Theta(0) < \theta(K)$, *then condition* (11) *is fulfilled and the plant exponentially converges to* X^r *in the sense that there exist* $R > 0$ *and* $\rho > 0$ *such that* $\Upsilon(t) \leq R\Upsilon(0)e^{-\rho t}$ *for all* $t \geq 0$.

This result has this relatively direct interpretation: the previously presented Theorem 1 requires the delay to vary sufficiently slowly, while on the other hand, the delay variations implicitly depend on the control input through the integral equation (3), with variations whose rapid swings are scaled by the gain K. Therefore, it would seem to be a natural requirement to restrict input variations by choosing initial conditions sufficiently near the desired equilibrium and which comply with the magnitude of the feedback gain.

The behavior of θ with respect to K may be investigated in future work. From the expression (31), this would involve a study of the solution of the Lyapunov equation and the behavior of its eigenvalues with respect to K. One would expect that $\theta(K) \to 0$ while $K \to -\infty$ (since $r(K)$ is expected in this case to tend to ∞). For this reason, for given initial conditions, the magnitude of the feedback gain should be chosen accordingly. For relatively large initial conditions, this would imply the use of a small feedback gain; therefore, this result can be interpreted as a small-gain condition. We now detail the proof of Theorem 2.

4 Proof of Theorem—Halanay-Like Inequalities

We prove Theorem 2. Taking a time-derivative of (13) and defining the error variable $\varepsilon = u - u^r$, one gets

$$\dot{D}(t) = \frac{\varphi(u(t-D(t))) - \varphi(u(t))}{\varphi(u(t-D(t)))} = \frac{\varphi(\varepsilon(t-D(t)) + u^r) - \varphi(\varepsilon(t) + u^r)}{\varphi(u(t-D(t)))}.$$

As φ is a polynomial function, it is locally Lipschitz with a constant $M(\kappa)$ on the interval $[-\kappa, \kappa]$ for any positive scalar κ. Then, since $\varphi \geq \underline{\varphi}$, we obtain

$$\dot{D}(t) \leq \frac{M(\kappa)|\varepsilon(t-D(t)) + u^r - (\varepsilon(t) + u^r)|}{\underline{\varphi}} = \frac{M(\kappa)|\varepsilon(t-D(t)) - \varepsilon(t)|}{\underline{\varphi}} \leq \frac{2M(\kappa)\max|\varepsilon_t|}{\underline{\varphi}},$$

if $\max|\varepsilon_t| \leq \kappa$. Hence, for any positive parameter $\kappa > 0$, condition (11) is satisfied if

$$\forall t \geq 0, \quad \max|\varepsilon_t| < \min\left\{\frac{\underline{\varphi}\Delta^*(K)}{2M(\kappa)}, \kappa\right\}. \tag{17}$$

This is the condition we now focus on. It yields the analysis of the dynamics of the variable ε. Before beginning this analysis in detail, we recall some well-known stability results for DDE and their extensions.

4.1 Halanay Inequality for Delay Equations of Order $n \geq 1$

We first recall the following result from [11, 13].[4]

Lemma 1 (Halanay inequality) *Consider a continuous* positive- *and real-valued function x such that, for some* $t_0 \in \mathbb{R}$,

$$\dot{x}(t) \leq -ax(t) + b\max x_t\,, \quad t \geq t_0$$

with $a \geq b \geq 0$. *Then, there exists* $\gamma \geq 0$ *such that*

$$\forall t \geq t_0\,, \quad x(t) \leq \max x_{t_0} e^{-\gamma(t-t_0)}$$

Straightforward extensions of this lemma are stated below.

Corollary 1 *Consider a* positive-valued *continuous function such that*

$$\begin{cases} \dot{x}(t) \leq -ax(t) + bh(t, x_t)\,, \quad t \geq t_0 \\ x_{t_0} = \psi \in \mathscr{C}^0([-\overline{D}, 0], \mathbb{R}_+), \end{cases} \tag{18}$$

where h is a continuous functional satisfying the sup-norm relation

$$h(t, x_t) \leq \max|x_t| \ \textit{if} \ \max|x_t| < \omega, \tag{19}$$

for a given $\omega > 0$. *If the initial condition is such that* $\max\psi < \omega$ *and if* $a \geq b \geq 0$, *then there exists* $\gamma \geq 0$ *(where* $\gamma = 0$ *if* $a = b$ *and* $\gamma > 0$ *otherwise) such that*

$$x(t) \leq \max x_{t_0} e^{-\gamma(t-t_0)} \tag{20}$$

holds for all $t \geq t_0$.

Proof Consider ψ such that x is a non-trivial continuous solution[5] of (18) which, using (19), satisfies the inequality

$$\dot{x}(t) \leq -ax(t) + b\max x_t \quad \text{if } \max x_t < \omega.$$

Following the seminal proof of [11], define $y(t) = ke^{-\gamma(t-t_0)}$, with $k > 0$ and γ chosen such that y satisfies the corresponding differential equation[6]

$$\dot{y}(t) = -ay(t) + b\max y_t, \quad t \geq t_0 \quad \text{and} \quad y_{t_0} = k.$$

---

[4] More precisely, in [11], this result is stated for $a > b > 0$.

[5] The case when x is identically 0 is trivial. The continuity (and even more) is obtained by assuming ψ is smooth enough.

[6] $\gamma \geq 0$ is the unique solution on $[0, \infty[$ of $a - \gamma = b\exp(\gamma\overline{D})$.

Now, we define the difference $z = y - x$, which is a continuous function, and we are interested in its sign change. We choose $k \in (\max x_{t_0}, \omega)$, which is a not empty set as $\max x_{t_0} = \max \psi < \omega$. This guarantees that $z(t) > 0$ for $t \in [t_0 - \overline{D}, t_0]$. The function z being continuous, we define

$$t_1 = \inf \{t > t_0 | z(t) = 0\} \in \mathbb{R} \cup \{\infty\} \ .$$

Assume that $t_1 < \infty$. From the definition of t_1, $z(t) > 0$ for $t \in [t_0, t_1]$ and, from the analytical expression of y and as both x and y are continuous, $x(t) < y(t) < k < \omega$ for $t \in [t_0, t_1)$. Therefore, the following inequality holds

$$\forall t \leq t_1 \, , \quad \dot{z}(t) \geq -az(t) + b\,(\max y_t - \max x_t)\, .$$

Then, $\dot{z}(t_1) \geq \max y_{t_1} - \max x_{t_1} > 0$, by definition of t_1. Yet, one has

$$\dot{z}(t_1) = \lim_{t \to t_1^-} \frac{z(t) - z(t_1)}{t - t_1} = \lim_{t \to t_1^-} \frac{z(t)}{t - t_1} \leq 0 \text{ as } z(t) \geq 0 \text{ on } [t_0, t_1].$$

We finally conclude that $t_1 = \infty$. Then, for $t \geq t_0$ and all $z(t) > 0$ and, for any $\varepsilon = k - \max x_{t_0} > 0$, we get

$$\forall t \geq t_0 \quad x(t) < (\max x_{t_0} + \varepsilon) e^{-\gamma(t - t_0)},$$

which gives the result.

Lemma 2 (stability of a nth order DDE) *Let x be a solution of the nth order DDE*

$$\begin{cases} x^{(n)}(t) + \alpha_{n-1} x^{(n-1)}(t) + \ldots + \alpha_0 x(t) = c\ell(t, x_t, \ldots x_t^{(n-1)}) \, , \; t \geq t_0 \\ X_{t_0} = \psi \in \mathscr{C}^0([-\overline{D}, 0], \mathbb{R}) \quad \text{with} \quad \max|\psi| < \omega, \end{cases} \tag{21}$$

where the left-hand side of the differential equation defines a polynomial whose roots have only strictly negative real parts, $c > 0$, $\omega > 0$ and ℓ is a continuous functional. Then, there exist $c_* > 0$ and $r_* > 0$ ($r_* = 1$ and $c_* = \alpha_0$ if $n = 1$) such that, if

- $c \leq c_*$;
- *ℓ satisfies the following sup norm relation, with $X = [x \;\; \dot{x} \; \ldots \; x^{(n-1)}]^T$,*

$$|\ell(t, x_t, \ldots, x_t^{(n-1)})| \leq \max|X_t|, \;\; \text{for } t \geq t_0, \;\; \max|X_t| < r\omega \text{ with } r \geq r_*,$$

then, there exists $\gamma \geq 0$ such that $|X(t)| \leq r_* \max|X_{t_0}| e^{-\gamma(t-t_0)}$ holds for all $t \geq 0$.

Proof Using the scalar result of Corollary 1, define the scalar positive valued function $m(t) = X(t)^T P X(t)$ where P is, as defined in the statement of the Lemma, the symmetric positive definite matrix solution of the Lyapunov equation $A_0^T P + PA_0 = -Q$, for some given symmetric positive definite matrix Q and

$$A_0 = \begin{pmatrix} 0 & 1 & & \\ \vdots & & \ddots & \\ 0 & & & 1 \\ -\alpha_0 & -\alpha_1 & \dots & -\alpha_{n-1}. \end{pmatrix}$$

is the companion matrix. Taking a time-derivative of m, we obtain

$$\dot{m}(t) = -X(t)^T Q X(t) + 2X(t)^T P \begin{pmatrix} 0 \\ \vdots \\ 0 \\ c\ell(t, x_t, \dots, x_t^{(n-1)}) \end{pmatrix}.$$

Therefore, defining $a \triangleq \frac{\underline{\lambda}(Q)}{\overline{\lambda}(P)}$ and $b \triangleq 2c\frac{\overline{\lambda}(P)}{\underline{\lambda}(P)}$, one obtains

$$\dot{m}(t) \leq -am(t) + bh(t), \text{ where} \tag{22}$$

$$\begin{aligned} h(t) &= \frac{\underline{\lambda}(P)}{\overline{\lambda}(P)} X(t)^T P \begin{pmatrix} 0 \\ \vdots \\ 0 \\ \ell(t, x_t, \dots, x_t^{(n-1)}) \end{pmatrix} \\ &\leq \underline{\lambda}(P)|X(t)||\ell(t, x_t, \dots, x_t^{(n-1)})| \leq \sqrt{\underline{\lambda}(P)}\sqrt{m(t)}|\ell(t, x_t, \dots, x_t^{(n-1)})|, \end{aligned} \tag{23}$$

which satisfies the following for $\max|X_t| < r\omega$ (and therefore, in particular, for $\max m_t < r^2\omega^2\underline{\lambda}(P)$): $|h(t)| \leq \sqrt{m(t)}\max\sqrt{m_t} \leq \max m_t$. Define the condition

$$\max m_{t_0} < r^2\omega^2\underline{\lambda}(P) \tag{24}$$

If (24) holds then we conclude by applying Corollary 1 to (22), that, if $a > b$, then there exists $\gamma > 0$ such that

$$\forall t \geq t_0, \quad m(t) \leq \max m_{t_0} e^{-2\gamma(t-t_0)}.$$

A sufficient condition for $\max m_{t_0} < r^2\omega^2\underline{\lambda}(P)$ is that

$$\max|X_{t_0}|^2 < r^2\omega^2\underline{\lambda}(P)/\overline{\lambda}(P). \tag{25}$$

By assumption $|\max X_{t_0}| < \omega$. Hence, to guarantee that (24) holds, we simply have to pick

$$r \geq r_* \triangleq \sqrt{\overline{\lambda}(P)/\underline{\lambda}(P)}. \tag{26}$$

Finally, the condition $a > b$ can be reformulated as

$$c < \frac{\underline{\lambda}(P)\underline{\lambda}(Q)}{2\overline{\lambda}(P)^2} \triangleq c_* \tag{27}$$

and one obtains

$$\forall t \geq t_0\,, \quad |X(t)| \leq \sqrt{\frac{\overline{\lambda}(P)}{\underline{\lambda}(P)}} \max|X_{t_0}|e^{-\gamma(t-t_0)},$$

which concludes the proof.

4.2 Application to Dynamical Equation Defined Through the Predictor-Based Control Law

We now focus on the DDE governing ε, which is given in the following lemma, the proof of which follows exactly the same lines as the one provided in [6]. This lemma can be extended without difficulty to the case of a piecewise continuous initial condition u_0. However, we prefer to restrict our exposition for the sake of simplicity.

Lemma 3 *Provided that $u_0 \in \mathscr{C}^n([-\overline{D}, 0], \mathbb{R})$, the error variable $\varepsilon = u - u^r$ with u defined in* (14) *satisfies the following differential equation for $t \geq 0$*

$$\begin{aligned}\varepsilon^{(n)}(t) + (a_{n-1} + b_0 k_{n-1})\varepsilon^{(n-1)}(t) + \ldots + (a_0 + b_0 k_0)\varepsilon(t) \\ = \pi_1\left(\dot{D}(t), \ldots, D^{(n)}(t), \varepsilon_t, \ldots, \varepsilon_t^{(n-1)}, \frac{1}{1+\dot{D}(t)}\right),\end{aligned} \tag{28}$$

where $[-k_0 \ldots - k_{n-1}] \stackrel{\Delta}{=} K$ and π_1 is a polynomial function which is at least quadratic in the variables $\dot{D}, \ldots, D^{(n)}, \varepsilon_t, \ldots, \varepsilon_t^{(n-1)}$.

Now that we are equipped with this last lemma, it is possible to use Lemma 2 to guarantee that the stability condition (17) holds.

Lemma 4 *Consider the functional Θ defined in* (15)*. Then, there exists a function $\theta : \mathbb{R}^n \mapsto \mathbb{R}_+^*$ such that, if $\Theta(0) < \theta(K)$, then* (17) *is fulfilled, which implies that* (11) *holds.*

Proof The input error ε satisfies dynamics (28), which is compliant with the assumptions of Lemma 2. In detail, first, the left-hand side of (28) is stable, as it represents the last line of the Hurwitz companion matrix $A + BK$. Second, by observing that φ and hence its derivatives are polynomial functions and that

$$\begin{aligned}
\dot{D}(t) &= \tfrac{\varphi(\varepsilon(t-D(t))+u^r)-\varphi(\varepsilon(t)+u^r)}{\varphi(\varepsilon(t-D(t))+u^r)} \\
\ddot{D}(t) &= \tfrac{1}{\varphi(\varepsilon(t-D(t))+u^r)^3}\big[\dot{\varepsilon}(t-D(t))\dot{\varphi}(\varepsilon(t-D(t))+u^r)\varphi(\varepsilon(t)+u^r) \\
&\quad -\dot{\varepsilon}(t)\dot{\varphi}(\varepsilon(t)+u^r)\varphi(\varepsilon(t-D(t))+u^r)\big] \\
D^{(3)}(t) &= \ldots,
\end{aligned}$$

we obtain by induction that, for $m \geq 1$, $D^{(m)}$ is a polynomial function in

$$\varepsilon_t, \ldots, \varepsilon_t^{(m-1)}, \frac{1}{\varphi(\varepsilon(t-D)+u^r)} \tag{29}$$

without terms of order 0 or 1. Therefore, π_1 is directly a polynomial function of the variables

$$\varepsilon_t, \ldots, \varepsilon_t^{(n-1)}, \frac{1}{1+\dot{D}}, \frac{1}{\varphi(\varepsilon(t-D)+u^r)}, \tag{30}$$

which is *at least quadratic* in the variables $\varepsilon_t, \ldots, \varepsilon_t^{(n-1)}$. Observing that

$$\frac{1}{1+\dot{D}(t)} = \frac{\varphi(\varepsilon(t-D)+u^r)}{2\varphi(\varepsilon(t-D)+u^r)-\varphi(\varepsilon(t)+u^r)},$$

we conclude that this term is continuous and therefore bounded on a given neighborhood of the origin. Hence, as π_1 is *at least quadratic*, it is possible to properly define a scalar $\omega^* > 0$ such that

$$\left|\pi_1\left(\varepsilon_t, \ldots, \varepsilon_t^{(n-1)}\right)\right| \leq c_* \max|E_t|, \ \max|E_t| < \omega^*,$$

in which c_* is introduced in Lemma 2. This neighborhood depends on the functional ℓ and on the constant c_* and therefore on the feedback gain K. Finally, define

$$c\ell(t, \varepsilon_t, \ldots, \varepsilon_t^{(n-1)}) = \pi_1\left(\dot{D}, \ldots, D^{(n)}, \varepsilon_t, \ldots, \varepsilon_t^{(n-1)}, \frac{1}{1+\dot{D}}\right).$$

Then, for $\max|E_t([-\overline{D}, 0])| < \omega^*(K)$, we obtain

$$\left|c\ell(t, \varepsilon_t, \ldots, \varepsilon_t^{(n-1)})\right| \leq c_* \max|E_t|.$$

Therefore, Lemma 2 guarantees the existence of $r_*(K) > 0$ and $\gamma \geq 0$ such that

$$\forall t \geq 0, \quad |E(t)| \leq r_*(K) \max|E_0| e^{-\gamma t}.$$

Hence, by choosing

$$\max|E_0| \leq \frac{1}{r_*(K)} \min \left\{ \frac{\underline{u}\Delta^*(K)}{2M(\kappa)}, \kappa, \omega^*(K) \right\} \triangleq \theta(K) \tag{31}$$

we ensure that this condition is fulfilled for any $t \geq 0$, that the initial condition lies in the neighborhood Ω and that

$$|E(t)| \leq \min \left\{ \frac{\underline{u}\Delta^*(|K|)}{2M(\kappa)}, \kappa \right\}, \ t \geq 0. \tag{32}$$

In particular, condition (17) is also fulfilled. Finally, the choice $\max|E_0| \leq \theta(K)$ can be expressed in terms of Θ. This gives the conclusion.

The proof of Theorem 2 directly follows from Lemma 4.

5 Conclusion

In this study we have extended some of our recent work to the compensation of input-dependent delay input defined through an integral equation and considered a wide class of kernels (positive polynomial functions of the input). This class of systems is representative of a large number of processes involving transport of material, such as the low-cost rocket design presented above. We prove that robust compensation is achieved by using the current value of the delay as prediction horizon, provided that the initial conditions are in accordance with the magnitude of the feedback gain. The proposed approach is based on a two-step methodology which, first, requires the delay variations to be bounded and, second, relates these variations to input fluctuations. Quantitative comparison of the practical performances of the proposed approach with those of a prediction-based controller using the (constant) final reference delay $D = 1/\varphi(u^r)$ is the natural next step in this approach. Extension to explicitly time-dependent delay-integral kernels is also a direction of future work.

References

1. Artstein, Z.: Linear systems with delayed controls: a reduction. IEEE Trans. Autom. Control **27**(4), 869–879 (1982)
2. Bekiaris-Liberis, N., Krstic, M.: Compensation of state-dependent input delay for nonlinear systems. IEEE Trans. Autom. Control **58**(2), 275–289 (2013)
3. Bresch-Pietri, D., Chauvin, J., Petit, N.: Adaptive control scheme for uncertain time-delay systems. Automatica **48**(8), 1536–1552 (2012)
4. Bresch-Pietri, D., Chauvin, J., Petit, N.: Invoking Halanay inequality to conclude on closed-loop stability of a process with input-varying delay. In: Proceedings of the 10th IFAC Workshop on Time Delay Systems, pp. 266–271 (2012)

5. Bresch-Pietri, D., Chauvin, C., Petit, N.: Prediction-based feedback control of a class of processes with input-varying delay. In: Proceedings of the American Control Conference, pp. 1991–1997 (2012)
6. Bresch-Pietri, D., Chauvin, J., Petit, N.: Sufficient condition for prediction-based stabilization of linear system subject to input-dependent input-delay. In: Proceedings of the American Control Conference, pp. 144–151 (2013)
7. Bresch-Pietri, D., Chauvin, J., Petit, N.: Prediction-based stabilization of linear systems subject to input-dependent input delay of integral-type. IEEE Trans. Autom. Control **59**(9), 2385–2399 (2014)
8. Chebre, M., Creff, Y., Petit, N.: Feedback control and optimization for the production of commercial fuels by blending. J. Process Control **20**(4), 441–451 (2010)
9. Depcik, C., Assanis, D.: One-dimensional automotive catalyst modeling. Prog. Energy Combust. Sci. **31**(4), 308–369 (2005)
10. Gu, K., Niculescu, S.-I.: Survey on recent results in the stability and control of time-delay systems. J. Dyn. Syst. Meas. Control **125**(2), 158–165 (2003)
11. Halanay, A.: Differential Equations: Stability, Oscillations, Time Lags. Academic Press, New York (1966)
12. Harmand, J., Dochain, D.: The optimal design of two interconnected (bio) chemical reactors revisited. Comput. Chem. Eng. **30**(1), 70–82 (2005)
13. Ivanov, A., Liz, E., Trofimchuk, S.: Halanay inequality, yorke 3/2 stability criterion, and differential equations with maxima. Tohoku Math. J. **54**(2), 277–295 (2002)
14. Jankovic, M.: Recursive predictor design for linear systems with time delay. In: Proceedings of the American Control Conference, pp. 4904–4909 (2008)
15. Krstic, M.: Boundary Control of PDEs: A Course on Backstepping Designs. Society for Industrial and Applied Mathematics, Philadelphia (2008)
16. Krstic, M.: Delay Compensation for Nonlinear, Adaptive, and PDE Systems. Birkhauser, Boston (2009)
17. Manitius, A., Olbrot, A.: Finite spectrum assignment problem for systems with delays. IEEE Trans. Autom. Control **24**(4), 541–552 (1979)
18. Michiels, W., Niculescu, S.-I.: Stability and Stabilization of Time-Delay Systems. Society for Industrial and Applied Mathematics, Philadelphia (2007)
19. Moon, Y., Park, P., Kwon, W.: Robust stabilization of uncertain input-delayed systems using reduction method. Automatica **37**(2), 307–312 (2001)
20. Nihtila, M.: Finite pole assignment for systems with time-varying input delays. In: Proceedings of the 30th IEEE Conference on Decision and Control, pp. 927–928 (1991)
21. Perry, R., Green, D., Maloney, J.: Perry's Chemical Engineers' Handbook. McGraw-Hill, New York (1984)
22. Petit, N., Creff, Y., Rouchon, P.: Motion planning for two classes of nonlinear systems with delays depending on the control. In: Proceedings of the 37th IEEE Conference on Decision and Control, pp. 1007–1011 (1998)
23. Richard, J.-P.: Time-delay systems: an overview of some recent advances and open problems. Automatica **39**(10), 1667–1694 (2003)
24. Sbarciog, M., De Keyser, R., Cristea, S., De Prada, C.: Nonlinear predictive control of processes with variable time delay. A temperature control case study. In Proceedings of the IEEE International Conference on Control Applications, pp. 1001–1006 (2008)
25. Smith, O.: A controller to overcome dead time. ISA J. **6**(2), 28–33 (1959)
26. Sutton, G., Biblarz, O.: Rocket Propulsion Elements. Wiley, New York (2011)
27. Witrant, E.: Stabilisation des systemes commandes par rseaux. Ph.D. thesis, Laboratoire d'Automatique de Grenoble (2005)
28. Yue, D., Han, Q.: Delayed feedback control of uncertain systems with time-varying input delay. Automatica **41**(2), 233–240 (2005)

State Estimation and Control of Nonlinear Systems with Large and Variable Measurement Delays

Filippo Cacace, Alfredo Germani and Costanzo Manes

Abstract This chapter deals with the problem of output feedback control of nonlinear systems affected by time-varying measurement delay. A control law is presented, which is made of an observer-controller cascade where the controller is a classic state-linearizing scheme, and the observer is a high-gain observer of chain-type. It is shown that under suitable conditions on the system, the observer is globally exponentially convergent, and the replacement of the true state with the observer state in the control law results in an exponentially stabilizing feedback scheme. The main limitation with a single observer is the presence of a delay bound that depends on the Lipschitz constant of the nonlinear system. To overcome this limitation it is possible to resort to a chain of observers that, at the cost of a growing realization space and convergence time, can in principle allow to compensate any delay. This design is straightforward when the delay is known and constant but its extension to time-varying delays requires special attention, in particular when the delay is not continuous with respect to time, as it frequently happens in the applications. We therefore introduce a classification of delay functions with respect to the available output information and illustrate how to design the cascade of elementary observers to solve the state reconstruction problem. We also characterize the class of delay functions for which this approach fails to provide a viable implementation.

F. Cacace (✉)
Università Campus Biomedico di Roma, Rome, Italy
e-mail: f.cacace@unicampus.it

A. Germani · C. Manes
Università dell'Aquila, L'Aquila, Italy
e-mail: alfredo.germani@univaq.it

C. Manes
e-mail: costanzo.manes@univaq.it

I. Karafyllis et al. (eds.), *Recent Results on Nonlinear Delay Control Systems*,
Advances in Delays and Dynamics 4, DOI 10.1007/978-3-319-18072-4_5

1 Overview

The problem of state estimation and control for nonlinear systems with delayed output measurements appears in many control systems areas, for example in networked control systems or when the measurement process intrinsically causes a non negligible time delay, as for example in biochemical reactors. For this reason the issue of state reconstruction in the presence of time delays in the system equations and/or in the measurement process is receiving increasing attention.

In this chapter, we extend the classical approach for non delayed systems, based on state estimation and state feedback, to the case of systems with large and variable measurement delays. Since the viability of the approach strongly depends on the features of the delay function, in Sect. 2 we describe a classification of the delay functions from the point of view of the state reconstruction.

The problem of state estimation under delays dates back to the prediction of the state evolution for stable linear systems with delays, solved by the Smith predictor [32]. Open loop predictors for nonlinear systems were subsequently developed [21, 29]. In recent years, several works have addressed the state observation problem under constant or variable delays [5, 7, 20, 28, 33, 34]. In Sect. 3, we describe an observer of high-gain type which guarantees exponential convergence of the estimation error to zero in the presence of uniformly bounded variable delays without any restriction on the delay function. Section 4 investigates conditions under which the separation principle holds and when it is possible to design a global stabilizing output feedback via the state estimation provided by the observer. This section also provides some references to this problem in the context of systems with output delay. It turns out that the corresponding sufficient conditions are more restrictive than in the non-delayed case.

The problem common to the approaches mentioned so far is the existence of a bound on the delay, that usually depends on the nonlinearities of the system. This is obviously a serious drawback, particularly when the bound of the delay is not known with precision. The idea of introducing a cascade of two or more observers to achieve the convergence of state estimation in presence of arbitrarily large, but constant, delays was first proposed in [13]. Each observer in the chain is in charge of predicting the system state for a suitable fraction of the total delay. The basic observer in the chain was the high-gain observer for non delayed measurements of [9, 10]. A similar approach has been used in [20], where some restrictions of the chain-observer in [13] have been overcome. In [1] another predictor for nonlinear triangular systems with delayed output, based on a cascade of observers, has been proposed. All these proposals deal with large but constant delays. In Sect. 5 we discuss the extension of the approach to time-varying delays. It turns out that chain observers can be extended to variable delays under some assumptions on the delay function. We describe a chain configuration that allows us to deal with discontinuous delay functions, which is an original contribution of this chapter.

2 Classification of Delay Functions

In this section, we analyze the time-varying delay functions from the point of view of the available output information. We argue that several situations, such as transmission delays, bufferization, sampling and data losses can be modeled by means of suitable delay functions.

The systems considered here have the form

$$\dot{x}(t) = F\big(x(t), u(t)\big), \qquad t \geq -\Delta, \tag{1}$$

$$\bar{y}(t) = h\big(x(t - \delta(t))\big), \quad t \geq 0, \tag{2}$$

$$x(-\Delta) = \bar{x} \in \mathbb{R}^n, \tag{3}$$

where $x(t) \in \mathbb{R}^n$ is the system state, $u(t) \in \mathbb{R}^p$ is a known input, and the function $F : \mathbb{R}^n \times \mathbb{R}^p \mapsto \mathbb{R}^n$ is affine in the input, i.e.,

$$F(x, u) = f(x) + G(x)u = f(x) + \sum_{k=1}^{p} g_k(x)u_k, \tag{4}$$

where $f(x)$ and $g_k(x)$ are C^∞ vector fields. $\bar{y}(t) \in \mathbb{R}$ is the measured output available at time t, which is a delayed function of the state, and $\delta(t)$ is a time-varying measurement delay. The function $h : \mathbb{R}^n \to \mathbb{R}$ is C^∞. For the sake of clarity we assume a scalar output, but the framework can be readily extended to vector output functions with specific delays on each component (see for example [7, 8]).

The problem we are facing for the system (1)–(2) is the dynamical reconstruction of the state at time t by exploiting the knowledge of the input $u(\tau)$, $\tau \in [-\Delta, t]$, and the output $\bar{y}(\tau)$, $\tau \in [0, t]$, and the simultaneous system stabilization through a function of the reconstructed state.

We denote as $y(t) = h\big(x(t)\big)$ the un-delayed output, so that $\bar{y}(t) = y\big(t - \delta(t)\big)$. We assume that the delay is bounded by a known upper bound, i.e. $\delta(t) \in [0, \Delta]$ with known Δ. We do not need to assume the a priori knowledge of the function $\delta(t)$: we only assume that $\delta(t)$ is known at the same time t at which $\bar{y}(t)$ is available. Thus, we say that the "fresh" information on the system at time t is given by the pair $\big(\bar{y}(t), \delta(t)\big)$. Moreover, we do not need to assume the continuity of $\delta(t)$, although the piecewise continuity is needed for the existence and uniqueness of the solution of a state observer in a differential form.

Note that when the delay function $\delta(t)$ is bounded and continuous the information flux is never interrupted, in that at time t all the measurements previous to $y(t - \delta(t)\big)$ are available.

Proposition 1 *If $\delta(t) \in [0, \Delta]$ is continuous, then at any $t > \Delta$ the measurement $y(t - \sigma)$ is available, if $\sigma > \delta(t)$. That is, there exists $t^* < t$ such that $\bar{y}(t^*) = y(t - \sigma)$.*

This implies that the measurements are eventually available after a time interval of at most Δ, and the information is never lost. Thus, the case of continuous delay is not too different from the constant delay case. In particular, any state predictor designed for the constant delay case can cope with continuous delay functions by artificially retarding the measurement $\bar{y}(t)$ with a delay $\theta(t) = \Delta - \delta(t)$, obtaining $\bar{y}\big(t - \theta(t)\big) = y(t - \Delta)$ (constant delay). Of course this choice in general is not efficient, but it highlights that continuous delays are similar to constant delays in the sense that output information $y(t)$ is entirely available after a certain time. This property is not limited to continuous delay functions, and it can be formally characterized with the notion of lossless delay functions (LDF) as follows:

Definition 1 A delay function $\delta(t)$ is LDF if $\forall t \geq 0, \exists t^* \geq t$ such that $t = t^* - \delta(t^*)$.

The above definition implies that if a delay function $\delta(t)$ is LDF, then $\forall t \geq 0$, $\exists t^* \geq t$ such that $\bar{y}(t^*) = y(t)$.

The LDF property implies that the output function can be fully reconstructed after a certain time interval. Notice that any continuous $\delta(t)$ is LDF, but the converse is not true, in that not all discontinuous delay functions cause information loss. An example of discontinuous LDF is the transmission delay over a network when no data packet is lost but the ordering of packets is not maintained; see Fig. 1. Sampled measurements can be modeled by discontinuous delay functions, as originally suggested by Fridman and co-authors in [30]. The delay function used when the sampling interval is T can be written as $\delta(t) = \mod(t, T)$.

In this case, $\bar{y}(t)$ is defined for any $t \geq 0$ using a piecewise constant function consisting of the last available measurements (Fig. 2, left), so that $\delta(t)$ has a saw-tooth shape. Is it immediate to see that delay functions that model sampling are not LDF.

This idea can be extended easily to the case of missing output measurements. Suppose that the measurements are available with a constant delay $\bar{\delta}$ (possibly zero), and are lost in the interval $[t_1, t_2]$. This can be modeled with a delay function $\delta(t) = \bar{\delta}$ for $t \notin [t_1, t_2)$ and $\delta(t) = t - t_1 + \bar{\delta}$ for $t \in [t_1, t_2)$, as shown in Fig. 2 (right). It is evident that the difficulty of estimating the state in presence of non LDF delays

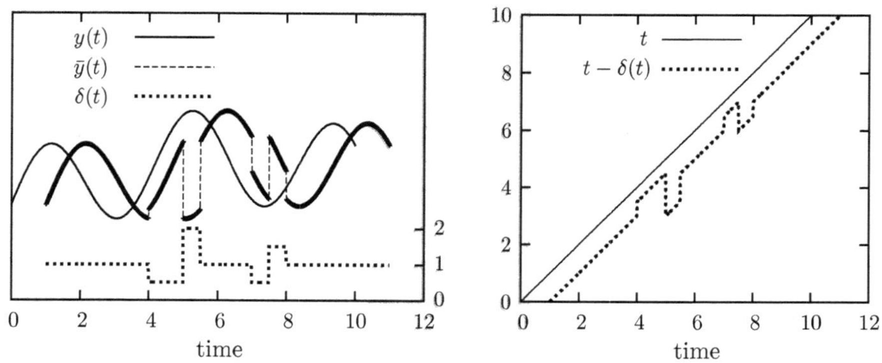

Fig. 1 Delayed output with discontinuous LDF $\delta(t)$ *(left)* and the corresponding function $t - \delta(t)$ *(right)*

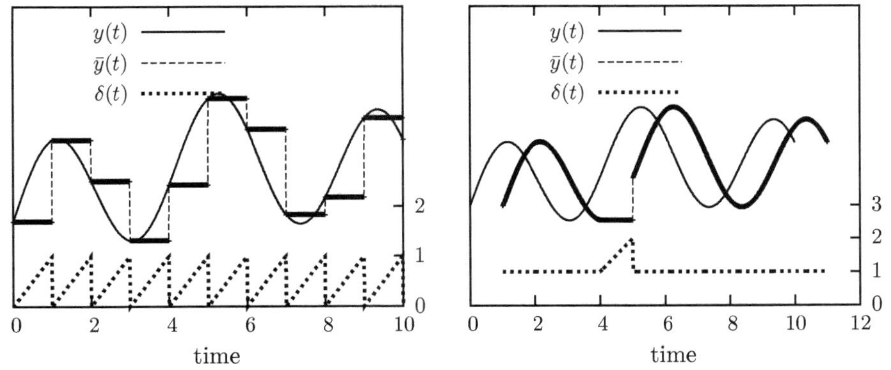

Fig. 2 Sampled output ($T = 1$) as discontinuous delay function (*left*) and packet loss represented as a discontinuous variable delay (*right*)

depends on the frequency and amplitude of the loss of information. A short blackout in the measurement process causes no substantial problem, whereas output sampling makes the measured output finite dimensional and can be more challenging.

We can define the largest time interval for which the output information is lost as follows.

Definition 2 The *maximal loss interval* of a delay function $\delta(t)$ is the least upper bound of the Lebesgue measures of time intervals I_τ such that $\forall \tau \in I_\tau,\ \forall t > 0,\ \tau \neq t - \delta(t)$.

For example, the sampling interval T is the maximal loss interval for $\delta(t) = \mathrm{mod}\ (t, T)$, whereas the jump $J_\delta(t_3)$ is the maximal loss interval for the delay function $\delta(t)$ in Fig. 3.

In the sequel, we shall see that the design of a chain of observers for non LDF delay functions is possible only when the maximal loss interval is less than the delay bound for the elementary observers that make up the chain.

Summarizing, delay functions $\delta(t) \leq \Delta$ can be characterized according to the output information available at time t:

- *continuous*: the output is completely available up to $t - \delta(t)$;
- *LDF*: the output is completely available up to $t - \Delta$;
- *loss bounded*: the output is available up to $t - \Delta$ except for regions not larger than the maximal loss interval;
- *unrestricted*: the output is not, in general, fully available.

We may conclude that variable output delays can be seen as a unifying framework to model buffering, discretization, sampling, transmission delays and packet losses.

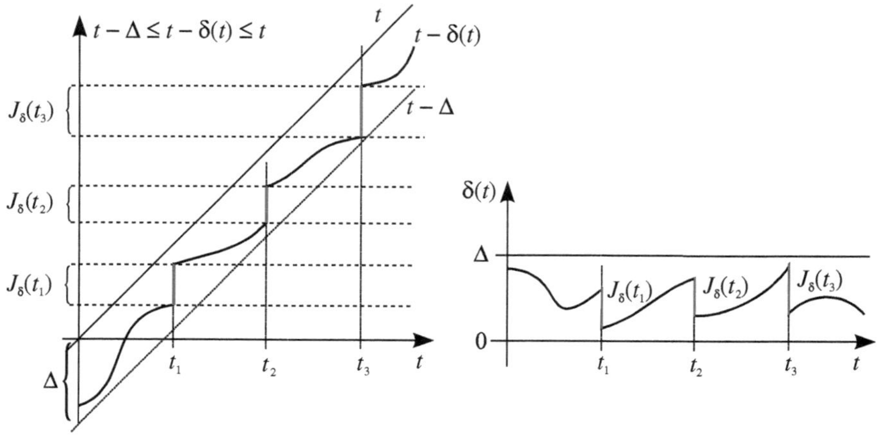

Fig. 3 Delay function $\delta(t)$ with discontinuities (Information Loss) in t_1, t_2 and t_3

Consequently, solving the problem of state estimation and control in presence of general delay functions encompasses a possible solution to all the above scenarios and to their combination.

3 Single-Step Exponential Observer

This section presents a single-step observer of system (1) and (2) and the related convergence result. In order to design a chain of observers it is crucial that the estimation error of the single-step observer is exponentially stable. The high-gain observer presented here is based on the drift-observability map and it was originally proposed in [6]. It is an exponentially convergent modification of the observer presented in [5] and it shares the essential properties of other recent proposals such as [2, 34].

In what follows we need to recall the definition of the Lie derivative of a C^∞ function $\lambda(x)$ with respect to the C^∞ vector field $\varphi : \mathbb{R}^n \to \mathbb{R}^n$, $L_\varphi \lambda(x) = d\lambda/dx \cdot \varphi(x)$ is a function from $\mathbb{R}^n$ to $\mathbb{R}$. The symbol $L_\varphi^k \lambda(x)$ is the k-times repeated iteration of $L_\varphi \lambda(x)$. For a C^∞ matrix function $G : \mathbb{R}^n \to \mathbb{R}^{n\times p}$, the Lie derivative $L_G \lambda(x) = [L_{g_1}\lambda(x) \ \dots \ L_{g_p}\lambda(x)]$ is a function from $\mathbb{R}^n$ to $\mathbb{R}^p$.

For system (4) the drift-observability map $z = \Phi(x)$ is defined as

$$z = \Phi(x) = \Big[h(x)\ L_f h(x)\ \dots\ L_f^{n-1} h(x)\Big]^T. \tag{5}$$

Definition 3 The system (1) is said to be globally drift-observable if the function $z = \Phi(x)$ is a diffeomorphism in all $\mathbb{R}^n$. A system is said to be globally uniformly Lipschitz drift-observable (GULDO) if it is globally drift-observable and the maps Φ and Φ^{-1} are uniformly Lipschitz.

When the system is globally drift-observable the map $z = \Phi(x)$ defines a global change of coordinates, and the Jacobian

$$Q(x) = \frac{d\Phi(x)}{dx} \tag{6}$$

is nonsingular for all $x \in \mathbb{R}^n$.

Definition 4 The triple $(f(x), G(x), h(x))$ is said to have *observation relative degree* r in a set $\Omega \subseteq \mathbb{R}^n$ if

$$\begin{aligned} L_G L_f^k h(x) &= 0, \quad k = 0, \dots, r-2, \ \forall x \in \Omega \\ L_G L_f^{r-1} h(x) &\neq 0_{1\times p}, \quad \text{for some } x \in \Omega. \end{aligned} \tag{7}$$

If $\Omega = \mathbb{R}^n$ the triple is said to have *observation* relative degree r.

If the system (1) is globally drift-observable and the observation relative degree in $\mathbb{R}^n$ is n, the following function is well defined

$$p(z, u) = \left(L_f^n h(x) + L_G L_f^{n-1} h(x) u \right)_{x=\Phi^{-1}(z)}, \tag{8}$$

and, in analogy with the non delayed nonlinear case [10], the representation in z-coordinates is

$$\begin{aligned} \dot{z}(t) &= A_b z(t) + B_b p(z(t), u(t)), \quad && t \geq -\Delta, \\ \bar{y}(t) &= C_b z(t - \delta(t)), && t \geq 0, \end{aligned} \tag{9}$$

with $z(-\Delta) = \Phi(\bar{x})$, where (A_b, B_b, C_b) is a Brunowsky triple of order n.

The hypotheses needed for the construction of the proposed observer of system (1) and (2) are summarized below.

$\mathscr{H}_1$ The nonlinear delay-free system described by the triple (f, G, h) is GULDO.

$\mathscr{H}_2$ The function $p(z, u)$ defined in (8) is globally uniformly Lipschitz with respect to z, and the Lipschitz coefficient γ_p is a (non decreasing) function of $\|u\|$, i.e. $\forall z_1, z_2 \in \mathbb{R}^n$

$$\|p(z_1, u) - p(z_2, u)\| \leq \gamma_p(\|u\|)\|z_1 - z_2\|. \tag{10}$$

$\mathscr{H}_3$ The triple (f, G, h) has uniform observation degree at least equal to n.

The proposed observer is the following for all $t > 0$,

$$\begin{aligned} \dot{\hat{x}}(t) &= f(\hat{x}(t)) + G(\hat{x}(t))u(t) + Q^{-1}(\hat{x}(t))K_\delta(t)\Big(\bar{y}(t) - h(\hat{x}(t - \delta(t)))\Big), \\ \hat{x}(\tau) &= \chi(\tau), \quad \tau \in [-\Delta, 0], \quad \text{where} \quad K_\delta(t) = e^{-\eta\delta(t)} K_0. \end{aligned} \tag{11}$$

$\chi : [-\Delta, 0] \to \mathbb{R}^n$ is a vector function used to initialize the observer, and is chosen continuous and bounded, i.e. $\chi \in \mathscr{C}_\Delta^n \cap L_\infty^n$. The vector $K_0 \in \mathbb{R}^n$ and the constant

$\eta \geq 0$ are design parameters. In particular, the constant η is a desired exponential decay rate for the observation error. The observer equation (11) in z-coordinates for $t > 0$ is

$$\dot{\hat{z}}(t) = A_b\hat{z}(t) + B_b p(\hat{z}(t), u(t)) + K_\delta(t)\left(\bar{y}(t) - C_b\hat{z}(t-\delta(t))\right). \tag{12}$$

Let $\tilde{z}(t) = z(t) - \hat{z}(t)$ be the observation error in $z-$coordinates. Subtracting equations (9) and (12) and defining the function

$$\tilde{p}(t,\tilde{z}) = p\big(z(t), u(t)\big) - p\big(z(t) - \tilde{z}, u(t)\big), \tag{13}$$

where $p(\cdot,\cdot)$ is defined in (8), we have that $\tilde{z}(t)$ satisfies for $t > 0$

$$\dot{\tilde{z}}(t) = A_b\tilde{z}(t) + B_b\tilde{p}\big(t,\tilde{z}(t)\big) - e^{-\eta\delta(t)}K_0 C_b\tilde{z}(t-\delta(t)), \tag{14}$$

Note that by the Lipschitz assumption (10) in $\mathscr{H}_2$ we have that if $\|u(t)\| \leq \bar{u}$, then

$$\|\tilde{p}\big(t,\tilde{z}(t)\big)\| \leq \gamma_p(\bar{u})\|\tilde{z}(t)\|. \tag{15}$$

Before giving the main convergence theorem, we recall [8, Lemma 4]. In the statement, we define the Vandermonde matrix associated with a vector $v \in \mathbb{R}^n$ by

$$V(v) = \left[v^{(n-1)} \;\dots\; v^{(2)} \;\; v \;\; v^{(0)}\right],$$

where $v^{(k)}$ is the vector of component-wise kth powers of v, i.e.

$$V(v) = \begin{bmatrix} v_1^{n-1} & \dots & v_1 & 1 \\ \vdots & \vdots\vdots\vdots & \vdots & \vdots \\ v_n^{n-1} & \dots & v_n & 1 \end{bmatrix}. \tag{16}$$

Lemma 1 [8] *Given a Brunowski triple (A_b, B_b, C_b) of order n and any given constants $a > 0$ and $b > 0$, there exist a vector $K_0 \in \mathbb{R}^n$ and a symmetric positive definite matrix P such that the following inequality holds with $\bar{A} = A_b - K_0 C_b$:*

$$\bar{A}^T P + P\bar{A} + aP + bI_n\left(B_b^T P B_b\right) \leq 0. \tag{17}$$

Moreover, given a vector $v \in \mathbb{R}^n$ with distinct and negative components, the pair $(K_0(\rho), P(\rho))$ defined by

$$K_0(\rho) = -\mathrm{diag}_{h=1}^{n}\{\rho^h\}\, V^{-1}(v)\, v^{(n)} \tag{18}$$

$$P(\rho) = \mathrm{diag}_{h=1}^{n}\{\rho^{-h}\}\, V^T(v)\, V(v)\, \mathrm{diag}_{h=1}^{n}\{\rho^{-h}\}\, \rho^{2n} \tag{19}$$

is a solution of (17) *if*

$$\rho > \max\left\{1,\ \frac{a + bn\|V(v)\|^2}{2w}\right\}, \tag{20}$$

where $w = -\max\{v\}$ *is the smallest component of* $-v$.

We now state the convergence theorem for the single-step exponential observer:

Theorem 1 [6] *For system* (1) *and* (2)*, let conditions* $\mathscr{H}_1$, $\mathscr{H}_2$, *and* $\mathscr{H}_3$ *be satisfied and assume that there is a constant* $\bar{u} > 0$ *such that* $\|u(t)\| \le \bar{u}$ *for all* $t \ge 0$. *Then, for any assigned decay rate* $\eta > 0$, *there exist a vector* $K_0 \in \mathbb{R}^n$ *and a positive value* $\bar{\Delta}$ *such that* (11) *is an* η*-exponential observer for system* (1) *provided that* $\Delta < \bar{\Delta}$, *in the following sense: For any given* $\chi \in \mathscr{C}^n_\Delta \cap L^n_\infty$ *and initial state* $x(-\Delta) = \bar{x} \in \mathbb{R}^n$, *we have*

$$\|x(t) - \hat{x}(t)\| \ \le\ \bar{c}e^{-\eta t}, \quad t \ge 0, \tag{21}$$

for some constant $\bar{c}$ *that depends on* $\bar{x}$ *and* χ. *In particular, if the vector* K_0 *and a matrix* P *are chosen to satisfy the inequality* (17) *with* $a = 2\eta + \alpha + 1$, *where* $\alpha > 0$ *is arbitrarily chosen and* $b = \gamma_p^2(\bar{u})$, *then the* η*-exponential convergence is ensured with a delay bound*

$$\bar{\Delta} = \frac{\alpha}{2 + \beta}, \quad \text{where } \beta = (K_0^T P K_0)\|P^{-1}\|\left(1 + \eta^2 + K_{0,1}^2\right). \tag{22}$$

Proof The proof, developed in [6], is obtained by rewriting the delay system (14) in a new coordinate system $\varepsilon(t)$ defined by means of the time-varying change of coordinates $\varepsilon(t) = e^{\eta t}\tilde{z}(t)$, and proving the asymptotic stability of the trivial solution $\varepsilon(t) = 0$ by using the Razumikhin Theorem. Then, from the property $\|\tilde{z}(t)\| = e^{-\eta t}\|\varepsilon(t)\|$ and by the Lipschitz property of Φ and Φ^{-1} due to $\mathscr{H}_1$, it is easily shown that the asymptotic stability of the equilibrium $\varepsilon(t) = 0$ implies the exponential convergence (21).

Notice that Theorem 1 only requires $\delta(t)$ to be bounded. In particular the continuity of $\delta(t)$ is not needed. Moreover, if the system (1) is autonomous, i.e. $u(t) = 0$, only assumptions $\mathscr{H}_1$ and $\mathscr{H}_2$ are required.

Remark 1 The global convergence result of Theorem 1 has been obtained under the global Lipschitz and observability assumptions $\mathscr{H}_1$ and $\mathscr{H}_2$. However, weaker local results can be obtained if local Lipschitz and observability assumptions are adopted instead. For those systems that admit compact invariant subsets of the state space, the assumptions $\mathscr{H}_1$ and $\mathscr{H}_2$ need to be satisfied only in such sets. The convergence of the observer (11) in such invariant sets can be proved following the same lines of the proof of Corollary 1 in [9] and of Theorem 3.9 in [10].

4 Sufficient Conditions for Global Regulation from Delayed Output

In this section we investigate conditions under which the single-step observer described in the previous section can be used to design a state-feedback control guaranteeing global stability. In other words, we are looking for sufficient conditions for a separation theorem in the case of drift-observable and full relative degree nonlinear systems with variable output delay. Even though this restriction somewhat reduces the generality of the approach, this class of nonlinear systems is relevant in many application areas, and a global regulation result with nonlinear delayed output feedback is interesting in its own.

In the non delayed case, the problem of output feedback control for nonlinear systems has received considerable attention in the literature. It is not possible to mention here the fundamentals developments in the theory of nonlinear output regulation (see for example [4, 17, 26] for necessary conditions for the existence of global and semi-global solutions). Having realized that the design of observers is of great relevance in the design of controllers for nonlinear systems, several alternative approaches have been investigated in [3, 22, 23]. State variables approaches for non uniformly completely observable systems have also been proposed [25]. Global stabilization of nonlinear systems with unmeasured state variables have already been obtained for some classes of nonlinear systems [11].

Many results are available in the literature for the input/output exact linearization problem and for the stabilization problem of retarded nonlinear systems with time-delay in the state (see, for instance, [12, 14, 16, 18, 19, 24, 27, 31, 35]). These papers assume the full knowledge of the system state. Comparable fewer works deal with output feedback stabilizing control laws in presence of delays (see [15, 36]).

In this chapter, we achieve global stability of the controlled system with a feedback computed using the state estimate provided by the nonlinear observer described in Sect. 3. The following hypotheses on (1) and (2) are needed:

$\mathscr{H}_1^c$ The nonlinear delay-free system described by the triple (f, G, h) is GULDO.
$\mathscr{H}_2^c$ The function $L_f^n h(\Phi^{-1}(z))$ is uniformly Lipschitz with respect to z.
$\mathscr{H}_3^c$ The function $L_G L_f^{n-1} h(x)$ is a constant row vector $\gamma^T \neq 0_{1\times p}$.
$\mathscr{H}_4^c$ The triple (f, G, h) has uniform observation degree n.

Comparing these with assumptions $\mathscr{H}_1$–$\mathscr{H}_3$ needed for the observer convergence, we note that the restrictive $\mathscr{H}_3^c$ condition has been introduced for ensuring the stabilizing action from the estimated state feedback.

Theorem 2 *Consider system* (1) *and* (2) *and a constant* $\eta_c > 0$, *and let the conditions* $\mathscr{H}_1^c$–$\mathscr{H}_4^c$ *be satisfied. Choose a constant* $\eta > \eta_c$, *and consider the observer* (11), *where the gain* $K_0 \in \mathbb{R}^n$ *and the delay bound* Δ *satisfy the condition* (22) *of* Theorem 1, *where the chosen* η *is used. Then there exists a feedback gain* $K_c \in \mathbb{R}^n$ *such that the trivial solution* $(x(t), \hat{x}(t)) = 0$ *of the closed loop system*

$$\dot{x}(t) = f\big(x(t)\big) + G\big(x(t)\big)u(t), \quad t \geq -\Delta, \quad x(-\Delta) = \bar{x}, \tag{23}$$
$$\bar{y}(t) = h\big(x\big(t - \delta(t)\big)\big), \qquad t \geq 0, \tag{24}$$
$$\begin{aligned}\dot{\hat{x}}(t) &= f(\hat{x}(t)) + G(\hat{x}(t))u(t) \\ &\quad + Q^{-1}(\hat{x}(t))K_\delta(t)\Big(\bar{y}(t) - h\big(\hat{x}(t - \delta(t))\big)\Big), \quad t \geq 0,\end{aligned} \tag{25}$$
$$\hat{x}(\tau) = \chi(\tau), \qquad u(\tau) = 0, \qquad \tau \in [-\Delta, 0], \tag{26}$$
$$u(t) = -\big(L_f^n h(\hat{x}(t)) + K_c^T \Phi(\hat{x})\big)\bar{\gamma}, \quad t \geq 0, \tag{27}$$

where $K_\delta(t) = e^{-\eta\delta(t)}K_0$, $\bar{x} \in \mathbb{R}^n$, $\chi \in \mathscr{C}_\Delta^n \cap L_\infty^n$, *and* $\bar{\gamma} \in \mathbb{R}^p$ *is such that* $\gamma^T\bar{\gamma} = 1$, *is globally exponentially stable with rate* η_c, *i.e. there exist* $\mu > 0$ *and* $\bar{\mu} > 0$, *both depending on* $\bar{x}$ *and* χ, *such that*

$$\|x(t)\| \leq \mu\, e^{-\eta_c t}, \qquad \|\hat{x}(t)\| \leq \bar{\mu}\, e^{-\eta_c t}, \quad t \geq 0. \tag{28}$$

Proof In the proof we will use the z-coordinates, $z = \Phi(x)$ and $\hat{z} = \Phi(\hat{x})$ and the global Lipschitz property contained in $\mathscr{H}_1^c$ on both Φ and Φ^{-1}. Defining $\pi(z) = L_f^n h(\Phi^{-1}(z))$, by assumption $\mathscr{H}_2^c$ we have $\|\pi(z) - \pi(\hat{z})\| \leq \gamma_\pi \|z - \hat{z}\|$, for some $\gamma_\pi > 0$. Thanks to assumption $\mathscr{H}_3^c$, we can write the function $p(z, u)$ defined in (8) as

$$p(z, u) = \pi(z) + \gamma^T u, \tag{29}$$

and the input $u(t)$ defined in (27) as

$$u(t) = -\Big(\pi\big(\hat{z}(t)\big) + K_c^T\hat{z}(t)\Big)\bar{\gamma}. \tag{30}$$

Defining the observation error $\tilde{z}(t) = z(t) - \hat{z}(t)$, the system (23) and the observer (25) can be written in z-coordinates as

$$\dot{z}(t) = A_b z(t) + B_b\Big(\pi\big(z(t)\big) + \gamma^T u(t)\Big) \tag{31}$$
$$\dot{\hat{z}}(t) = A_b\hat{z}(t) + B_b\Big(\pi\big(\hat{z}(t)\big) + \gamma^T u(t)\Big) + K_\delta(t)C_b\tilde{z}\big(t - \delta(t)\big), \tag{32}$$

where the identity $\bar{y}(t) - h\big(\hat{x}(t - \delta(t))\big) = C_b\tilde{z}\big(t - \delta(t)\big)$ has been used. From these, the observation error dynamics is obtained

$$\dot{\tilde{z}}(t) = A_b\tilde{z}(t) + B_b\Big(\pi\big(z(t)\big) - \pi\big(\hat{z}(t)\big)\Big) - K_\delta(t)C_b\tilde{z}\big(t - \delta(t)\big). \tag{33}$$

Since $\|\pi(z) - \pi(\hat{z})\| \leq \gamma_\pi\|\tilde{z}\|$, all the hypotheses needed to prove the exponential convergence to 0 of $\tilde{z}(t)$ following the lines of Theorem 1 are satisfied, and therefore there exists $\kappa > 0$, depending on $\bar{x}$ and χ, such that $\|\tilde{z}(t)\| \leq \kappa\, e^{-\eta t}$.

Substitution of $u(t)$ given by (27) in (32) yields

$$\dot{\hat{z}}(t) = (A_b - B_b K_c^T)\hat{z}(t) + K_\delta C_b \tilde{z}\big(t - \delta(t)\big). \tag{34}$$

By choosing K_c such that all the eigenvalues of $A_b - B_b K_c^T$ have the real part less than $-\eta_c$ and recalling that $-\eta < -\eta_c$ by assumption, we can easily prove that there exists $\bar{\kappa}$, depending on $\kappa, \bar{x}$ and χ, such that $\|\hat{z}\| \le \bar{\kappa}\, e^{-\eta_c t}$. From this and the Lipchitz property of $\Phi^{-1}(\cdot)$, the second inequality of (28) follows. Since $z(t) = \hat{z}(t) + \tilde{z}(t)$, we have that $\|z(t)\| \le (\kappa + \bar{\kappa})\, e^{-\eta_c t}$. Again, from the Lipschiz assumption on $\Phi^{-1}(\cdot)$ the first of (28) follows, and the thesis is proved.

Remark 2 An analogous separation result for nonlinear systems with non delayed output can be established under the weaker condition that $L_G L_f^{n-1} h(x)$ depends on x only through $h(x)$ (output injection). The case of delayed output can be handled only with the stronger requirement of a constant $L_G L_f^{n-1} h(x)$ (assumption $\mathscr{H}_3^c$). This is the price to pay to have a separation result when measurements are available with some delay. When $\mathscr{H}_3^c$ holds, the gain matrices K_0 and K_c can be chosen independently, in the same way as in the linear non delayed case (classical separation result). Of course, $\mathscr{H}_3^c$ is a sufficient condition. Weaker conditions may be derived in this framework for both the global and local stabilization case (see [15] for some results in this direction).

5 Chain Observers

It may happen that the maximum measurement delay Δ in system (1) and (2) is *too large* for a single-step observer of the type (11) to achieve η-exponential convergence for some chosen $\eta > 0$. This happens if the condition (22) of Theorem 1 can not be satisfied for any $\bar{\Delta} > \Delta$ and gain $K_0 \in \mathbb{R}^n$. In this case we can resort to a *chain observer*, in which, roughly speaking, he actual delay bound Δ is split in smaller sub-delays that satisfy the convergence condition (22). or a precise description of the operations of a chain observer the following definition is useful:

Definition 5 Given a delay $\Delta > 0$ and an integer $m > 1$, an *m-partition* of Δ is a strictly increasing sequence $\bar{\sigma} = \{\sigma_j\}_{j=0}^m$, such that $\sigma_0 = 0$ and $\sigma_m = \Delta$, so that $\Delta = \sum_{j=1}^m d_j$, where $d_j = \sigma_j - \sigma_{j-1}$.

As a general statement, given any integer $m > 1$ and an m-partition $\bar{\sigma}$ of the maximum delay Δ, a chain observer is a set of m (or $m+1$) interconnected observers, each one devoted to the observation of the state at time $t - \sigma_j$. Previously proposed chain observers (such as those in [2, 13, 20]) only considered the case of a single constant measurement delay, i.e. $\delta(t) \equiv \Delta$, and used a uniform m-partition of Δ. Here, we extend this framework to time-dependent delays, using a non necessarily uniform m-partition. In the proposed approach, we consider a cascade of m observers, numbered with $j = 1, \ldots, m$, where the output of the jth observer, denoted $\hat{x}_j(t)$, estimates $x(t-\sigma_{j-1})$. The output of the first observer, $\hat{x}_1(t)$, is devoted to the observation of the *current* state $x(t)$ (i.e., $x(t-\sigma_0)$), and is the output of the chain observer.

Notice that the design of chain observers for variable delays is considerably more difficult than for the constant delay case. When $\delta(t)$ is LDF (according to Definition 1), the output measurements previous to $t-\Delta$ are all available. It is therefore possible to buffer the measurements and use a chain observer designed for the constant delay case. Of course it is more effective to use in the observer of the latest available measurement, avoiding the introduction of an additional buffering delay just with the aim of emulating a constant delay.

5.1 Chain Observers for Variable LDF Delays

The idea behind a chain observer is to build a cascade of observers, each one aimed at estimating the state at time instants with fixed delays with respect to the current time t, so that the last observer of the chain provides the estimate of the state at time t. The set of fixed delays is defined in the m-partition $\bar{\sigma}$ of the overall delay Δ. By the assumption that $\delta(t) \in [0, \Delta]$, at each time t we have that $\delta(t)$ belongs to one of the intervals defined by the m-partition. If $\delta(t) \in (\sigma_j, \sigma_{j-1}]$, then the jth observer of the chain can use the current measurement $\bar{y}(t)$ to compute $\hat{x}_j(t)$, which is an estimate of $x(t-\sigma_{j-1})$. The measurement $\bar{y}(t) = y\big(t-\delta(t)\big)$ is seen by the jth observer as a delayed measurement of $h\big(x(t-\sigma_{j-1})\big)$, where the delay is $\delta(t)-\sigma_{j-1}$ (note that the maximum output delay seen by the jth observer is $d_j = \sigma_j - \sigma_{j-1}$). At time t the jth observer is the only observer in the chain that is fed with the current measurement $\bar{y}(t)$, and can provide an estimate of the measurement $y(t-\sigma_{j-1})$, computed as $h\big(\hat{x}_j(t)\big)$, to the adjacent observer in the chain (the $(j-1)$th observer), for which no real measurement is available. In a similar way, all observers with index k lower than $j-1$ (the *following* sub-chain) are fed by an estimate of the unavailable output $y(t-\sigma_k)$, computed by the observer with index $k+1$ as $h\big(\hat{x}_{k+1}(t)\big)$. As for the observers with index k higher than j (the *leading* sub-chain), these can be driven by artificially retarded measurements or by estimated measurements, depending on the assumption on the delay function $\delta(t)$.

If $\delta(t)$ is continuous, all the output values preceding $y\big(t-\delta(t)\big)$ are available, and therefore it is possible to feed the kth observers in the leading sub-chain ($k > j$ and $\delta(t) < \sigma_k$) with the un-delayed measurements $y(t-\sigma_k)$ obtained by introducing an artificial delay $\tau_k(t) = \sigma_k - \delta(t)$ to the available output $\bar{y}(t)$, thus obtaining $\bar{y}\big(t-\tau_k(t)\big) = y\big(t-\delta(t)-\tau_k(t)\big) = y(t-\sigma_k)$.

In the case of discontinuous delay $\delta(t)$, the values $y(t-\sigma_k)$, for $k > j$, may be unavailable. However, under the LDF assumption (see Definition 1), the value of $y(t-\Delta)$ is always available, and therefore we can use this as the forcing term of the last observer in the chain (the mth observer). Note that $y(t-\Delta)$ is obtained by artificially retarding the current measurement $\bar{y}(t)$ introducing the delay $\tau_m(t) = \Delta - \delta(t)$, so that $\bar{y}\big(t-\tau_m(t)\big) = y\big(t-\delta(t)-\tau_m(t)\big) = y(t-\Delta)$. All the other observers in the

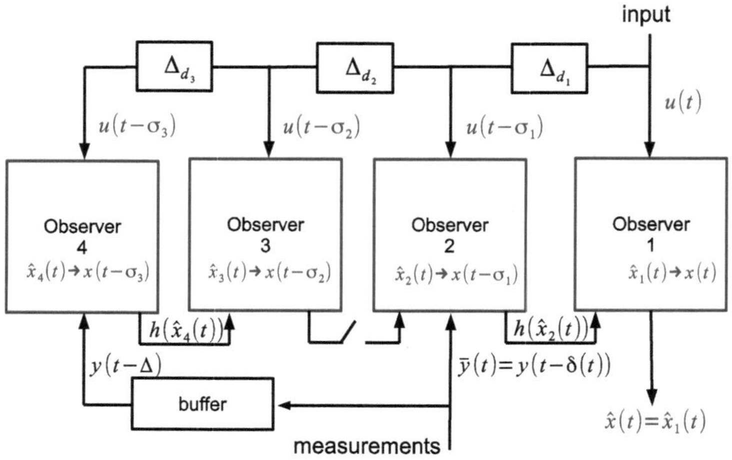

Fig. 4 Configuration of a chain of 4 observers when $\sigma_1 < \delta(t) < \sigma_2$ for a LDF $\delta(t)$

leading sub-chain can be driven by estimated outputs, i.e., the kth observer is fed by the estimate of $y(t-\sigma_k)$, computed by the $k+1$th observer as $h\big(\hat{x}_{k+1}(t)\big)$.

The situation is represented in Fig. 4 for the case $m = 4$, when $\delta(t) \in (\sigma_2, \sigma_1]$. Several variants are possible (for instance, in the leading sub-chain, the true measurement can be used instead of the estimated measurement when it is available). The description and convergence analysis of these variants are essentially the same, thus we limit ourselves to describe the LDF case for the design represented in the Fig. 4. The jth observers in the chain, $j = 1, \ldots, m$, the one aimed at estimating $x(t-\sigma_{j-1})$ through the variable $\hat{x}_j(t)$, obeys the following equations

$$\dot{\hat{x}}_j(t) = F(\hat{x}_j(t), \bar{u}_j(t)) + Q^{-1}(\hat{x}_j(t))e^{-\eta_j\delta_j(t)}K_j v_j(t), \ t \geq 0 \tag{35}$$

$$\bar{u}_j(t) = u(t-\sigma_{j-1}), \tag{36}$$

$$v_j(t) = \bar{y}_j(t) - h\big(\hat{x}_j(t-\delta_j(t))\big), \tag{37}$$

$$\hat{x}_j(\tau) = \varphi(\tau-\sigma_{j-1}), \quad \tau \in [-d_j, 0] \tag{38}$$

where η_j and K_j are design parameters, $d_j = \sigma_{j-1} - \sigma_j$, $\varphi \in \mathscr{C}^n_\Delta \cap L^n_\infty$, and the pairs $(\bar{y}_j(t), \delta_j(t))$, together with $\bar{u}_j(t)$, are the driving terms of the observers in the chain, and are suitably computed exploiting $u(t)$ and the pair $\big(\bar{y}(t), \delta(t)\big)$. The terms $\bar{u}_j(t)$ are computed as in (36), by artificially introducing delays σ_{j-1} in the input function $u(t)$. The computation of the pairs $(\bar{y}_j(t), \delta_j(t))$ corresponds to the description given at the beginning of this section, where the operations of a chain observers are detailed, and is synthetically reported in Table 1.

Intuitively, $\delta_j(t)$ is the delay of the local measurement function with respect to $t-\sigma_j$. The local measurement function corresponds to the actual delayed measurement when $\delta(t)$ is in the delay range of the observer, otherwise it is provided by the previous observer in the chain. The last observer has a special status, since it uses $y(t-\Delta)$, which is always available because $\delta(t)$ is LDF.

Table 1 Definition of the local delay and measurement functions for a chain of observers when $\delta(t)$ is LDF

		$\bar{y}_j(t)$	$\delta_j(t)$
$j = 1, \ldots, m-1$	$\delta(t) \in [\sigma_{j-1}, \sigma_j]$	$\bar{y}(t)$	$\delta(t) - \sigma_{j-1}$
	$\delta(t) \in [0, \sigma_{j-1}) \cup (\sigma_j, \Delta]$	$h\left(\hat{x}_{j+1}(t)\right)$	d_j
$j = m$	$\delta(t) \in [0, \sigma_{m-1})$	$\bar{y}\big(t - \Delta + \delta(t)\big)$	d_m
	$\delta(t) \in [\sigma_{m-1}, \Delta]$	$\bar{y}(t)$	$\delta(t) - \sigma_{m-1}$

Note that in the table $\bar{y}\big(t - \Delta + \delta(t)\big) = y(t - \Delta)$, and $[0, \sigma_0) = \emptyset$

Note that the bound on the local measurement delay $\delta_j(t)$ of the jth observer is d_j, i.e. $\delta(t) \in [0, d_j]$ (recall that $d_j = \sigma_j - \sigma_{j-1}$).

The convergence properties of the chain of observers have been investigated in [8], for the more general case of multi-output systems, were each output component has its own delay. An important result in [8] is the following Theorem, that is the single-output version of Theorem 9 in [8].

Theorem 3 *Consider system* (1) *and* (2), *with* $\delta(t) : \mathbb{R}_+ \to [0, \Delta]$ *a LDF function. Let conditions* $\mathcal{H}_1$, $\mathcal{H}_2$, $\mathcal{H}_3$ *be satisfied, and assume that* $\|u(t)\| \leq \bar{u}$, $\forall t \geq -\Delta$, *for some* $\bar{u} > 0$. *For a given integer* $m > 1$, *consider an* m*-partition* $\bar{\sigma}$ *of* Δ. *Let* $\eta > 0$ *be a given desired error decay rate. Consider a strictly increasing sequence of* m *positive numbers* $\{\eta_j\}_0^m$, *with* $\eta_0 = \eta$. *Consider* m *pairs* $(K_j, P_j) \in \mathbb{R}^n \times \mathbb{R}^{n \times n}$, P_j *symmetric positive definite, that satisfy* m *inequalities of the type* (17), *with* $a = 2\eta_j + \alpha + 1$, *with arbitrary* $\alpha > 0$. *For* $j = 1, \ldots, m$ *let*

$$\bar{\Delta}_j = \frac{\alpha}{2 + \beta_j}, \quad \text{with } \beta_j = (K_j^T P_j K_j)\|P_j^{-1}\| \left(1 + \eta_j^2 + K_{j,1}^2\right). \tag{39}$$

Then, if $d_j < \bar{\Delta}_j$ *the system* (35)–(38) *is a global* η*-exponential chain observer for system* (1) *and* (2).

A further theorem in [8] (Th. 10) states that, under the same assumptions of Theorem 3, for any desired convergence rate η, there exist an integer m, an m-partition $\bar{\sigma}$ of Δ, and a set of m observer gains K_j, such that the conditions of η-exponential convergence given in Theorem 3 are satisfied, and therefore the system (35)–(38) is a global η-exponential observer of system (1) and (2).

Notice that, if $\mathcal{H}_3^c$ of Sect. 4 holds, Theorem 2 can be immediately modified by replacing the single-step observer (25) with the chain observer (35)–(38). Thus in principle it is possible to design a globally stabilizing control of a nonlinear system affected by any variable measurement delay when $\delta(t)$ is LDF and $\mathcal{H}_1^c$–$\mathcal{H}_4^c$ hold.

5.2 Chain Observers for Variable Non-LDF Delays

The analysis of discontinuous delay functions of Sect. 2 leads us to conclude that discontinuities of $\delta(t)$ may cause a loss of information about the output. This affects the chain structure in Fig. 4 because there is no guarantee that $y(t - \Delta)$ is available for the correction term of mth observer. Therefore, it is not possible to define the correction term of the observers in the left part of the chain. Although this chain structure is not suited for non-LDF delay functions, it may be extended with slight modifications to compensate information losses that last less than the delay range of the last observer in the chain. To achieve this, we recall the notion of *maximal loss interval* of a delay function $\delta(t)$, given in Definition 2 and introduce:

Definition 6 A delay function $\delta(t)$ is said to be compatible with respect to a delay partition $\bar{\sigma}$ if its maximal loss interval b satisfies $b \le d_m \equiv \Delta - \sigma_{m-1}$.

When $\delta(t)$ is compatible with respect to $\bar{\sigma}$, the amplitudes of the losses are contained within the delay interval of the last observer. As a consequence, when $y(t - \sigma_{m-1})$ – the output measurement that carries no delay relatively to $\hat{x}_m(t)$ – is not available, it is still possible to use the most recent previous measurement. This measurement will have a delay in the range of the last observer, and it is possible to replace $y(t - \Delta)$ with it. This can be formalized as follows.

Let $\delta^o(t)$ be the function

$$\delta^o(t) \;=\; \min_{\tau \in [0,t]} \left\{ ((t - \sigma_{m-1}) \;-\; (\tau - \delta(\tau)))^+ \right\}, \tag{40}$$

where $(.)^+$ denotes the positive part, i.e. the maximum between the argument and 0. Informally, $\delta^o(t) \ge 0$ is the delay of the available measurement closest in time to $y(t - \sigma_{m-1})$. Notice that if $\delta(t)$ is continuous, $\delta^o(t) = 0$ for any t, since, by Proposition 1, $\exists t^* < t : t^* - \delta_{t^*} = t - \sigma_{m-1}$. Let $t^o \le t$ be the time value such that $\delta^o(t) = (t - \sigma_{m-1}) - (t^o - \delta_{t^o})$. It is easy to prove the following.

Lemma 2 *If $\delta(t)$ is compatible with respect to $\bar{\sigma}$ then $\bar{y}(t^o) = y(t^o - \delta_{t^o})$ is available at time t, and $\delta^o(t) \le d_m$.*

Using these definitions, the chain for compatible $\delta(t)$ can be defined as in Table 2. Theorem 3 can be immediately extended to this definition of a chain observer.

For example, when $\delta(t) = \bmod(t, T)$, a delay function that corresponds to sampling, the requirement of Definition 2 is equivalent to $T \le d_m$, that is, the sampling interval must be smaller than the delay range of the last observer. Therefore, a chain observer has the same maximum tolerable sampling interval of a single observer. However, if a continuous variable delay is added to the delay introduced by sampling measurements, a chain observer can tolerate larger total delays.

Table 2 Definition of the local delay and measurement functions for a chain of observers when $\delta(t)$ is compatible with $\bar{\sigma}$

		$\bar{y}_j(t)$	$\delta_j(t)$
$j = 1, \ldots, m-1$	$\delta(t) \in [\sigma_{j-1}, \sigma_j]$	$\bar{y}(t)$	$\delta(t) - \sigma_{j-1}$
	$\delta(t) \in [0, \sigma_{j-1}) \cup (\sigma_j, \Delta]$	$h(\hat{x}_{j+1}(t))$	d_j
$j = m$	$\delta(t) \in [0, \sigma_{m-1})$	$\bar{y}(t^o) = y(t^o - \delta_{t^o})$	$\delta^o(t)$
	$\delta(t) \in [\sigma_{m-1}, \Delta]$	$\bar{y}(t)$	$\delta(t) - \sigma_{m-1}$

6 Conclusions

We addressed the problem of state estimation and control of drift-observable nonlinear systems that have time varying measurement delays. We argued that many situations affecting the output measurements, such as buffering, sampling, losses and transmission delays, can be modeled by delays. To handle this generalized notion of delay, it is important to introduce some classification of the delay function, whose essential metric is the amount of information lost. This is evident when the estimation problem is solved by a structured observer composed of a chain of elementary observers. In principle, this approach allies to delays of any size. However, we showed that in order to guarantee exponential convergence to zero of the estimation error, the possible information blackout cannot last more than the delay range of the elementary observers in the chain. An open question to investigate is whether there are structured observer configurations that allow us to avoid this restriction.

References

1. Ahmed-Ali, T., Cherrier, E., M'Saad, M.: Cascade high gain observers for nonlinear systems with delayed output. In: Proceedings of the 48th Conference on Decision and Control, pp. 8226–8231 (2009)
2. Ahmed-Ali, T., Cherrier, E., Lamnabhi-Lagarrigue, F.: Cascade high gain predictor for a class of nonlinear systems. IEEE Trans. Autom. Control **57**(1), 221–226 (2012)
3. Andrieu, V., Praly, L.: On the existence of a Kazantis-Kravaris/Luemberger observer. SIAM J. Control Optim. **45**(2), 432–456 (2007)
4. Byrnes, C.I., Isidori, A.: Nonlinear internal models for output regulation. IEEE Trans. Autom. Control **49**(12), 2244–2247 (2004)
5. Cacace, F., Germani, A., Manes, C.: An observer for a class of nonlinear systems with time varying observation delay. Syst. Control Lett. **59**(5), 305–312 (2010)
6. Cacace, F., Germani, A., Manes, C.: An exponential observer with delay-dependent gain for a class of nonlinear systems with time-varying measurement delay. In: Proceedings of the IEEE Conference on Decision and Control, pp. 2364–2369 (2013)
7. Cacace, F., Germani, A., Manes, C.: A chain approach for state observation of a class of MIMO nonlinear systems with time-varying output delays. In: Proceedings of the 11th IFAC Workshop on Time-Delay Systems, pp. 541–546 (2013)
8. Cacace, F., Germani, A., Manes, C.: A chain observer for nonlinear systems with multiple time-varying measurement delays. SIAM J. Control Optim. **52**(3), 1862–1885 (2014)
9. Ciccarella, G., Dalla Mora, M., Germani, A.: A Luenberger-like observer for nonlinear systems. Int. J. Control **57**(3), 537–556 (1993)
10. Dalla Mora, M., Germani, A., Manes, C.: Design of state observers from a drift-observability property. IEEE Transac. Autom. Control **45**(8), 1536–1540 (2000)
11. Ding, Z.: Output feedback stabilization of systems with nonlinearity of unmeasured states. In: Proceedings of the IEEE Conference on Decision and Control, pp. 5414–5419 (2008)
12. Fridman, E.: Output regulation of nonlinear systems with delay. Syst. Control Lett. **50**(2), 81–93 (2003)
13. Germani, A., Manes, C., Pepe, P.: A new approach to state observation of nonlinear systems with delayed output. IEEE Trans. Autom. Control **47**(1), 96–101 (2002)
14. Germani, A., Manes, C., Pepe, P.: Input-output linearization with delay cancellation for nonlinear delay systems: the problem of the internal stability. Int. J. Robust Nonlinear Control **13**(9), 909–937 (2003)

15. Germani, A., Manes, C., Pepe, P.: Observer-based stabilizing control for a class of nonlinear retarded systems. In: Sipahi, R., Vyhlidal, T., Niculescu, S.-I., Pepe, P. (eds.) Time Delay Systems: Methods, Applications and New Trends, pp. 331–342. Springer, Berlin Heidelberg (2012)
16. Hua, C., Guan, X., Shi, P.: Robust stabilization of a class of nonlinear time-delay systems. Appl. Math. Comput. **155**(3), 737–752 (2004)
17. Isidori, A.: Nonlinear Control Systems: An Introduction. Springer-Verlag, Berlin (1995)
18. Jankovic, M.: Control Lyapunov-Razumikhin functions and robust stabilization of time delay systems. IEEE Trans. Autom. Control **46**(7), 1048–1060 (2001)
19. Karafyllis, I., Jiang, Z-P.: Necessary and sufficient Lyapunov-like conditions for robust nonlinear stabilization. ESAIM: Control, Optim. Calc. Var. **16**(4), 887–928 (2010)
20. Kazantzis, N., Wright, R.: Nonlinear observer design in the presence of delayed output measurements. Syst. Control Lett. **54**(9), 877–886 (2005)
21. Kravaris, C., Wright, R.: Dead time compensation for nonlinear chemical processes. AlChE J. **35**(9), 1535–1542 (1989)
22. Kreisselmeier, G., Engel, R.: Nonlinear observer for autonomous Lipschitz continuous systems. IEEE Trans. Autom. Control **48**(3), 451–464 (2003)
23. Krener, A., Xiao, M.: Erratum: Nonlinear observer design in the Siegel domain. SIAM J. Control Optim. **43**(1), 377–378 (2004)
24. Lien, C.-H.: Global exponential stabilization for several classes of uncertain nonlinear systems with time-varying delay. Nonlinear Dyn. Syst. Theor. **4**(1), 15–30 (2004)
25. Maggiore, M., Passino, K.: A separation principle for a class of non uniformly completely observable systems. IEEE Trans. Autom. Control **48**(7), 1122–1133 (2003)
26. Marconi, L., Praly, L., Isidori, A.: Output stabilization via nonlinear Luenberger observers. SIAM J. Control Optim. **45**(6), 2277–2298 (2007)
27. Márquez-Martinez, L., Moog, C.: Input-output feedback linearization of time-delay systems. IEEE Trans. Autom. Control **49**(5), 781–785 (2004)
28. Márquez-Martinez, L., Moog, C., Velasco-Villa, M.: Observability and observers for nonlinear systems with time-delays. In: Proceedings of the 2nd IFAC Workshop on Time Delay Systems, pp. 52–58 (2000)
29. Meadows, E., Rawlings, J.: Model predictive control. In: Henson, M., Seborg, D. (eds.) Nonlinear Process Control, pp. 233–310. Prentice Hall, Englewood Cliffs, NJ (1997)
30. Mikheev, Y., Sobolev, V., Fridman, E.: Asymptotic analysis of digital control systems. Autom. Remote Control **49**(9), 1175–1180 (1988)
31. Oguchi, T., Watanabe, A., Nakamizo, T.: Input-output linearization of retarded non-linear systems by using an extension of Lie derivative. Int. J. Control **75**(8), 582–590 (2002)
32. Smith, O.: Closer control of loops with dead-time. Chem. Eng. Prog. **53**(5), 217–219 (1957)
33. Tatiraju, S., Soroush, M., Ogunnaike, B.: Multirate nonlinear state estimation with application to a polymerization reactor. AlChE J. **45**(4), 769–776 (1999)
34. Van Assche, V., Ahmed-Ali, T., Hann, C., Lamnabhi-Lagarrigue, F.: High gain observer design for nonlinear systems with time varying delayed measurement. In: Proceedings of the 18th IFAC World Congress, pp. 692–696 (2011)
35. Zhang, X., Cheng, Z.: Global stabilization of a class of time-delay nonlinear systems. Int. J. Syst. Sci. **36**(8), 461–468 (2005)
36. Zhang, X., Zhang, C., Cheng, Z.: Asymptotic stabilization via output feedback for nonlinear systems with delayed output. Int. J. Syst. Sci. **37**(9), 599–607 (2006)

On the Existence of the Normal Form for Nonlinear Delay Systems

Claudia Califano and Claude H. Moog

Abstract The normal form is discussed for nonlinear systems affected by constant commensurate delays. Two different forms are argued. In particular, necessary and sufficient conditions are given under which a nonlinear time-delay system can be decomposed into a (weakly) observable subsystem and a non observable subsystem. Whenever such a decomposition exists, additional conditions are required to ensure the feedback linearization of the weakly observable subsystem. Finally, a full characterization is derived for the nonlinear time delay system to have an unobservable subsystem not directly affected by the input and a weakly observable subsystem which is linearizable by feedback. The performed analysis is carried out within a new geometric framework recently introduced in the literature.

1 Introduction

In this chapter the conditions are investigated under which a given nonlinear single input- single output continuous time system, affected by constant commensurate delays, can be represented up to a feedback as the connection of two subsystems, a weakly observable subsystem, and an unobservable one. The latter naturally yields the notion of zero dynamics of the system. It is also investigated under which conditions, the weakly observable subsystem can be rendered linear. The interest in such a canonical form arises in various control problems including the study of stability and stabilization [14].

As well known, in the continuous delay-free case, under the assumption of defined relative degree, such a decomposition is always possible, and it is referred to as

---

C. Califano (✉)
Dipartimento di Ingegneria Informatica Automatica e Gestionale "Antonio Ruberti", Università di Roma La Sapienza, Via Ariosto 25, 00185 Rome, Italy
e-mail: califano@dis.uniroma1.it

C.H. Moog
L'UNAM, IRCCyN, UMR C.N.R.S. 6597, 1 rue de la Noë, BP 92101, 44321 Nantes Cedex 3, France
e-mail: moog@ieee.org

I. Karafyllis et al. (eds.), *Recent Results on Nonlinear Delay Control Systems*, Advances in Delays and Dynamics 4, DOI 10.1007/978-3-319-18072-4_6

normal form [10]. In this case the input-output behavior can always be rendered linear through an appropriate choice of the feedback law and can be represented as a chain of integrators. Furthermore for single output systems, it has also been shown that the unobservable dynamics is independent of the control, a property which is in general lost in the multi-output case. This property which has some interesting implications on the study of the zero dynamics of the system is not guaranteed for nonlinear discrete time single-input single-output systems [3]. In fact the assumption of defined relative degree allows to achieve the desired decomposition, but in general the unobservable subsystem is influenced by the control.

The case of single-input systems of retarded type is even more involved since the assumption of well defined relative degree does not guarantee anymore the existence of such a decomposition. Furthermore in order to get a linear input-output behavior some additional conditions are needed, and even if these conditions are satisfied, a representation of such a behavior as a chain of integrators can be achieved only under very strong conditions. It is also shown that the unobservable dynamics is in general influenced by the control. Some preliminary results were presented in [4].

The outline of this book chapter is as follows. The statements of the problems under investigation are given in Sect. 2, while some recalls on the algebraic and geometric tools used in the chapter for dealing with nonlinear time-delay systems are given in Sect. 3. Section 4 is devoted to the main results. Some concluding remarks are given in Sect. 5.

2 Preliminaries and Problem Statements

A class of nonlinear single-input single-output systems is considered whose state variables and control are subject to constant and commensurate delays. Without loss of generality, and after an eventual preliminary sampling (as explained in [9, 12]), the dynamics reads:

$$\Sigma : \begin{aligned} \dot{x}_{[0]} &= F(\mathbf{x}_{[s]}) + \sum_{j=0}^{s} G_j(\mathbf{x}_{[s]}) u_{[0]}(-j) \\ y_{[0]} &= H(\mathbf{x}_{[s]}) \end{aligned} \tag{1}$$

where

$$\mathbf{x}_{[s]}^T = (x^T(t), \cdots x^T(t-s)) \in \mathbb{R}^{(s+1)n},$$

represents the first $(s+1)n$ components of the state of the infinite dimensional system (1). The notations $x_{[0]} = x(t) \in \mathbb{R}^n$, $u_{[0]} = u(t) \in \mathbb{R}$ and $y_{[0]} = y(t) \in \mathbb{R}$ stand for the instantaneous values of the state, the input and the output. The functions $F(\mathbf{x}_{[s]})$, $G_j(\mathbf{x}_{[s]})$ for $j \in [0, s]$ and $H(\mathbf{x}_{[s]})$ are assumed analytic in their arguments. Additional assumptions such as dealing with local/global Lipschitz functions or forward completeness may be introduced depending on the requirements on the existence/uniqueness of the solution of (1). These assumptions play a key role when dealing with stability problems.

Denote by

$$\mathbf{x}_{[s]}^T(-p) = (x^T(t-p), \cdots x^T(t-s-p));$$

in a similar vein, we define $\mathbf{u}_{[s]}$ and $\mathbf{u}_{[s]}(-p)$. When there is no confusion, the subscript will be dropped so that $\mathbf{x}$ represents $\mathbf{x}_{[s]}$ and $\mathbf{x}(-p)$ represents $\mathbf{x}_{[s]}(-p)$. On the contrary if a function $\tau(\cdot)$ depends not only on $\mathbf{x}_{[s]}$, but also on some future values of the variable $x(t)$, in order to put in evidence that such a function is not causal, we will say that $\tau(\cdot) = \tau(\mathbf{x}_{[*]})$. Given a function $\tau(\mathbf{x}_{[s]})$, its rth derivative with respect to time will be denoted as

$$\tau^{(r)}(\mathbf{x}_{[s]}) := \frac{d^r \tau(\mathbf{x}_{[s]})}{dt^r}.$$

The notation $\mathbf{x}_{[\gamma]}^0$ represents an equilibrium point.

In the following, $\mathscr{K}$ will be used to denote the field of meromorphic functions of a *finite* number of variables in

$$\left\{\mathbf{x}_{[0]}(-i), \mathbf{u}_{[0]}(-i), \cdots, \mathbf{u}_{[0]}^{(k)}(-i),\ k, i \in \mathbb{N}\right\},$$

where $\mathbb{N}$ is the set of natural numbers. The notation "d" will represent the differential operator, while "δ" will represent the backward time-shift operator: for $a(\mathbf{x}_{[\alpha]}), f(\mathbf{x}_{[\beta]}) \in \mathscr{K}$:

$$\delta[a(\mathbf{x}_{[\alpha]})\, df(\mathbf{x}_{[\beta]})] = a(\mathbf{x}_{[\alpha]}(-1))\delta df(\mathbf{x}_{[\beta]}) = a(\mathbf{x}_{[\alpha]}(-1))df(\mathbf{x}_{[\beta]}(-1)).$$

The left ring of polynomials in δ with coefficients in $\mathscr{K}$ is denoted $\mathscr{K}(\delta]$.

In this framework a one-form is written as

$$\omega(\mathbf{x}, \delta)dx_{[0]} = \sum_{i=1}^{n} \omega_i(\mathbf{x}, \delta)dx_{i,[0]},\ \ \omega_i(\mathbf{x}, \delta) \in \mathscr{K}(\delta],\ i \in [1, n].$$

If there exists $f(\mathbf{x})$ such that $df(\mathbf{x}) = \omega(\mathbf{x}, \delta)dx_{[0]}$, then the one-form $\omega(\mathbf{x}, \delta)dx_{[0]}$ is exact. A set of one-forms $\{\omega_i(\mathbf{x}, \delta)dx_{[0]}, i \in [1, j]\}$ are independent over $\mathscr{K}(\delta]$ if

$$\sum_{i=1}^{j} \alpha_i(\mathbf{x}, \delta)\omega_i(\mathbf{x}, \delta) = 0$$

only if $\alpha_i(\mathbf{x}, \delta) = 0, \forall i \in [1, j]$. Finally $\Delta = \mathrm{span}_{\mathscr{K}}\{r_1(\mathbf{x}), \cdots, r_j(\mathbf{x})\}$ denotes the distribution spanned by the $r_i(\mathbf{x})$'s. Any element of Δ can be expressed as a linear combination of the $r_i(\mathbf{x})$'s with coefficient in $\mathcal{K}$. $\bar{\Delta}$ will denote its involutive closure. Analogously $\Delta(\mathbf{x}, \delta) = \text{span}_{\mathcal{K}(\delta]}\{r_1(\mathbf{x}, \delta), \cdots, r_j(\mathbf{x}, \delta)\}$ denotes the submodule spanned by the $r_i(\mathbf{x}, \delta)$'s. Any element of $\Delta(\mathbf{x}, \delta)$ can be expressed as

$$r(\mathbf{x}, \delta) = \sum_{i=0}^{j} r_i(\mathbf{x})\alpha_i(\mathbf{x}, \delta)$$

with $\alpha_i(\mathbf{x}, \delta) \in \mathcal{K}(\delta]$.

Before stating the different addressed problems, we will recall the notions of relative degree of a given output function, observability of the given system, bicausality of a change of coordinates and regular static state feedback, which play a key role. More precisely let us recall that the relative degree r of the output $y_{[0]}$ is defined as the smallest integer r, such that the jth derivative of the output function $y_{[0]}^{(j)}$ is not influenced by the control, for $0 \leq j \leq r - 1$, while $y_{[0]}^{(r)}$ is influenced by the (eventually delayed) control. For the class of system considered, algebraic conditions are given in Definition 6, which characterize the relative degree through the use of the extended Lie derivative (Definition 5).

Definition 1 (*Observability*, [11]) System Σ with instantaneous state variable $x_{[0]} \in \mathbb{R}^n$ is said to be weakly observable if the observability matrix $\mathbf{O}(\mathbf{x}, \delta)$ satisfying

$$\begin{pmatrix} dy_{[0]} \\ \vdots \\ dy_{[0]}^{(n-1)} \end{pmatrix} = \mathbf{O}(\mathbf{x}, \delta)dx_{[0]},$$

is a full rank matrix.The system is strongly observable if $\mathbf{O}(\mathbf{x}, \delta)$ is unimodular, i.e. it admits a polynomial inverse.

Since we will study the equivalence of systems linked through a change of coordinates, recall the following definition of bicausal change of coordinates adapted from [15].

Definition 2 (*Bicausal change of coordinates*) Consider system Σ in the state coordinates x. The mapping $z_{[0]} = \varphi(\mathbf{x}_{[\alpha]})$, where $\alpha \in \mathbb{N}$ and $\varphi \in \mathcal{K}^n$, is a local *bicausal* change of coordinates for Σ if there exists an integer $\ell \in \mathbb{N}$ and a function $\psi(\mathbf{z}_{[\ell]}) \in \mathcal{K}^n$ such that, assuming $z_{[0]}$ and $x_{[0]}$ defined for $t \geq -(\alpha + \ell)$, then $\psi(\varphi(\mathbf{x}_{[\alpha]}), \cdots, \varphi(\mathbf{x}_{[\alpha]}(-\ell)) = x_{[0]}$ for $t \geq 0$.

Definition 3 (*Bicausal regular static state feedback*) Consider the input $w_{[0]} \in \mathrm{R}^m$. The feedback

$$w_{[0]} = \alpha(\mathbf{x}) + \sum_{j=0}^{i} \beta_j(\mathbf{x}) v_{[0]}(-j)$$

is a local *bicausal* regular static state feedback if there exists an integer $\ell \in \mathbb{N}$ and functions $\bar{\alpha}(\mathbf{x})$ and $\bar{\beta}_\mu(\mathbf{x})$ such that,

$$\alpha(\mathbf{x}) + \sum_{j=0}^{i} \beta_j(\mathbf{x}) \left(\bar{\alpha}(\mathbf{x}(-j)) + \sum_{\mu=0}^{\ell} \bar{\beta}_\mu(\mathbf{x}(-j)) w_{[0]}(-\mu - j) \right) = w_{[0]}.$$

According to the definition, the regularity and bicausality of the feedback is equivalent to the unimodularity of the matrix

$$\beta(\mathbf{x}, \delta) = \sum_{j=0}^{i} \beta_j(\mathbf{x}) \delta^j. \tag{2}$$

In the single input case, considered in the present chapter, it thus reduces to the independence of δ, that is $\beta(\mathbf{x}, \delta) = \beta(\mathbf{x}) \neq 0$, or equivalently $\beta_0(\mathbf{x}) \neq 0$ and $\beta_j(\mathbf{x}) = 0$ for all $j \in [1, i]$.

The following problems can now be set.

Problem Statement 1. [Existence of the Generalized Normal Form] Given system (1), with output $y_{[0]}$ characterized by the well defined relative degree r, find, if possible, a regular bicausal static state feedback

$$u_{[0]} = \alpha(\mathbf{x}) + \beta(\mathbf{x}) v_{[0]},$$

and a bicausal change of coordinates

$$z_{[0]} = \begin{pmatrix} z_{1,[0]} \\ z_{2,[0]} \end{pmatrix} = \varphi(\mathbf{x})$$

with $z_{1,[0]} \in \mathbb{R}^r$, such that in the new coordinates the closed-loop system reads as:

$$\dot{z}_{1,[0]} = \theta_1(\mathbf{z}_{1,[\ell]}) + \sum_{j=0}^{\ell} \theta_{2j}(\mathbf{z}_{1,[\ell]}) v_{[0]}(-j),$$

$$\dot{z}_{2,[0]} = \eta_1(\mathbf{z}_{[\ell]}) + \sum_{j=0}^{\ell} \eta_{2j}(\mathbf{z}_{[\ell]}) v_{[0]}(-j), \quad y_{[0]} = \tilde{H}(\mathbf{z}_{1,[\ell]}), \tag{3}$$

where the subsystem

$$\dot{z}_{1,[0]} = \theta_1(\mathbf{z_{1,[\ell]}}) + \sum_{j=0}^{\ell} \theta_{2j}(\mathbf{z_{1,[\ell]}}) v_{[0]}(-j), \quad y_{[0]} = \tilde{H}(\mathbf{z_{1,[\ell]}})$$

is weakly observable. ◁

Such a problem can be further developed by requiring that the input-output behavior is described by a linear system, which leads to the second problem examined in this chapter:

Problem Statement 2. [Existence of the Normal Form] Given system (1), with output $y_{[0]}$ characterized by the well defined relative degree r, find, if possible, a regular bicausal static state feedback $u_{[0]} = \alpha(\mathbf{x}) + \beta(\mathbf{x})v_{[0]}$, and a bicausal change of coordinates

$$z_{[0]} = \begin{pmatrix} z_{1,[0]} \\ z_{2,[0]} \end{pmatrix} = \varphi(\mathbf{x})$$

with $z_{1,[0]} \in \mathbb{R}^r$, such that in the new coordinates the closed-loop system reads as (3) with a linear (weakly) observable input-output behavior, that is with

$$\dot{z}_{1,[0]} = \theta_1(\mathbf{z_{1,[\ell]}}) + \sum_{j=0}^{\ell} \theta_{2j}(\mathbf{z_{1,[\ell]}}) v_{[0]}(-j) = \sum_{j=0}^{\ell} A_j z_{1,[0]}(-j) + \sum_{j=0}^{\ell} B_j v_{[0]}(-j),$$

$$y_{[0]} = \tilde{H}(\mathbf{z_{1,[\ell]}}) = \sum_{j=0}^{\ell} C_j z_{1,[0]}(-j),$$

where the polynomial matrices

$$A(\delta) = \sum_{j=0}^{\ell} A_j \delta^j, \quad B(\delta) = \sum_{j=0}^{\ell} B_j \delta^j, \quad \text{and } C(\delta) = \sum_{j=0}^{\ell} C_j \delta^j$$

define a linear (weakly) observable system. ◁

Finally, the conditions under which the unobservable dynamics defined by the derivative of $z_{2,[0]}$ in (3) is not affected by the control v are investigated; that is the bicausal change of coordinates

$$z_{[0]} = \begin{pmatrix} z_{1,[0]} \\ z_{2,[0]} \end{pmatrix} = \varphi(\mathbf{x})$$

can be chosen to guarantee that

$$\dot{z}_{2,[0]} = \eta_1(\mathbf{z}_{[\ell]}).$$

This has interesting consequences when studying the stability properties of the whole system as discussed in Sect. 5.

Remark 1 System (1) is defined for $t \geq 0$ once the initial condition $x(t) = \varphi(t)$ is set on the interval $[-s, 0]$. For the equivalence to the form (3) to hold true for $t \geq 0$, it may be necessary to consider the initial condition $x(t) = \varphi(t)$ set on a greater interval $[-\bar{s}, 0]$, to guarantee that the change of coordinates and the static state feedback are well defined for $t \geq 0$, as well as the initial condition $z(t) = \bar{\varphi}(t)$ which must be well defined on the interval $[-\ell, 0]$.

This study is performed within a new geometric framework which was recently introduced in [1], and appeared to be successful to derive various canonical forms [1, 2]. More precisely, the conditions are stated in terms of extended Lie brackets of vector fields which are defined on appropriate finite dimensional manifolds.

3 Recalling Results on Algebraic and Geometric Tools

This Section is devoted to recall some basic tools and results which will be used to address the problems stated in Sect. 2. The approach has shown to be efficient for solving various problems for nonlinear time delay systems, including the identification of the delay [16].

3.1 Differential Representation

Given system Σ defined by (1), its differential representation is

$$\Sigma_L : \begin{aligned} d\dot{x}_{[0]} &= f(\mathbf{x}_{[s]}, \mathbf{u}_{[s]}, \delta)dx_{[0]} + g_1(\mathbf{x}_{[s]}, \delta)du_{[0]} \\ dy_{[0]} &= h(\mathbf{x}_{[s]}, \delta)dx_{[0]} \end{aligned} \tag{4}$$

with

$$\begin{aligned} f(\mathbf{x}_{[s]}, \mathbf{u}_{[s]}, \delta) &= \sum_{i=0}^{s} \frac{\partial F(\mathbf{x}_{[s]})}{\partial x_{[0]}(-i)} \delta^i + \sum_{j=0}^{s} u_{[0]}(-j) \sum_{i=0}^{s} \frac{\partial G_j(\mathbf{x}_{[s]})}{\partial x_{[0]}(-i)} \delta^i \\ g_1(\mathbf{x}_{[s]}, \delta) &= \sum_{j=0}^{s} G_j(\mathbf{x}_{[s]})\delta^j = \sum_{j=0}^{s} g_1^j(\mathbf{x}_{[s]})\delta^j, \\ h(\mathbf{x}_{[s]}, \delta) &= \sum_{j=0}^{s} \frac{\partial H(\mathbf{x}_{[s]})}{\partial x_{[0]}(-j)} \delta^j \end{aligned}$$

3.2 Left- and Right-Annihilators

- Let $T(\mathbf{x}, \delta)$ be a $(n \times r)$ polynomial matrix of rank r. Its left-annihilator $\Omega(\mathbf{x}, \delta)$ consists of all row vectors $\omega(\mathbf{x}, \delta)$ such that $\omega(\mathbf{x}, \delta)T(\mathbf{x}, \delta) = 0$. By definition, the submodule $\Omega(\mathbf{x}, \delta)$ is closed in the following sense of left-closure:

$$P(\mathbf{x}, \delta)\pi \in \Omega \Rightarrow \pi \in \Omega$$

  where $P(\mathbf{x}, \delta)$ stands for a polynomial operator.
- Let $T(\mathbf{x}, \delta)$ be a $(r \times n)$ polynomial matrix of rank r. Its right-annihilator $\Delta(\mathbf{x}_{[*]}, \delta)$ consists of all column vectors $v(\mathbf{x}_{[*]}, \delta)$ such that $T(\mathbf{x}, \delta)v(\mathbf{x}_{[*]}, \delta) = 0$. By definition, the right-kernel is closed in the following sense of right-closure:

$$v(\mathbf{x}_{[*]}, \delta)P(\mathbf{x}_{[*]}, \delta) \in \Delta(\mathbf{x}_{[*]}, \delta) \Rightarrow v(\mathbf{x}_{[*]}, \delta) \in \Delta(\mathbf{x}_{[*]}, \delta)$$

- The left closure of the polynomial matrix $T(\mathbf{x}, \delta)$ is obtained computing the left-annihilator of the right-annihilator of $T(\mathbf{x}, \delta)$.
- The right closure of the polynomial matrix $T(\mathbf{x}, \delta)$ is obtained computing the right-annihilator of the left-annihilator of $T(\mathbf{x}, \delta)$.
- Note that starting from the causal $(n \times r)$ polynomial matrix $T(\mathbf{x}, \delta)$ of rank r, its left-annihilator can always be expressed by causal generators. On the contrary starting from the causal $(r \times n)$ polynomial matrix $T(\mathbf{x}, \delta)$ of rank r, it may not be possible to express its right-annihilator through causal generators, that is the entries of the right-annihilator are polynomials in δ whose coefficients are functions which may depend on future time instants $t + k$ for some $k > 0$. For instance, consider

$$T(\mathbf{x}, \delta) = \begin{pmatrix} x_1(-1) + x_1\delta \\ x_2\delta \end{pmatrix}.$$

  Then its left annihilator is

$$\omega = \left(x_2\delta,\ -x_1(-2) - x_2\frac{x_1(-1)}{x_2(-1)}\delta\right),$$

  which is causal. Now consider the transpose of $T(\mathbf{x}, \delta)$, namely,

$$\bar{T}(\mathbf{x}, \delta) = \left(x_1(-1) + x_1\delta,\ x_2\delta\right).$$

  Its right annihilator is

$$v = \begin{pmatrix} x_2\delta \\ -x_1 - \frac{x_1(+1)}{x_2(+1)}x_2\delta \end{pmatrix},$$

  which is not causal.

3.3 The Action of a Bicausal Change of Coordinates

Given the bicausal change of coordinates $z_{[0]} = \varphi(\mathbf{x})$, let us consider the associated differential representation $d\mathbf{z}_{[0]} = T(\mathbf{x}, \delta)dx_{[0]}$. Then, $T(\mathbf{x}_{[\alpha]}, \delta)$ is unimodular. Moreover, its inverse has a polynomial degree $\gamma \leq \alpha(n-1)$. Moreover, under such a bicausal change of coordinates, the differential representation (4) is transformed into

$$d\dot{z}_{[0]} = \tilde{f}(\mathbf{z}, \mathbf{u}, \delta)dz_{[0]} + \tilde{g}_1(\mathbf{z}, \delta)du_{[0]}, \quad dy_{[0]} = \tilde{h}(\mathbf{z}, \delta)dz_{[0]}, \tag{5}$$

where

$$\tilde{f}(\mathbf{z}, \mathbf{u}, \delta) = \left[\left(T(\mathbf{x}, \delta)f(\mathbf{x}, \mathbf{u}, \delta) + \dot{T}(\mathbf{x}, \delta)\right)T^{-1}(\mathbf{x}, \delta)\right]_{\varphi^{-1}(\mathbf{z})}$$

$$\tilde{g}_1(\mathbf{z}, \delta) = (T(\mathbf{x}, \delta)g_1(\mathbf{x}, \delta))_{\varphi^{-1}(\mathbf{z})}$$

Remark 2 The rank of $T(\mathbf{x}, \delta)$ may drop around some singular function $\varphi(\mathbf{x})$, which may eventually be a trajectory for the system. In this case the equivalence does not hold true around such a singular trajectory.

3.4 Extended Lie Bracket

We next recall the definition of the extended Lie bracket introduced in [1] to tackle nonlinear time-delay systems and which is used to characterize the integrability of 1-forms defined not only from state variables at time t but also from their time shifts.

Definition 4 (*Extended Lie Bracket*) Let

$$r_\beta(\mathbf{x}, \delta) = \sum_{j=0}^{s} r_\beta^j(\mathbf{x})\delta^j,$$

where $\beta = 1, 2$. The extended Lie bracket

$$[r_1^k(\mathbf{x}), r_2^l(\mathbf{x})]_{E_i}$$

is defined on $\mathbb{R}^{(i+1)n}$, $i \geq 0$, by

$$\left[r_1^k(\cdot), r_2^l(\cdot)\right]_{E_i} = \sum_{j=0}^{\bar{k}} \left([r_1^{k-j}(\cdot), r_2^{l-j}(\cdot)]_{E_0}\right)^T_{|(\mathbf{x}(-j),\mathbf{u}(-j))} \frac{\partial}{\partial x_{[0]}(-j)}, \tag{6}$$

with $\bar{k} = min(k, l, i)$, and

$$[r_1^k(\cdot), r_2^l(\cdot)]_{E_0} = \sum_{i=0}^{k} \frac{\partial r_2^l(\mathbf{x})}{\partial x_{[0]}(-i)} r_1^{k-i}(\mathbf{x}(-i)) - \sum_{i=0}^{l} \frac{\partial r_1^k(\mathbf{x})}{\partial x_{[0]}(-i)} r_2^{l-i}(\mathbf{x}(-i)). \quad (7)$$

The extended Lie bracket makes sense for the infinite dimensional system

$$\begin{aligned} \dot{x}_{[0]} &= F(\mathbf{x}_{[s]}) + \sum_{j=0}^{s} G_j(\mathbf{x}_{[s]}) u_{[0]}(-j) \\ \dot{x}_{[0]}(-1) &= F(\mathbf{x}_{[s]}(-1)) + \sum_{j=0}^{s} G_j(\mathbf{x}_{[s]}(-1)) u_{[0]}(-j-1) \\ &\vdots \end{aligned} \quad (8)$$

which is associated to the time-delay system (1). The computation of the extended Lie bracket can be implemented in some computer algebra software as done in [6].

As for delay-free systems, it is useful to introduce an extended Lie derivative whose definition is given below and differs slightly from the one in [7, 13].

Definition 5 (*Extended Lie Derivative*) Given a function $\lambda(\mathbf{x}_{[s]})$ and the vector

$$r_i(\mathbf{x}, \delta) = \sum_{j=0}^{\bar{s}} r_i^j(\mathbf{x}) \delta^j,$$

the extended Lie derivative $L_{r_i^j(\mathbf{x})} \lambda(\mathbf{x}_{[s]})$ of $\lambda(\mathbf{x}_{[s]})$ in the direction r_i^j is

$$L_{r_i^j(\mathbf{x})} \lambda(\mathbf{x}_{[s]}) = \sum_{l=0}^{j} \frac{\partial \lambda(\mathbf{x}_{[s]})}{\partial x_{[0]}(-l)} r_i^{j-l}(\mathbf{x}(-l)). \quad (9)$$

One thus gets

$$\left[r_1^k(\cdot), r_2^l(\cdot)\right]_{E_0} = \left(L_{r_1^k(\mathbf{x})} r_2^l(\mathbf{x}) - L_{r_2^l(\mathbf{x})} r_1^k(\mathbf{x})\right)^T \frac{\partial}{\partial x_{[0]}}.$$

This allows to recover standard definitions of Lie derivatives and Lie brackets which are used in the delay free case. From (1), consider now the polynomial vector

$$\mathbf{F}(\mathbf{x}_{[s]}, \delta) = \sum_{j=0}^{ns} F^j(\mathbf{x}_{[s]}) \delta^j = \sum_{j=0}^{ns} F(\mathbf{x}_{[s]}) \delta^j. \quad (10)$$

The derivative of $\lambda(\mathbf{x}_{[s]})$ of order i which is computed for $\mathbf{u} = 0$, is given by

$$\lambda^{(i)}(\mathbf{x}, 0) = L^i_{F^{ns}(\mathbf{x})}\lambda(\mathbf{x}_{[s]}),$$

for $i \in [1, n]$. The definition of relative degree can then be stated as follows.

Definition 6 The function $\lambda(\mathbf{x}_{[s]})$ has relative degree $k > 0$ if

$$L_{g_1^j} L^i_{F^{ns}}\lambda(\mathbf{x}) \equiv 0 \ \ \forall j \geq 0, \forall 0 \leq i < k-1,$$

and if there exists an integer $j \geq 0$ such that

$$L_{g_1^j} L^{k-1}_{F^{ns}}\lambda(\mathbf{x}) \neq 0 \tag{11}$$

The relative degree is said to be strong if (11) is fulfilled for $j = 0$.

3.5 Integrability

Next, let us recall a fundamental result on integrability of a submodule, giving conditions under which the left annihilator of the submodule is spanned by exact one-forms. The following is a more elegant revisited version of a result in [1]. Consider

$$P_j(\mathbf{x}, \delta) = \left[r_1(\mathbf{x}, \delta), \cdots, r_j(\mathbf{x}, \delta)\right] = \sum_{l=0}^{s} P_{jl}(\mathbf{x})\delta^l$$

where $P_{j0}(\mathbf{x})$ has full column rank j and

$$r_k(\mathbf{x}, \delta) = \sum_{l=0}^{s} r_k^l(\mathbf{x})\delta^l, \ \ k \in [1, j],$$

and let s be the maximum delay in the $\mathbf{x}$ variable. Denoting, with some abuse of notations, in compact form

$$r^l = r_1^l, \cdots, r_j^l$$

and by I_n the $n \times n$ identity matrix, consider the distributions Δ_i, $i \geq 0$ which are defined on $\mathbb{R}^{(i+s+1)n}$ as follows:

$$\Delta_0 = \text{span}_{\mathcal{K}} \left\{ \begin{matrix} r^0 & \cdots & r^s & 0 & \cdots & \\ & \cdots & & I_n & & \\ & \cdots & & & \ddots & \ddots \\ & \cdots & & & 0 & I_n \end{matrix} \right\}$$

$$\Delta_1 = \text{span}_{\mathcal{K}} \left\{ \begin{bmatrix} r^0 & \cdots & r^s & 0 & \cdots & & \\ 0 & r^0(-1) & \cdots & r^s(-1) & 0 & \cdots & \\ & & & & I_n & & \\ & & \cdots & & & \ddots & \ddots \\ & & \cdots & & & 0 & I_n \end{bmatrix} \right\} \quad (12)$$

$$\vdots$$

$$\Delta_i = \text{span}_{\mathcal{K}} \left\{ \begin{bmatrix} r^0 & \cdots & r^s & 0 & \cdots & & & \\ 0 & \ddots & \cdots & \ddots & 0 & \cdots & & \\ & & r^0(-i) & \cdots & r^s(-i) & 0 & \cdots & \\ & & & & & I_n & & \\ & & & \cdots & & & \ddots & \ddots \\ & & & \cdots & & & 0 & I_n \end{bmatrix} \right\}$$

Let $\rho_i = \dim(\bar{\Delta}_i)$ in the neighborhood of $\mathbf{x}^0$.

Theorem 1 *Consider the submodule*

$$\Delta(\mathbf{x},\delta) = \text{span}_{\mathcal{K}(\delta]} \left\{ r_1(\mathbf{x},\delta), \cdots, r_j(\mathbf{x},\delta) \right\}, \quad \text{where} \quad r_i(\mathbf{x},\delta) = \sum_{l=0}^{s} r_i^l(\mathbf{x}_{[s]})\delta^l,$$

and such that the matrices

$$P_j(\mathbf{x},\delta) = \left(r_1(\mathbf{x},\delta), \cdots, r_j(\mathbf{x},\delta) \right) = \sum_{l=0}^{s} P_{jl}(\mathbf{x})\delta^l$$

and $P_{j0}(\mathbf{x})$ have rank j. Let Δ_i be the set of distributions defined by (12)*, which have by assumption constant dimension $\rho_i = \dim(\bar{\Delta}_i)$ locally around $\mathbf{x}^0_{[i]}$, with $\rho_{-1} = ns$ by definition. Then $\Delta(\mathbf{x},\delta)$ is completely integrable if and only if there exists an integer γ such that $\rho_\gamma - \rho_{\gamma-1} = j$.*

If the given submodule $\Delta(\mathbf{x},\delta)$ is not completely integrable, one may wish to compute the largest number of exact one-forms independent over $\mathcal{K}(\delta]$ which are in the left-kernel of $\Delta(\mathbf{x},\delta)$. The following result holds:

Theorem 2 *Consider the submodule*

$$\Delta(\mathbf{x},\delta) = \text{span}_{\mathcal{K}(\delta]} \left\{ r_1(\mathbf{x},\delta), \cdots, r_j(\mathbf{x},\delta) \right\}$$

with

$$r_i(\mathbf{x},\delta) = \sum_{l=0}^{s} r_i^l(\mathbf{x}_{[s]})\delta^l,$$

and such that matrices

$$P_j(\mathbf{x},\delta) = \big(r_1(\mathbf{x},\delta),\cdots,r_j(\mathbf{x},\delta)\big) = \sum_{l=0}^{s} P_{jl}(\mathbf{x})\delta^l$$

and $P_{j0}(\mathbf{x})$ have rank j. Let Δ_i be the set of distributions defined by (12)*, which have by assumption constant dimension $\rho_i = \dim(\bar{\Delta}_i)$ locally around $\mathbf{x}^0_{[i]}$, where $\rho_{-1} = ns$ by definition. Let γ be the smallest index such that $\rho_{\gamma+k} - \rho_{\gamma+k-1} = \ell \geq j$, $\forall k \geq 0$. The smallest completely integrable submodule $\bar{\Delta}(\mathbf{x},\delta) \supset \Delta(\mathbf{x},\delta)$ has rank ℓ. Consequently, there exist $n-\ell$ exact one-forms independent over $\mathscr{K}(\delta]$ which generate the left kernel of $\Delta(\mathbf{x},\delta)$.*

We end this section by stating the following result from [5], which is fundamental for developing the results in the next section.

Theorem 3 *Let $\Delta(\mathbf{x},\delta)$ be a given submodule and let $\bar{\Delta}(\mathbf{x},\delta)$ denote the smallest completely integrable submodule such that $\bar{\Delta}(\mathbf{x},\delta) \supset \Delta(\mathbf{x},\delta)$. Let j be the rank of $\bar{\Delta}(\mathbf{x},\delta)$, then there exist $n-j$ exact one forms $(d\lambda_1,\cdots,d\lambda_{n-j})$ which generate the left annihilator of $\bar{\Delta}(\mathbf{x},\delta)$. Furthermore there exist j exact one forms $(d\lambda_{n-j+1},\cdots,d\lambda_n)$, which can be taken to define a basis completion, that is*

$$\begin{pmatrix} d\lambda_1 \\ \vdots \\ d\lambda_n \end{pmatrix} = T(\mathbf{x},\delta)dx_{[0]}$$

with $T(\mathbf{x},\delta)$ unimodular.

4 Main Results

In this section we address the problem of characterizing the different geometrical structures analyzed in the Introduction. The conditions are linked to two different properties of the system, basically, what can be achieved under bicausal change of coordinates, and what can be achieved under regular bicausal static state feedback.

4.1 Generalized Normal Form

Hereafter we analyze the conditions under which there exist a bicausal change of coordinates and a regular bicausal static state feedback thanks to which the closed-loop system can be decomposed in the *generalized normal form*

$$\begin{aligned}
\dot{z}_{1,[0]} &= \theta_1(\mathbf{z}_{1,[\ell]}) + \sum_{j=0}^{\ell} \theta_{2j}(\mathbf{z}_{1,[\ell]}) v_{[0]}(-j), \\
\dot{z}_{2,[0]} &= \eta_1(\mathbf{z}_{[\ell]}) + \sum_{j=0}^{\ell} \eta_{2j}(\mathbf{z}_{[\ell]}) v_{[0]}(-j) \\
y_{[0]} &= \tilde{H}(\mathbf{z}_{1,[\ell]}),
\end{aligned} \tag{13}$$

with $z_{1,[0]} \in \mathbb{R}^r$, r being the relative degree of the output.

Before enouncing the necessary and sufficient conditions, the following result is required, which enlightens under which conditions it is possible to define a bicausal change of coordinates such that the output function and its derivative up to the order $r-1$ can be expressed as functions of the variable $z_{1,[0]} \in \mathbb{R}^r$ and its delayed values. This condition, which is necessary to solve Problem 1, does not follow from the assumption of defined relative degree, as it happens in the delay-free case.

The following result holds:

Lemma 1 *Given system* (1), *with output $y_{[0]}$ characterized by the well defined relative degree r, there exists a bicausal change of coordinates*

$$z_{[0]} = \begin{pmatrix} z_{1,[0]} \\ z_{2,[0]} \end{pmatrix} = \begin{pmatrix} \varphi_1(\mathbf{x}) \\ \varphi_2(\mathbf{x}) \end{pmatrix} = \varphi(\mathbf{x})$$

with $z_{1,[0]} \in \mathbb{R}^r$, such that in the new coordinates

$$dy_{[0]}^{(i)} = \varphi_{i+1}(\mathbf{z}_1) \qquad i \in [0, r-1] \tag{14}$$

if and only if setting

$$\begin{pmatrix} dy_{[0]} \\ \vdots \\ dy_{[0]}^{(r-1)} \end{pmatrix} = \mathbf{Y}(\mathbf{x}, \delta) dx_{[0]},$$

the right annihilator $\Delta(\cdot, \delta)$ of $\mathbf{Y}(\mathbf{x}, \delta)$ is causal.

Proof Necessity. Assume for instance that there exists a bicausal change of coordinates

$$z_{[0]} = \begin{pmatrix} z_{1,[0]} \\ z_{2,[0]} \end{pmatrix} = \begin{pmatrix} \varphi_1(\mathbf{x}) \\ \varphi_2(\mathbf{x}) \end{pmatrix} = \varphi(\mathbf{x})$$

with $z_{1,[0]} \in \mathbb{R}^r$, such that in the new coordinates the output and its derivative up to the order $r-1$ are expressed as functions of the variable $\mathbf{z}_1$ only. Let

$$dz_{[0]} = \begin{pmatrix} dz_{1,[0]} \\ dz_{2,[0]} \end{pmatrix} = \begin{pmatrix} T_1(\mathbf{x}, \delta) \\ T_2(\mathbf{x}, \delta) \end{pmatrix} dx_{[0]}$$

Then one gets immediately that

$$\begin{pmatrix} dy_{[0]} \\ \vdots \\ dy_{[0]}^{(r-1)} \end{pmatrix} = \Psi(\mathbf{z}_1, \delta) dz_{1,[0]} = \Psi(\mathbf{z}_1, \delta)_{z_{1,[0]}=\varphi_1(\mathbf{x})} T_1(\mathbf{x}, \delta) dx_{[0]} = \mathbf{Y}(\mathbf{x}, \delta) dx_{[0]}.$$

Consider now the matrix $T^{-1}(\mathbf{x}, \delta)$. Since $T(\mathbf{x}, \delta)$ is unimodular by assumption, $T^{-1}(\mathbf{x}, \delta)$ is causal. Furthermore setting $T^{-1}(\mathbf{x}, \delta) = (L_1(\mathbf{x}, \delta), L_2(\mathbf{x}, \delta))$, by construction $T_1(\mathbf{x}, \delta) L_2(\mathbf{x}, \delta) = 0$. Consequently, setting $\bar{\Psi}(\mathbf{x}, \delta) = \Psi(\mathbf{z}_1, \delta)_{z_{1,[0]}=\varphi_1(\mathbf{x})}$,

$$\mathbf{Y}(\mathbf{x}, \delta) L_2(\mathbf{x}, \delta) = \bar{\Psi}(\mathbf{x}, \delta) T_1(\mathbf{x}, \delta) L_2(\mathbf{x}, \delta) = 0.$$

It follows that the right annihilator $\Delta(\cdot, \delta)$ of $\mathbf{Y}(\mathbf{x}, \delta)$ is given by $\Delta(\cdot, \delta) = L_2(\mathbf{x}, \delta)$, has rank $n - r$ and is causal.

Sufficiency. By assumption $\Delta(\cdot, \delta)$ is causal, has rank $n - r$ and there exist r exact differentials, independent over $\mathcal{K}(\delta]$ which lay in the left annihilator. Consequently the rank of the involutive closure $\bar{\Delta}(\mathbf{x}, \delta)$ of $\Delta(\mathbf{x}, \delta)$ is still $n - r$ and there exists r exact differentials $d\lambda_1(\mathbf{x}), \cdots, d\lambda_r(\mathbf{x})$ which generate the left annihilator, so that for $i \in [0, r-1]$,

$$dy_{[0]}^{(i)} = \sum_{j=1}^{r} \alpha_{ij}(\mathbf{x}, \delta) d\lambda_j(\mathbf{x}) \tag{15}$$

Due to Theorem 3 it is also possible to compute $n - r$ functions $\lambda_{r+1}(\mathbf{x}), \cdots, \lambda_n(\mathbf{x})$, so that

$$\begin{pmatrix} d\lambda_1(\mathbf{x}) \\ \vdots \\ d\lambda_n(\mathbf{x}) \end{pmatrix} = T(\mathbf{x}, \delta) dx_{[0]}$$

with $T(\mathbf{x}, \delta)$ unimodular. As a consequence, the transformation

$$z_{[0]} = \begin{pmatrix} \lambda_1(\mathbf{x}) \\ \vdots \\ \lambda_n(\mathbf{x}) \end{pmatrix}$$

defines a bicausal change of coordinates. Denoting by $z_{1,[0]}$ the first r-components, and by $z_{2,[0]}$ the last $n - r$ components of $z_{[0]}$, in these new coordinates, we have that, due to Eq. (15), for $i \in [0, r-1]$

$$dy_{[0]}^{(i)} = \sum_{j=1}^{n} \alpha_{ij}(\mathbf{x}, \delta) d\lambda_j(\mathbf{x}) = \sum_{j=1}^{r} \alpha_{ij}(\mathbf{x}, \delta) d\lambda_j(\mathbf{x}) = \sum_{j=1}^{r} \bar{\alpha}_{ij}(\mathbf{z}, \delta) dz_{1,[0]}$$

Since on the left hand-side one has exact differentials, then necessarily $\bar{\alpha}_{ij}(\mathbf{z}, \delta) = \bar{\alpha}_{ij}(\mathbf{z}_1, \delta)$, which ends the proof.

Theorem 4 *Given system* (1)*, with output $y_{[0]}$ characterized by defined relative degree r, there exist a regular bicausal static state feedback*

$$u_{[0]} = \hat{\alpha}(\mathbf{x}) + \hat{\beta}(\mathbf{x}) v_{[0]},$$

and a bicausal change of coordinates

$$z_{[0]} = \begin{pmatrix} z_{1,[0]} \\ z_{2,[0]} \end{pmatrix} = \varphi(\mathbf{x})$$

with $z_{1,[0]} \in \mathbb{R}^r$, such that in the new coordinates the closed-loop system reads as the Generalized Normal Form (13)*, if and only if the following conditions hold:*

(i) Let

$$\begin{pmatrix} dy_{[0]} \\ \vdots \\ dy_{[0]}^{(r-1)} \end{pmatrix} = \mathbf{Y}(\mathbf{x}, \delta) dx_{[0]}. \tag{16}$$

The right annihilator $\Delta(\cdot, \delta)$ of $\mathbf{Y}(\mathbf{x}, \delta)$ has rank $n - r$ and is causal.

(ii) $dy_{[0]}^{(r)}$, the differential of the rth derivative of the output $y^{(r)}(\mathbf{x}_{[s]}, \mathbf{u}_{[s]})$, given by

$$dy_{[0]}^{(r)} = a(\mathbf{x}, \mathbf{u}, \delta) dx_{[0]} + b(\mathbf{x}, \delta) du_{[0]}, \tag{17}$$

satisfies the following properties:

(iia) $b(\mathbf{x}, \delta)$ in (17) *can be factorized as*

$$b(\mathbf{x}, \delta) = \tilde{b}(\mathbf{x}, \delta) \beta(\mathbf{x}) \tag{18}$$

with $\tilde{b}(\mathbf{x}, \delta) = \sum_{j=0}^{\ell} \tilde{b}_j(\mathbf{x}) \delta^{\ell}$ such that

$$d\tilde{b}_j(\mathbf{x}) \in \Delta(\mathbf{x}, \delta)^{\perp}, \quad \forall j \in [0, \ell] \tag{19}$$

(iib) there exists a function $\alpha(\mathbf{x})$ such that denoting by

$$\gamma(\mathbf{x}, \delta) dx_{[0]} = d\alpha(\mathbf{x}),$$

then, $a(\mathbf{x}, \mathbf{u}, \delta)$ in (17) *computed for $\mathbf{u} = 0$ and $\tilde{b}(\mathbf{x}, \delta)$ in* (18) *satisfy*

$$\left(a(\mathbf{x},0,\delta)-\tilde{b}(\mathbf{x},\delta)\gamma(\mathbf{x},\delta)\right)dx_{[0]}\in\Delta(\mathbf{x},\delta)^{\perp} \tag{20}$$

If conditions (i) and (ii) are satisfied, the coordinates $z_{[0]}=\varphi(\mathbf{x})$, *which transform the closed-loop system in the form* (13), *can be chosen in such a way that* $\theta_{2j}(\mathbf{z}_1)$ *has only the last component different from* 0, *that is*

$$\theta_{2j}(\mathbf{z}_1)=\begin{pmatrix}0\\ \vdots\\ 0\\ \vartheta_{2j}(\mathbf{z}_1)\end{pmatrix},\quad \forall i\in[0,\ell] \tag{21}$$

Proof Assume that there exist a regular bicausal static state feedback

$$u_{[0]}=\hat{\alpha}(\mathbf{x})+\hat{\beta}(\mathbf{x})v_{[0]}$$

and a bicausal change of coordinates,

$$z_{[0]}=\begin{pmatrix}z_{1,[0]}\\ z_{2,[0]}\end{pmatrix}=\varphi(\mathbf{x})$$

with $z_{1,[0]}\in\mathbb{R}^r$, such that the closed-loop system in the new coordinates reads (13). Then

$$dy_{[0]}=d\tilde{H}(\mathbf{z}_1)\text{ and }dy^{(i)}_{[0]}=d\psi_i(\mathbf{z}_1)$$

for $i\in[1,r-1]$. Let

$$dz_{[0]}=\begin{pmatrix}dz_{1,[0]}\\ dz_{2,[0]}\end{pmatrix}=\begin{pmatrix}T_1(\mathbf{x},\delta)\\ T_2(\mathbf{x},\delta)\end{pmatrix}dx_{[0]}=T(\mathbf{x},\delta)dx_{[0]}$$

be the differential representation of the bicausal change of coordinates. Then, by assumption,

$$\begin{pmatrix}dy_{[0]}\\ \vdots\\ dy^{(r-1)}_{[0]}\end{pmatrix}=Q(\mathbf{z}_1,\delta)dz_{1,[0]}=Q(\mathbf{z}_1,\delta)|_{z_{1,[0]}=\varphi_1(\mathbf{x})}T_1(\mathbf{x},\delta)dx_{[0]}.$$

so that $\mathbf{Y}(\mathbf{x},\delta)$ in (16) is given by $\mathbf{Y}(\mathbf{x},\delta)=\bar{Q}(\mathbf{x},\delta)T_1(\mathbf{x},\delta)$. It is immediately seen that denoting by $\Delta(\mathbf{x},\delta)$ the last $n-r$ columns of $T^{-1}(\mathbf{x},\delta)$, then $\Delta(\mathbf{x},\delta)$ is the right annihilator of $\mathbf{Y}(\mathbf{x},\delta)$, has rank $n-r$ and is causal, which proves the necessity of (i).

As for (ii), by assumption in the z-coordinates, and after the regular bicausal static state feedback

$$u_{[0]} = \hat{\alpha}(\mathbf{x}) + \hat{\beta}(\mathbf{x})v_{[0]}$$

which in the z-coordinates reads

$$u_{[0]} = \tilde{\alpha}(\mathbf{z}) + \tilde{\beta}(\mathbf{z})v_{[0]},$$

the differential representation of the rth derivative of the output of the closed-loop system is given by

$$dy_{[0]}^{(r)} = \tilde{a}(\mathbf{z}_1, \mathbf{v}, \delta)dz_{1,[0]} + \tilde{b}(\mathbf{z}_1, \delta)dv_{[0]}.$$

Let

$$v_{[0]} = \bar{\alpha}(\mathbf{z}) + \bar{\beta}(\mathbf{z})u_{[0]}$$

be the inverse feedback of

$$u_{[0]} = \tilde{\alpha}(\mathbf{z}) + \tilde{\beta}(\mathbf{z})v_{[0]}.$$

Then

$$\bar{\alpha}(\mathbf{z}) = -\tilde{\beta}^{-1}(\mathbf{z})\tilde{\alpha}(\mathbf{z}) \text{ and } \bar{\beta}(\mathbf{z}) = \tilde{\beta}^{-1}(\mathbf{z}).$$

Let $dv_{[0]} = \tilde{\gamma}(\mathbf{z}, \mathbf{u}, \delta)dz_{[0]} + \bar{\beta}(\mathbf{z})du_{[0]}$ be its differential representation and note that $d\bar{\alpha}(\mathbf{z}) = \tilde{\gamma}(\mathbf{z}, 0, \delta)dz_{[0]}$. Then for the open loop system, one has that

$$\begin{aligned} dy_{[0]}^{(r)} &= \bar{a}(\mathbf{z}, \mathbf{u}, \delta)dz_{[0]} + \bar{b}(\mathbf{z}, \delta)du_{[0]} \\ &= \tilde{a}(\mathbf{z}_1, \mathbf{v}, \delta)|_{v_{[0]}=\bar{\alpha}(\mathbf{z})+\bar{\beta}(\mathbf{z})u_{[0]}}dz_{1,[0]} + \tilde{b}(\mathbf{z}_1, \delta)\tilde{\gamma}(\mathbf{z}, \mathbf{u}, \delta)dz_{[0]} \\ &\quad + \tilde{b}(\mathbf{z}_1, \delta)\bar{\beta}(\mathbf{z})du_{[0]} \\ &= \hat{a}(\mathbf{z}, \mathbf{u}, \delta)dz_{1,[0]} + \tilde{b}(\mathbf{z}_1, \delta)\tilde{\gamma}(\mathbf{z}, \mathbf{u}, \delta)dz_{[0]} + \tilde{b}(\mathbf{z}_1, \delta)\bar{\beta}(\mathbf{z})du_{[0]} \end{aligned}$$

which shows that

$$\begin{aligned} \bar{b}(\mathbf{z}, \delta) &= \tilde{b}(\mathbf{z}_1, \delta)\bar{\beta}(\mathbf{z}) \\ \bar{a}(\mathbf{z}, \mathbf{u}, \delta)dz_{[0]} &= \hat{a}(\mathbf{z}, \mathbf{u}, \delta)dz_{1,[0]} + \tilde{b}(\mathbf{z}_1, \delta)\tilde{\gamma}(\mathbf{z}, \mathbf{u}, \delta)dz_{[0]}, \end{aligned}$$

so that

$$\bar{a}(\mathbf{z}, 0, \delta)dz_{[0]} = \hat{a}(\mathbf{z}, 0, \delta)dz_{1,[0]} + \tilde{b}(\mathbf{z}_1, \delta)\tilde{\gamma}(\mathbf{z}, 0, \delta)dz_{[0]}. \qquad (22)$$

Recall that $b(\mathbf{x}, \delta)$ in (17) is linked to $\bar{b}(\mathbf{z}, \delta)$ computed above through the relation $b(\mathbf{x}, \delta) = \bar{b}(\mathbf{z}, \delta)|_{z[0]=\varphi(\mathbf{x})}$, so that one gets:

$$b(\mathbf{x}, \delta) = \bar{b}(\mathbf{z}, \delta)|_{z[0]=\varphi(\mathbf{x})} = \bar{b}(\mathbf{z}_1, \delta)|_{z_{1,[0]}=\varphi_1(\mathbf{x})}\bar{\beta}(\mathbf{z})|_{z[0]=\varphi(\mathbf{x})} = \tilde{b}(\mathbf{x}, \delta)\beta(\mathbf{x})$$

where $\tilde{b}_j(\mathbf{x}) = \bar{b}_j(\mathbf{z}_1)|_{z_{1,[0]}=\varphi_1(\mathbf{x})}$ so that for all $j \in [0, \ell]$, $d\tilde{b}_j(\mathbf{x}) \in \Delta(\mathbf{x}, \delta)^{\perp}$, thus proving (18) and (19) in (iia).

Analogously $a(\mathbf{x}, 0, \delta)$ and $\gamma(\mathbf{x}, \delta)$ used in (20) are linked respectively to $\bar{a}(\mathbf{z}, 0, \delta)$ and $\tilde{\gamma}(\mathbf{z}, \delta)$ defined above through the relations $a(\mathbf{x}, 0, \delta) = \bar{a}(\mathbf{z}, 0, \delta)|_{z[0]=\varphi(\mathbf{x})} T(\mathbf{x}, \delta)$ and

$$\gamma(\mathbf{x}, \delta) = \tilde{\gamma}(\mathbf{z}, \delta)|_{z[0]=\varphi(\mathbf{x})} T(\mathbf{x}, \delta).$$

It follows from (22), that

$$\begin{aligned}
&\left(a(\mathbf{x}, 0, \delta) - \tilde{b}(\mathbf{x}, \delta)\gamma(\mathbf{x}, \delta)\right) dx_{[0]} \\
&= \left((\hat{a}(\mathbf{z}, 0, \delta),\ 0) - \bar{b}(\mathbf{z}_1, \delta)\tilde{\gamma}(\mathbf{z}, \delta)\right) |_{z[0]=\varphi(\mathbf{x})} T(\mathbf{x}, \delta) dx_{[0]} \\
&= \bar{a}(\mathbf{z}, 0, \delta)|_{z[0]=\varphi(\mathbf{x})} d\varphi_1(\mathbf{x})
\end{aligned}$$

which proves (20).

As for the sufficiency, assume that the conditions are satisfied and let the rth derivative of the output be

$$y_{[0]}^{(r)} = \bar{a}(\mathbf{x}) + \sum_{j=0}^{\ell} b_j(\mathbf{x}) u_{[0]}(-j).$$

Consider the regular bicausal static state feedback

$$u_{[0]} = \hat{\alpha}(\mathbf{x}) + \hat{\beta}(\mathbf{x}) v_{[0]} = \beta^{-1}(\mathbf{x}) \left(-\alpha(\mathbf{x}) + v_{[0]}\right),$$

where $\beta(\mathbf{x})$ is computed from (18), while $\alpha(\mathbf{x})$ is computed starting from $\gamma(\mathbf{x}, \delta)$ satisfying (20), with

$$d\alpha(\mathbf{x}) = \gamma(\mathbf{x}, \delta) dx_{[0]}.$$

For the closed-loop system, the rth derivative of the output becomes

$$y_{[0]}^{(r)} = \bar{a}(\mathbf{x}) + \sum_{j=0}^{\ell} b_j(\mathbf{x}) \beta^{-1}(\mathbf{x}(-j))(-\alpha(\mathbf{x}(-j)) + v_{[0]}(-j))$$

Due to (18), each coefficient $b_j(\mathbf{x}) = \tilde{b}_j(\mathbf{x})\beta(\mathbf{x}(-j))$, so that

$$y_{[0]}^{(r)} = \bar{a}(\mathbf{x}) + \sum_{j=0}^{\ell} \tilde{b}_j(\mathbf{x})(-\alpha(\mathbf{x}(-j)) + v_{[0]}(-j))$$

From (i) choose a bicausal change of coordinates $z_{[0]} = \varphi(\mathbf{x})$, such that $dz_{1,0} \perp \Delta(\mathbf{x}, \delta)$ with $z_{1,[0]} \in \mathrm{R}^r$. Then, by Lemma 1,

$$dy_{[0]}^{(i)} \in \mathrm{span}_{\mathcal{K}(\delta]}\{dz_{1,[0]}\},$$

for $i \in [0, r-1]$. Furthermore in these new coordinates the rth derivative of the output of the closed-loop system is

$$y_{[0]}^{(r)} = \bar{a}(\mathbf{x})|_{x[0]=\varphi^{-1}(\mathbf{z})} - \sum_{j=0}^{\ell} (\tilde{b}_j(\mathbf{x})\alpha(\mathbf{x}(-j)))|_{x[0]=\varphi^{-1}(\mathbf{z})}$$
$$+ \sum_{j=0}^{\ell} \tilde{b}_j(\mathbf{x})|_{x[0]=\varphi^{-1}(\mathbf{z})} v_{[0]}(-j)$$

Due to (19), $d\tilde{b}_j(\mathbf{x}) \in \Delta^{\perp}(\mathbf{x}, \delta)$, so that $\tilde{b}_j(\mathbf{x})|_{x[0]=\varphi^{-1}(\mathbf{z})} = \tilde{b}_j(\mathbf{z}_1)$. Finally due to (20), we get

$$d\bar{a}(\mathbf{x}) - \sum_{j=0}^{\ell} (\tilde{b}_j(\mathbf{x})\delta^j d\alpha(\mathbf{x})) \in \Delta(\mathbf{x}, \delta)^{\perp},$$

so that also

$$d\bar{a}(\mathbf{x}) - \sum_{j=0}^{\ell} d(\tilde{b}_j(\mathbf{x})\alpha(\mathbf{x}(-j))) \in \Delta(\mathbf{x}, \delta)^{\perp},$$

which proves that for the closed-loop system in the new coordinates

$$y_{[0]}^{(r)} = \tilde{a}(\mathbf{z}_1) + \sum_{j=0}^{\ell} \tilde{b}_j(\mathbf{z}_1) v_{[0]}(-j)$$

Due to Lemma 1, in the z-coordinates one has that

$$\begin{pmatrix} dy_{[0]} \\ \vdots \\ dy_{[0]}^{(r-1)} \end{pmatrix} = Q(\mathbf{z}_1, \delta) dz_{1,[0]} \tag{23}$$

Consider the derivative of both sides of (23). Then one gets that for the closed-loop system

$$\begin{pmatrix} d\dot{y}_{[0]} \\ \vdots \\ dy_{[0]}^{(r)} \end{pmatrix} = \dot{Q}(\mathbf{z}_1, \delta)dz_{1,[0]} + Q(\mathbf{z}_1, \delta)d\dot{z}_{1,[0]} \tag{24}$$
$$= P_1(\mathbf{z}_1, \mathbf{v}, \delta)dz_{1,[0]} + P_2(\mathbf{z}_1, \delta)dv_{[0]},$$

that is

$$Q(\mathbf{z}_1, \delta)d\dot{z}_{1,[0]} = \bar{P}_1(\mathbf{z}, \mathbf{v}, \delta)dz_{1,[0]} + P_2(\mathbf{z}_1, \delta)dv_{[0]}. \tag{25}$$

The proof is carried out by contradiction, by showing that $d\dot{z}_{1,[0]}$ cannot depend on $dz_{2,[0]}$. Assume in fact that this is not the case. Then,

$$d\dot{z}_{1,[0]} = \hat{P}_{11}(\mathbf{z}, \delta)dz_{1,[0]} + \hat{P}_{12}(\mathbf{z}, \delta)dz_{2,[0]} + \hat{P}_{13}(\mathbf{z}, \delta)dv_{[0]}$$

Thus, premultiplying both sides by $Q(\mathbf{z}_1, \delta)$, one gets that

$$Q(\mathbf{z}_1, \delta)d\dot{z}_{1,[0]} = Q(\mathbf{z}_1, \delta)\hat{P}_{11}(\mathbf{z}, \delta)dz_{1,[0]} + Q(\mathbf{z}_1, \delta)\hat{P}_{12}(\mathbf{z}, \delta)dz_{2,[0]}$$
$$+ Q(\mathbf{z}_1, \delta)\hat{P}_{13}(\mathbf{z}, \delta)dv_{[0]}$$

which compared with (25) implies that $Q(\mathbf{z}_1, \delta)\hat{P}_{12}(\mathbf{z}, \delta) = 0$. Since by assumption $Q(\mathbf{z}_1, \delta)$ is a full rank matrix, this can only happen if $\hat{P}_{12}(\mathbf{z}, \delta) = 0$, which thus proves the thesis. As a consequence the system can be put in the form (13). To end the proof we have to show that the $z_{1,[0]}$ coordinates can always be chosen in order to guarantee that the θ_{2j}'s in (13) are of the form (21). This follows immediately by considering the subsystem

$$\dot{z}_{1,[0]} = \theta_1(\mathbf{z}_{1,[\ell]}) + \sum_{j=0}^{\ell} \theta_{2j}(\mathbf{z}_{1,[\ell]})v_{[0]}(-j),$$
$$y_{[0]} = \tilde{H}(\mathbf{z}_{1,[\ell]}),$$

whose associated differential representation (4) is characterized by

$$g_{11}(\mathbf{z}_{1,[\ell]}, \delta) = \sum_{j=0}^{\ell} \theta_{2j}(\mathbf{z}_{1,[\ell]})\delta^j$$

Since $dy_{[0]}^{(i)}\, g_{11}(\mathbf{z}_1, \delta) = 0$ for $i \in [0, r-2]$, then

$$\Delta(\mathbf{z}_1, \delta) = \operatorname{span}_{\mathscr{K}(\delta]}\{g_{11}(\mathbf{z}_1, \delta)\}$$

is completely integrable and since rank $\Delta(\mathbf{x}, \delta) = 1$, there exists $r - 1$ exact differentials which lay in the left-annihilator. Due to Lemma 1 there exists a bicausal change of coordinates $\tilde{z}_{1,[0]} = \bar{\varphi}_1(\mathbf{z}_1)$, such that in the new coordinates

$$\tilde{g}_{11}(\tilde{\mathbf{z}}_{1,[\bar{\ell}]}, \delta) = \begin{pmatrix} 0 \\ \vdots \\ 0 \\ * \end{pmatrix}$$

that is

$$\begin{aligned} \dot{\tilde{z}}_{1,[0]} &= \tilde{\theta}_1(\tilde{\mathbf{z}}_{1,[\bar{\ell}]}) + \sum_{j=0}^{\bar{\ell}} \tilde{\theta}_{2j}(\tilde{\mathbf{z}}_{1,[\bar{\ell}]}) v_{[0]}(-j), \\ y_{[0]} &= \tilde{H}(\tilde{\mathbf{z}}_{1,[\bar{\ell}]}), \end{aligned}$$

with

$$\tilde{\theta}_{2j}(\tilde{\mathbf{z}}_{1,[\bar{\ell}]}) = \begin{pmatrix} 0 \\ \vdots \\ 0 \\ \tilde{\vartheta}_{2j}(\tilde{\mathbf{z}}_1) \end{pmatrix}, \quad \forall j \in [0, \bar{\ell}]$$

which ends the proof.

We next give two examples which show how to apply the previous result.

Example 1 Consider the dynamics

$$\begin{aligned} \dot{x}_{1,[0]} &= x_{1,[0]} x_{1,[0]}(-1) + x_{2,[0]} - x_{1,[0]} + v_{[0]} \\ \dot{x}_{2,[0]} &= -(x_{1,[0]} x_{1,[0]}(-1) + x_{2,[0]})^3 \\ &\quad - (x_{1,[0]} x_{1,[0]}(-1) + x_{2,[0]} - x_{1,[0]} + v_{[0]}) x_{1,[0]}(-1) \\ &\quad - (x_{1,[0]}(-1) x_{1,[0]}(-2) + x_{2,[0]}(-1) - x_{1,[0]}(-1) + v_{[0]}(-1)) x_{1,[0]} \\ &\quad + v_{[0]}(-1) \end{aligned}$$

with output function

$$y_{[0]} = x_{1,[0]} x_{1,[0]}(-1) + x_{2,[0]}$$

Then we get that the output has relative degree $r = 1$ and we have that

$$dy_{[0]} = \big(x_{1,[0]}(-1) + x_{1,[0]}\delta,\ 1\big)\, dx_{[0]}.$$

Accordingly, the right annihilator is

$$\Delta(\mathbf{x},\delta) = \begin{pmatrix} -1 \\ x_{1,[0]}(-1) + x_{1,[0]}\delta \end{pmatrix}$$

which is causal. Since $dy_{[0]}$ is closed we have that $d\lambda_1(\mathbf{x}) = dy_{[0]}$ and a possible completion to get a bicausal change of coordinates is $dz_{2,[0]} = dx_{1,[0]}$. One thus has that in the coordinates

$$dz_{[0]} = \begin{pmatrix} x_{1,[0]}(-1) + x_{1,[0]}\delta & 1 \\ 1 & 0 \end{pmatrix} dx_{[0]} \quad \rightarrow z_{[0]} = \begin{pmatrix} x_{1,[0]}x_{1,[0]}(-1) + x_{2,[0]} \\ x_{1,[0]} \end{pmatrix}$$

the system reads

$$\dot{z}_{1,[0]} = -z_{1,[0]}^3 + v_{[0]}(-1), \quad \dot{z}_{2,[0]} = z_{1,[0]} - z_{2,[0]} + v_{[0]}, \quad y_{[0]} = z_{1,[0]}.$$

Example 2 Consider again the dynamics (1), but with output function

$$y_{[0]} = x_{1,[0]}x_{1,[0]}(-1) + x_{2,[0]}x_{2,[0]}(-1)$$

In this case,

$$dy_{[0]} = \left(x_{1,[0]}(-1) + x_{1,[0]}\delta \;\; x_{2,[0]}(-1) + x_{2,[0]}\delta\right),$$

which is still closed. However the right annihilator in this case is

$$\Delta(\mathbf{x},\delta) = \begin{pmatrix} -\frac{x_{2,[0]}(-1)}{x_{1,[0]}(-1)} b_0 - x_{2,[0]}(+1)\delta \\ b_0 + x_{1,[0]}(+1)\delta \end{pmatrix},$$

with

$$b_0 = \frac{x_{1,[0]}x_{2,[0]}(+2) - x_{2,[0]}x_{1,[0]}(+2)}{x_{2,[0]}(+1)x_{1,[0]}(-1) - x_{1,[0]}(+1)x_{2,[0]}(-1)} x_{1,[0]}(-1).$$

Obviously the generator of $\Delta(\mathbf{x},\delta)$ can never be chosen causal, so that there is no causal function $\lambda(\mathbf{x})$ such that $d\mathrm{x}_{[0]} \in \mathrm{span}_{\mathcal{K}(\delta]}\{dy_{[0]}, d\lambda\}$.

We end this section by giving the conditions under which the unobservable dynamics in (13) can be chosen in order to be independent of the control. The following result holds true.

Theorem 5 *Given system* (1), *with output* $y_{[0]}$ *characterized by the well defined relative degree* r, *there exist a regular bicausal static state feedback* $u_{[0]} = \alpha(\mathbf{x}) + \beta(\mathbf{x})v_{[0]}$, *and a bicausal change of coordinates*

$$z_{[0]} = \begin{pmatrix} z_{1,[0]} \\ z_{2,[0]} \end{pmatrix} = \varphi(\mathbf{x})$$

with $z_{1,[0]} \in \mathbb{R}^r$, such that in the new coordinates the closed-loop system reads as (13) *with $\eta_{2j} = 0$, $j \in [0, \ell]$, if and only if the conditions of Theorem* 4 *are satisfied, and additionally,*
(a) given the differential representation (4)*, associated with the system,*

$$\Delta(\mathbf{x}, \delta) = \text{span}_{\mathscr{K}(\delta]}\{g_1(\mathbf{x}, \delta)\}$$

is completely integrable, that is, there exists an index γ such that $\rho_\gamma - \rho_{\gamma-1} = 1$;
(b) in the z-coordinates in which the system reads (13) *with θ_{2j} of the form* (21)*, denoting by $\eta_2(\mathbf{z}, \delta) = \sum_{j=0}^{\ell} \eta_{2j}(\mathbf{z})\delta^j$ and by $\vartheta_2(\mathbf{z}_1, \delta) = \sum_{j=0}^{\ell} \vartheta_{2j}(\mathbf{z}_1)\delta^j$, then*

$$\eta_2(\mathbf{z}, \delta) = \begin{pmatrix} m_1(\mathbf{z}, \delta) \\ \vdots \\ m_{n-r}(\mathbf{z}, \delta) \end{pmatrix} \vartheta_2(\mathbf{z}_1, \delta)$$

Proof The proof of the necessity is immediate if one considers that, condition (a) ensures that there exist $n-1$ exact differentials in the left kernel of $g_1(\mathbf{x}, \delta)$, which are the candidates for the bicausal change of coordinates. The second condition is necessary to guarantee that one can find a bicausal change of coordinates which achieves the result by transforming only the last $n-r$ components, since the first r ones are related to the input output behavior.

As for the sufficiency, suppose for simplicity that the system has already been transformed in the form (13), with θ_{2j} of the form (21). Since (a) holds true one can find $n-1$ exact differentials which lie in the left kernel of

$$\Delta(\mathbf{x}, \delta) = \text{span}_{\mathscr{K}(\delta]}\{g_1(\mathbf{z}, \delta)\}.$$

Starting with the computation of such exact differentials at the first step one gets

$$dz_{11,[0]}, \cdots, dz_{1r-1,[0]}.$$

We now have to show the form of the last $n-r$ exact differentials. To this end note that, due to b), a basis for $\Delta(\mathbf{x}, \delta)^\perp$ is

$$\Delta(\mathbf{x}, \delta)^\perp = \text{span}_{\mathscr{K}(\delta]}\{dz_{1i,[0]}, i \in [1, r-1], dz_{2j,[0]} - m_j(\mathbf{z}, \delta)dz_{1r,[0]}, j \in [1, n-r]\}$$

Since the previous collection of one-forms is closed, then any other collection which generates $\Delta(\mathbf{x}, \delta)^\perp$ must be obtained through a unimodular matrix. Let $d\varphi_{2j}(\mathbf{z})$, $j \in [1, \cdots, n-r]$, be the set of exact differentials such that

$$\Delta(\mathbf{x}, \delta)^\perp = \text{span}_{\mathscr{K}(\delta]}\{dz_{1i,[0]},\ i \in [1, r-1], d\varphi_{2j}(\mathbf{z}),\ j \in [1, n-r]\}$$

then

$$\begin{pmatrix} dz_{11,[0]} \\ \vdots \\ dz_{1\,r-1,[0]} \\ d\varphi_{21}(\mathbf{z}) \\ \vdots \\ d\varphi_{2\,n-r}(\mathbf{z}) \end{pmatrix} = \begin{pmatrix} I & 0 \\ T_{21}(\mathbf{z},\delta) & T_{22}(\mathbf{z},\delta) \end{pmatrix} \begin{pmatrix} dz_{11,[0]} \\ \vdots \\ dz_{1\,r-1,[0]} \\ dz_{21,[0]} - m_1(\mathbf{z},\delta)dz_{1\,r,[0]} \\ \vdots \\ dz_{2\,n-r,[0]} - m_{n-r}(\mathbf{z},\delta)dz_{1\,r,[0]} \end{pmatrix}$$

where $T_{22}(\mathbf{z},\delta)$ is unimodular. Consequently the change of coordinates

$$\begin{pmatrix} dz_{11,[0]} \\ \vdots \\ dz_{1\,r,[0]} \\ d\varphi_{21}(\mathbf{z}) \\ \vdots \\ d\varphi_{2\,n-r}(\mathbf{z}) \end{pmatrix}$$

is bicausal.

4.2 Normal Form

In this section we analyze the conditions under which there exist a bicausal change of coordinates and a regular bicausal static state feedback thanks to which the closed-loop system can be decomposed in the form (13), with a linear input output behavior, that is the closed-loop system displays a linear input-output relation and reads in the new coordinates

$$\begin{aligned} \dot{z}_{1,[0]} &= \sum_{j=0}^{\ell} A_j z_{1,[0]}(-j) + \sum_{j=0}^{\ell} B_j v_{[0]}(-j), \\ \dot{z}_{2,[0]} &= \eta_1(\mathbf{z}_{[\ell]}) + \sum_{j=0}^{\ell} \eta_{2j}(\mathbf{z}_{[\ell]}) v_{[0]}(-j) \\ y_{[0]} &= \sum_{j=0}^{\ell} C_j z_{1,[0]}(-j), \end{aligned} \tag{26}$$

with $z_{1,[0]} \in \mathbb{R}^r$, r being the relative degree of the output.

Of course, in this case the conditions of the previous Theorem 4 must be satisfied together with some new additional ones. The following result holds true.

Theorem 6 *Given system* (1)*, with output* $y_{[0]}$ *characterized by the well defined relative degree r, there exist a regular bicausal static state feedback*

$$u_{[0]} = \hat{\alpha}(\mathbf{x}) + \hat{\beta}(\mathbf{x})v_{[0]},$$

and a bicausal change of coordinates

$$z_{[0]} = \begin{pmatrix} \mathbf{z}_{1,[0]} \\ \mathbf{z}_{2,[0]} \end{pmatrix} = \varphi(\mathbf{x})$$

with $z_{1,[0]} \in \mathbb{R}^r$*, such that in the new coordinates the closed-loop system is in the* Normal Form (26) *if and only if the conditions of Theorem* 4 *are satisfied and additionally*

(i') $\mathbf{Y}(\mathbf{x}, \delta)$ *in* (16) *can be factorized as* $\mathbf{Y}(\mathbf{x}, \delta) = Q(\delta)T_1(\mathbf{x}, \delta)$*, with* $Q(\delta)$ *a matrix with coefficients in* $R(\delta]$*, and the submodule generated by* $T_1(\mathbf{x}, \delta)d\mathrm{x}_{[0]}$ *is left-closed.*

(ii') $\tilde{b}(\mathbf{x}, \delta)$ *in* (18) *is independent of* $\mathbf{x}$ *that is* $\tilde{b}(\mathbf{x}, \delta) = \tilde{b}(\delta)$*. Accordingly,* (19) *is always satisfied.*

(iii') *the exact differential* $\gamma(\mathbf{x}, 0, \delta)dx_{[0]} = d\alpha(\mathbf{x})$ *must satisfy*

$$\Big(a(\mathbf{x}, 0, \delta) - \tilde{b}(\delta)\gamma(\mathbf{x}, \delta)\Big)\, dx_{[0]} \in \mathrm{span}_{R(\delta]}\{d\lambda_1, \cdots, d\lambda_r\} \tag{27}$$

The coordinates $z_1 = \big(\lambda_1, \cdots, \lambda_r\big)^T$ *can be chosen in such a way that the* B_j*'s in the closed-loop system* (26)*, only have the last component different from* 0*, that is,*

$$B_j = \begin{pmatrix} 0 \\ \vdots \\ 0 \\ b_{jr} \end{pmatrix}, \quad j \in [0, \ell]. \tag{28}$$

Proof Of course the conditions of Theorem 4 must be satisfied, since the structure (26) represents a special case of the structure (13). The necessity and sufficiency of the additional conditions are proved hereafter. As for the necessity, if the system can be put in the form (26), then one gets that in the good coordinates

$$\begin{pmatrix} dy_{[0]} \\ \vdots \\ dy_{[0]}^{(r-1)} \end{pmatrix} = \begin{pmatrix} C(\delta) \\ \vdots \\ C(\delta)A^{r-1}(\delta) \end{pmatrix} dz_{1,[0]} = Q(\delta)T_1(\mathbf{x}, \delta)dx_{[0]},$$

which proves the necessity of (i'). Furthermore, for the closed-loop system one must have that

$$dy_{[0]}^{(r)} = \tilde{a}(\delta)dz_{1,[0]} + \tilde{b}(\delta)dv_{[0]}$$

For any regular bicausal static state feedback $v_{[0]} = \bar{\alpha}(\mathbf{z}) + \bar{\beta}(\mathbf{z})u_{[0]}$, one has that

$$dv_{[0]} = \bar{\gamma}(\mathbf{z}, \mathbf{u}, \delta)dz_{[0]} + \bar{\beta}(\mathbf{z})du_{[0]},$$

so that

$$\begin{aligned} dy_{[0]}^{(r)} &= \tilde{a}(\delta)dz_{1,[0]} + \tilde{b}(\delta)\bar{\gamma}(\mathbf{z}, \mathbf{u}, \delta)dz_{[0]} + \tilde{b}(\delta)\bar{\beta}(\mathbf{z})du_{[0]} \\ &= \tilde{a}(\mathbf{z}, \mathbf{u}, \delta)dz_{[0]} + \tilde{b}(\delta)\bar{\beta}(\mathbf{z})du_{[0]} \end{aligned}$$

which immediately proves (ii') and (iii').

As for the sufficiency, since the conditions of Theorem 4 are satisfied, in the new coordinates the system reads as (13).

Due to (i'), we can take as $z_{1,[0]}$ coordinates $z_{1,[0]} = (\lambda_1(\mathbf{x}), \cdots, \lambda_r(\mathbf{x}))^T$.

Taking $u_{[0]} = \beta^{-1}(\mathbf{x})(-\alpha(\mathbf{x}) + v_{[0]})$ as the feedback law, where $\beta(\mathbf{x})$ and $\alpha(\mathbf{x})$ are chosen as in Theorem 4 one gets that for the closed loop system, the rth derivative of the output becomes

$$y_{[0]}^{(r)} = \bar{a}(\mathbf{x}) + \sum_{j=0}^{\ell} b_j(\mathbf{x})\beta^{-1}(\mathbf{x}(-j))(-\alpha(\mathbf{x}(-j)) + v_{[0]}(-j))$$

Due to (ii') and (iii'), each coefficient $b_j(\mathbf{x}) = \tilde{b}_j\beta(\mathbf{x}(-j))$, so that

$$y_{[0]}^{(r)} = \bar{a}(\mathbf{x}) + \sum_{j=0}^{\ell} \tilde{b}_j(-\alpha(\mathbf{x}(-j)) + v_{[0]}(-j))$$

Consequently, its differential representation is

$$\begin{aligned} dy_{[0]}^{(r)} &= d\bar{a}(\mathbf{x}) - \textstyle\sum_{j=0}^{\ell} \tilde{b}_j d\alpha(\mathbf{x}(-j)) + \sum_{j=0}^{\ell} \tilde{b}_j dv_{[0]}(-j) \\ &= \tilde{a}(\delta)dz_{1,[0]} + \tilde{b}(\delta)dv_{[0]} \end{aligned}$$

Furthermore, from (i'),

$$\begin{pmatrix} dy_{[0]} \\ \vdots \\ dy_{[0]}^{(r-1)} \end{pmatrix} = Q(\delta)T_1(\mathbf{x}, \delta)dx_{[0]} = Q(\delta)dz_{1,[0]}.$$

As a consequence,

$$\begin{pmatrix} d\dot{y}_{[0]} \\ \vdots \\ dy_{[0]}^{(r)} \end{pmatrix} = Q(\delta)d\dot{z}_{1,[0]} = P_1(\delta)dz_{1,[0]} + P_2(\delta)du_{[0]}.$$

Multiply both sides on the left by the adjugate matrix $[Q(\delta)]^{(a)}$ of $Q(\delta)$. Since

$$[Q(\delta)]^{(a)}Q(\delta) = q_o(\delta)I,$$

where q_0 is the determinant of $Q(\delta)$ and I is the identity matrix, we get

$$q_o(\delta)d\dot{z}_{1,[0]} = [Q(\delta)]^{(a)}P_1(\delta)dz_{1,[0]} + [Q(\delta)]^{(a)}P_2(\delta)du_{[0]}$$

which immediately implies (from the identity of polynomials) that

$$d\dot{z}_{1,[0]} = A(\delta)dz_{1,[0]} + B(\delta)du_{[0]}$$

as well as that $dy_{[0]} = C(\delta)dz_{1,[0]}$. This ends the proof.

In this case, one may also investigate under which conditions the unobservable dynamics is independent of the control variable. The generalization of Theorem 5 is straightforward. The following result, whose proof is omitted since it follows the same lines as the proof of Theorem 5, holds:

Theorem 7 *Given system* (1), *with output* $y_{[0]}$ *characterized by the well defined relative degree* r, *there exist a regular bicausal static state feedback*

$$u_{[0]} = \hat{\alpha}(\mathbf{x}) + \hat{\beta}(\mathbf{x})v_{[0]},$$

and a bicausal change of coordinates

$$z_{[0]} = \begin{pmatrix} z_{1,[0]} \\ z_{2,[0]} \end{pmatrix} = \varphi(\mathbf{x})$$

with $z_{1,[0]} \in \mathbb{R}^r$, *such that in the new coordinates the closed-loop system reads as* (26) *with* $\eta_{2j} = 0$, $j \in [0, \ell]$, *if and only if the conditions of Theorem* 6 *are satisfied and additionally*
(a) given the differential representation (4), *associated to the system,* $\Delta(\mathbf{x}, \delta) = \text{span}_{\mathcal{K}(\delta]}\{g_1(\mathbf{x}, \delta)\}$ *is completely integrable, that is there exists an index* γ *such that* $\rho_\gamma - \rho_{\gamma-1} = 1$; *and*
(b) in the z *coordinates in which the system is* (26) *with* B_j *of the form* (28), *we have*

$$\eta_2(\mathbf{z},\delta) = \begin{pmatrix} m_1(\mathbf{z},\delta) \\ \vdots \\ m_{n-r}(\mathbf{z},\delta) \end{pmatrix} b(\delta),$$

where $\eta_2(\mathbf{z},\delta) = \sum_{j=0}^{\ell} \eta_{2j}\delta^j$ *and* $b(\delta) = \sum_{j=0}^{\ell} b_j\delta^j$.

5 Some Concluding Remarks

Various normal forms have been derived for single-output time-delay systems. Some of those normal forms include candidates for defining the so-called zero dynamics. However, as already underlined in [8], the extension of the concept of zero dynamics to nonlinear systems with delays is not straightforward. As a matter of fact, in the delay-free case, zeroing the output, which has by assumption relative degree r, leads to the computation of a residual dynamics, called zero dynamics, which has dimension $n-r$, and in the case of single-input single-output systems, is independent of the control u. As shown in [10], this property has some important consequences on the stabilization procedure, since if the zero dynamics is locally asymptotically stable, then one may choose the input in order to satisfy some requirements on the input-output behavior. In the case of delay systems, instead, the zero dynamics may have a dimension greater than $n-r$ as shown in [8]. However if the conditions of Theorem 4 (respectively Theorem 6) hold, then the system can be split, as already noted, into two subsystems, the first one representing the input-output behavior, and the second one influenced by the control and the state variable of the first subsystem. The stability properties of this last subsystem become important when designing feedback laws, to guarantee a certain required input-output behavior.

References

1. Califano, C., Marquez-Martinez, L., Moog, C.H.: Extended lie brackets for nonlinear time-delay systems. IEEE Trans. Autom. Control **56**(9), 2213–2218 (2011)
2. Califano, C., Marquez-Martinez, L., Moog, C.H.: Linearization of time-delay systems by input-output injection and output transformation. Automatica **49**(6), 1932–1940 (2013)
3. Califano, C., Monaco, S., Normand-Cyrot, D.: On the discrete-time normal form. IEEE Trans. Autom. Control **43**(11), 1654–1658 (1998)
4. Califano, C., Moog, C.H.: De l'existence de la forme normale pour les systemes non lineaires a retards. In: Proceedings of Conference Internationale Francophone d'Automatique, pp. 37–42 (2012)
5. Califano, C., Moog, C.H.: Coordinates transformations in nonlinear time-delay systems. In: Proceedings of the 53rd IEEE Conference on Decision and Control, pp. 475–480 (2014)
6. Garate-Garcia, A., Marquez-Martinez, L., Cuesta-Garcia, J., Garcia-Ramirez, E.: A computer algebra system for analysis and control of nonlinear time-delay systems. Adv. Eng. Softw. **65**, 138–148 (2013)

7. Germani, A., Manes, C., Pepe, P.: Linearization of input-output mapping for nonlinear delay systems via static state feedback. In: Proceedings of the IEEE-IMACS Conference on Computer Engineering in System Applications, pp. 599–602 (1996)
8. Germani, A., Manes, C., Pepe, P.: Input-output linearization with delay cancellation for nonlinear delay systems: the problem of the internal stability. Int. J. Robust Nonlinear Control **13**(9), 909–937 (2003)
9. Gu, K., Kharitonov, V., Chen, J.: Stability of Time-Delay Systems. Birkhauser, Boston (2003)
10. Isidori, A.: Nonlinear Control Systems, 3rd edn. Springer, New York (1995)
11. Marquez-Martinez, L., Moog, C.H., Velasco-Villa, M.: Observability and observers for nonlinear systems with time delay. Kybernetika **38**(4), 445–456 (2002)
12. Michiels, W., Niculescu, S-I.: Stability and Stabilization of Time-Delay Systems. An Eigenvalue-Based Approach. SIAM, Philadelphia (2007). (Advances in Design and Control, 12)
13. Oguchi, T.: A finite spectrum assignment for retarded non-linear systems and its solvability condition. Int. J. Control **80**(6), 898–907 (2007)
14. Pepe, P., Jiang, Z.-P.: A Lyapunov Krasovskii methodology for ISS and iISS of time-delay systems. Syst. Control Lett. **55**(12), 1006–1014 (2006)
15. Xia, X., Marquez-Martinez, L., Zagalak, P., Moog, C.H.: Analysis of nonlinear time-delay systems using modules over non-commutative rings. Automatica **38**(9), 1549–1555 (2002)
16. Zheng, G., Barbot, J., Boutat, D.: Identification of the delay parameter for nonlinear time-delay systems with unknown inputs. Automatica **49**(6), 1755–1760 (2013)

Compensating for Fatigue-Induced Time-Varying Delayed Muscle Response in Neuromuscular Electrical Stimulation Control

R. Downey, R. Kamalapurkar, N. Fischer and W. Dixon

Abstract Neuromuscular electrical stimulation (NMES), often called functional electrical stimulation (FES), is a prescribed treatment for various neuromuscular disorders. When applied to articulate a person's limb, the respective skeletal muscle groups are known to rapidly fatigue compared to muscles activated by the nervous system. Recent results have shown that muscles have a delayed response to electrical stimulation, and more recent results indicate that this delayed response increases as the muscle fatigues. A NMES control method is developed in this chapter as a means to compensate for the varying input delay for the uncertain nonlinear dynamics for the lower limb. Experimental results are provided to demonstrate the performance of the developed controller.

1 Introduction

Neuromuscular electrical stimulation (NMES) is the use of electric current to activate skeletal muscle, typically applied by electrodes placed on the surface of the skin. NMES is commonly used in rehabilitative settings where the goal is to increase muscle size, strength, and function [65, 73, 74, 80] and may also be used to produce functional tasks (e.g., standing, stepping, reaching, grasping, cycling) [19, 21, 36, 61, 64] where it is termed functional electrical stimulation (FES). Various feedback-based NMES controllers have been developed [1, 4, 18, 20, 22, 30, 31, 35, 43, 60, 68–70]; however, results that consider the muscle's delayed response to electrical

R. Downey (✉) · R. Kamalapurkar · N. Fischer · W. Dixon
Department of Mechanical and Aerospace Engineering,
231 MAE-A Building, 116250, 32611-6250 Gainesville, FL, USA
e-mail: DowneyRyanJ@gmail.com

R. Kamalapurkar
e-mail: RKamalapurkar@ufl.edu

N. Fischer
e-mail: nic.r.fischer@gmail.com

W. Dixon
e-mail: wdixon@ufl.edu

I. Karafyllis et al. (eds.), *Recent Results on Nonlinear Delay Control Systems*,
Advances in Delays and Dynamics 4, DOI 10.1007/978-3-319-18072-4_7

stimulation, known as electromechanical delay (EMD), are not common. EMD may lead to degraded performance and instability, motivating the need for control designs that compensate for delay.

Time-delays are prevalent in many engineering systems and have been well documented in literature (cf. [26, 62, 71, 78] and recent monographs such as [14, 25, 39, 45, 47, 57]). Few mathematical tools exist that can be used to develop controllers that compensate for input delays. Of these tools (namely Smith predictors [72], Artstein model reduction [2], and finite spectrum assignment [50]), few variations have been developed that can compensate for uncertain nonlinear systems. Methods which solve the input delay problem for uncertain nonlinear systems with known and unknown constant time-delays have been studied in [7, 9, 13, 33, 40, 42, 52, 67, 68].

Based on the development in [67] for general Euler-Lagrange systems, the results in [68] developed a model-free robust controller that enabled the leg shank of a healthy normal volunteer to track a desired angular trajectory about the knee with a uniformly ultimately bounded error despite a known constant EMD. Motivated by [68], a state predictive hybrid control approach was developed in [33] that considered sampled state measurements and limb constraints in addition to the known constant EMD. Assuming exact knowledge of the limb dynamics, the result in [33] yields exponential tracking. Although results such as [33, 68] provide insights on NMES in the presence of EMD, significant changes in muscle EMD were reported in [10, 11, 58, 81] during voluntary fatiguing exercises. Further, NMES is well known to induce significant fatigue in contrast to volitional contractions. There are a number of suggested causes of NMES-induced fatigue [5, 23] and efforts have been made to prevent or slow the onset of fatigue [6, 15–17, 24, 29, 34, 48, 49, 53, 56, 59, 75, 79, 82]. However, NMES-induced fatigue and the resulting time-varying EMD is of key importance when developing NMES controllers.

Control methods for time-varying input-delayed systems with linear plant models have been studied extensively. Discrete predictor-based techniques have been developed for linear systems with time-varying input delay in [46], where small bounded uncertainties in the system parameters, delay, and sampling instants are considered. A delayed feedback controller was developed in [83] for uncertain linear systems with a time-varying input delay based on a reduction method. A robust control method for uncertain linear systems with time-varying input delays was developed in [76], which combines a novel Lyapunov-Krasovskii (LK) functional and a neutral transformation to obtain sufficient conditions for closed-loop robustness. Predictive controllers have also been developed under the assumption that input delay systems can be represented by hyperbolic partial differential equations (cf. [25, 62] and references therein). This fact is exploited in [41] to design controllers for actuator delayed linear systems where the time delayed system is modeled as an ordinary differential equation (ODE)—partial differential equation (PDE) cascade using an infinite dimensional transformation where the non-delayed input acts at the PDE boundary. Linearized controllers have been developed for nonlinear systems [27, 44], but because the stability of the closed loop system is only valid within a region around the point of linearization, a complete nonlinear control solution to the time-varying input delay problem is still motivated.

A finite-time stabilizing controller developed in [32] compensates for time-varying input delays in nonlinear systems with triangular structures using an integrator backstepping technique. More recently, Bekiaris-Liberis and Krstic [3] extended the results in [40, 41] to develop a control method for forward complete nonlinear systems with time-varying input delays. Under the assumption of the existence of a stabilizing controller in the absence of the input delay, an invertible infinite dimensional backstepping transformation is used to yield an asymptotically stable system in the presence of a time-varying input delay. While these results have been successful for certain classes of nonlinear input-delayed systems, the applicability of the methods to general uncertain Euler-Lagrange dynamics is not clear. Motivated by this issue, [51] provided a transformation to convert an Euler-Lagrange system into a forward-complete system, but such a transformation requires exact model knowledge of the Euler-Lagrange dynamics; thus, the technique is not applicable when the system parameters are unknown or the dynamics are uncertain. This implies that methods developed for forward-complete systems with input delays may not be applicable to uncertain Euler-Lagrange systems.

In this chapter, a control method is developed to compensate for time-varying EMD during NMES where the muscle dynamics are uncertain, nonlinear, and contain additive disturbances, under the assumption that the known time-delay is bounded and slowly varying. As in our previous work, LK functionals are used to facilitate the design and analysis of a control method that can compensate for the input delay. Since the LK functionals contain time-varying delay terms, additional complexities are introduced into the analysis. Techniques used to compensate for the time-varying delay result in new sufficient control conditions that depend on the length of the delay as well as the rate of delay. The developed controller achieves semi-global uniformly ultimately bounded tracking despite the time-varying input delay, parametric uncertainties and additive bounded disturbances in the dynamics. Experiments are provided to examine the performance of the developed controller.

2 Knee Joint Dynamics

The knee-joint dynamics are modeled as [70]

$$M_I(\ddot{q}) + M_e(q) + M_g(q) + M_v(\dot{q}) + \bar{d} = \mu, \tag{1}$$

where $M_I : \mathbb{R} \to \mathbb{R}$ denotes the inertial effects of the shank-foot complex about the knee-joint, $M_e : \mathbb{R} \to \mathbb{R}$ denotes the elastic effects due to joint stiffness, $M_g : \mathbb{R} \to \mathbb{R}$ denotes the gravitational component, $M_v : \mathbb{R} \to \mathbb{R}$, denotes the viscous effects due to damping in the musculotendon complex, $\bar{d} \in \mathbb{R}$ is an unknown bounded time-varying disturbance from unmodeled dynamics, $\mu \in \mathbb{R}$ denotes the torque produced at the knee-joint due to stimulation, and $\tau \in \mathbb{R}$ denotes the EMD. The inertial and gravitational effects in (1) are modeled as

$$M_I(\ddot{q}) \triangleq J\ddot{q} \quad \text{and} \quad M_g(q) \triangleq mgl\sin(q), \tag{2}$$

where $J, m, g, l \in \mathbb{R}$ are positive constants and $q, \dot{q}, \ddot{q} \in \mathbb{R}$ denote the angular position, velocity, and acceleration of the shank about the knee-joint, respectively. The terms J, m, and l denote the unknown inertia of the combined shank and foot, the unknown combined mass of the shank and foot, and the unknown distance between the knee-joint and the lumped center of mass of the shank and foot, respectively, while g denotes the gravitational acceleration. The elastic and viscous effects are modeled as

$$M_e(q) \triangleq k_1(\exp(-k_2 q))(q - k_3), \tag{3}$$

where $k_1, k_2, k_3 \in \mathbb{R}$ are unknown positive constants and

$$M_v(\dot{q}) \triangleq -B_1\tanh(-B_2\dot{q}) + B_3\dot{q}, \tag{4}$$

where $B_1, B_2, B_3 \in \mathbb{R}$ are unknown positive constants. The subsequent development is based on the assumption that q and $\dot{q}$ are measurable outputs. Throughout the paper, a time-dependent delayed function is denoted as

$$(\cdot)_\tau(t) \triangleq \begin{cases} (\cdot)(t - \tau(t)) & t - \tau(t) \geq t_0 \\ 0 & t - \tau(t) < t_0 \end{cases}, \tag{5}$$

where $t_0 \in \mathbb{R}$ is the initial time. Additionally, let $\|\cdot\|$ denote the Euclidean norm of a vector.

The torque produced about the knee is controlled through muscle forces that are elicited by NMES/FES. For simplicity (and without loss of generality), the subsequent development focuses on producing knee torque through muscle tendon forces generated by electrical stimulation of the quadriceps. The total muscle force is a net sum of active force generated by contractile, elastic, and viscous elements [70]. The muscle force generated at the tendon is the projection of net sum of these elements along the line parallel to the tendon. The force development in the muscle is delayed due to the finite propagation time of chemical ions such as Ca^{2+} and action potential along the T-tubule system, cross-bridge formation between actin and myosin filaments, the subsequent tension development, and the stretching of the series elastic components by the contractile components in the muscle [10, 63, 84]. The EMD is influenced by factors such as fatigue, rate of force production, and types of muscle contractions. The total muscle force generated at the tendon, denoted by $F \in \mathbb{R}$, is defined as

$$F \triangleq \xi(q, \dot{q})u_\tau. \tag{6}$$

In (6), $\xi : \mathbb{R} \times \mathbb{R} \to \mathbb{R}$ denotes an unknown nonlinear function of the muscle length and velocity, and the applied voltage potential across the quadriceps muscle, denoted by $u_{\tau} \in \mathbb{R}$, includes the time-delay to capture the latency that is present between the application of voltage and force production [31, 66]. The introduction of the unknown nonlinear function ξ enables the muscle contraction to be considered under general dynamic conditions in the subsequent control development. The uncertain and unknown function ξ captures the dynamic characteristics of muscle recruitment (approximated by a continuously differentiable function), muscle force-length and muscle force-velocity relationships, and active and passive muscle characteristics [70]. The knee torque is related to the muscle tendon force as

$$\mu = \zeta(q)F, \tag{7}$$

where $\zeta : \mathbb{R} \to \mathbb{R}$ denotes a positive moment arm that changes with the extension and flexion of the leg [8, 38].

The model developed in (1)–(7) is used to examine the stability of the subsequently developed controller, but the controller does not explicitly depend on these models. The following assumptions and notations are used to facilitate the subsequent control development and stability analysis.

Assumption 1 The moment arm ζ is assumed to be a non-zero, positive, bounded function [8, 38] whose first two time derivatives exist and are bounded. Based on the empirical data [55, 77], the function ξ is assumed to be a non-zero, positive, and bounded function with bounded first and second time derivatives.

For notational brevity, an auxiliary non-zero unknown scalar function $\Omega : \mathbb{R} \times \mathbb{R} \to \mathbb{R}$ is defined as

$$\Omega\left(q, \dot{q}\right) \triangleq \zeta\left(q\right) \xi\left(q, \dot{q}\right). \tag{8}$$

From Assumption 1, the first and second time derivatives of Ω exist and are bounded.

Assumption 2 The unknown disturbance $\bar{d}$ is bounded and its first and second derivatives with respect to time exist and are bounded. Based on Assumption 1, the ratio $\bar{d}/\Omega(q, \dot{q})$ denoted by d is also bounded and its first and second derivatives with respect to time exist and are bounded.

Assumption 3 The time delay and its first and second time derivatives are bounded such that $0 \leq \tau(t) \leq \varphi_1$, $|\dot{\tau}(t)| < \varphi_2 < 1$, and $|\ddot{\tau}(t)| \leq \varphi_3$, for all $t \in \mathbb{R}_{\geq 0}$, where $\varphi_1, \varphi_2, \varphi_3 \in \mathbb{R}$ are known positive constants.

The implications of Assumptions 2 and 3 are that the disturbance and delay are sufficiently smooth and that the delay is sufficiently slow. The development of an input-delayed controller for arbitrarily fast time-varying delays remains an open problem for NMES with unknown dynamics.

Combining (1)–(8), the knee joint dynamics can be expressed as

$$M(q,\dot{q})\ddot{q} + f(q,\dot{q}) + d = u_{\tau}, \tag{9}$$

where $f : \mathbb{R} \times \mathbb{R} \to \mathbb{R} \triangleq \frac{M_e + M_g + M_v}{\Omega}$, $M : \mathbb{R} \times \mathbb{R} \to \mathbb{R} \triangleq \frac{J}{\Omega}$, and $d \triangleq \frac{\bar{d}}{\Omega}$. Based on Assumption 1, M can be bounded as

$$\underline{m} \leq |M(x_1, x_2)| \leq \overline{m} \tag{10}$$

for all $x_1, x_2 \in \mathbb{R}$, where $\underline{m}, \overline{m} \in \mathbb{R}$ are known positive constants.

3 Control Objective

The objective is to design a continuous controller that will ensure the generalized state q of the input-delayed system in (9) tracks a desired trajectory despite uncertainties and additive bounded disturbances in the dynamic model. To quantify the control objective, a tracking error denoted by $e \in \mathbb{R}$ is defined as

$$e \triangleq q_d - q, \tag{11}$$

where $q_d \in \mathbb{R}$ denotes the desired trajectory and is designed such that $q_d, \dot{q}_d, \ddot{q}_d \in \mathcal{L}_{\infty}$. To facilitate the subsequent analysis, a measurable auxiliary tracking error, denoted by $r \in \mathbb{R}$, is defined as

$$r \triangleq \dot{e} + \alpha e - B e_z, \tag{12}$$

where $\alpha \in \mathbb{R}^+$ is a known constant control gain, and $B \in \mathbb{R}^+$ is a known constant best guess estimate of M^{-1}. In (12), $e_z \in \mathbb{R}$ is an auxiliary signal containing the time-delay in the system, defined as

$$e_z \triangleq \int_{t-\tau(t)}^{t} u(\theta)\, \mathrm{d}\theta. \tag{13}$$

The error between B and M^{-1} is denoted by $\eta : \mathbb{R} \times \mathbb{R} \to \mathbb{R}$ and is defined as

$$\eta(q,\dot{q}) \triangleq B - \frac{1}{M(q,\dot{q})} \tag{14}$$

and satisfies

$$|\eta(x_1, x_2)| \leq \overline{\eta} \tag{15}$$

for all $x_1, x_2 \in \mathbb{R}$, where $\overline{\eta} \in \mathbb{R}^+$ is a known constant.

4 Control Development

The open-loop error system can be obtained by multiplying the time derivative of (12) by M and utilizing the expressions in (9), (11), (13) and (14) to yield

$$\begin{aligned} M(q,\dot{q})\dot{r} &= M(q,\dot{q})\ddot{q}_d + f(q,\dot{q}) + d + \alpha M(q,\dot{q})\dot{e} \\ &\quad - M(q,\dot{q})\eta(q,\dot{q})(u - u_\tau + u_\tau\dot{\tau}) - u - u_\tau\dot{\tau}. \end{aligned} \tag{16}$$

For notational brevity, the dependance of M and η on q and $\dot{q}$ is omitted henceforth. Based on the error system formulation in (12) and (13), the open-loop error system in (16) contains a delay-free control input. From (16) and the subsequent stability analysis, the control input is designed as

$$u = k_b r, \tag{17}$$

where $k_b \in \mathbb{R}$ is a known positive constant control gain. To facilitate the subsequent stability analysis, an auxiliary signal $N_d \in \mathbb{R}^n$ is defined as

$$N_d \triangleq M_d\ddot{q}_d + f_d, \tag{18}$$

where $M_d \triangleq M(q_d, \dot{q}_d)$ and $f_d \triangleq f(q_d, \dot{q}_d)$.

The closed-loop error system is obtained by adding and subtracting N_d and e to (16) and utilizing (12) and (17) to yield

$$\begin{aligned} M\dot{r} &= -\tfrac{1}{2}\dot{M}r + \chi + S - k_b r - k_b M\eta(r - r_\tau + r_\tau\dot{\tau}) \\ &\quad -k_b r_\tau\dot{\tau} - e, \end{aligned} \tag{19}$$

where the auxiliary terms $\chi, S \in \mathbb{R}^n$ are defined as

$$\begin{aligned} \chi &\triangleq \frac{1}{2}\dot{M}r + e + (M - M_d)\ddot{q}_d + f(q,\dot{q}) - f_d \\ &\quad + \alpha M(r - \alpha e + Be_z) \end{aligned} \tag{20}$$

$$S \triangleq N_d + d. \tag{21}$$

Using Assumption 2, the following inequality can be shown, based on the expression in (21)

$$\|S\| \leq \bar{s}, \tag{22}$$

where $\bar{s} \in \mathbb{R}^+$ is a known constant. The structure of (19) is motivated by the desire to segregate terms that can be upper bounded by state-dependent terms and terms that can be upper bounded by constants. Using the Mean Value Theorem, the expression in (20) can be upper bounded as

$$\|\chi\| \leq \rho\left(\|z\|\right)\|z\|, \tag{23}$$

where $\rho\left(\|z\|\right)$ is a positive and strictly increasing function and $z \in \mathbb{R}^3$ is defined as

$$z \triangleq \left[e\ r\ e_z\right]^T. \tag{24}$$

5 Stability Analysis

To facilitate the subsequent stability analysis, let $y \in \mathbb{R}^4$ be defined as

$$y \triangleq \left[e\ r\ \sqrt{P}\ \sqrt{Q}\right]^T, \tag{25}$$

where the signals $P, Q \in \mathbb{R}$ are defined as

$$P \triangleq \omega \int_{t-\tau(t)}^{t} \left(\int_{s}^{t} u^2\left(\theta\right) \mathrm{d}\theta \right) \mathrm{d}s, \tag{26}$$

$$Q \triangleq \frac{k_b\left(2\overline{m}\,\overline{\eta} + \varphi_2\right)}{2\left(1-\dot{\tau}\right)} \int_{t-\tau(t)}^{t} \|r\left(\theta\right)\|^2 \mathrm{d}\theta, \tag{27}$$

where $\omega \in \mathbb{R}$ is a known positive constant. Let the auxiliary constants σ, δ, and γ be defined as

$$\sigma = \min\left\{\frac{\alpha}{4}, \frac{k_b}{8}, \frac{\omega\left(1-\varphi_2\right)}{8\varphi_1}\right\}, \tag{28}$$

$$\delta = \min\left\{\frac{\alpha}{4}, \frac{k_b}{8}, \frac{\left(1-\varphi_2\right)}{4\varphi_1}, \frac{\omega k_b\left(1-\varphi_2\right)^2}{\left(4\overline{m}\,\overline{\eta} + 2\varphi_2\right)}\right\}, \tag{29}$$

$$\gamma = \max\left\{1, \sqrt{\frac{4k_b\varphi_1}{2\overline{m}\,\overline{\eta} + \varphi_2}}\right\}, \tag{30}$$

and let

$$\mathscr{D} \triangleq \left\{x \in \mathbb{R}^4 \mid \|x\| \leq \frac{1}{\gamma} \inf\left\{\rho^{-1}\left[\sqrt{k_b\sigma}, \infty\right)\right\}\right\},$$

$$\mathscr{S}_{\mathscr{D}} \triangleq \left\{x \in \mathscr{D} \mid \|x\| < \sqrt{\frac{\phi_1}{\gamma^2\phi_2}} \inf\left\{\rho^{-1}\left[\sqrt{k_b\sigma}, \infty\right)\right\}\right\}.$$

Theorem 1 *Given the dynamics in* (9)*, provided the control gains are selected based on the sufficient conditions*

$$\alpha > B, \quad \omega > \frac{4B\varphi_1}{(1-\varphi_2)}, \tag{31}$$

and the input delay τ*, its time derivatives* $\dot{\tau}$ *and* $\ddot{\tau}$*, and the inertia estimate mismatch* η *are small enough so that there exists a positive gain* $k_b \in \mathbb{R}$ *that satisfies*

$$k_b > \frac{2\varphi_3\left(2\overline{m}\,\overline{\eta} + \varphi_2\right)}{\omega\left(1-\varphi_2\right)^3}, \tag{32}$$

$$\varphi_1 < \frac{1}{\omega k_b}\left(\frac{1}{4} - \frac{\left(2\overline{\eta}\overline{m} + \varphi_2\right)}{1-\varphi_2}\right), \tag{33}$$

$$\frac{\phi_2 \bar{s}^2 \gamma^2}{\phi_1 \delta k_b} < \left(\inf\left\{\rho^{-1}\left[\sqrt{k_b\sigma}, \infty\right)\right\}\right)^2, \tag{34}$$

the controller in (17) *ensures uniformly ultimately bounded tracking in the sense that*

$$\limsup_{t\to\infty} \|y(t)\| \le \sqrt{\frac{\phi_2 \bar{s}^2}{\phi_1 \delta k_b}},$$

for all $y(t_0) \in \mathscr{S}_{\mathscr{D}}$*, and the convergence to the ultimate bound is exponential.*

Proof Let $V_L : \mathscr{D} \times [0, \infty) \to \mathbb{R}$ be a continuously differentiable, positive-definite function defined as

$$V_L \triangleq \frac{1}{2}e^T e + \frac{1}{2}r^T M r + P + Q, \tag{35}$$

which can be bounded as

$$\phi_1 \|y\|^2 \le V_L \le \phi_2 \|y\|^2 \tag{36}$$

where the constants $\phi_1, \phi_2 \in \mathbb{R}$ are defined as

$$\phi_1 \triangleq \frac{1}{2}\min\left[\underline{m}, 1\right], \quad \phi_2 \triangleq \max\left[\frac{1}{2}\overline{m}, 1\right]. \tag{37}$$

After utilizing (12) and (19), applying the Leibniz Rule to determine the time derivative of (26) and (27), and by canceling similar terms, the time derivative of (35) can be expressed as

$$\begin{aligned}\dot{V}_L = &-\alpha e^2 - k_b r^2 + B e e_z + r\chi + rS - k_b \eta M r^2 \\ &+ k_b\left(1-\dot{\tau}\right)\eta M r r_\tau - k_b r r_\tau \dot{\tau} + \omega\tau k_b^2 r^2 \\ &- \omega\left(1-\dot{\tau}\right)\int_{t-\tau(t)}^{t} u^2(\theta)\, d\theta\end{aligned}$$

$$+\frac{\ddot{\tau}\left(2\overline{m}\,\overline{\eta}+\varphi_2\right)}{2k_b\left(1-\dot{\tau}\right)^2}\int_{t-\tau(t)}^{t}u^2\left(\theta\right)\mathrm{d}\theta$$
$$+\frac{k_b\left(2\overline{m}\,\overline{\eta}+\varphi_2\right)}{2\left(1-\dot{\tau}\right)}r^2-\frac{k_b\left(2\overline{m}\,\overline{\eta}+\varphi_2\right)}{2}r_\tau^2. \tag{38}$$

By utilizing Young's inequality, Assumption 3, (15), (17), (22), and (23), (38) can be expanded, regrouped and upper bounded as

$$\begin{aligned}\dot{V}_L \leq& -\alpha e^2-k_b r^2+\rho\left(\|z\|\right)\|z\|\,|r|+|r|\,\overline{s}\\ &+k_b\frac{\left(2\overline{\eta}\overline{m}+\varphi_2\right)}{1-\varphi_2}r^2+\omega\varphi_1 k_b^2 r^2\\ &+\frac{B}{2}e^2+\frac{B}{2}e_z^2-\omega\left(1-\varphi_2\right)\int_{t-\tau(t)}^{t}u^2\left(\theta\right)d\theta\\ &+\frac{\varphi_3\left(2\overline{m}\,\overline{\eta}+\varphi_2\right)}{2k_b\left(1-\varphi_2\right)^2}\int_{t-\tau(t)}^{t}u^2\left(\theta\right)\mathrm{d}\theta.\end{aligned} \tag{39}$$

Utilizing the Cauchy-Schwartz inequality and (13) yields

$$\|e_z\|^2\leq\tau\int_{t-\tau(t)}^{t}\|u\left(\theta\right)\|^2\mathrm{d}\theta. \tag{40}$$

Using (40) and the inequality [67]

$$\int_{t-\tau(t)}^{t}\left(\int_{s}^{t}\|u\left(\theta\right)\|^2\mathrm{d}\theta\right)\mathrm{d}s\leq\tau\int_{t-\tau(t)}^{t}\|u\left(\theta\right)\|^2\mathrm{d}\theta$$

the expression in (39) can be bounded as

$$\begin{aligned}\dot{V}_L \leq& -\frac{\alpha}{4}e^2-\frac{k_b}{8}r^2-\frac{\omega\left(1-\varphi_2\right)}{8\varphi_1}e_z^2+\frac{\rho^2\left(\|z\|\right)}{k_b}\|z\|^2\\ &-\frac{\alpha}{4}e^2-\frac{k_b}{8}r^2-\frac{\left(1-\varphi_2\right)}{4\varphi_1}P-\frac{\omega k_b\left(1-\varphi_2\right)^2}{\left(4\overline{m}\,\overline{\eta}+2\varphi_2\right)}Q+\frac{\overline{s}^2}{k_b}\\ &-\left(\frac{\omega\left(1-\varphi_2\right)}{4}-\frac{\varphi_3\left(2\overline{m}\,\overline{\eta}+\varphi_2\right)}{2k_b\left(1-\varphi_2\right)^2}\right)\int_{t-\tau(t)}^{t}u^2\left(\theta\right)\mathrm{d}\theta\\ &-\left(\frac{\alpha}{2}-\frac{B}{2}\right)e^2-\left(\frac{\omega\left(1-\varphi_2\right)}{8\varphi_1}-\frac{B}{2}\right)e_z^2\\ &-k_b\left(\frac{1}{4}-\frac{\left(2\overline{\eta}\overline{m}+\varphi_2\right)}{1-\varphi_2}-\omega\varphi_1 k_b\right)r^2.\end{aligned} \tag{41}$$

Provided the sufficient gain conditions in (31)–(33) are satisfied, the inequality $\|z\|\leq\gamma\|y\|$ can be used to bound the expression in (41) as

$$\dot{V}_L \leq -\left(\sigma - \frac{\rho^2\left(\gamma \left\|y\right\|\right)}{k_b}\right)\left\|z\right\|^2 - 2\delta\left\|y\right\|^2 + \frac{\bar{s}^2}{k_b},$$
$$\leq -\delta\left\|y\right\|^2,\ \forall y \in \mathscr{D},\ \left\|y\right\| > \frac{\bar{s}}{\sqrt{\delta k_b}}, \tag{42}$$

where σ and δ were introduced in (28) and (29), respectively. Using (34)–(37) and (42), Theorem 4.18 in [37] can be invoked to conclude that

$$y\left(t\right) \in \mathscr{D},\ \forall t \in \left[t_0, \infty\right)\ \forall y\left(t_0\right) \in \mathscr{S}_{\mathscr{D}},$$
$$\limsup_{t\to\infty} \left\|y(t)\right\| \leq \sqrt{\frac{\phi_2 \bar{s}^2}{2\phi_1 \delta k_b}},\ \forall y\left(t_0\right) \in \mathscr{S}_{\mathscr{D}}. \tag{43}$$

Thus, $e, r, P, Q \in \mathscr{L}_\infty$, hence, using $\left\|z\right\| \leq \gamma\left\|y\right\|$, $\left\|z\right\| \in \mathscr{L}_\infty$, and hence, $e_u \in \mathscr{L}_\infty$. The closed-loop error system can be used to conclude that the remaining signals are bounded.

Using (37), for all $y\left(t_0\right) \in \mathscr{S}_{\mathscr{D}}$, the Lyapunov derivative in (42) can be bounded as

$$\dot{V}_L \leq -\frac{2\delta}{\phi_2} V_L + \frac{\bar{s}^2}{k_b}. \tag{44}$$

Solving the differential inequality, (44), the Lyapunov function can be bounded as

$$V_L\left(y\left(t\right), t\right) \leq \left(V_L\left(y\left(t_0\right), t_0\right) - \frac{\phi_2 \bar{s}^2}{2\delta k_b}\right) e^{-\frac{2\delta\left(t_0 - t\right)}{\phi_2}} + \frac{\phi_2 \bar{s}^2}{2\delta k_b}. \tag{45}$$

Substituting for the Lyapunov function in (45) using the bound (37) yields

$$\left\|y\left(t\right)\right\|^2 \leq \left(\frac{\phi_2\left\|y\left(t_0\right)\right\|^2}{\phi_1} - \frac{\phi_2 \bar{s}^2}{2\phi_1 \delta k_b}\right) e^{-\frac{2\delta\left(t_0 - t\right)}{\phi_2}} + \frac{\phi_2 \bar{s}^2}{2\phi_1 \delta k_b},$$

establishing exponential convergence of the tracking error to the ultimate bound.

6 Experiments

One able-bodied male (age 26) participated in the study to examine the performance of the developed delay compensation controller. The electrical stimulation responses of healthy subjects have been reported to be similar to those of paraplegic subjects [12, 28, 31, 43]. Therefore, a healthy subject was used as a substitute for affected individuals. Prior to participation, written informed consent was obtained from the individual, as approved by the institutional review board at the University of Florida.

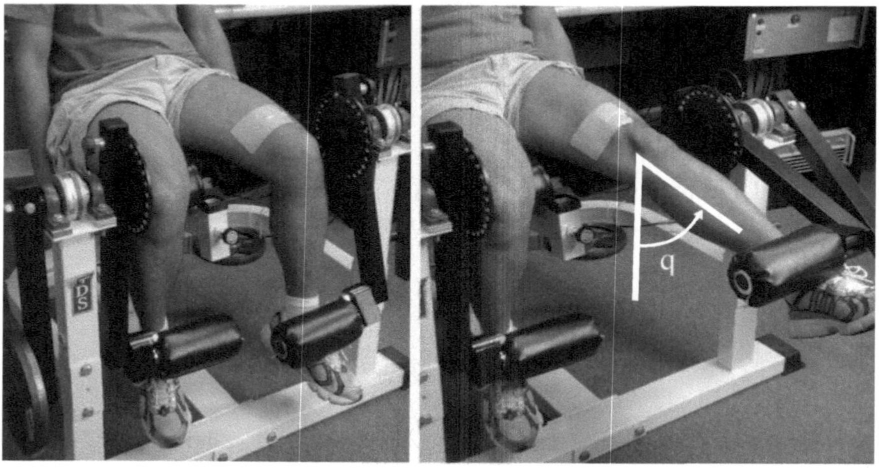

Fig. 1 The experimental setup includes a leg extension exercise machine, encoders to determine the subject's leg angle $q(t)$, and a computer to control stimulation and gather data

All testing was performed using an apparatus that consists of a custom computer-controlled stimulation circuit and a leg extension machine (LEM; Fig. 1). The LEM includes optical encoders to measure the angle between the femur and the tibia. The LEM allows seating adjustments to ensure that the rotation of the knee is about the encoder axis and a mechanical stop was used to prevent hyperextension. A computer was used to collect data from the encoders and execute the closed-loop control algorithms. Voltage was applied with a pair of 3″ by 5″ oval PALS® surface electrodes placed over the distal–medial and proximal–lateral portions of the quadriceps femoris muscle group. Surface electrodes for the study were provided compliments of Axelgaard Manufacturing Co., Ltd. To better understand the effect of the delay compensation term Be_z in (12) the developed proportional-derivative delay compensation (PDDC) controller was compared to a controller of similar form that did not include the delay compensation term. In other words, the PDDC controller was compared to a proportional-derivative (PD) controller. During testing, the individual was instructed to relax as much as possible and to allow the stimulation to control the limb motion (i.e., the subject was not supposed to influence the leg motion voluntarily and was not allowed to see the desired trajectory). The desired trajectory was a sinusoid ranging from 10° to 45° with a period of 2.5 s.

The desired trajectory also included a smooth step function (0°–30° lasting 1 s) at equally spaced intervals to measure the EMD during the course of the experiment. The EMD was calculated as the difference between the time at the onset of stimulation and the time at which the leg angle increased by 0.005 rad. Measurement of the EMD during the PD controller trials allowed for modeling of the delay during the PDDC controller trials. As such, the PD trials were always performed prior to the PDDC trials. The EMD measurements were curve-fit as a function of time with a model of the form $\tau(t) = a\exp(bt) + c\exp(dt)$ and this model was used to calculate the control term e_z in (12).

Eight trials were done where the PD and PDDC controls were examined twice for each leg. The PD control was first examined where the control gains were tuned in pretrial tests to reduce tracking error. After gain tuning, two PD trials were completed where 5 min of rest was allowed between the trials and the same control gains were used in each trial. The individual was then allowed to rest for 1 h before two PDDC trials were completed, where the same gain tuning and resting procedures were followed as in the PD trials. One-way, paired t-tests were used to compute statistical differences between the two controls in terms of the measured RMS and peak errors. The significance level was set at $\alpha = 0.05$ for the t-tests.

7 Results

The RMS error and peak error were calculated during steady state and are in Tables 1 and 2, respectively. Steady state was defined by removing the first period of the desired sinusoid to be tracked (i.e., the transient period). T-tests indicate that the mean RMS and peak errors are statistically less for the PDDC controller with P-values of 0.001 and 0.009, respectively. An example trial run which compares the PD and PDDC controllers is shown in Fig. 2 with a more detailed view in Fig. 3. Example measurements and curve fit of the time-varying EMD are in Fig. 4.

Table 1 Steady state RMS error (degrees) of the two examined controllers

Leg–Trial	PD	PDDC
Left–1st	7.22	4.61
Left–2nd	7.84	4.10
Right–1st	7.40	3.01
Right–2nd	6.91	3.70
Mean	7.34	3.86[a]
SD	0.39	0.68

PD indicates a proportional-derivative controller while PDDC indicates the developed proportional-derivative controller with delay compensation
[a] Indicates statistically significant difference in the means (P-value = 0.001)

Table 2 Peak steady state errors (degrees) of the two examined controllers

Leg–Trial	PD	PDDC
Left–1st	18.78	10.17
Left–2nd	24.30	9.68
Right–1st	18.93	14.22
Right–2nd	19.32	9.40
Mean	20.33	10.86[a]
SD	2.65	2.26

PD indicates a proportional-derivative controller while PDDC indicates the developed proportional-derivative controller with delay compensation
[a] Indicates statistically significant difference in the means (P-value = 0.009)

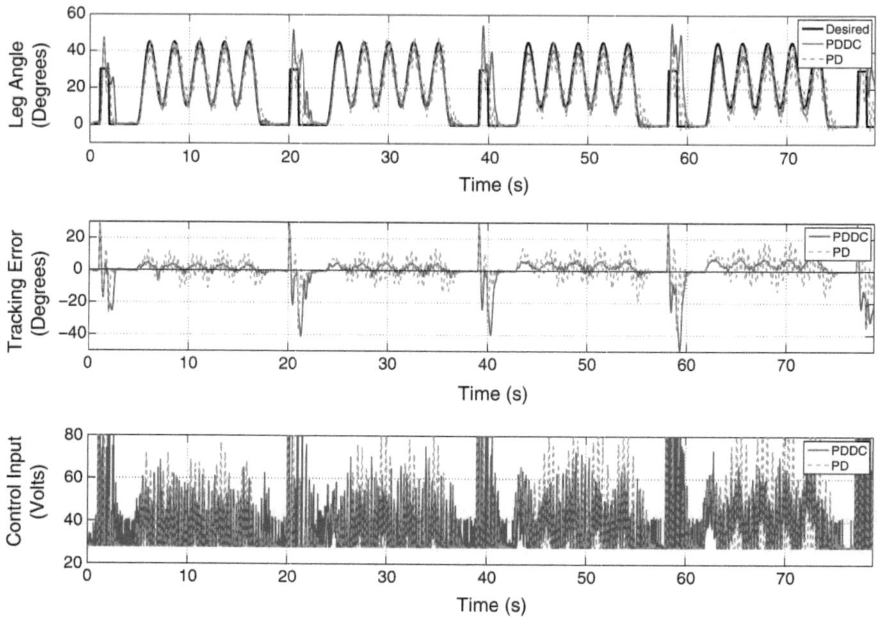

Fig. 2 Example tracking performance of the PD and PDDC controllers

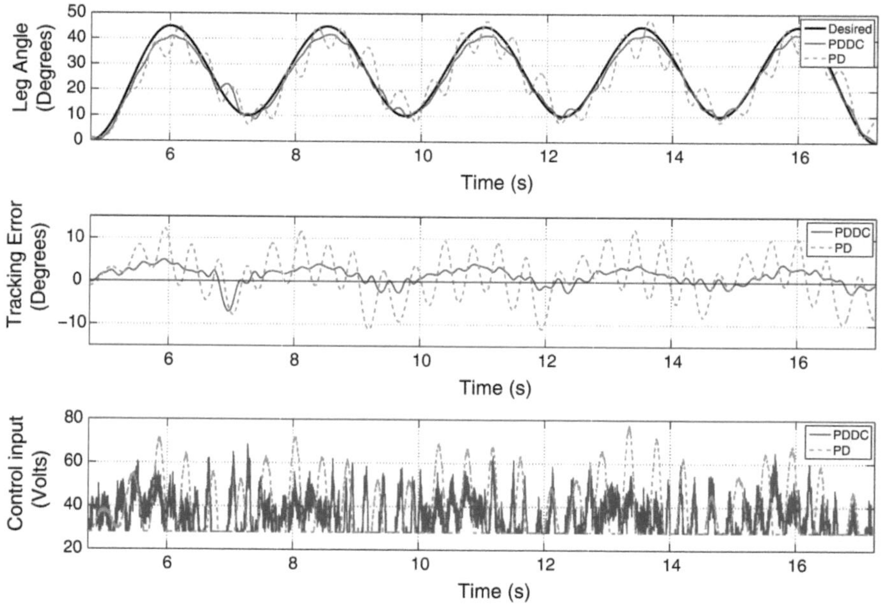

Fig. 3 Cropped example of the tracking performance of the PD and PDDC controllers

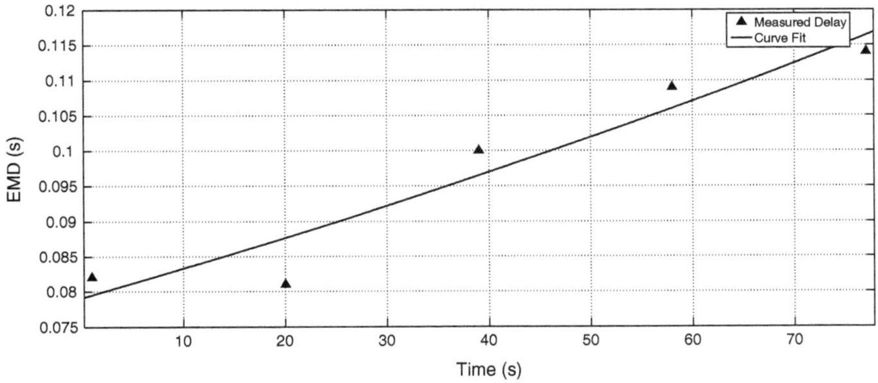

Fig. 4 Example of the time-varying EMD and curve fit

8 Discussion

The experimental results indicate that our PDDC controller results in statistically improved tracking performance compared to the PD controller. Overall, the PDDC control resulted a 47 % reduction in RMS error and a 46 % reduction in peak error; see Tables 1 and 2. While delay compensation was able to improve the tracking performance, there are some limitations to the controller. The primary limitation to the PDDC controller is that it requires the time-varying EMD to be known. In the experiments, an estimate of the time-varying EMD was computed by periodically placing step functions throughout the desired trajectory. If the controller is implemented during a functional activity (e.g., walking or cycling), then an estimate of the delay would need to be calculated without interrupting the desired trajectory. One potential solution may be to use electromyography to estimate muscle fatigue as stimulus pulses are delivered [54] and subsequently estimate the EMD as a function of fatigue. Another potential solution is to develop a controller which does not require knowledge of the time-varying EMD. RISE-based and neural network-based NMES controls have demonstrated better performance without considering the EMD [69, 70] for short duration experiments; however, it is unclear how these controllers can be modified to compensate for EMD. Future efforts should examine methods to estimate the time-varying EMD, methods to compensate for known time-varying EMD in alternative control structures, and methods to compensate for unknown time-varying EMD.

References

1. Abbas, J., Chizeck, H.: Feedback control of coronal plane hip angle in paraplegic subjects using functional neuromuscular stimulation. IEEE Trans. Biomed. Eng. **38**(7), 687–698 (1991)
2. Artstein, Z.: Linear systems with delayed controls: A reduction. IEEE Trans. Autom. Control **27**(4), 869–879 (1982)

3. Bekiaris-Liberis, N., Krstic, M.: Compensation of time-varying input delay for nonlinear systems. In: Proceedings of the Mediterranean Conference on Control and Automation, pp. 1040–1045 (2011)
4. Bellman, M., Cheng, T.-H., Downey, R., Dixon, W.: Stationary cycling induced by switched functional electrical stimulation control. In: Proceedings of the American Control Conference, pp. 4802–4809 (2014)
5. Bickel, C., Gregory, C., Dean, J.: Motor unit recruitment during neuromuscular electrical stimulation: a critical appraisal. Eur. J. Appl. Physiol. **111**(10), 2399–2407 (2011)
6. Binder-Macleod, S., Lee, S., Baadte, S.: Reduction of the fatigue induced force decline in human skeletal muscle by optimized stimulation trains. Archives Phys. Med. Rehabil. **78**(10), 1129–1137 (1997)
7. Bresch-Pietri, D., Krstic, M.: Adaptive trajectory tracking despite unknown input delay and plant parameters. Automatica **45**(9), 2074–2081 (2009)
8. Buford W. Jr., Ivey F. Jr., Malone, J., Patterson, R., Peare, G., Nguyen, D., Stewart, A.: Muscle balance at the knee—moment arms for the normal knee and the ACL—minus kne. IEEE Trans. Rehabitat. Eng. **5**(4), 367–379 (1997)
9. Castillo-Toledo, B., Di Gennaro, S., Castro, G.: Stability analisys for a class of sampled nonlinear systems with time-delay. In: Proceedings of the IEEE Conference on Decision and Control, pp. 1575–1580 (2010)
10. Cavanagh, P., Komi, P.: Electromechanical delay in human skeletal muscle under concentric and eccentric contractions. Eur. J. Appl. Physiol. Occup. Physiol. **42**(3), 159–163 (1979)
11. Ce, E., Rampichini, S., Agnello, L., Veicsteinas, A., Esposito, F.: Effects of temperature and fatigue on the electromechanical delay components. Muscle Nerve **47**(4), 566–576 (2013)
12. Chang, G.-C., Lub, J.-J., Liao, G.-D., Lai, J.-S., Cheng, C.-K., Kuo, B.-L., Kuo, T.-S.: A neuro-control system for the knee joint position control with quadriceps stimulation. IEEE Trans. Rehabitat. Eng. **5**(1), 2–11 (1997)
13. Chen, B., Liu, X., Tong, S.: Robust fuzzy control of nonlinear systems with input delay. Chaos, Solitons, Fractals **37**(3), 894–901 (2008)
14. Chiasson, J., Loiseau, J.: Applications of Time Delay Systems. Series Lecture Notes in Control and Information Sciences. Springer, New York (2007)
15. Doucet, B., Griffin, L.: Maximal versus submaximal intensity stimulation with variable patterns. Muscle Nerve **37**(6), 770–777 (2008)
16. Downey, R., Ambrosini, E., Ferrante, S., Pedrocchi, A., Dixon, W., Ferrigno, G.: Asynchronous stimulation with an electrode array reduces muscle fatigue during FES cycling. In: Proceedings of the International Functional Electrical Stimulation Society Conference, pp. 154–157 (2012)
17. Downey, R., Bellman, M., Sharma, N., Wang, Q., Gregory, C., Dixon, W.: A novel modulation strategy to increase stimulation duration in neuromuscular electrical stimulation. Muscle Nerve **44**(3), 382–387 (2011)
18. Downey, R., Cheng, T.-H., Dixon, W.: Tracking control of a human limb during asynchronous neuromuscular electrical stimulation. In: Proceedings of the IEEE Conference on Decision and Control, pp. 139–144 (2013)
19. Duffell, L., Donaldson, N., Newham, D.: Power output during functional electrically stimulated cycling in trained spinal cord injured people. Neuromodulation: technology at the neural. Interface **13**(1), 50–57 (2010)
20. Ebrahimpour, M., Erfanian, A.: Comments on sliding mode closed-loop control of FES: controlling the shank movement. IEEE Trans. Biomed. Eng. **55**(12), 2842 (2008)
21. Ferrante, S., Ambrosini, E., Ferrigno, G., Pedrocchi, A.: Biomimetic NMES controller for arm movements supported by a passive exoskeleton. In: Proceedings of the EMBC Annual International Conference, pp. 1888–1891 (2012)
22. Gollee, H., Hunt, K., Wood, D.: New results in feedback control of unsupported standing in paraplegia. IEEE Trans. Neural Syst. Rehabil. Eng. **12**(1), 73–91 (2004)
23. Gregory, C., Bickel, C.: Recruitment patterns in human skeletal muscle during electrical stimulation. Phys. Therapy **85**(4), 358–364 (2005)

24. Gregory, C., Dixon, W., Bickel, C.: Impact of varying pulse frequency and duration on muscle torque production and fatigue. Muscle Nerve **35**(4), 504–509 (2007)
25. Gu, K., Kharitonov, V., Chen, J.: Stability of Time-Delay Systems. Birkhauser, Boston (2003)
26. Gu, K., Niculescu, S.-I.: Survey on recent results in the stability and control of time-delay systems. J. Dyn. Syst. Meas. Control **125**(2), 158–165 (2003)
27. Guo, L.: H infinity output feedback control for delay systems with nonlinear and parametric uncertainties. IEEE Proc. Control Theory Appl. **149**(3), 226–236 (2002)
28. Hausdorff, J., Durfee, W.: Open-loop position control of the knee joint using electrical stimulation of the quadriceps and hamstrings. Med. Biol. Eng. Comput. **29**(3), 269–280 (1991)
29. Hughes, A., Guo, L., De Weerth, S.: Interleaved multichannel epimysial stimulation for eliciting smooth contraction of muscle with reduced fatigue. In: Proceedings of the Annual International Conference of the IEEE Engineering in Medicine and Biology Society, pp. 6226–6229 (2010)
30. Hunt, K., Munih, M.: Feedback control of unsupported standing in paraplegia-part 1: optimal control approach. IEEE Trans. Rehabil. Eng. **5**(4), 331–340 (1997)
31. Jezernik, S., Wassink, R., Keller, T.: Sliding mode closed-loop control of FES: controlling the shank movement. IEEE Trans. Biomed. Eng. **51**(2), 263–272 (2004)
32. Karafyllis, I.: Finite-time global stabilization by means of time-varying distributed delay feedback. SIAM J. Control Optim. **45**(1), 320–342 (2006)
33. Karafyllis, I., Malisoff, M., de Queiroz, M., Krstic, M., Yang, R.: Predictor-based tracking for neuromuscular electrical stimulation. Intern. J. Robust Nonlinear Control, to appear. doi:10.1002/rnc.3211
34. Karu, Z., Durfee, W., Barzilai, A.: Reducing muscle fatigue in FES applications by stimulating with N-let pulse trains. IEEE Trans. Biomed. Eng. **42**(8), 809–817 (1995)
35. Kawai, H., Bellman, M., Downey, R., Dixon, W.: Tracking control for FES-cycling based on force direction efficiency with antagonistic bi-articular muscles. In: Proceedings of the American Control Conference, pp. 5484–5489 (2014)
36. Kesar, T., Perumal, R., Reisman, D., Jancosko, A., Rudolph, K., Higginson, J., Binder-Macleod, S.: Functional electrical stimulation of ankle plantarflexor and dorsiflexor muscles: effects on poststroke gait. Stroke **40**(12), 3821–3827 (2009)
37. Khalil, H.: Nonlinear Systems, 3rd edn. Prentice-Hall, Englewood Cliffs, NJ (2002)
38. Krevolin, J., Pandy, M., Pearce, J.: Moment arm of the patellar tendon in the human knee. J. Biomech. **37**(5), 785–788 (2004)
39. Krstic, M.: Delay Compensation for Nonlinear, Adaptive, and PDE Systems. Springer, New York (2009)
40. Krstic, M.: Input delay compensation for forward complete and strict-feedforward nonlinear systems. IEEE Trans. Autom. Control **55**(2), 287–303 (2010)
41. Krstic, M.: Lyapunov stability of linear predictor feedback for time-varying input delay. IEEE Trans. Autom. Control **55**(2), 554–559 (2010)
42. Krstic, M., Smyshlyaev, A.: Backstepping boundary control for first-order hyperbolic PDEs and application to systems with actuator and sensor delays. Syst. Control Lett. **57**(9), 750–758 (2008)
43. Kurosawa, K., Futami, R., Watanabe, T., Hoshimiya, N.: Joint angle control by FES using a feedback error learning controller. IEEE Trans. Neural Syst. Rehabil. Eng. **13**(3), 359–371 (2005)
44. Li, W., Dong, Y., Wang, X.: Robust H^{∞} control of uncertain nonlinear systems with state and input time-varying delays. In: Proceedings of the Chinese Control and Decision Conference, pp. 317–321 (2010)
45. Loiseau, J., Michiels, W., Niculescu, S.-I., Sipahi, R. (eds.): Topics in Time Delay Systems: Analysis, Algorithms, and Control. Springer, New York (2009)
46. Lozano, R., Castillo, P., Garcia, P., Dzul, A.: Robust prediction-based control for unstable delay systems: application to the yaw control of a mini-helicopter. Automatica **40**(4), 603–612 (2004)
47. Mahmoud, M.: Robust Control and Filtering for Time-Delay Systems. Marcel Dekker, New York (2000)

48. Malesevic, N., Popovic, L., Schwirtlich, L., Popovic, D.: Distributed low-frequency functional electrical stimulation delays muscle fatigue compared to conventional stimulation. Muscle Nerve **42**(4), 556–562 (2010)
49. Maneski, L., Malesevic, N., Savic, A., Keller, T., Popovic, D.: Surface-distributed low-frequency asynchronous stimulation delays fatigue of stimulated muscles. Muscle Nerve **48**(6), 930–937 (2013)
50. Manitius, A., Olbrot, A.: Finite spectrum assignment problem for systems with delays. IEEE Trans. Autom. Control **24**(4), 541–552 (1979)
51. Mazenc, F., Bowong, S.: Tracking trajectories of the cart-pendulum system. Automatica **39**(4), 677–684 (2003)
52. Mazenc, F., Mondie, S., Francisco, R.: Global asymptotic stabilization of feedforward systems with delay in the input. IEEE Trans. Autom. Control **49**(5), 844–850 (2004)
53. McDonnall, D., Clark, G., Normann, R.: Interleaved, multisite electrical stimulation of cat sciatic nerve produces fatigue-resistant, ripple-free motor responses. IEEE Trans. Neural Syst. Rehabil. Eng. **12**(2), 208–215 (2004)
54. Mizrahi, J., Levy, M., Ring, H., Isakov, E., Liberson, A.: EMG as an indicator of fatigue in isometrically FES-activated paralyzed muscles. IEEE Trans. Rehabitat. Eng. **2**(2), 57–65 (1994)
55. Nathan, R., Tavi, M.: The influence of stimulation pulse frequency on the generation of joint moments in the upper limb. IEEE Trans. Biomed. Eng. **37**(3), 317–322 (1990)
56. Nguyen, R., Masani, K., Micera, S., Morari, M., Popovic, M.: Spatially distributed sequential stimulation reduces fatigue in paralyzed triceps surae muscles: a case study. Artif. Organs **35**(12), 1174–1180 (2011)
57. Niculescu, S.-I., Gu, K.: Advances in Time-Delay Systems. Springer, New York (2004)
58. Paasuke, M., Ereline, J., Gapeveva, H.: Neuromuscular fatigue during repeated exhaustive submaximal static contractions of knee extensor muscles in endurance-trained, power-trained and untrained men. Acta Physiologica Scandinavica **166**(4), 319–326 (1999)
59. Popovic, L., Malesevic, N.: Muscle fatigue of quadriceps in paraplegics: Comparison between single vs. multi-pad electrode surface stimulation. In: Proceedings of the Annual International Conference of the IEEE Engineering in Medicine and Biology Society, pp. 6785–6788 (2009)
60. Previdi, F., Ferrarin, M., Savaresi, S., Bittanti, S.: Gain scheduling control of functional electrical stimulation for assisted standing up and sitting down in paraplegia: a simulation study. Intern. J. Adapt. Control Signal Process. **19**(5), 327–338 (2005)
61. Prochazka, A., Gauthier, M., Wieler, M., Kenwell, Z.: The bionic glove: an electrical stimulator garment that provides controlled grasp and hand opening in quadriplegia. Archives Phys. Med. Rehabil. **78**(6), 608–614 (1997)
62. Richard, J.-P.: Time-delay systems: an overview of some recent advances and open problems. Automatica **39**(10), 1667–1694 (2003)
63. Riener, R., Fuhr, T.: Patient-driven control of FES-supported standing up: a simulation study. IEEE Trans. Rehabitat. Eng. **6**(2), 113–124 (1998)
64. Sabut, S., Lenka, P., Kumar, R., Mahadevappa, M.: Effect of functional electrical stimulation on the effort and walking speed, surface electromyography activity, and metabolic responses in stroke subjects. J. Electromyogr. Kinesiol. **20**(6), 1170–1177 (2010)
65. Sabut, S., Sikdar, C., Mondal, R., Kumar, R., Mahadevappa, M.: Restoration of gait and motor recovery by functional electrical stimulation therapy in persons with stroke. Disabil. Rehabil. **32**(19), 1594–1603 (2010)
66. Schauer, T., Negard, N., Previdi, F., Hunt, K., Fraser, M., Ferchland, E., Raisch, J.: Online identification and nonlinear control of the electrically stimulated quadriceps muscle. Control Eng. Pract. **13**(9), 1207–1219 (2005)
67. Sharma, N., Bhasin, S., Wang, Q., Dixon, W.: Predictor-based control for an uncertain Euler-Lagrange system with input delay. Automatica **47**(11), 2332–2342 (2011)
68. Sharma, M., Gregory, C., Dixon, W.: Predictor-based compensation for electromechanical delay during neuromuscular electrical stimulation. IEEE Trans. Neural Syst. Rehabil. Eng. **19**(6), 601–611 (2011)

69. Sharma, N., Gregory, C., Johnson, M., Dixon, W.: Closed-loop neural network-based NMES control for human limb tracking. IEEE Trans. Control Syst. Technol. **20**(3), 712–725 (2012)
70. Sharma, N., Stegath, K., Gregory, C., Dixon, W.: Nonlinear neuromuscular electrical stimulation tracking control of a human limb. IEEE Trans. Neural Syst. Rehabil. Eng. **17**(6), 576–584 (2009)
71. Sipahi, R., Niculescu, S.-I., Abdallah, C., Michiels, W., Gu, K.: Stability and stabilization of systems with time delay: limitations and opportunities. IEEE Control Syst. Mag. **31**(1), 38–65 (2011)
72. Smith, O.: A controller to overcome deadtime. ISA J. **6**(2), 28–33 (1959)
73. Snyder-Mackler, L., Delitto, A., Stralka, S., Bailey, S.: Use of electrical stimulation to enhance recovery of quadriceps femoris muscle force production in patients following anterior cruciate ligament reconstruction. Phys. Ther. **74**(10), 901–907 (1994)
74. Stackhouse, S., Binder-Macleod, S., Stackhouse, C., McCarthy, J., Prosser, L., Lee, S.: Neuromuscular electrical stimulation versus volitional isometric strength training in children with spastic diplegic cerebral palsy: A preliminary study. Neurorehabil. Neural Repair **21**(6), 475–485 (2007)
75. Thomsen, M., Veltink, P.: Influence of synchronous and sequential stimulation on muscle fatigue. Med. Biol. Eng. Comput. **35**(3), 186–192 (1997)
76. Wang, Z., Goldsmith, P., Tan, D.: Improvement on robust control of uncertain systems with time-varying input delays. IET Control Theory Appl. **1**(1), 189–194 (2007)
77. Watanabe, T., Futami, R., Hoshimiya, N., Handa, Y.: An approach to a muscle model with a stimulus frequency-force relationship for FES applications. IEEE Trans. Rehabil. Eng. **7**(1), 12–17 (1999)
78. Watanabe, K., Nobuyama, E., Kojima, A.: Recent advances in control of time delay systems-a tutorial review. In: Proceedings of the IEEE Conference on Decision and Control, pp. 2083–2089 (1996)
79. Wise, A., Morgan, D., Gregory, J., Proske, U.: Fatigue in mammalian skeletal muscle stimulated under computer control. J. Appl. Physiol. **90**(1), 189–197 (2001)
80. Yan, T., Hui-Chan, C., Li, L.: Functional electrical stimulation improves motor recovery of the lower extremity and walking ability of subjects with first acute stroke: A randomized placebo-controlled trial. Stroke **36**(1), 80–85 (2005)
81. Yavuz, S., Sendemir-Urkmez, A., Turker, K.: Effect of gender, age, fatigue and contraction level on electromechanical delay. Clin. Neurophysiol. **121**(10), 1700–1706 (2010)
82. Yoshida, K., Horch, K.: Reduced fatigue in electrically stimulated muscle using dual channel intrafascicular electrodes with interleaved stimulation. Ann. Biomed. Eng. **21**(6), 709–714 (1993)
83. Yue, D., Han, Q.-L.: Delayed feedback control of uncertain systems with time-varying input delay. Automatica **41**(2), 233–240 (2005)
84. Zhou, S., Lawson, D., Morrison, W., Fairweather, I.: Electromechanical delay in isometric muscle contractions evoked by voluntary, reflex and electrical stimulation. Eur. J. Appl. Physiol. Occup. Physiol. **70**(2), 138–145 (1995)

Global and Local Weighted Homogeneity for Time-Delay Systems

Denis Efimov, Wifrid Perruquetti and Jean-Pierre Richard

Abstract The notion of weighted homogeneity is extended to time-delay systems. It is shown that the stability (resp., instability) of homogeneous functional systems on a sphere implies the global stability (resp., instability) of the system. The notion of local homogeneity is introduced, and a relation between stability and instability of the locally approximating dynamics and the original time-delay system is established using a Lyapunov-Razumikhin approach. An implication between homogeneity and input-to-state stability is investigated. Examples of applications of the proposed theory are given.

1 Introduction

For homogeneous ordinary differential equations (ODEs), the global behavior of trajectories can be evaluated based on their behavior on a suitably defined sphere around the origin [36]. Thus, the local and global behaviors of homogeneous systems are the same. This property has been found to be useful for stability analysis [2, 4, 25, 27, 35], approximation of system dynamics [24, 30], stabilization [7, 18, 28, 32, 38], and estimation [2, 30]. It has been shown that for stability and instability analysis, Lyapunov functions of a homogeneous system can be chosen to be homogeneous [13, 35, 40]. Thus, the numerical analysis and design of homogeneous systems may

D. Efimov (✉) · W. Perruquetti · J.-P. Richard
Non-A Team @ Inria, 40 Avenue Halley, 59650 Villeneuve d'ascq, France
e-mail: Denis.Efimov@inria.fr

D. Efimov
Department of Control Systems and Informatics, Saint Petersburg State University of Information Technologies Mechanics and Optics (ITMO), Kronverkskiy av. 49, Saint Petersburg 197101, Russia

W. Perruquetti · J.-P. Richard
LAGIS (UMR-CNRS 8146), Ecole Centrale de Lille, 59651 Villeneuve-d'ascq, France
e-mail: wilfrid.perruquetti@inria.fr

J.-P. Richard
e-mail: jean-pierre.richard@ec-lille.fr

I. Karafyllis et al. (eds.), *Recent Results on Nonlinear Delay Control Systems*, Advances in Delays and Dynamics 4, DOI 10.1007/978-3-319-18072-4_8

be simpler, since for example, a Lyapunov function must only be constructed on a sphere (and it can be extended to the whole state space using homogeneity).

The class of nonlinear systems with a globally (stable or unstable) behavior is rather restricted. That is why the notion of local homogeneity has been defined and widely used [2, 13, 40]. This concept implies that the system is similar to a homogeneous one, but only on a sphere. In this case, the theory of homogeneous systems can be applied to locally analyze the behavior of the original nonlinear system (around that sphere).

The theory of homogeneous systems has been developed for time-invariant ODEs. Another important class of systems can be modeled by functional differential equations (FDEs), and in particular, systems with time delays. Delays are typically present in many applications [9, 16, 34]. For example, networked systems suffer from various sources of delays (such as access time delays, or communication or packet dropouts), as do embedded real-time systems (where the asynchronous sampling effect can be modeled by delays, as noted in [17] and the references therein). Appearance of a delay critically impacts the system stability and behavior complexity [19, 23, 29], which is why most of the examples in the literature are modeled by linear time-delay systems. As a matter of fact, the analysis for such models is well developed (and even converse Lyapunov-Krasovskii theorems in the case of constant delays have been proposed [19]). By contrast, for nonlinear time-delay systems, the design of a Lyapunov-Krasovskii functional or a Lyapunov-Razumikhin function is a difficult problem. As has been shown in the ODE case, homogeneity may simplify the analysis and design of nonlinear systems, and this is a motivation for the development of homogeneity theory for FDEs.

The main obstruction for such a development is that time-delay systems operate in an infinite dimensional state space. An extension of the standard results [2, 13, 35, 40] to this class of systems needs a complete revision of the arguments and tools of the theory of homogeneous system developed for ODEs. There exist a few works that deal with time-delay systems using homogeneity theory [1, 3, 8, 11]. In all of these works, homogeneity theory for ODEs has been applied to the delay-free part of the systems, and the concept of homogeneity has not been extended to the functional spaces. In the recent works [14, 15], homogeneity theory has been extended to FDEs; the present chapter mainly follows these works, and some new results and examples are considered.

The outline of this chapter is as follows. Preliminary definitions and the system equations are given in Sect. 2. The homogeneous norm, an extended definition of homogeneity for time-delay systems, and sufficient stability or instability conditions are presented in Sect. 3. Local homogeneity theory is studied in Sect. 4. In [5, 26, 37], it has been shown that for ODEs, homogeneity implies some kind of robustness with respect to external disturbances, and a corresponding link between input-to-state stability (ISS) and homogeneity for FDEs is established in Sect. 5. Examples are presented in Sect. 6.

2 Preliminaries

Consider an autonomous functional differential equation of retarded type of the form [29]:

$$dx(t)/dt = f(x_t),\ t \geq 0, \tag{1}$$

where $x \in \mathbb{R}^n$ and $x_t \in C_{[-\tau,0]}$ is the state function and defined by $x_t(s) = x(t+s)$ when $-\tau \leq s \leq 0$ (where $C_{[-\tau,0]}$ for $0 < \tau < +\infty$ denotes the Banach space of all continuous functions $\varphi : [-\tau, 0] \to \mathbb{R}^n$ with the uniform norm $||\varphi|| = \sup_{-\tau \leq \varsigma \leq 0} |\varphi(\varsigma)|$, where $|\cdot|$ is the standard Euclidean norm), and $f : C_{[-\tau,0]} \to \mathbb{R}^n$ is a locally Lipschitz continuous function such that $f(0) = 0$. The representation (1) includes pointwise or distributed retarded systems with either constant or variable time delay $\tau(t) \in [0, \tau]$. We assume that the solutions of the system (1) satisfy the initial functional condition $x_0 \in C_{[-\tau,0]}$. It is known from the theory of functional differential equations [29] that under the above assumptions, the system (1) has a unique solution $x(t, x_0)$ satisfying the initial condition x_0, which is defined on some finite time interval $[-\tau, T)$ (and we will use the notation $x(t)$ to reference $x(t, x_0)$ if the choice of x_0 is clear).

For a locally Lipschitz continuous function $V : \mathbb{R}^n \to \mathbb{R}_+$, the lower or upper directional Dini derivatives are defined as follows:

$$\begin{aligned} D^- V[x_t(0)]f(x_t) &= \lim_{h \to 0^+} \inf \tfrac{V[x_t(0)+hf(x_t)]-V[x_t(0)]}{h}, \\ D^+ V[x_t(0)]f(x_t) &= \lim_{h \to 0^+} \sup \tfrac{V[x_t(0)+hf(x_t)]-V[x_t(0)]}{h}. \end{aligned}$$

A continuous function $\sigma : \mathbb{R}_+ \to \mathbb{R}_+$ is of class $\mathcal{K}$ provided it is strictly increasing and $\sigma(0) = 0$; it is of class $\mathcal{K}_\infty$ provided it is also radially unbounded. A continuous function $\beta : \mathbb{R}_+ \times \mathbb{R}_+ \to \mathbb{R}_+$ is of class $\mathcal{KL}$ provided $\beta(\cdot, r) \in \mathcal{K}$ and $\beta(r, \cdot)$ is asymptotically strictly decreasing to zero for any fixed $r \in \mathbb{R}_+$. The symbol $\overline{1, m}$ denotes the sequence of integers $1, ..., m$ for each m.

3 Homogeneity

For any $r_i > 0$ where $i \in \overline{1, n}$ and $\lambda > 0$, define the dilation linear operator $\Lambda_r(\lambda) = \text{diag}\{\lambda^{r_i}\}_{i=1}^n$ and the vector of weights $r = [r_1, ..., r_n]^T$. For any $r_i > 0$ where $i \in \overline{1, n}$ and $x \in \mathbb{R}^n$, the homogeneous norm can be defined as follows:

$$|x|_r = \left(\sum_{i=1}^n |x_i|^{\rho/r_i} \right)^{1/\rho}, \quad \rho = \prod_{i=1}^n r_i.$$

For all $x \in \mathbb{R}^n$, the Euclidean norm $|x|$ is related to the homogeneous one as follows:

$$\underline{\sigma}_r(|x|_r) \leq |x| \leq \bar{\sigma}_r(|x|_r),$$
$$\bar{\sigma}_r(s) = \max_{|x|_r \leq s} |x|, \ \underline{\sigma}_r^{-1}(s) = \max_{|x| \leq s} |x|_r,$$

where $\underline{\sigma}_r$ and $\bar{\sigma}_r$ are in $\mathcal{K}_\infty$, which defines the Euclidean norm deviations with respect to the homogeneous norm. The homogeneous norm has an important property that $|\Lambda_r(\lambda)x|_r = \lambda|x|_r$ for all $x \in \mathbb{R}^n$. Define $\mathbb{S}_r = \{x \in \mathbb{R}^n : |x|_r = 1\}$.

Since the transformation by the dilation operator $\Lambda_r(\lambda)$ is linear, it can be applied to functional arguments. Indeed, for any $r_i > 0$ and all $i \in \overline{1,n}$ and $\varphi \in C_{[a,b]}$, if $0 \leq a < b \leq +\infty$, then the homogeneous norm can be defined as follows:

$$||\varphi||_r = \left(\sum_{i=1}^{n} ||\varphi_i||^{\rho/r_i}\right)^{1/\rho}, \ \rho = \prod_{i=1}^{n} r_i.$$

Lemma 1 [15] *There exist two functions $\underline{\rho}_r$ and $\bar{\rho}_r$ in $\mathcal{K}_\infty$ such that for all $\varphi \in C_{[a,b]}$, we have*

$$\underline{\rho}_r(||\varphi||_r) \leq ||\varphi|| \leq \bar{\rho}_r(||\varphi||_r).$$

Therefore, the proposed homogeneous norm is equivalent to the uniform norm in $C_{[a,b]}$. The homogeneous norm in the Banach space has the same important property that $||\Lambda_r(\lambda)\varphi||_r = \lambda||\varphi||_r$ for all $\varphi \in C_{[a,b]}$. Define the corresponding unit sphere by $\mathcal{S}_r = \{\varphi \in C_{[-\tau,0]} : ||\varphi||_r = 1\}$.

Definition 1 A function $g : C_{[-\tau,0]} \to \mathbb{R}$ is called r-homogeneous (where $r_i > 0$ for all $i \in \overline{1,n}$) provided for any $\varphi \in C_{[-\tau,0]}$, the relation

$$\lambda^{-d} g(\Lambda_r(\lambda)\varphi) = g(\varphi)$$

holds for some $d \in \mathbb{R}$ and all $\lambda > 0$. The function $f : C_{[-\tau,0]} \to \mathbb{R}^n$ is called r-homogeneous (where $r_i > 0$ for all $i \in \overline{1,n}$) provided for any $\varphi \in C_{[-\tau,0]}$, the relation

$$\lambda^{-d} \Lambda_r^{-1}(\lambda) f(\Lambda_r(\lambda)\varphi) = f(\varphi)$$

holds for some $d \geq -\min_{1 \leq i \leq n} r_i$ and all $\lambda > 0$. In both cases, the constant d is called the degree of homogeneity.

The notion of homogeneity introduced in $C_{[-\tau,0]}$ is reduced to the standard one in $\mathbb{R}^n$ [35] under a vector argument substitution. An advantage of homogeneous systems described by nonlinear ordinary differential equations is that analysis of their stability can be performed on only the unit sphere $\mathbb{S}_r$ [35] (and the homogeneous system trajectories have a similar behavior on any other sphere defined by the norm $|\cdot|_r$). This conclusion is based on the property that any solution of a homogeneous system can be obtained from another solution under a dilation rescaling and a suitable time re-parametrization. A similar property holds for functional homogeneous systems.

Proposition 1 *Let* (1) *be an* r*-homogeneous system with degree* d *and* $x(t, x_0)$ *be its solution for any initial condition* $x_0 \in C_{[-\tau,0]}$, $\tau \in (0, +\infty)$. *For any* $\lambda > 0$, *the functional differential equation*

$$dy(t)/dt = f(y_t), \ t \geq 0 \tag{2}$$

with $y_t \in C_{[-\lambda^{-d}\tau,0]}$ *has a solution* $y(t, y_0) = \Lambda_r(\lambda)x(\lambda^d t, x_0)$ *for all* $t \geq 0$ *with the initial condition*

$$y_0 \in C_{[-\lambda^{-d}\tau,0]},$$

namely, $y_0(s) = \Lambda_r(\lambda)x_0(\lambda^d s)$ *for all* $s \in [-\lambda^{-d}\tau, 0]$.

Proof By definition, $x_{\lambda^d t}(s) = x(\lambda^d t + s) \in \mathbb{R}^n$ and $y_t(s) = \Lambda_r(\lambda)x_{\lambda^d t}(s)$ for any $-\lambda^{-d}\tau \leq s \leq 0$, so

$$\dot{y}(t) = \frac{d}{dt}\left(\Lambda_r(\lambda)x(\lambda^d t)\right) = \lambda^d \Lambda_r(\lambda) f(x_{\lambda^d t}) = f(\Lambda_r(\lambda)x_{\lambda^d t}) = f(y_t)$$

and $y(t)$ is a solution of (2).

In order to better explain this result, consider this particular case of the system (1) with the pointwise delay:

$$f(x_\tau) = F[x(t), x(t-\tau)]$$

Then the homogeneity condition reads $F[\Lambda_r(\lambda)x, \Lambda_r(\lambda)z] = \lambda^d \Lambda_r(\lambda)F[x, z]$ for any x and z in $\mathbb{R}^n$ and $\lambda \in (0, +\infty)$. Select a solution $x(t, x_0)$ of this system for the initial condition $x_0 \in C_{[-\tau,0]}$ for some $\tau > 0$ as before (i.e., $\dot{x}(t, x_0) = F[x(t, x_0), x(t-\tau, x_0)]$). Take some $\lambda \in (0, +\infty)$ and define $y_0 \in C_{[-\lambda^{-d}\tau,0]}$ by

$$y_0(s) = \Lambda_r(\lambda)x_0(\lambda^d s)$$

for $s \in [-\lambda^{-d}\tau, 0]$, then consider $y(t, y_0) = \Lambda_r(\lambda)x(\lambda^d t, x_0)$. Taking its derivative with respect to time, we obtain

$$\begin{aligned}
\frac{dy(t, y_0)}{dt} &= \frac{d\Lambda_r(\lambda)x(\lambda^d t, x_0)}{dt} \\
&= \lambda^d \Lambda_r(\lambda)F[x(\lambda^d t, x_0), x(\lambda^d t - \tau, x_0)] \\
&= F[\Lambda_r(\lambda)x(\lambda^d t, x_0), \Lambda_r(\lambda)x(\lambda^d\{t - \lambda^{-d}\tau\}, x_0)] \\
&= F[y(t, y_0), y(t - \lambda^{-d}\tau, y_0)].
\end{aligned}$$

Therefore, $y(t, y_0)$ is the system solution with initial conditions y_0 for another delay $\lambda^{-d}\tau$.

Corollary 1 *Let the origin be locally asymptotically stable for a r-homogeneous system* (1) *with the degree $d = 0$. Then it is globally asymptotically stable.*

Proof Assume that the origin is locally attractive for (1) with an open domain of attraction $\mathscr{A} \subset C_{[-\tau,0]}$, i.e., for any $\varepsilon > 0$ and $x_0 \in \mathscr{A}$, there is a $T_{\varepsilon,x_0} \geq 0$ such that $||x(t, x_0)||_r \leq \varepsilon$ for all $t \geq T_{\varepsilon,x_0}$; by Lemma 1 the norms $|| \cdot ||$ and $|| \cdot ||_r$ can be replaced. Take a $\mu > 0$ such that $\mathscr{S}_r^\mu \subset \mathscr{A}$, where $\mathscr{S}_r^\mu = \mu\mathscr{S}_r$. Then for any $\xi \in C_{[-\tau,0]}$, there is an $x_0 \in \mathscr{S}_r^\mu$ such that $\xi = \Lambda_r(\lambda)x_0$ where $\lambda = \mu^{-1}||\xi||_r$, and the corresponding unique solution is $x(t, \xi) = \Lambda_r(\lambda)x(\lambda^d t, x_0)$, by Proposition 1. Obviously, if $x(t, x_0) \to 0$ for all $x_0 \in \mathscr{S}_r^\mu$ with $t \to +\infty$, then so is $x(t, \xi) = \Lambda_r(\lambda)x(\lambda^d t, x_0)$, and the claims about global attractiveness and forward completeness follow.

To prove that local stability of the origin implies global one in this case, assume that $\sup_{t\geq 0} ||x(t, x_0)||_r \leq \sigma(||x_0||_r)$ for all $x_0 \in \mathscr{A}$ and some $\sigma \in \mathscr{K}$. Take any $\xi \in C_{[-\tau,0]}$, then there is $x_0 \in \mathscr{S}_r^\mu \subset \mathscr{A}$ such that $\xi = \Lambda_r(\lambda)x_0$ for $\lambda = \mu^{-1}||\xi||_r$ with the corresponding unique solution

$$x(t, \xi) = \Lambda_r(\lambda)x(\lambda^d t, x_0),$$

by Proposition 1. Therefore

$$\begin{aligned}\sup_{t\geq 0} ||x(t, \xi)||_r &= \sup_{t\geq 0} ||\Lambda_r(\lambda)x(\lambda^d t, x_0)||_r = \lambda \sup_{t\geq 0} ||x(\lambda^d t, x_0)||_r \\ &\leq ||\xi||_r \sigma(||x_0||_r)/\mu = \sigma(\mu)/\mu||\xi||_r,\end{aligned}$$

and the system is Lyapunov stable [19, 23, 29].

Thus, we have proven that for homogeneous time-delay systems with $d = 0$, any local stability/attractivity properties at the origin for a given delay τ hold globally. The case of homogeneous systems with $d = 0$ becomes important for stability analysis using local homogeneous approximating dynamics, which are considered in Sect. 4 below (where the approximating dynamics degree can be assigned to be zero). For the case $d \neq 0$ even more interesting conclusions can be obtained for homogeneous systems.

Lemma 2 *Let the system* (1) *be r-homogeneous with degree $d \neq 0$ and globally asymptotically stable for some delay $0 < \tau_0 < +\infty$. Then, it is globally asymptotically stable for any delay $0 < \tau < +\infty$.*

Proof In this case, for all $x_0 \in C_{[-\tau_0,0]}$, there is a function $\sigma \in \mathscr{K}$ such that $|x(t, x_0)|_r \leq \sigma(||x_0||_r)$ holds for all $t \geq 0$ and $\lim_{t\to+\infty} |x(t, x_0)|_r = 0$. Take some $\tau \in (0, +\infty)$ and select an initial condition $y_0 \in C_{[-\tau,0]}$. Then for

$$\lambda = \left(\frac{\tau_0}{\tau}\right)^{1/d},$$

where λ is well defined since $d \neq 0$, there exists $x_0 \in C_{[-\tau_0,0]}$ such that $y_0(s) = \Lambda_r(\lambda)x_0(\lambda^d s)$ for all $s \in [-\tau, 0]$, and

$$y(t, y_0) = \Lambda_r(\lambda)x(\lambda^d t, x_0)$$

holds for all $t \geq 0$, by Proposition 1. Thus,

$$\lim_{t\to+\infty} |y(t, y_0)|_r = \lambda \lim_{t\to+\infty} |x(\lambda^d t, x_0)|_r = \lambda \lim_{t\to+\infty} |x(t, x_0)|_r = 0,$$

and the solution $y(t, y_0)$ is converging asymptotically to the origin. In addition, $|y(t, y_0)|_r = \lambda|x(\lambda^d t, x_0)|_r \leq \lambda\sigma(||x_0||_r) = \lambda\sigma(\lambda^{-1}||y_0||_r)$ for all $t \geq 0$, which implies stability of the system (1) for the delay τ. The proven convergence to the origin and stability give the global asymptotic stability of the system for an arbitrary delay $\tau \in (0, +\infty)$.

It is a well known fact for linear systems (which are homogeneous systems of degree $d = 0$) that their stability for a sufficiently small delay does not imply stability for all $\tau \in (0, +\infty)$; Corollary 1 has been proven for a fixed delay τ. For nonlinear homogeneous systems with degree $d \neq 0$, this is not the case, since according to the result of Lemma 2, if they are globally stable for some delay, then they can preserve their stability for an arbitrary delay $\tau \in (0, +\infty)$. This is a surprising advantage of this class of "nonlinear" time-delay systems. We next consider several useful consequences of Proposition 1 and Lemma 2.

Corollary 2 *Let the system* (1) *be* r*-homogeneous with degree* d *and asymptotically stable into the set* $\Omega = B_\rho^\tau = \{\varphi \in C_{[-\tau,0]} : ||\varphi||_r \leq \rho\}$ *for some* $0 < \rho < +\infty$ *for any value of delay* $0 \leq \tau < +\infty$. *Then it is globally asymptotically stable independently of delay.*

Proof For any $\tau > 0$, take $y_0 \in C_{[-\tau,0]}$, $y_0 \notin B_\rho^\tau$. Then, according to Proposition 1, there is a positive $\lambda < +\infty$ (with $\lambda = \rho^{-1}||y_0||_r$) and

$$x_0 \in B_\rho^{\lambda^d \tau}$$

such that $y_0(s) = \Lambda_r(\lambda)x_0(\lambda^d s)$ for all $s \in [-\tau, 0]$ and $y(t, y_0) = \Lambda_r(\lambda)x(\lambda^d t, x_0)$ for all $t \geq 0$. Since $x(t)$ converges asymptotically to the origin, the same property is satisfied for $y(t)$ and it enters the set B_ρ^τ in a finite time.

Thus, for the case $d \neq 0$, the result of Corollary 1 can be obtained if local stability holds independently of delay, and then (1) is globally asymptotically stable.

For ordinary differential equations, it has also been shown that asymptotically stable/unstable homogeneous systems always have homogeneous Lyapunov functions [2, 13, 35]. In this work, we would like to prove a similar (sufficient only) result for time-delay homogeneous systems. For time-delay systems, there exist two main techniques for stability analysis based on the Lyapunov approach. The first one is based on Lyapunov-Krasovskii functionals, and another one on Lyapunov-Razumikhin functions [19, 23, 29]. Unfortunately, due to peculiarities of the directional derivatives for functionals, the Lyapunov-Krasovskii approach is hard to develop using homogeneity in general; the derivative of a homogeneous functional may be non-homogeneous [14, 15].

The Razumikhin approach is based on Lyapunov-Razumikhin functions defined on $\mathbb{R}^n$ [19, 23, 29], which give a pointwise sufficient criteria for stability (and not a functional one). It allows us to develop this approach using homogeneous arguments. If a function $V : \mathbb{R}^n \to \mathbb{R}_+$ is positive definite and radially unbounded, then there are functions α_1 and α_2 in $\mathscr{K}_\infty$ such that $\alpha_1(|x|) \le V(x) \le \alpha_2(|x|)$ for all $x \in \mathbb{R}^n$. If V is r-homogeneous with a degree ν, then for any $x \in \mathbb{R}^n$, we have $V(x) = V(\Lambda_r(|x|_r)y) = |x|_r^\nu V(y)$ for some $y \in \mathbb{S}_r$, so in this case, $\alpha_1(s) = [\bar{\sigma}_r^{-1}(s)]^\nu \min_{y\in\mathbb{S}_r} V(y)$ and $\alpha_2(s) = [\underline{\sigma}_r^{-1}(s)]^\nu \max_{y\in\mathbb{S}_r} V(y)$, and ν should be strictly positive for radial unboundedness of V. In addition, if V is continuously differentiable at the origin, then ν should be bigger than 1.

Theorem 1 *Assume that the function f in* (1) *is r-homogeneous of some degree $d \ge -\min_{1\le i\le n} r_i$ and that there exists a locally Lipschitz continuous r-homogeneous Lyapunov-Razumikhin function $V : \mathbb{R}^n \to \mathbb{R}_+$ with degree $\nu > \max\{0, -d\}$ such that*

(i) $\min_{y\in\mathbb{S}_r} V(y) > 0$ *and there exist functions $\alpha, \gamma \in \mathscr{K}$ such that for all $\varphi \in \mathscr{S}_r$*

$$\max_{\theta\in[-\tau,0]} V[\varphi(\theta)] < \gamma\{V[\varphi(0)]\} \Rightarrow D^+V[\varphi(0)]f(\varphi) \le -\alpha(|\varphi(0)|);$$

(ii) there exists a function $\wp \in \mathscr{K}$ such that $\lambda s < \wp(\lambda s) \le \lambda\gamma(s)$ for all s and λ in $\mathbb{R}_+ \setminus \{0\}$.

Then the origin is globally asymptotically stable for the system (1).

Proof The dilation transformation $\varphi = \Lambda_r(\lambda)\varphi$ connects any $\varphi \in C_{[-\tau,0]} \setminus \{0\}$ with some $\varphi \in \mathscr{S}_r$ for properly chosen $\lambda > 0$. Let us multiply the left-hand part of the implication (i) by λ^ν and the right-hand part by $\lambda^{\nu+d}$:

$$\max_{\theta\in[-\tau,0]} \lambda^\nu V[\varphi(\theta)] < \lambda^\nu\gamma\{V[\varphi(0)]\} \Rightarrow \lambda^{\nu+d} D^+V[\varphi(0)]f(\varphi) \le -\lambda^{\nu+d}\alpha(|\varphi(0)|),$$

where the inequalities and relations should not be changed for $\lambda > 0$. By the properties of the functions γ and $\wp$, the inequality $\wp\{\lambda^\nu V[\varphi(0)]\} \le \lambda^\nu\gamma\{V[\varphi(0)]\}$ holds. Formally $\inf_{\varphi\in\mathscr{S}_r}\{\alpha(|\varphi(0)|)\} = 0$, however having in mind that φ is a continuous function and a solution of (1), there exists

$$a = \inf_{\varphi\in\mathscr{S}_r, \max_{\theta\in[-\tau,0]} V[\varphi(\theta)]<\gamma\{V[\varphi(0)]\}} \alpha(|\varphi(0)|)$$

and $a > 0$. Indeed, for the case $\max_{\theta\in[-\tau,0]} V[\varphi(\theta)] < \gamma\{V[\varphi(0)]\}$,we have

$$\alpha_1(\underline{\rho}_r(1)) \le \alpha_1(||\varphi||) = \alpha_1(\max_{\theta\in[-\tau,0]} |\varphi(\theta)|)$$
$$= \max_{\theta\in[-\tau,0]} \alpha_1(|\varphi(\theta)|) \le \max_{\theta\in[-\tau,0]} V(\varphi(\theta)) < \gamma[V(\varphi(0))],$$

so $\alpha_2^{-1} \circ \gamma^{-1} \circ \alpha_1(\underline{\rho}_r(1)) < |\varphi(0)|$ for all such $\varphi \in \mathcal{S}_r$ and $a \ge \alpha_2^{-1} \circ \gamma^{-1} \circ \alpha_1(\underline{\rho}_r(1)) > 0$. Therefore, due to homogeneity of the functions f and V, we have

$$\max_{\theta\in[-\tau,0]} V[\Lambda_r(\lambda)\varphi(\theta)] < \wp\{V[\Lambda_r(\lambda)\varphi(0)]\} \Rightarrow$$
$$D^+V[\Lambda_r(\lambda)\varphi(0)]f(\Lambda_r(\lambda)\varphi) \le -\lambda^{\nu+d}a,$$

or equivalently, (since $\lambda = ||\varphi||_r \ge |\varphi(0)|_r \ge \bar{\sigma}_r^{-1}(|\varphi(0)|)$),

$$\max_{\theta\in[-\tau,0]} V[\varphi(\theta)] < \wp\{V[\varphi(0)]\} \Rightarrow D^+V[\varphi(0)]f(\varphi) \le -\alpha'(|\varphi(0)|),$$

where the function $\alpha'(s) = (\bar{\sigma}_r^{-1}(s))^{\nu+d}a$ is from class $\mathcal{K}$ since $\nu + d > 0$. Therefore, if Razumikhin arguments are true for $\varphi \in \mathcal{S}_r$, then they are valid for any $\varphi \in C_{[-\tau,0]}$, which implies the global asymptotic stability of the origin for (1) [19, 23, 29].

The condition (i) imposed in Theorem 1 on the system (1) behavior is the conventional Razumikhin condition (except that in the homogeneous case it can be verified on the sphere $\mathcal{S}_r$ only). The constraint (ii) on existence of the function $\wp$ is new. For example, it is satisfied for any $\gamma(s) > s > 0$.

Remark 1 Note that the derivative $D^+V[\varphi(0)]f(\varphi)$ is a function of $\varphi \in \mathcal{S}_r$, however under the condition

$$\max_{\theta\in[-\tau,0]} V[\Lambda_r(\lambda)\varphi(\theta)] < \wp\{V[\Lambda_r(\lambda)\varphi(0)]\},$$

the inequality

$$D^+V[\varphi(0)]f(\varphi) < m(\varphi(0))$$

holds for some $m : \mathbb{R}^n \to \mathbb{R}$. If the function m is r-homogeneous, then the property $m(\varphi(0)) \le -\alpha(|\varphi(0)|)$ in (ii) need only be verified for $\varphi(0) \in \mathbb{S}_r$.

Instability conditions in the Lyapunov-Razumikhin framework have been formulated in [21, 22]. We will say that the system (1) is unstable at the origin provided for any $\delta > 0$, there exist $\varepsilon > 0$, $||x_0|| \le \delta$, and $t'_{x0,\varepsilon} \ge 0$ such that

$$||x_{t'_{x0,\varepsilon}}|| > \varepsilon.$$

Given a function $V:\mathbb{R}^n\to\mathbb{R}_+$, define these two sets:

$$P_M^V=\{\varphi\in\mathcal{S}_r:\ V[\varphi(0)]=\max\nolimits_{\theta\in[-\tau,0]}V[\varphi(\theta)]\},$$
$$P_m^V=\{\varphi\in\mathcal{S}_r:\ V[\varphi(0)]=\min_{\theta\in[-\tau,0]}V[\varphi(\theta)]\}.$$

Theorem 2 *Let f in* (1) *be r-homogeneous of some degree $d\geq-\min_{1\leq i\leq n}r_i$ and there exist a locally Lipschitz continuous r-homogeneous Lyapunov-Razumikhin function $V:\mathbb{R}^n\to\mathbb{R}_+$ with degree $\nu>\max\{0,-d\}$ such that $\min_{y\in\mathbb{S}_r}V(y)>0$ and one of the following properties is satisfied:*

(i) $D^-V[\varphi(0)]f(\varphi)>0$ for all $\varphi\in P_M^V$;
(ii) $D^-V[\varphi(0)]f(\varphi)>0$ for all $\varphi\in P_m^V$.

Then the system (1) *is unstable at the origin.*

Proof Select a $\lambda>0$ such that the dilation transformation $\varphi=\Lambda_r(\lambda)\varphi$ connects a $\varphi\in C_{[-\tau,0]}\setminus\{0\}$ with some $\varphi\in\mathcal{S}_r$. The sets

$$\Pi_M^V=\{\varphi\in C_{[-\tau,0]}:\ V[\varphi(0)]=\max\nolimits_{\theta\in[-\tau,0]}V[\varphi(\theta)]\}$$

and

$$\Pi_m^V=\{\varphi\in C_{[-\tau,0]}:\ V[\varphi(0)]=\min_{\theta\in[-\tau,0]}V[\varphi(\theta)]\}$$

are well defined by the dilation transformation of P_M^V and P_m^V, as follows:

$$\begin{aligned}\cup_{\lambda>0}\Lambda_r(\lambda)P_M^V&=\{\varphi=\Lambda_r(\lambda)\varphi,\lambda>0,\varphi\in\mathcal{S}_r:V[\varphi(0)]=\max\nolimits_{\theta\in[-\tau,0]}V[\varphi(\theta)]\}\\&=\{\varphi=\Lambda_r(\lambda)\varphi,\lambda>0,\varphi\in\mathcal{S}_r:\lambda^\nu V[\varphi(0)]=\lambda^\nu\max\nolimits_{\theta\in[-\tau,0]}V[\varphi(\theta)]\}\\&=\Pi_M^V,\\ \cup_{\lambda>0}\Lambda_r(\lambda)P_m^V&=\{\varphi=\Lambda_r(\lambda)\varphi,\lambda>0,\varphi\in\mathcal{S}_r:V[\varphi(0)]=\min_{\theta\in[-\tau,0]}V[\varphi(\theta)]\}\\&=\{\varphi=\Lambda_r(\lambda)\varphi,\lambda>0,\varphi\in\mathcal{S}_r:\lambda^\nu V[\varphi(0)]=\lambda^\nu\min_{\theta\in[-\tau,0]}V[\varphi(\theta)]\}\\&=\Pi_m^V.\end{aligned}$$

Take $\varphi\in\Pi_M^V$ or $\varphi\in\Pi_m^V$. Then

$$D^-V[\varphi(0)]f(\varphi)=\lambda^{\nu+d}D^-V[\varphi(0)]f(\varphi)>0.$$

Therefore, $D^-V[\varphi(0)]f(\varphi)>0$ for all $\varphi\in\Pi_M^V$ or $\varphi\in\Pi_m^V$, which are the conditions of instability of (1) into the sets Π_M^V or Π_m^V from [21, 22].

Theorems 1 and 2 mean that by using homogeneous Lyapunov-Razumikhin functions, the global stability or instability of a homogeneous system at the origin can be checked on the sphere $\mathcal{S}_r$ only. These facts may simplify the search for the function V and the system analysis, using an application of a numerical routine. The drawback is that in the space $C_{[-\tau,0]}$, the sphere $\mathcal{S}_r$ is a rather complex object.

4 Local Homogeneity

A disadvantage of the global homogeneity introduced so far is that such systems possess the same behavior "globally". Thus, homogeneous systems are not really "nonlinear", in that they have similar diversity of operating modes as linear systems. In fact, from the analysis and design points of view, homogeneous systems are a generalization of linear ones. Compared with other nonlinear systems, it may be easier to find a (homogeneous) Lyapunov function for homogeneous systems. That is why finding a possibility to apply this approach for a broader class of nonlinear systems is very important.

An approach to resolve this issue consists of introducing a local version of homogeneity as in [13]:

Definition 2 The function $g : C_{[-\tau,0]} \to \mathbb{R}$ is called (r,λ_0,g_0)-homogeneous (where $r_i > 0$ for all $i \in \overline{1,n}$ and $g_0 : C_{[-\tau,0]} \to \mathbb{R}$) provided for any $\varphi \in \mathcal{S}_r$, the relation

$$\lim_{\lambda\to\lambda_0} \lambda^{-d_0} g(\Lambda_r(\lambda)\varphi) - g_0(\varphi) = 0$$

is satisfied uniformly on $\mathcal{S}_r$ for $\lambda_0 \in \{0, +\infty\}$, for some $d_0 \in \mathbb{R}$.

The system (1) is called (r,λ_0,f_0)-homogeneous (with $r_i > 0$ for all $i \in \overline{1,n}$, and $f_0 : C_{[-\tau,0]} \to \mathbb{R}^n$) provided for any $\varphi \in \mathcal{S}_r$, the relation

$$\lim_{\lambda\to\lambda_0} \lambda^{-d_0} \Lambda_r^{-1}(\lambda) f(\Lambda_r(\lambda)\varphi) - f_0(\varphi) = 0$$

is satisfied uniformly on $\mathcal{S}_r$ for $\lambda_0 \in \{0, +\infty\}$, for some $d_0 \geq -\min_{1\leq i\leq n} r_i$.

For a given λ_0, g_0 and f_0 are called approximating functions.

For any $0 < \lambda_0 < +\infty$, the following formulas give a variant of homogeneous approximating functions g_0 and f_0:

$$g_0(\varphi) = ||\varphi||_r^d \lambda_0^{-d_0} g(\Lambda_r(\lambda_0)\Lambda_r^{-1}(||\varphi||_r)\varphi), \ d \geq 0,$$
$$f_0(\varphi) = ||\varphi||_r^d \lambda_0^{-d_0} \Lambda_r(||\varphi||_r)\Lambda_r^{-1}(\lambda_0) f(\Lambda_r(\lambda_0)\Lambda_r^{-1}(||\varphi||_r)\varphi), \ d \geq -\min_{1\leq i\leq n} r_i.$$

This property is called *local* homogeneity [13], and allows us to analyze local stability or instability of the system (1) on the basis of a simplified system

$$dy(t)/dt = f_0[y_\tau(t)], t \geq 0, \tag{3}$$

called the *local approximating dynamics* for (1).

Theorem 3 [15] *Let the system* (1) *be* (r,λ_0,f_0)*-homogeneous for some* $r_i > 0$ *for all* $i \in \overline{1,n}$, *and the function* f_0 *be continuous and* r*-homogeneous with the degree* d_0. *Assume that there exists a locally Lipschitz continuous* r*-homogeneous Lyapunov-Razumikhin function* $V_0 : \mathbb{R}^n \to \mathbb{R}_+$ *with the degree* $\nu_0 > \max\{0, -d_0\}$ *such that* $\alpha_1(|x|) \leq V_0(x) \leq \alpha_2(|x|)$ *holds for all* $x \in \mathbb{R}^n$ *and some* $\alpha_1, \alpha_2 \in \mathcal{K}_\infty$, *and that*

(i) there exist functions α *and* γ *in* $\mathcal{K}$ *such that for all* $\varphi \in \mathcal{S}_r$, *we have*

$$\max_{\theta\in[-\tau,0]} V_0[\varphi(\theta)] < \gamma\{V_0[\varphi(0)]\} \Rightarrow D^+ V_0[\varphi(0)] f_0(\varphi) \leq -\alpha(|\varphi(0)|);$$

(ii) there exists a function $\wp \in \mathcal{K}$ *such that* $\lambda s < \wp(\lambda s) \leq \lambda\gamma(s)$ *holds for all* $s, \lambda \in \mathbb{R}_+ \setminus \{0\}$.

Then

(1) if $\lambda_0 = 0$, *then there exists a* $\bar{\lambda}_\varepsilon$ *satisfying* $0 < \bar{\lambda}_\varepsilon$ *such that* (1) *is locally asymptotically stable to the origin with the domain of attraction containing the set*

$$X_0 = \{\varphi \in C_{[-\tau,0]} : ||\varphi|| \leq \alpha_1^{-1} \circ \alpha_2 \circ \bar{\rho}_r(\bar{\lambda}_\varepsilon)\};$$

(2) if $\lambda_0 = +\infty$, *then there exists an* $\underline{\lambda}_\varepsilon$ *satisfying* $0 < \underline{\lambda}_\varepsilon < +\infty$ *such that the system* (1) *is globally asymptotically stable with respect to forward invariant set*

$$X_\infty = \{\varphi \in C_{[-\tau,0]} : ||\varphi|| \leq \alpha_1^{-1} \circ \alpha_2 \circ \underline{\rho}_r(\underline{\lambda}_\varepsilon)\};$$

(3) if $0 < \lambda_0 < +\infty$, *then there exist an* $\underline{\lambda}_\varepsilon$ *and* $\bar{\lambda}_\varepsilon$ *satisfying* $0 < \underline{\lambda}_\varepsilon \leq \lambda_0 \leq \bar{\lambda}_\varepsilon < +\infty$ *such that* (1) *is asymptotically stable with respect to the forward invariant set* X_∞ *with region of attraction*

$$X = \{\varphi \in C_{[-\tau,0]} : \alpha_1^{-1} \circ \alpha_2 \circ \underline{\rho}_r(\underline{\lambda}_\varepsilon) < ||\varphi|| < \alpha_1^{-1} \circ \alpha_2 \circ \bar{\rho}_r(\bar{\lambda}_\varepsilon)\}$$

provided that the set X *is connected and nonempty.*

Theorem 4 [15] *Let the system* (1) *be* (r,λ_0,f_0)*-homogeneous for some* $r_i > 0$ *for all* $i \in \overline{1,n}$, *and the function* f_0 *be continuous and* r*-homogeneous with the degree* d_0. *Assume that there exists a locally Lipschitz continuous* r*-homogeneous Lyapunov-Razumikhin function* $V_0 : \mathbb{R}^n \to \mathbb{R}_+$ *with some degree* $\nu_0 > \max\{0, -d_0\}$ *that satisfies* $\alpha_1(|x|) \leq V_0(x) \leq \alpha_2(|x|)$ *for all* $x \in \mathbb{R}^n$ *for some* α_1 *and* $\alpha_2 \in \mathcal{K}_\infty$, *and that admits a constant* a *such that*

$$D^- V_0[\varphi(0)] f_0(\varphi) \geq a > 0$$

for all $\varphi \in P_m^{V_0} \cup P_M^{V_0}$. *Then:*

(1) if $\lambda_0 = 0$*, then there exists* $\bar{\lambda}_\varepsilon > 0$ *such that for the system* (1)*, the set*

$$X_0 = \{\varphi \in C_{[-\tau,0]} : ||\varphi|| \leq \alpha_1^{-1} \circ \alpha_2 \circ \bar{\rho}_r(\bar{\lambda}_\varepsilon)\}$$

is unstable;

(2) if $\lambda_0 = +\infty$*, then there exists* $\underline{\lambda}_\varepsilon \in (0, +\infty)$ *such that for the system* (1)*, the set*

$$X_\infty = \{\varphi \in C_{[-\tau,0]} : ||\varphi|| \leq \alpha_1^{-1} \circ \alpha_2 \circ \underline{\rho}_r(\underline{\lambda}_\varepsilon)\}$$

is unstable;

(3) if $0 < \lambda_0 < +\infty$*, then there exist an* $\underline{\lambda}_\varepsilon$ *satisfying* $0 < \underline{\lambda}_\varepsilon \leq \lambda_0 \leq \bar{\lambda}_\varepsilon < +\infty$ *such that for the system* (1)*, the set* X_∞ *is unstable provided that the set*

$$X = \{\varphi \in C_{[-\tau,0]} : \alpha_1^{-1} \circ \alpha_2 \circ \underline{\rho}_r(\underline{\lambda}_\varepsilon) < ||\varphi|| < \alpha_1^{-1} \circ \alpha_2 \circ \bar{\rho}_r(\bar{\lambda}_\varepsilon)\}$$

is connected and nonempty.

These results establish links between different variants of local homogeneity with stable or unstable approximating dynamics (3) and the stability or instability properties of the original system (1), similarly to [13].

Corollary 3 [15] *Let the system* (1) *be* (r,λ_0, f_0)*-homogeneous for some* $r_i > 0$ *for all* $i \in \overline{1,n}$*, the function* $f_0 : R^n \to \mathbb{R}^n$ *be continuous and r-homogeneous with the degree* d_0 *and there exist an r-homogeneous Lyapunov-Razumikhin function* $V_0 : \mathbb{R}^n \to \mathbb{R}_+$ *with the degree* $v_0 > \max\{0, -d_0\}$ *such that there exists a* $\alpha \in \mathscr{K}$ *such that* $D^+V_0(x) f_0(x) \leq -\alpha(|x|)$ *(resp.,* $D^-V_0(x) f_0(x) \geq \alpha(|x|)$*) for all* $x \in \mathbb{S}_r$*. Then all conclusions of Theorem* 3 *(resp., Theorem* 4*) hold.*

Similarly to [13], one can use these conditions to detect Yakubovich oscillations for (1) [12].

5 ISS Property of Time-Delay Homogeneous Systems

Consider the system (1) with inputs:

$$dx(t)/dt = f[x_t, u(t)], \; t \geq 0, \tag{4}$$

where $x \in \mathbb{R}^n$, $x_t \in C_{[-\tau,0]}$ is the state function as before, $u : \mathbb{R}_+ \to \mathbb{R}^m$ is an essentially bounded (Lebesgue) measurable input, $||u||_\infty = ess.\sup_{t\geq 0} |u(t)|$ (and we will let $\mathscr{L}_\infty$ denote the set of all inputs $u : \mathbb{R}_+ \to \mathbb{R}^m$ such that $||u||_\infty < +\infty$), and $f : C_{[-\tau,0]} \times \mathbb{R}^m \to \mathbb{R}^n$ is a continuous function which is locally Lipschitz with respect to x_t and which satisfies $f(0, 0) = 0$. Under these conditions, the system (4) has a unique solution $x(t, x_0, u)$ for any $u \in \mathscr{L}_\infty$ and $x_0 \in C_{[-\tau,0]}$ defined on some interval $[-\tau, T)$. Lyapunov-Razumikhin theory has been extended to ISS [39], as follows.

Definition 3 The system (4) is called ISS provided for all $x_0 \in C_{[-\tau,0]}$ and $u \in \mathscr{L}_\infty$, there exist $\beta \in \mathscr{KL}$ and $\gamma \in \mathscr{K}$ such that $|x(t, x_0, u)| \le \beta(||x_0||, t) + \gamma(||u||_\infty)$ holds for all $t \ge 0$.

Theorem 5 *Let there exist a locally Lipschitz continuous ISS Lyapunov-Razumikhin function $V : \mathbb{R}^n \to \mathbb{R}_+$ for* (4)*, i.e., there are functions $\alpha_i \in \mathscr{K}_\infty$ for $i = 1, 2, 3$ and γ and χ in $\mathscr{K}$ such that $\gamma(s) > s$ for all $s \in \mathbb{R}_+$ and such that for all $x \in \mathbb{R}^n$, $\varphi \in C_{[-\tau,0]}$, and $u \in \mathbb{R}^m$, we have $\alpha_1(|x|) \le V(x) \le \alpha_2(|x|)$ and*

$$\begin{aligned}&\max\left(\max_{\theta\in[-\tau,0]} V[\varphi(\theta)], \chi(|u|)\right) < \gamma\{V[\varphi(0)]\} \Rightarrow \\ &D^+V[\varphi(0)]f(\varphi, u) \le -\alpha_3(|\varphi(0)|).\end{aligned}$$

Then (4) *is ISS.*

In [6, 26, 37], it has been shown that if a nonlinear dynamical system is homogeneous, then it is also ISS with respect to an input (e.g., an additive disturbance or measurement noise). A similar link for time-delay system (4) is established in the theorem below. Define $\tilde{f}(x_t, u) = [f(x_t, u)^T \ 0_m]^T \in \mathbb{R}^{n+m}$, which is an extended auxiliary vector field for the system (4), where 0_m is the zero vector with m elements.

Theorem 6 [15] *Let the vector field $\tilde{f}$ be homogeneous with the weights $r = [r_1, \ldots, r_n] > 0$ and $\tilde{r} = [\tilde{r}_1, \ldots, \tilde{r}_m] > 0$ with a degree*

$$d \ge -r_{\min}, \ r_{\min} = \min_{1\le i\le n} r_i, \ \textit{i.e.,}$$

$$f(\Lambda_r(\lambda)x_t, \Lambda_{\tilde{r}}(\lambda)u) = \lambda^d \Lambda_r(\lambda) f(x_t, u) \quad \forall \lambda > 0.$$

Assume that for the system (4) *for $u = 0$, there exists a continuously differentiable r-homogeneous Lyapunov-Razumikhin function $V : \mathbb{R}^n \to \mathbb{R}_+$, $V(0) = 0$ with degree $\nu > r_{\max} = \max_{1\le i\le n} r_i$ such that:*

(i) $\min_{y\in\mathbb{S}_r} V(y) > 0$ and there exist functions α and γ in $\mathscr{K}$ such that for all $\varphi \in \mathscr{S}_r$, we have

$$\max_{\theta\in[-\tau,0]} V[\varphi(\theta)] < \gamma\{V[\varphi(0)]\} \Rightarrow D^+V[\varphi(0)]f(\varphi, 0) \le -\alpha(|\varphi(0)|);$$

(ii) there is a $\wp \in \mathscr{K}$ such that $\lambda s < \wp(\lambda s) \le \lambda\gamma(s)$ for all s and λ in $\mathbb{R}_+ \setminus \{0\}$. Then the system (4) *is ISS.*

Theorem 6 says that if the conditions of Theorem 1 are satisfied for the case $u = 0$ for the system (4), then it is ISS. This conclusion highlights an additional importance of the homogeneity concept we introduced for time-delay systems, namely, under additional algebraic restrictions on the system equations and its Lyapunov-Razumikhin function (i.e., homogeneity) we gain system robustness.

6 Examples

In this section, we consider several examples of homogeneous systems to illustrate various aspects of Proposition 1 and Theorems 3, 4 and 6.

6.1 Homogeneous Time-Delay System with Degree d = 0

Consider the system $\dot{x}(t) = -2|x(t)|^{\alpha}\mathrm{sign}[x(t)] + |x(t-\tau)|^{\alpha}\mathrm{sign}[x(t-\tau)]$, where $x(t) \in \mathbb{R}$, $\tau > 0$ is a fixed time delay, and $\alpha > 0$ is a parameter. Applying the Lyapunov-Razumikhin approach with $V(x) = 0.5x^2$, it is straightforward to show that the system is stable for any $\tau > 0$. The system is homogeneous for $r = 1$ and $d = \alpha - 1$. According to Proposition 1, if $x(t, x_0)$ is a solution of the system (1) with initial condition $x_0 \in C_{[-\tau,0]}$, then $y(t, y_0) = \Lambda_r(\lambda)x(\lambda^d t, x_0)$ is a solution of the system (2) with the initial condition

$$y_0 \in C_{[-\lambda^{-d}\tau,0]}, \text{ where } y_0(s) = \Lambda_r(\lambda)x_0(\lambda^d s)$$

for $s \in [-\lambda^{-d}\tau, 0]$ for any $\lambda > 0$. For $\tau = 10$ and two values $\alpha = 0.5$ and $\alpha = 1.5$ of α, the system (1) trajectory $x(t, x_0)$ with $x_0(s) = 2$ for all $s \in [-\tau, 0]$, the system (2) trajectory $z(t, z_0)$ for $z_0(s) = 2$ for $s \in [-3^{1-\alpha}\tau, 0]$, and $y(t) = 3x(3^{\alpha-1}t, x_0)$ are shown in Fig. 1. The results of these simulations confirm the scaling property established in Proposition 1 since the variables $y(t)$ and $z(t)$ coincide.

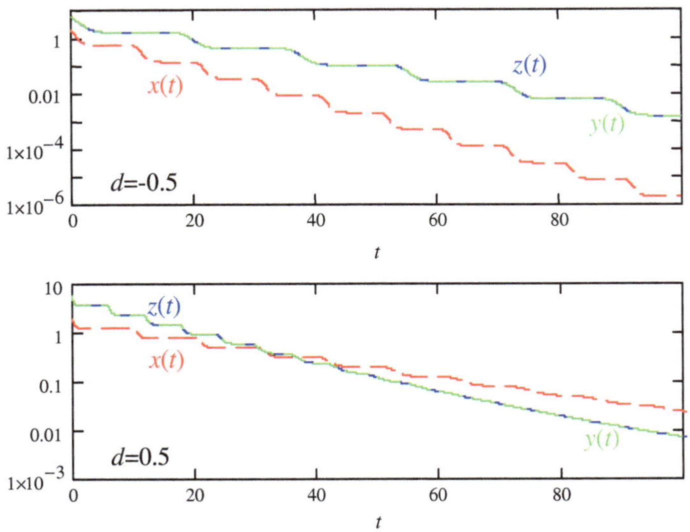

Fig. 1 Scaling of trajectories of a homogeneous system with different degrees

6.2 Nicholson's Blowfly Population Dynamics

Consider the recruitment-delayed model $\dot{x}(t) = bx(t-\tau)e^{-cx(t-\tau)} - ax(t)$ of a population [20, 33], where $x(t) \in \mathbb{R}_+$ is the adult population size; a, b, and c are positive parameters; and $\tau > 0$ is a fixed delay. The first term on the right-hand side is the birth rate, and the second is the death rate. This equation was introduced by Nicholson [33] to model a laboratory fly population. The delay models the maturation period after birth. This system is locally homogeneous in the bi-limit:

$$\lambda_1 = 0, \ r_1 = 1, \ f_1(x_\tau) = bx(t-\tau) - ax(t), \ d_1 = 0;$$
$$\lambda_2 = +\infty, \ r_2 = 1, \ f_2(x_\tau) = -ax(t), \ d_2 = 0.$$

Usually, it is assumed that $b > a$. Then, the approximation at the origin is unstable, and there is a nonzero equilibrium $\bar{x} = c^{-1}\ln(b/a)$. First approximation arguments show that for $b > ae^2$, the steady state $\bar{x}$ is also unstable for all $\tau \geq \tau_0$ for a sufficiently big critical delay $\tau_0 > 0$. Thus, (3) is unstable at $\lambda_1 = 0$. Then, Theorem 4 implies instability of the system around the origin. For $\lambda_2 = +\infty$, the system (3) is asymptotically stable. From Theorem 3, this ensures global convergence of the trajectories to a vicinity of the origin. Approximate behavior of the system is defined by the delay-free dynamics (by Corollary 3). Since a nontrivial equilibrium $\bar{x}$ is also unstable, this population dynamics is oscillating in the Yakubovich sense [12]. Simulations for $b = 15$, $a = 1$, $\tau = 10$, and $c = 0.5$ are presented in Fig. 2 (where two trajectories $x(t)$ are shown for different initial conditions).

6.3 ISS Academic Example of Degree $d \neq 0$

Consider the following system, which a modification of quasi-polynomial model widely used in the systems biology: [10, 31]):

$$\dot{x}_1(t) = -a_1x_1(t)[x_1(t) + u_1(t)] + b_1x_2(t-\tau) + u_2(t),$$
$$\dot{x}_2(t) = -a_2x_2^{1.5}(t) + [b_2x_2(t) + u_3(t)]x_1(t-\tau),$$

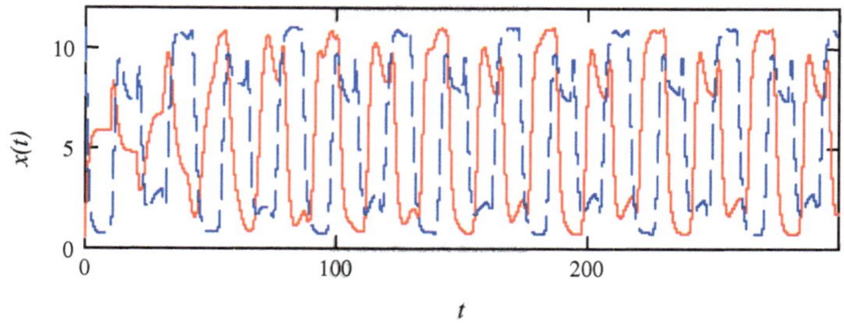

Fig. 2 Simulation for Nicholson's population model

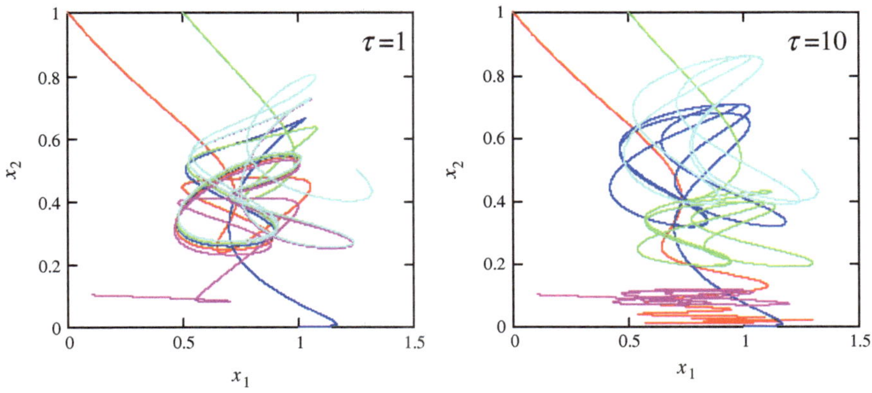

Fig. 3 Trajectories of an ISS system

where $x_1(t) \in \mathbb{R}_+$ and $x_2(t) \in \mathbb{R}_+$ represent some concentrations in a genetic network, the inputs $u_i(t) \in \mathbb{R}_+, i \in \overline{1,3}$ are the model uncertainties, $\tau > 0$ represents the transition delay in the network; and a_1, a_2, b_1 and b_2 are positive parameters. It is straightforward to check that the system is positive and homogeneous for $r = [1\ 2]$ and $\tilde{r} = [1\ 2\ 2]$ with degree $d = 1$. For $u_i = 0$ and $i \in \overline{1,3}$, consider the Lyapunov function $V(x_1, x_2) = x_1^2 + x_2$, which is r-homogeneous and. Note that $\max\{x_1^2, x_2\} \le V(x_1, x_2)$ and $V^{1.5}(x_1, x_2) \le \frac{3}{2}[x_1^3 + x_2^{1.5}]$. Moreover,

$$\begin{aligned}\dot{V}(t) &= -2a_1x_1^3(t) + 2b_1x_1(t)x_2(t-\tau) - a_2x_2^{1.5}(t) + b_2x_1(t-\tau)x_2(t)\\ &\le -\min\{2a_1, a_2\}[x_1^3(t) + x_2^{1.5}(t)]\\ &\quad + \max\{2b_1, b_2\}[x_1(t)x_2(t-\tau) + x_1(t-\tau)x_2(t)]\\ &\le -\frac{2}{3}\min\{2a_1, a_2\}V^{1.5}(t) + \max\{2b_1, b_2\}[\sqrt{V(t)}V(t-\tau) + \sqrt{V(t-\tau)}V(t)].\end{aligned}$$

Assume that $\max\{2b_1, b_2\}[p + \sqrt{p}] + \varepsilon \le \frac{2}{3}\min\{2a_1, a_2\}$ holds for some $p > 1$ and $\varepsilon > 0$. Then, $pV(t) \ge V(t-\tau) \Rightarrow \dot{V}(t) \le -\varepsilon V^{1.5}(t)$ and V is a Lyapunov-Razumikhin function for the unperturbed system. Then, by Theorem 6, the system is ISS with respect to inputs $u_i(t)$, $i = \overline{1,3}$. This conclusion is confirmed by the results of numerical simulation presented in Fig. 3 for $a_1 = 1$, $a_2 = 2$, $b_1 = 0.25$, $b_2 = 0.5$, $p = 0.9$, $u_1(t) = \sin^2(0.5t)$, $u_2(t) = 2\cos^2(2t)$, $u_3(t) = \sin^2(t)$ and different values of τ.

7 Conclusions

We presented a way to develop homogeneity notions for time-delay systems. We showed that a local stability notion for homogeneous time-delay systems holds globally, which simplifies the analysis of such systems by applying linearization at the origin, for instance. Sufficient conditions for stability and instability of homogeneous systems are presented. The conditions are based on Razumikhin stability arguments. It is also shown that if a homogeneous system has a homogeneous Lyapunov-Razumikhin function, then under a mild structural condition (involving the degree of homogeneity), it is ISS. The definition of local homogeneity is proposed, and relations between stability and instability of locally approximating dynamics and the original system are established. Efficiency of the proposed approach is demonstrated with numerical experiments.

References

1. Aleksandrov, A., Zhabko, A.: On the asymptotic stability of solutions of nonlinear systems with delay. Siberian Math. J. **53**(3), 393–403 (2012)
2. Andrieu, V., Praly, L., Astolfi, A.: Homogeneous approximation, recursive observer design, and output feedback. SIAM J. Control Optim. **47**(4), 1814–1850 (2008)
3. Asl, F., Ulsoy, A.: Analytical solution of a system of homogeneous delay differential equations via the Lambert function. In: Proceedings of the American Control Conference, pp. 2496–2500 (2000)
4. Bacciotti, A., Rosier, L.: Liapunov Functions and Stability in Control Theory. Springer, Berlin (2001)
5. Bernuau, E., Polyakov, A., Efimov, D., Perruquetti, W.: Verification of ISS, iISS and IOSS properties applying weighted homogeneity. Syst. Control Lett. **62**(12), 1159–1167 (2013)
6. Bernuau, E., Polyakov, A., Efimov, D., Perruquetti, W.: On ISS and iISS properties of homogeneous systems. In: Proceedings of the European Control Conference, pp. 1705–1710 (2013)
7. Bhat, S., Bernstein, D.: Geometric homogeneity with applications to finite-time stability. Math. Control Signals Syst. **17**(2), 101–127 (2005)
8. Bokharaie, V., Mason, O., Verwoerd, M.: D-stability and delay-independent stability of homogeneous cooperative systems. IEEE Trans. Autom. Control **55**(12), 2882–2885 (2010)
9. Chiasson, J., Loiseau, J. (eds.): Applications of Time Delay Systems. Springer, Berlin (2007)
10. Chowdhury, A., Chetty, M., Vinh, N.: Incorporating time-delays in S-system model for reverse engineering genetic networks. BMC Bioinf. **14**, 196 (2013)
11. Dublik, J.: Asymptotic equilibrium for homogeneous delay linear differential equations with l-perturbation term. Nonlinear Anal. Theory Methods Appl. **30**(6), 3927–3933 (1997)
12. Efimov, D., Fradkov, A.: Oscillatority conditions for nonlinear systems with delays. J. Appl. Math. **2007**(2007), 1–12 (2007)
13. Efimov, D., Perruquetti, W.: Oscillations conditions in homogeneous systems. In: Proceedings of the 8th IFAC Symposium on Nonlinear Control Systems, pp. 1379–1384 (2010)
14. Efimov, D., Perruquetti, W.: Homogeneity for time-delay systems. In: Proceedings of the 18th IFAC World Congress, pp. 3861–3866 (2011)
15. Efimov, D., Perruquetti, W., Richard, J.-P.: Development of homogeneity concept for time-delay systems. SIAM J. Control Optim. **52**(3), 1547–1566 (2014)
16. Erneux, T.: Applied Delay Differential Equations. Springer, New York (2009)
17. Fridman, E.: A refined input delay approach to sampled-data control. Automatica **46**(2), 421–427 (2010)
18. Grune, L.: Homogeneous state feedback stabilization of homogeneous systems. SIAM J. Control Optim. **38**(4), 1288–1314 (2000)
19. Gu, K., Kharitonov, K., Chen, J.: Stability of Time-Delay Systems. Birkhauser, Boston (2003)
20. Gurney, W., Blythe, S., Nisbet, R.: Nicholsons blowfly revisited. Nature **287**, 17–21 (1980)

21. Haddock, J., Ko, Y.: Lyapunov-Razumikhin functions and an instability theorem for autonomous functional differential equations with finite delay. Rocky Mt. J. Math. **25**(1), 261–267 (1995)
22. Haddock, J., Zhao, J.: Instability for autonomous and periodic functional differential equations with finite delay. Funkcialaj Ekvacioj **39**(3), 553–570 (1996)
23. Hale, J.: Theory of Functional Differential Equations. Springer-verlag, New York (1977)
24. Hermes, H.: Nilpotent and high-order approximations of vector field systems. SIAM Rev. **33**(2), 238–264 (1991)
25. Hermes, H.: Homogeneous coordinates and continuous asymptotically stabilizing feedback controls. Differential Equations (Colorado Springs. CO, 1989), pp. 249–260. Marcel Dekker, New York (1991)
26. Hong, Y.: H_∞ control, stabilization, and input-output stability of nonlinear systems with homogeneous properties. Automatica **37**(7), 819–829 (2001)
27. Hong, Y.: Finite-time stabilization and stabilizability of a class of controllable systems. Syst. Control Lett. **46**(4), 231–236 (2002)
28. Kawski, M.: Homogeneous feedback stabilization. In: Conte, G., Perdon, A., Wyman, B. (eds.) New Trends in Systems Theory, pp. 464–471. Birkhauser, Boston (1991)
29. Kolmanovsky, V., Nosov, V.: Stability of Functional Differential Equations. Kluwer Academic, Dordrecht (1986)
30. Meenard, T., Moulay, E., Perruquetti, W.: Homogeneous approximations and local observer design. ESAIM Control Optim. Calculus Var. **19**(3), 906–929 (2013)
31. Motee, N., Bamieh, B., Khammash, M.: Stability analysis of quasi-polynomial dynamical systems with applications to biological network models. Automatica **48**(6), 2945–2950 (2012)
32. Moulay, E., Perruquetti, W.: Finite time stability and stabilization of a class of continuous systems. J. Math. Anal. Appl. **323**(2), 1430–1443 (2006)
33. Nicholson, A.: An outline of the dynamics of animal populations. Aust. J. Zool. **2**(1), 9–65 (1954)
34. Richard, J.-P.: Time-delay systems: an overview of some recent advances and open problems. Automatica **39**(10), 1667–1694 (2003)
35. Rosier, L.: Homogeneous Lyapunov function for homogeneous continuous vector fields. Syst. Control Lett. **19**(6), 467–473 (1992)
36. Rothschild, L., Stein, E.: Hypoelliptic differential operators and nilpotent groups. Acta Math. **137**(1), 247–320 (1976)
37. Ryan, E.: Universal stabilization of a class of nonlinear systems with homogeneous vector fields. Syst. Control Lett. **26**(3), 177–184 (1995)
38. Sepulchre, R., Aeyels, D.: Stabilizability does not imply homogeneous stabilizability for controllable systems. SIAM J. Control Optim. **34**(5), 1798–1813 (1996)
39. Teel, A.: Connections between Razumikhin-type theorems and the ISS nonlinear small gain theorem. IEEE Trans. Autom. Control **43**(7), 960–964 (1998)
40. Zubov, V.: Systems of ordinary differential equations with generalized-homogeneous right-hand sides. Izvestija Vyssih Ucebnyh Zavedenii Matematika **1**(2), 80–88 (1958)

Sliding Mode Observer for Robust Fault Reconstruction of Time Delay Systems

Xiaoran Han, Emilia Fridman and Sarah K. Spurgeon

Abstract A sliding mode observer in the presence of known output delay and its application to robust fault reconstruction is studied. The observer is designed using a singular perturbation method for which sufficient conditions are given in the form of linear matrix inequalities (LMIs) to guarantee ultimate boundedness of the error dynamics. Though an ideal sliding motion cannot be achieved in the observer when the outputs are delayed, ultimately bounded solutions can be obtained, provided the delay is sufficiently small. The bound on the solution is proportional to the delay and the magnitude of the switching gain. The proposed observer design is applied to the problem of fault reconstruction under delayed outputs and system uncertainties. It is shown that actuator or sensor faults can be reconstructed reliably from the output error dynamics. An example of observer design for an inverted pendulum system is used to demonstrate the merit of the proposed methodology, compared with existing sliding mode observer design approaches.

1 Introduction

A sliding mode observer is a category of robust observer which facilitates the complete rejection of a class of uncertainty between the system and observer [25]. In most cases, the sliding surface is set to be the difference between the observer outputs and system outputs, which is therefore forced to zero [6, 28]. A discontinuous injection term is designed and applied to drive the observer so that the error between the outputs

X. Han (✉)
Aeronautical and Automotive Engineering, Loughborough University,
Loughborough LE11 3TU, UK
e-mail: ttxh@lboro.ac.uk

E. Fridman
School of Electrical Engineering, Tel Aviv University, 69978 Tel Aviv, Israel
e-mail: emilia@eng.tau.ac.il

S.K. Spurgeon
School of Engineering and Digital Arts, University of Kent, Canterbury CT2 7NZ, UK
e-mail: S.K.Spurgeon@kent.ac.uk

I. Karafyllis et al. (eds.), *Recent Results on Nonlinear Delay Control Systems*,
Advances in Delays and Dynamics 4, DOI 10.1007/978-3-319-18072-4_9

of the observer and the outputs of the plant will move onto this surface within the error space and then remain there. In terms of implementation, delays exist in many applications, such as those caused by transmission delay and computational delay. If performance levels are to be optimized in the presence of such delays, it is necessary to consider the development of methodologies which incorporate knowledge of the delay in the design framework. There have been many results that investigate the effect of state delay on observer design [4, 5]. However, observer design with delays in the output measurements has begun to receive more interest. Observer design for nonlinear systems with delayed outputs has been considered in [1, 2, 13, 14], and [18]. It has been studied for linear systems in [26]. In terms of work that considers the effect of time-delay in sliding mode observers, the literature is very sparse [19] and is strongly aligned with observer based control, rather than fault detection and estimation with an emphasis on state delay rather than measurement delay [23, 24]. Since the switching term in a sliding mode observer depends on the output measurement, which may be subject to delay in practice, the resulting discontinuous injection applied to the observer has the potential to cause chattering of large amplitude. This behavior may limit the magnitude of the discontinuous signal that it is possible to apply with the observer.

There has been a great deal of interest in the application of sliding mode observers to the problem of model based fault detection and isolation [8, 12, 30]. The merit of the approach lies in the application of the so-called *equivalent output injection* to explicitly reconstruct fault signals. The results obtained to date mostly require that an ideal sliding motion is attained in finite time before the appearance of faults, and that no delay is present in the output measurement used to drive the observer. The aim in this chapter is to develop a general framework for sliding mode observer design and fault reconstruction under multiple delayed outputs. The error dynamics is forced to exhibit a bound proportional to the delay period of the outputs and the magnitude of the discontinuous switching gain employed in the observer. The observer, which is designed using a singular perturbation approach, possesses a sufficiently small perturbation parameter μ such that faults are reliably constructed despite the presence of the delayed output. The observer synthesis is formulated in terms of LMIs, the feasibility of which is guaranteed for small enough μ. The effect of uncertainties on the fault reconstruction is minimized by incorporating H_∞ concepts within the observer design framework. In Sect. 2, the problem of sliding mode observer design with delayed outputs is formulated, in terms of a system representation with known varying delay. Section 3 develops an observer design approach which ensures the ultimate boundedness of the error dynamics. By using the singular perturbation method, Section 4 shows that approximate fault reconstruction can be achieved. The sensor fault reconstruction is demonstrated in Sect. 5, where new measurable states are augmented to the original faulty system so that the results developed in the previous sections can be applied as the sensor fault is now transformed into input fault. In Sect. 6, the effectiveness of the result is demonstrated using a linearized model of the inverted pendulum. Some preliminary results from this chapter in the context of the input delay problem were presented in [16].

Notation: Throughout the chapter, the superscript "T" stands for matrix transposition, $\mathbb{R}^n$ denotes the n-dimensional Euclidean space with vector norm $\|\cdot\|$, $\mathbb{R}^{n\times m}$ is the set of all $n \times m$ real matrices, and the notation $P > 0$, for $P \in \mathbb{R}^{n\times n}$ means that P is symmetric and positive definite. The symmetric elements of the symmetric matrix are denoted by $*$. The symbol $\|\cdot\|_\infty$ stands for essential supremum.

2 Problem Statement

Consider the following linear, uncertain system with a known measurement delay:

$$\begin{aligned} \dot{x}(t) &= Ax(t) + Bu(t) + Df_i(t) + M\zeta(t, y, u) \\ y(t) &= Cx(t - \tau(t)) \end{aligned} \tag{1}$$

where $x \in \mathbb{R}^n$ and $u \in \mathbb{R}^m$ are the state and the input vector, respectively, and $A,\ B,\ C,\ D$ and M are constant and known system matrices of appropriate dimensions. The unknown actuator faults $f_i \in \mathbb{R}^q$ are assumed to be bounded, i.e., $\sup_t \|f_i(t)\| \le \Delta$, where Δ is known. The signal $\zeta(:)\mathbb{R}_+ \times \mathbb{R}^p \times \mathbb{R}^m \to \mathbb{R}^k$ encapsulates the uncertainty in the system. It is assumed to be unknown but that it admits a uniform bound $\|\zeta(t, y, u)\| \le \beta$ where the positive scalar β is known. For real systems, methods to establish an appropriate fixed matrix M so that $M\zeta$ captures the uncertainty are developed in [3]. The measurement delay is assumed to be known and bounded, i.e., $\tau(t) \in [0, h]$ for all t, where h is a known bound. It is assumed that the delay is either fast varying (without any constraints on the delay derivative) or slowly varying, where the delay-derivative satisfies a bound of the form $\dot{\tau} \le d < 1$. It is assumed that $q \le p < n$ and that:

1. rank $(CD) = q$; and
2. any invariant zeros of (A, D, C) lie in the left half plane.

Invariant zeros in the sliding mode dynamics are those eigenvalues of A for which Rosenbrock's system matrix loses normal rank [7]. Under these assumptions, there exists a linear change of coordinates

$$\begin{bmatrix} x_1 \\ x_2 \end{bmatrix} = T_0 x,$$

where T_0 is non-singular, such that the transformed system has the following form:

$$\begin{aligned} \dot{x}_1(t) &= A_{11}x_1(t) + A_{12}x_2(t) + B_1u(t) + M_1\zeta(t, y, u) \\ \dot{x}_2(t) &= A_{21}x_1(t) + A_{22}x_2(t) + B_2u(t) + D_1f_i(t) + M_2\zeta(t, y, u) \\ y(t) &= Tx_2(t - \tau(t)) \end{aligned} \tag{2}$$

(see [7]), where $x_1 \in \mathbb{R}^{n-p},\ x_2 \in \mathbb{R}^p$,

$$D_1 = \begin{bmatrix} 0 \\ \bar{D}_1 \end{bmatrix},$$

$\bar{D}_1 \in \mathbb{R}^{q\times q}$, and T is an orthogonal matrix. An observer will be designed which for sufficiently large t, induces motion proportional to the $h(\Delta + \beta)$-neighbourhood of the surface

$$\mathscr{E} = \{x_2, \hat{x}_2 \in \mathbb{R}^p \ : \ s_e(t) = T\big(x_2(t-\tau(t)) - \hat{x}_2(t-\tau(t))\big) = 0\}$$

where $\hat{x}_2(t-\tau(t))$ is the corresponding component of the estimated states from an observer to be designed; see our main theorem below. An ideal sliding mode can be achieved with $h = 0$ under Assumptions 1 and 2.

3 Observer Design

Noting that (A_{11}, A_{21}) is detectable (by Assumptions 1 and 2), choose a matrix $L \in \mathbb{R}^{(n-p)\times p}$ of the form $L = \big[\bar{L}\ 0\big]$ with $\bar{L} \in \mathbb{R}^{(n-p)\times(p-q)}$ such that $LD_1 = 0$ and $A_{11} + LA_{21}$ is stable. Consider the following observer of system (2):

$$\begin{aligned}
&\dot{\hat{x}}_1(t) = A_{11}\hat{x}_1(t) + A_{12}\hat{x}_2(t) + B_1u(t)\\
&-(\tfrac{1}{\mu}L + A_{11}L)(x_2(t-\tau(t)) - \hat{x}_2(t-\tau(t))) + LT^Tv(t-\tau(t))\\
&\dot{\hat{x}}_2(t) = A_{21}\hat{x}_1(t) + A_{22}\hat{x}_2(t) + B_2u(t)\\
&-(A_{21}L - \tfrac{1}{\mu}I_p)(x_2(t-\tau(t)) - \hat{x}_2(t-\tau(t)) - T^Tv(t-\tau(t))\\
&\hat{y}(t) = T\hat{x}_2(t-\tau(t))
\end{aligned} \tag{3}$$

where $\mu > 0$ is a scalar. Here the discontinuous injection term v is given by

$$v(t) = -M_\beta[\text{sign } \bar{e}_{2_1}(t), \ldots, \text{sign } \bar{e}_{2_p}(t)]^T, \tag{4}$$

where

$$M_\beta = \|TD_1\|\Delta + \delta_1\Delta + \delta_2\beta + \|TM_2\|\beta,$$

δ_1 and δ_2 are positive scalars, and the subscript p denotes the p-th component of $\bar{e}_2(t) = T(x(t) - \hat{x}(t))$. Our objective is to find appropriate design parameters δ_1, δ_2, and μ such that $\|\bar{e}_2(t)\|$ is minimized for a given bound on the delay h for all t. The linear change of coordinates

$$\hat{x} = T_0^{-1}\big[\hat{x}_1\ \hat{x}_2\big]^T$$

leads to the observer of (1) given by

$$\begin{aligned}
&\dot{\hat{x}}(t) = A\hat{x}(t) + Bu(t) - G_l\bar{e}_2(t-\tau(t)) + G_nv(t-\tau(t))\\
&\hat{y}(t) = C\hat{x}(t-\tau(t))
\end{aligned} \tag{5}$$

where observer gains $G_l \in \mathbb{R}^{n\times p}$ and $G_n \in \mathbb{R}^{n\times p}$ are designed to be in the form

$$G_l = T_0^{-1}\begin{bmatrix} \frac{1}{\mu}L + A_{11}L \\ A_{21}L - \frac{1}{\mu}I_p \end{bmatrix} \text{ and } G_n = T_0^{-1}\begin{bmatrix} LT^T \\ -T^T \end{bmatrix} \tag{6}$$

where $\mu > 0$ is a scalar. Defining the state estimation error as $e_1(t) = x_1(t) - \hat{x}_1(t)$ and $e_2(t) = x_2(t) - \hat{x}_2(t)$, we obtain

$$\begin{aligned}
&\dot{e}_1(t) = A_{11}e_1(t) + A_{12}e_2(t) + M_1\zeta(t, y, u) \\
&+ L\big(\tfrac{1}{\mu}e_2(t-\tau(t)) - T^T v(t-\tau(t))\big) + A_{11}Le_2(t-\tau(t)) \\
&\dot{e}_2(t) = A_{21}e_1(t) + A_{22}e_2(t) + D_1 f_i(t) + M_2\zeta(t, y, u) \\
&+ T^T v(t-\tau(t)) - (\tfrac{1}{\mu}I_p - A_{21}L)e_2(t-\tau(t)).
\end{aligned}$$

Changing variables using

$$\begin{bmatrix} \bar{e}_1(t) \\ \bar{e}_2(t) \end{bmatrix} = T_L \begin{bmatrix} e_1(t) \\ e_2(t) \end{bmatrix}, \quad \text{where } T_L = \begin{bmatrix} I_{n-q} & L \\ 0 & T \end{bmatrix},$$

one obtains

$$\begin{aligned}
\dot{\bar{e}}_1(t) &= (A_{11} + LA_{21})\bar{e}_1(t) - (A_{11}L + LA_{21}L - A_{12} - LA_{22})T^T\bar{e}_2(t) \\
&+ (A_{11} + LA_{21})LT^T\bar{e}_2(t-\tau(t)) + (M_1 + LM_2)\zeta(t, y, u)
\end{aligned} \tag{7}$$

$$\begin{aligned}
\dot{\bar{e}}_2(t) &= TA_{21}\bar{e}_1(t) - (TA_{21}LT^T - TA_{22}T^T)\bar{e}_2(t) + TA_{21}LT^T\bar{e}_2(t-\tau(t)) \\
&- \tfrac{1}{\mu}\bar{e}_2(t-\tau(t)) + v(t-\tau(t)) + TD_1 f_i(t) + TM_2\zeta(t, y, u)
\end{aligned} \tag{8}$$

with initial condition

$$\bar{e}(t_0) = \bar{e}_0, \quad \bar{e}(t) = 0, \quad t < t_0 \tag{9}$$

The dynamics of the switching manifold is governed by (7).

3.1 Input-to-State Stability of the Error Dynamics: Singular Perturbation Approach

The closed-loop system (7)–(8) can be expressed as

$$\dot{\bar{e}}_1(t) = \bar{A}_{11}\bar{e}_1(t) + \bar{A}_{12}\bar{e}_2(t) + \bar{A}_\tau\bar{e}_2(t-\mu\xi(t)) + \bar{M}_1\zeta(t, y, u) \tag{10}$$

$$\mu\dot{\bar{e}}_2(t) = \mu\bar{A}_{21}\bar{e}_1(t) + \mu\bar{A}_{22}\bar{e}_2(t) + (\mu\bar{A}_{d22} - I_p)\bar{e}_2(t-\mu\xi(t)) + \mu\bar{f}_i(t) \tag{11}$$

where $\mu\xi(t) = \tau(t)$, $0 \le \xi(t) \le \bar{\xi}$, $\mu\bar{\xi} = h$, $\bar{A}_{11} = A_{11} + LA_{21}$, $\bar{A}_{12} = -(A_{11}L + LA_{21}L - A_{12} - LA_{22})T^T$, $\bar{A}_\tau = (A_{11} + LA_{21})LT^T$, $\bar{M}_1 = (M_1 + LM_2)$, $\bar{A}_{21} = TA_{21}$, $\bar{A}_{22} = -(TA_{21}LT^T - TA_{22}T^T)$, $\bar{A}_{d22} = TA_{21}LT^T$, and

$$\bar{f}_i(t) = v(t - \mu\xi(t)) + TD_1 f_i(t) + TM_2\zeta(t, y, u),$$

i.e. $\|\bar{f}_i(t)\| \le M_f = M_\beta\sqrt{p} + \|TD_1\|\Delta + \|TM_2\|\beta$. Let $P_\mu \in \mathbb{R}^{n\times n}$ be a positive definite matrix with the structure [20]

$$P_\mu = \begin{bmatrix} P_1 & \mu P_2^T \\ * & \mu P_3 \end{bmatrix} > 0 \tag{12}$$

where $P_1 \in \mathbb{R}^{n-p}$, and choose the Lyapunov-Krasovskii functional

$$\begin{aligned} V(t) = &\int_{t-\mu\xi(t)}^t e^{\bar{\alpha}(s-t)}\bar{e}_2^T(s)S\bar{e}_2(s)ds + \int_{t-\mu\bar{\xi}}^t e^{\bar{\alpha}(s-t)}\bar{e}_2^T(s)G\bar{e}_2(s)ds \\ &+ \mu\bar{\xi}\int_{-\mu\bar{\xi}}^0 \int_{t+\theta}^t e^{\bar{\alpha}(s-t)}\dot{\bar{e}}_2^T(s)R\dot{\bar{e}}_2(s)dsd\theta + \bar{e}^T(t)P_\mu\bar{e}(t) \end{aligned} \tag{13}$$

with respect to the error dynamics (10)–(11), where G, R and $S \in \mathbb{R}^p$ are positive matrices. Then, the following lemma can be stated:

Lemma 1 *Given positive tuning scalars μ, $\bar{\xi}$, $\bar{\alpha}$, $\bar{b}$, and $\bar{b}_1$, let there exist $P_\mu > 0$ in (12) with $(n-p)\times(n-p)$ matrix $P_1 > 0$, $p\times(n-p)$-matrix P_2 and $p\times p$ positive matrices P_3, G, R, and S such that the LMI with entries*

$$\Theta_\mu = \begin{bmatrix} \tilde{\theta}_{1,1} & \cdots & \tilde{\theta}_{1,7} \\ * & \cdots & \tilde{\theta}_{7,7} \end{bmatrix} < 0 \tag{14}$$

$$\begin{aligned}
\tilde{\theta}_{1,1} &= P_1\bar{A}_{11} + \bar{A}_{11}^T P_1 + \mu P_2^T\bar{A}_{21} + \mu\bar{A}_{21}^T P_2 + \bar{\alpha}P_1, \\
\tilde{\theta}_{1,2} &= P_1\bar{A}_{12} + \mu\bar{A}_{21}^T P_3 + \mu\bar{A}_{11}^T P_2^T + \mu P_2^T\bar{A}_{22} + \bar{\alpha}\mu P_2^T, \\
\tilde{\theta}_{1,4} &= P_2^T(\mu\bar{A}_{d22} - I_p) + P_1\bar{A}_\tau,\ \tilde{\theta}_{1,5} = P_2^T,\ \tilde{\theta}_{1,6} = P_1\bar{M}_1,\ \tilde{\theta}_{1,7} = \mu\bar{\xi}\bar{A}_{21}^T R, \\
\tilde{\theta}_{2,2} &= \mu P_2\bar{A}_{12} + \mu\bar{A}_{12}^T P_2^T + \mu P_3\bar{A}_{22} + \mu\bar{A}_{22}^T P_3 + \bar{\alpha}\mu P_3 + G - e^{-\bar{\alpha}\mu\bar{\xi}}R + S, \\
\tilde{\theta}_{2,4} &= P_3(\mu\bar{A}_{d22} - I_p) + e^{-\bar{\alpha}\mu\bar{\xi}}R + \mu P_2\bar{A}_\tau, \\
\tilde{\theta}_{2,5} &= P_3,\ \tilde{\theta}_{2,6} = \mu P_2\bar{M}_1,\ \tilde{\theta}_{2,7} = \mu\bar{\xi}\bar{A}_{22}^T R, \\
\tilde{\theta}_{3,3} &= -e^{-\bar{\alpha}\mu\bar{\xi}}G - e^{-\bar{\alpha}\mu\bar{\xi}}R,\ \tilde{\theta}_{3,4} = e^{-\bar{\alpha}\mu\bar{\xi}}R, \\
\tilde{\theta}_{4,4} &= -2e^{-\bar{\alpha}\mu\bar{\xi}}R - (1-d)Se^{-\bar{\alpha}\mu\bar{\xi}},\ \tilde{\theta}_{4,7} = \bar{\xi}(\mu\bar{A}_{d22} - I_p)^T R, \\
\tilde{\theta}_{5,5} &= -\bar{b}I_p,\ \tilde{\theta}_{5,7} = \bar{\xi}R,\ \tilde{\theta}_{6,6} = -\bar{b}_1 I_k,\ \tilde{\theta}_{7,7} = -R
\end{aligned}$$

is feasible. Then all solutions of (7)–(9) *satisfy the bound*

$$\bar{e}^T(t)P_\mu\bar{e}(t) < e^{-\bar{\alpha}(t-t_0)}\bar{e}^T(t_0)P_\mu\bar{e}(t_0) + \tfrac{\mu^2\bar{b}}{\bar{\alpha}}\|\bar{f}_{i_{[t_0,t]}}\|_\infty^2 + \tfrac{\bar{b}_1}{\bar{\alpha}}\|\zeta_{[t_0,t]}\|_\infty^2 \tag{15}$$

when $\|\bar{e}_0\|^2 + \|\bar{f}_{i[t_0,t]}\|^2_\infty + \|\zeta_{[t_0,t]}\|^2_\infty > 0$ *and for all* $\xi(t) \in [0,\ \bar{\xi}]$ *satisfying* $\mu\dot{\xi}(t) \leq d < 1$ *(and thus* (7)–(8) *is input-to-state stable). Moreover, all solutions of* (7)–(8) *satisfy* (15) *for all fast-varying delays* $\xi(t) \in [0,\ \bar{\xi}]$ *if the LMI* (14) *is feasible with* $S = 0$.

Proof The inequality

$$W(t) = \tfrac{d}{dt}V(t) + \bar{\alpha}V(t) - \mu^2\bar{b}\bar{f}_i^{\,T}(t)\bar{f}_i(t) - \bar{b}_1\zeta^T(t,y,u)\zeta(t,y,u) < 0 \qquad (16)$$

along all trajectories of (7)–(8) guarantees (15) [10]. Differentiating V of the structure (12) and (13), along (10) and (11), it follows from (16) that

$$\begin{aligned}
&W(t) \leq 2\bar{e}_1^T(t)P_1\big([\bar{A}_{11}\ \bar{A}_{12}]\bar{e}(t) + \bar{A}_\tau\bar{e}_2(t-\mu\xi(t)) + \bar{M}_1\zeta(t,y,u)\big)\\
&+2\mu\bar{e}_2^T(t)P_2\big([\bar{A}_{11}\ \bar{A}_{12}]\bar{e}(t) + \bar{A}_\tau\bar{e}_2(t-\mu\xi(t)) + \bar{M}_1\zeta(t,y,u)\big)\\
&+2\bar{e}_1^T(t)P_2^T\big(\mu[\bar{A}_{21}\ \bar{A}_{22}]\bar{e}(t) + (\mu\bar{A}_{d22} - I_p)\bar{e}_2(t-\mu\xi(t)) + \mu\bar{f}_i(t)\big)\\
&+2\,\bar{e}_2^T(t)P_3\big(\mu[\bar{A}_{21}\ \bar{A}_{22}]\bar{e}(t) + (\mu\bar{A}_{d22} - I_p)\bar{e}_2(t-\mu\xi(t)) + \mu\bar{f}_i(t)\big)\\
&-\mu^2\bar{b}\bar{f}_i^T(t)\bar{f}_i(t) - \bar{b}_1\zeta^T(t,y,u)\zeta(t,y,u) + \bar{\alpha}\bar{e}_1^T(t)P_1\bar{e}_1(t) + \bar{\alpha}\mu\bar{e}_2^T(t)P_2\bar{e}_1(t)\\
&+\bar{\alpha}\mu\bar{e}_1^T(t)P_2^T\bar{e}_2(t) + \bar{\alpha}\mu\bar{e}_2^T(t)P_3\bar{e}_2(t) + \mu^2\bar{\xi}^2\dot{\bar{e}}_2^T(t)R\dot{\bar{e}}_2(t)\\
&-\mu\bar{\xi}\int_{t-\mu\bar{\xi}}^t e^{-\bar{\alpha}\mu\bar{\xi}}\dot{\bar{e}}_2^T(s)R\dot{\bar{e}}_2(s)ds + \bar{e}_2^T(t)G\bar{e}_2(t)\\
&-e^{-\bar{\alpha}\mu\bar{\xi}}\bar{e}_2^T(t-\mu\bar{\xi})G\bar{e}_2(t-\mu\bar{\xi}) + \bar{e}_2^T(t)S\bar{e}_2(t)\\
&-(1-d)e^{-\bar{\alpha}\mu\xi(t)}\bar{e}_2^T(t-\mu\xi(t))S\bar{e}_2(t-\mu\xi(t))\ .
\end{aligned}$$

Using the identity

$$\begin{aligned}
&\int_{t-\mu\bar{\xi}}^t e^{-\bar{\alpha}\mu\bar{\xi}}\dot{\bar{e}}_2^T(s)R\dot{\bar{e}}_2(s)ds = \int_{t-\mu\bar{\xi}}^{t-\mu\xi(t)} e^{-\bar{\alpha}\mu\bar{\xi}}\dot{\bar{e}}_2^T(s)R\dot{\bar{e}}_2(s)ds\\
&+\int_{t-\mu\xi(t)}^t e^{-\bar{\alpha}\mu\bar{\xi}}\dot{\bar{e}}_2^T(s)R\dot{\bar{e}}_2(s)ds,
\end{aligned}$$

we apply Jensen's inequality to obtain the following:

$$\begin{aligned}
&-\mu\bar{\xi}\int_{t-\mu\bar{\xi}}^{t-\mu\xi(t)} e^{-\bar{\alpha}\mu\bar{\xi}}\dot{\bar{e}}_2^T(s)R\dot{\bar{e}}_2(s)ds\\
&\leq -e^{-\bar{\alpha}\mu\bar{\xi}}[\bar{e}_2^T(t-\mu\xi(t)) - \bar{e}_2^T(t-\mu\bar{\xi})]R[\bar{e}_2(t-\mu\xi(t)) - \bar{e}_2(t-\mu\bar{\xi})]\\
&-\mu\bar{\xi}\int_{t-\mu\xi(t)}^t e^{-\bar{\alpha}\mu\bar{\xi}}\dot{\bar{e}}_2^T(s)R\dot{\bar{e}}_2(s)ds\\
&\leq -e^{-\bar{\alpha}\mu\bar{\xi}}[\bar{e}_2^T(t) - \bar{e}_2^T(t-\mu\xi(t))]R[\bar{e}_2(t) - \bar{e}_2(t-\mu\xi(t))]
\end{aligned}$$

Then, setting $\eta(t) = col\{\bar{e}_1(t), \bar{e}_2(t), \bar{e}_2(t-\mu\bar{\xi}), \bar{e}_2(t-\mu\xi(t)), \mu\bar{f}_i(t), \zeta(t,y,u)\}$ and applying Schur complements to the term $\mu^2\bar{\xi}^2\dot{\bar{e}}_2^T(t)R\dot{\bar{e}}_2(t)$, where $\dot{\bar{e}}_2(t)$ is substituted by the right-hand site of (11), it is established that $W(t) < 0$ if $\Theta_\mu < 0$.

3.2 LMIs for Switching Gain Design

Conditions will now be derived to guarantee the following bound for the solutions of (8):

$$
\begin{aligned}
&\limsup_{t\to\infty} \left\| \begin{bmatrix} \bar{A}_{21} \; \bar{A}_{22} \end{bmatrix} \bar{e}(t) \right\| \le k_1(\delta_1 \Delta + \delta_2 \beta) \text{ and} \\
&\limsup_{t\to\infty} \left\| \begin{bmatrix} 0 \; \bar{A}_{d22} \end{bmatrix} \bar{e}(t-\tau(t)) \right\| \le k_2(\delta_1 \Delta + \delta_2 \beta)
\end{aligned} \tag{17}
$$

for some $k_1 \ge 0$ and $k_2 \ge 0$ such that $k_1 + k_2 = 1$. Taking (15) into account, we can conclude that (17) holds if the following inequalities are satisfied for large enough t:

$$
\bar{e}^T(t)[\bar{A}_{21} \; \bar{A}_{22}]^T[\bar{A}_{21} \; \bar{A}_{22}]\bar{e}(t) < \frac{\bar{\alpha} k_1^2 (\delta_1 \Delta + \delta_2 \beta)^2 \bar{e}^T(t) P_\mu \bar{e}(t)}{\mu^2 \bar{b} M_f^2 + \bar{b}_1 \beta^2} \tag{18}
$$

$$
\bar{e}^T(t-\mu\xi(t))[0 \; \bar{A}_{d22}]^T[0 \; \bar{A}_{d22}]\bar{e}(t-\mu\xi(t)) < \frac{\bar{\alpha} k_2^2 (\delta_1 \Delta + \delta_2 \beta)^2 \bar{e}^T(t-\mu\xi(t)) P_\mu \bar{e}(t-\mu\xi(t))}{\mu^2 \bar{b} M_f^2 + \bar{b}_1 \beta^2} \tag{19}
$$

Hence, the inequalities

$$
\begin{bmatrix} -k_1^2 M_\Omega P_1 & -\mu k_1^2 M_\Omega P_2^T & \bar{A}_{21}^T \\ * & -\mu k_1^2 M_\Omega P_3 & \bar{A}_{22}^T \\ * & * & -I_p \end{bmatrix} < 0, \quad \begin{bmatrix} -k_2^2 M_\Omega P_1 & -\mu k_2^2 M_\Omega P_2^T & 0 \\ * & -\mu k_2^2 M_\Omega P_3 & \bar{A}_{d22}^T \\ * & * & -I_p \end{bmatrix} < 0 \tag{20}
$$

where

$$
M_\Omega = \frac{\bar{\alpha}(\delta_1 \Delta + \delta_2 \beta)^2}{\mu^2 \bar{b} M_f^2 + \bar{b}_1 \beta^2}
$$

guarantee that the solutions of (7) and (8) satisfy the bound (17). Matrix inequalities (12), (14) and (20) have been derived for finding parameters μ and δ_1 and δ_2 of the observer (5). It will now be shown that if the μ-dependent LMI

$$
\Theta_0 = \Theta_{\mu|\mu=0} < 0, \tag{21}
$$

is feasible, then for big enough $\delta_1 > 0$ and $\delta_2 > 0$, inequalities (12), (14) and (20) are feasible for small enough μ and $\bar{\xi}$. Let P_1, P_2, and P_3 satisfy the above conditions. Then, for small enough $\mu > 0$ and $\bar{\xi} > 0$, (12), (14) are feasible for the same μ-independent matrices P_1, P_2, and P_3. Hence, given big enough $\delta_1 > 0$ and $\delta_2 > 0$, (18) and (19) are feasible for small enough μ and $\bar{\xi}$. Note that feasibility of (21) guarantees exponential stability of the slow subsystem

$$
\dot{\bar{e}}_1(t) = \bar{A}_{11}\bar{e}_1(t)
$$

with decay rate $\bar{\alpha}/2$, and asymptotic stability of the fast subsystem

$$
\mu\dot{\bar{e}}_2(t) = -\bar{e}_2(t-\mu\xi(t)), \quad \xi \in [0, \; \bar{\xi}].
$$

Since $\bar{A}_{11}$ is Hurwitz by design, there exists a $P_1 > 0$ satisfying

$$P_1\bar{A}_{11} + \bar{A}_{11}^T P_1 + \bar{\alpha} P_1 < 0$$

for small enough $\bar{\alpha} > 0$. Next, choose, $P_2 = 0,\ G = S = 0$ and $R = P_3 = p_3 I_m$. By using Schur complements, it can be shown that $\Theta_0 < 0$ holds for big enough $p_3 > 0$, $\bar{b}$, and $\bar{b}_1$ and small enough $\bar{\xi}$.

Proposition 1 *(i) Given positive tuning scalars* $\bar{\alpha},\ \mu,\ \bar{\xi},\ \bar{b}$, *and* $\bar{b}_1$, *let there exist a* P_1 *such that* $0 < P_1 \in \mathbb{R}^{(n-p)\times(n-p)},\ P_2 \in \mathbb{R}^{p\times(n-p)}$ *and positive* $p \times p$-*matrices* $P_3,\ G,\ R$, *and* S *such that LMI* (21) *is feasible. Then, for all large enough* $\delta_1 > 0$ *and* $\delta_2 > 0$, *there exists small enough* $\mu(\delta) > 0$ *such that for all* $\mu \in (0, \mu(\delta)]$, *the LMIs* (12), (14) *and* (20) *are feasible, so all solutions of* (7) *and (8) satisfy the bound* (17).

(ii) The LMI in (21) *is feasible for small enough* $\bar{\alpha},\ \mu$, *and* $\bar{\xi}$ *and big enough* $\bar{b}$ *and* $\bar{b}_1$.

3.3 Ultimate Boundedness of the Error Dynamics

Let $\varphi(t, t_0, \mu)$ be the fundamental solution of the equation

$$\mu\dot{z}(t) = -z(t - \mu\xi(t)), \quad z(t) \in \mathbb{R},\ \xi(t) \in [0,\ \bar{\xi}] \tag{22}$$

with $\varphi(t_0, t_0, \mu) = 1$ and $\varphi(t, t_0, \mu) = 0$ for $t < t_0$. It is shown in [9] that (22) remains exponentially stable for all variable delays such that $\mu\xi(t) \le 1.99$ for all t. Then the bound

$$\|\varphi(t, t_0, \mu)\| \le e^{-\frac{\alpha_2(t-t_0)}{\mu}} \tag{23}$$

holds for small enough $\alpha_2 > 0$ and for all $\mu > 0$ that admit a constant d such that $\mu\xi(t) \le h$ and $\mu\dot{\xi}(t) \le d < 1$. Our results may now be stated (where the proof follows the arguments of [16] as shown in the appendix below):

Theorem 1 *Given positive tuning scalars* $\mu,\ \bar{\xi},\ \bar{\alpha},\ \bar{b},\ \bar{b}_1,\ \delta_1,\ \delta_2\ k_1$, *and* k_2, *let there exist a* $n \times n$-*matrix* $P_\mu > 0$ *with* $(n-p)\times(n-p)$ *matrix* $P_1 > 0$, $p \times (n-p)$ *matrix* P_2 *and* $p \times p$ *positive matrices* $P_3,\ G,\ R,\ S$ *such that LMIs* (12), (14) *and* (20) *are feasible. Let* $\bar{e}(t)$ *be a solution to* (7)–(8). *Then, for all* $\mu\xi(t) \in [0,\ h]$ *that admit a constant* d *such that* $\mu\dot{\xi}(t) \le d < 1$, *every component of* $\bar{e}_2(t)$ *satisfies the bound*

$$\limsup_{t\to\infty} |\bar{e}_{2_i}(t)| \le 2M_0\mu\bar{\xi} \tag{24}$$

where

$$M_0 = 2(\delta_1 \Delta + \delta_2 \beta + \|TD_1\| \Delta + \|TM_2\| \beta)$$

and where Δ *and* β *denote the known bound of* f_i *and* ζ*, respectively.*

Remark 1 If the condition $\|\zeta\| \leq k_0 + k_1\|y\| + k_2\|u\|$ on the disturbance is preferred, then it is always possible to find a large enough r_0 such that the terms in (17) are bounded by

$$\limsup_{t\to\infty}(\|[\bar{A}_{21}\ \bar{A}_{22}]\bar{e}(t)\| + \|[0\ \bar{A}_{d22}]\bar{e}(t-\tau(t))\|) \leq r_0. \tag{25}$$

Choosing the switching gain in (4) as $M_\beta = r_o + \|TD_1\|\Delta + \|TM_2\|(k_0 + k_1\|y\| + k_2\|u\|)$, Theorem 1 will still hold with $M_0 = 2M_\beta$.

Remark 2 It has been shown in [9, 11], and [22] that a sampled-data output can be represented as a continuous one with a fast varying delay. For the observer design under sampled outputs, please refer to [17].

4 Input Fault Reconstruction in the Presence of Uncertainty

The fault reconstruction properties of the observer designed above are now considered. This effectively extends the presentation in [30] to consider the effect of delays in outputs. For sufficiently small μ, (10) and (11) become:

$$\dot{\bar{e}}_1(t) = (A_{11} + LA_{21})\bar{e}_1(t) + (M_1 + LM_2)\zeta(t, y, u) \tag{26}$$

$$0 \approx \bar{A}_{21}\bar{e}_1(t) - \frac{1}{\mu}\bar{e}_2(t-\tau(t)) + v(t-\tau(t)) + TD_1 f_i(t) + TM_2\zeta(t, y, u) \tag{27}$$

Define

$$W := \left[\, W_1\ \bar{D}_1^{-1} \,\right],$$

where $W_1 \in \mathbb{R}^{q\times(p-q)}$ is a tuning matrix, and

$$\hat{f}_i(t) = WT^T(v(t-\tau(t)) - \frac{1}{\mu}\bar{e}_2(t-\tau(t))). \tag{28}$$

Then equation (27) can be rewritten as

$$0 \approx WA_{21}\bar{e}_1(t) + \hat{f}_i(t) + WD_1 f_i(t) + WM_2\zeta(t, y, u),$$

or equivalently,

$$-\hat{f}_i(t) = f_i(t) + \hat{G}(s)\zeta(t, y, u), \tag{29}$$

where

$$G(s) = WA_{21}(sI - (A_{11} + LA_{21}))^{-1}(M_1 + LM_2) + WM_2.$$

The effect of $\zeta(t, y, u)$ on fault reconstruction can be minimized by minimizing the H_∞ norm of the transfer function $\hat{G}(s)$ from $\zeta(t, y, u)$ to $\hat{f}_i(t)$. This is equivalent to ensuring that the following inequality is satisfied, whereby the H_∞ norm of the transfer function is less than some positive γ:

$$\begin{bmatrix} P(A_{11} + LA_{21}) + (A_{11} + LA_{21})^T P & -P(M_1 + LM_2) & -(WA_{21})^T \\ * & -\gamma I & (WM_2)^T \\ * & * & -\gamma I \end{bmatrix} < 0 \tag{30}$$

where $P \in \mathbb{R}^{(n-p)\times(n-p)} > 0$. The above inequality can be reduced to LMI by taking $Y = PL$.

Remark 3 Fault reconstruction using the sliding mode technique usually requires an ideal sliding motion to be attained in finite time [8, 29]. Due to model uncertainties and output delay effects for example, an ideal sliding motion in the observer does not usually appear. Instead, the motion is bounded within a region of the sliding surface. This chapter uses a singular perturbation approach for fault reconstruction under sampled outputs, where by choosing a sufficiently small μ, the fault can be approximated by an approximation that only depends on the outputs error.

5 Sensor Fault Reconstruction in the Presence of Uncertainty

In this case, the system under consideration is the following:

$$\begin{aligned} \dot{x}(t) &= Ax(t) + Bu(t) + M\zeta(t, y, u) \\ y(t) &= Cx(t - \tau(t)) + Nf_0(t), \quad \tau(t) \in [0,\ h] \end{aligned} \tag{31}$$

where $f_0 \in \mathbb{R}^r$ is the vector of sensor faults and satisfies a bound of the form $\|f_0\| \leq \Delta < \infty$, $N \in \mathbb{R}^{p\times r}$, and $r \leq p$. Consider new measurable states $z_f \in \mathbb{R}^p$ that satisfy

$$\dot{z}_f(t) = -A_f z_f(t) + A_f Cx(t - \tau(t)) + A_f Nf_0(t), \tag{32}$$

where $-A_f$ is a user-defined stable matrix. The system (31), (32) is

$$\begin{bmatrix} \dot{x}(t) \\ \dot{z}_f(t) \end{bmatrix} = \underbrace{\begin{bmatrix} A & 0 \\ 0 & -A_f \end{bmatrix}}_{\tilde{A}} \underbrace{\begin{bmatrix} x(t) \\ z_f(t) \end{bmatrix}}_{x_z(t)} + \underbrace{\begin{bmatrix} 0 & 0 \\ A_f C & 0 \end{bmatrix}}_{\tilde{A}_d} \begin{bmatrix} x(t-\tau(t)) \\ z_f(t-\tau(t)) \end{bmatrix} + \underbrace{\begin{bmatrix} B \\ 0 \end{bmatrix}}_{\tilde{B}} u(t)$$
$$+ \underbrace{\begin{bmatrix} 0 \\ A_f N \end{bmatrix}}_{\tilde{D}} f_0(t) + \underbrace{\begin{bmatrix} M \\ 0 \end{bmatrix}}_{\tilde{M}} \zeta(t, y, u), \quad z_f(t) = \underbrace{\begin{bmatrix} 0 & I_p \end{bmatrix}}_{\tilde{C}} x_z(t-\tau(t)). \tag{33}$$

Since (33) has a similar structure to (1) apart from the term $\tilde{A}_d x_z(t-\tau(t))$, the observer structure for system (33) will be similar to (5), where

$$v(t) = -(\|TD_1\|\Delta + \delta_1 \Delta + \delta_2 \beta)[\text{sign } \bar{e}_{2_1}(t), \ldots, \text{sign } \bar{e}_{2_p}(t)]^T$$

and $\hat{x}_z$ represents the observer states with an additional term $\tilde{A}_d \hat{x}_z(t-\tau(t))$. Denoting A, $-A_f$ in $\tilde{A}$ and $A_f C$ in $\tilde{A}_d$ in (33) by A_{11} and A_{22} and a new term A_{d21} in (2) respectively, and $A_f N$ and M in (33) by D_1 and M_1 in (2), then the error dynamics (10) becomes

$$\begin{aligned} \dot{\bar{e}}_1(t) = \bar{A}_{11}\bar{e}_1(t) + \bar{A}_{12}\bar{e}_2(t) + \bar{A}_{d11}\bar{e}_1(t-\mu\xi(t)) \\ + \bar{A}_\tau \bar{e}_2(t-\mu\xi(t)) + \bar{M}_1 \zeta(t, y, u) \end{aligned} \tag{34}$$

$$\begin{aligned} \mu\dot{\bar{e}}_2(t) = \mu\bar{A}_{22}\bar{e}_2(t) + \mu T A_{d21}\bar{e}_1(t-\mu\xi(t)) \\ + (\mu\bar{A}_{d22} - I_p)\bar{e}_2(t-\mu\xi(t)) + \mu\bar{f}_i(t) \end{aligned} \tag{35}$$

where $\bar{A}_{11} = A_{11}$, $\bar{A}_{d11} = LA_{d21}$, $\bar{A}_{12} = -(A_{11}L - LA_{22})T^T$, $\bar{A}_\tau = (A_{11} - LA_{d21})LT^T$, $\bar{M}_1 = M_1$, $\bar{A}_{22} = TA_{22}T^T$, $\bar{A}_{d22} = -TA_{d21}LT^T$, $\mu\xi(t) = \tau(t)$, $\mu\bar{\xi} = h$, $0 \le \xi(t) \le \bar{\xi}$ and $\bar{f}_i(t) = v(t-\mu\xi(t)) + TD_1 f_i(t)$, i.e. $\|\bar{f}_i(t)\| \le M_f = (\|TD_1\|\Delta + \delta_1\Delta + \delta_2\beta)\sqrt{p} + \|TD_1\|\Delta$. Choose the Lyapunov-Krasovskii functional

$$\begin{aligned} V(t) = \int_{t-\mu\xi(t)}^{t} e^{\bar{\alpha}(s-t)}\bar{e}^T(s)S\bar{e}(s)ds + \int_{t-\mu\bar{\xi}}^{t} e^{\bar{\alpha}(s-t)}\bar{e}^T(s)G\bar{e}(s)ds \\ + \mu\bar{\xi}\int_{-\mu\bar{\xi}}^{0}\int_{t+\theta}^{t} e^{\bar{\alpha}(s-t)}\dot{\bar{e}}^T(s)R\dot{\bar{e}}(s)dsd\theta + \bar{e}^T(t)P_\mu\bar{e}(t) \end{aligned} \tag{36}$$

where $P_\mu \in \mathbb{R}^{n+p}$, S, G, and $R \in \mathbb{R}^{n+p}$ are positive matrices. Set

$$2\bar{e}^T(t)P_\mu\dot{\bar{e}}(t) = 2\bar{e}^T(t)\underbrace{\begin{bmatrix} P_1 & P_2^T \\ \mu P_2 & P_3 \end{bmatrix}}_{P_{1\mu}}\begin{bmatrix} \dot{\bar{e}}_1(t) \\ \mu\dot{\bar{e}}_2(t) \end{bmatrix}, \tag{37}$$

where

$$\begin{bmatrix} \dot{\bar{e}}_1(t) \\ \mu\dot{\bar{e}}_2(t) \end{bmatrix} = \underbrace{\begin{bmatrix} \bar{A}_{11} & \bar{A}_{12} \\ 0 & \mu\bar{A}_{22} \end{bmatrix}}_{\bar{A}_1}\bar{e}(t) + \underbrace{\begin{bmatrix} \bar{A}_{d11} & \bar{A}_\tau \\ \mu TA_{d21} & \mu\bar{A}_{d22} - I_p \end{bmatrix}}_{\bar{A}_{d1}}\bar{e}(t-\mu\xi(t)) + \underbrace{\begin{bmatrix} \bar{M}_1 & 0 \\ 0 & I_p \end{bmatrix}}_{\bar{A}_D}\underbrace{\begin{bmatrix} \zeta(t, y, u) \\ \mu\bar{f}_i(t) \end{bmatrix}}_{\zeta_f(t)}. \tag{38}$$

Then a sufficient condition for $W = \dot{V} + \bar{\alpha}V - \bar{b}\zeta_f^T\zeta_f < 0$ along trajectories (34) and (35) is the feasibility of the LMI

$$\Theta_\mu = \begin{bmatrix} \prod & 0 & P_{1\mu}\tilde{A}_{d1} + Re^{-\bar{\alpha}\mu\bar{\xi}} & P_{1\mu}\tilde{A}_D & \mu\bar{\xi}\tilde{A}_1^T R \\ * & -(G+R)e^{-\bar{\alpha}\mu\bar{\xi}} & Re^{-\bar{\alpha}\mu\bar{\xi}} & 0 & 0 \\ * & * & -2Re^{-\bar{\alpha}\mu\bar{\xi}} - (1-d)Se^{-\bar{\alpha}\mu\bar{\xi}} & 0 & \mu\bar{\xi}\tilde{A}_{d1}^T R \\ * & * & * & -\bar{b} & \mu\bar{\xi}\tilde{A}_D^T R \\ * & * & * & * & -R \end{bmatrix} < 0, \tag{39}$$

where $\prod = P_{1\mu}\tilde{A}_1 + \tilde{A}_1^T P_{1\mu}^T + \bar{\alpha}P_\mu + S + G - Re^{-\bar{\alpha}\mu\bar{\xi}}$. Then the error dynamics (34) and (35) satisfy the bound

$$\bar{e}^T(t)P_\mu\bar{e}(t) < e^{-\bar{\alpha}(t-t_0)}\bar{e}^T(t_0)P_\mu\bar{e}(t_0) + \frac{\bar{b}}{\bar{\alpha}}\|\zeta_{f[t_0,t]}\|_\infty^2. \tag{40}$$

In the switching gain design, (17) becomes

$$\begin{aligned} &\limsup_{t\to\infty} \| \begin{bmatrix} 0 & \bar{A}_{22} \end{bmatrix} \bar{e}(t)\| \le k_1(\delta_1\Delta + \delta_2\beta), \\ &\limsup_{t\to\infty} \| \begin{bmatrix} TA_{d21} & \bar{A}_{d22} \end{bmatrix} \bar{e}(t-\tau(t))\| \le k_2(\delta_1\Delta + \delta_2\beta). \end{aligned} \tag{41}$$

Since $\bar{A}_{11} + \bar{A}_{d11}$ is Hurwitz by design, it can be shown that $\Theta_0 < 0$ holds for small enough $\bar{\alpha}$ and $\bar{\xi}$ and big enough $\bar{b}$. Furthermore, the bounds in (41) hold for large enough δ_1 and δ_2.

Proposition 2 *(i) Given positive tuning scalars $\bar{\alpha}$, μ, $\bar{\xi}$, $\bar{b}$, δ_1, and δ_2, let there exist an $n \times n$ matrix $P_1 > 0$, a $p \times p$ matrix $P_3 > 0$, a $p \times n$ matrix P_2 such that matrix $P_\mu > 0$ in* (12)*, and positive $(n+p) \times (n+p)$ matrices R, G, and S such that inequality* (39) *is feasible with $\mu = 0$. Then for big enough $\delta_1 > 0$ and $\delta_2 > 0$, there exist small enough $\bar{\alpha}$, μ, and $\bar{\xi} > 0$ such that the bounds in* (41) *are feasible. Then, all solutions of* (34) *and* (35) *satisfy the bound* (24)*, where $M_0 = 2(\delta_1\Delta + \delta_2\beta + \|TD_1\|\Delta)$.*
(ii) LMIs (39) *are feasible for small enough $\bar{\alpha}$, μ, $\bar{\xi}$ and big enough $\bar{b}$.*

Similarly to Sect. 4, for sufficiently small μ, equations (26) and (29) become

$$\begin{aligned} \dot{\bar{e}}_1(t) &= \bar{A}_{11}\bar{e}_1(t) + \bar{A}_{d11}\bar{e}_1(t-\mu\xi(t)) + \bar{M}_1\zeta(t) \\ -\hat{f}_i(t) &= f_i + WA_{d21}\bar{e}_1(t-\mu\xi(t)). \end{aligned}$$

Applying the result from [27] for delay systems to the above equation, the H_∞ norm from $\zeta(t, y, u)$ to $\hat{f}_i(t)$ will be less than a positive number γ if the following inequality is feasible:

$$\begin{bmatrix} \prod_1 & \hat{P}_1 - \hat{P}_2^T + \varepsilon(\bar{A}_{11}^T\hat{P}_2 + A_{d21}^T Y^T) & \mu\bar{\xi}YA_{d21} & \hat{P}_2^T\bar{M}_1 & A_{d21}^T W^T \\ * & -\varepsilon(\hat{P}_2 + \hat{P}_2^T) + \mu\bar{\xi}R & \mu\bar{\xi}\varepsilon YA_{d21} & \varepsilon\hat{P}_2^T\bar{M}_1 & 0 \\ * & * & -\mu\bar{\xi}R & 0 & \mu\bar{\xi}A_{d21}^T W^T \\ * & * & * & -\gamma^2 I_k & 0 \\ * & * & * & * & -I_r \end{bmatrix} < 0 \tag{42}$$

where $\prod_1 = \hat{P}_2^T \bar{A}_{11} + \bar{A}_{11}^T \hat{P}_2 + YA_{d21} + A_{d21}^T Y^T$, $\hat{P}_1$, $R \in \mathbb{R}^{n\times n} > 0$, $\hat{P}_3 = \varepsilon \hat{P}_2 \in \mathbb{R}^{n\times n}$, $Y = \hat{P}_2^T L \in \mathbb{R}^{n\times p}$, ε is a nonzero scalar, and $\gamma > 0$. To reconstruct the fault signals, continuous approximation of the discontinuous component $v(t)$, i.e.,

$$v_r = -(\|TD_1\| + \delta)\Delta \left[\frac{\bar{e}_{2_1}}{|\bar{e}_{2_1}| + r}, \ldots, \frac{\bar{e}_{2_p}}{|\bar{e}_{2_p}| + r} \right]^T, \tag{43}$$

where $r \geq 0$ as proposed in [8], will be adopted.

6 Example

Consider an inverted pendulum (as in [8]) which is linearized about its equilibrium at the origin:

$$A = \begin{bmatrix} 0 & 0 & 1 & 0 \\ 0 & 0 & 0 & 1 \\ 0 & -1.9333 & -1.9872 & 0.0091 \\ 0 & 36.9771 & 6.2589 & -0.1738 \end{bmatrix}, \quad B = D = \begin{bmatrix} 0 \\ 0 \\ 0.3205 \\ -1.0095 \end{bmatrix},$$

$$C = \begin{bmatrix} 1\,0\,0\,0 \\ 0\,1\,0\,0 \\ 0\,0\,1\,0 \end{bmatrix}, \quad M = \begin{bmatrix} 0 \\ 0 \\ 0 \\ 1 \end{bmatrix}, \quad \zeta = \begin{bmatrix} 0 \; 2.4 \; 2.4 \end{bmatrix} y$$

The sensor fault distribution matrix is

$$N = \begin{bmatrix} 0 \; 0 \; 1 \end{bmatrix}$$

in (31). A compensator approach from [15] is designed to stabilize the pendulum. It is assumed that an input fault is bounded by $\|f_i\| \leq \Delta = 0.6$.

6.1 Reconstruction of an Actuator Fault

The partitioned system (2) can be obtained by choosing an appropriate transformation matrix T_0 which yields $T = I_3$, an identity matrix. In deriving the observer gain, the LMI (30) is feasible under tuning parameters

$$\gamma = 0.02 \;\; \text{and} \;\; W_1 = \begin{bmatrix} 0.1 \; 0.001 \end{bmatrix},$$

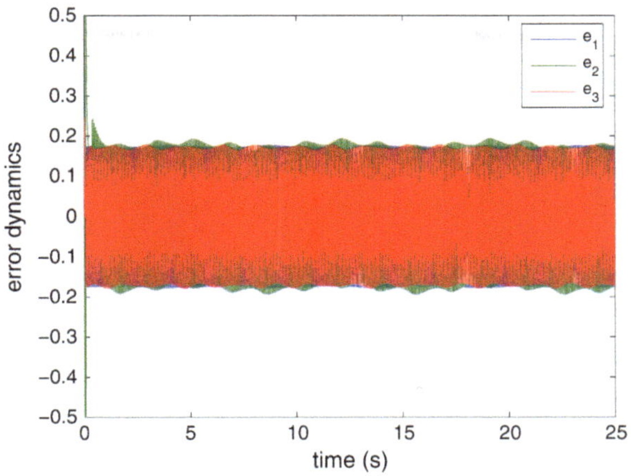

Fig. 1 Error dynamics e_2 without smoothing factor in control and with output delay $h = 13\,\text{ms}$

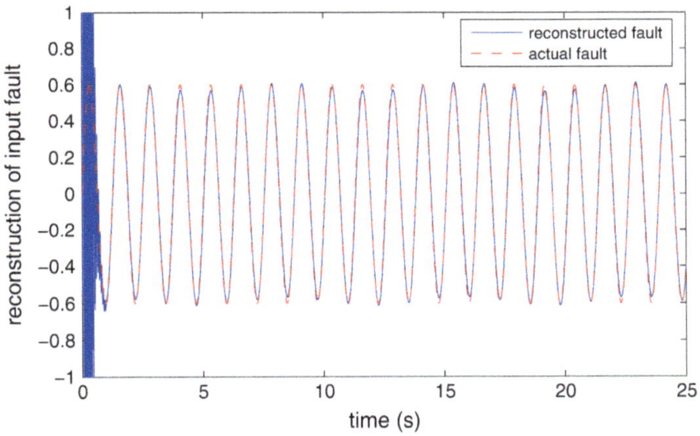

Fig. 2 Input fault reconstruction in the presence of uncertainty

which give the observer gain $L = [0\ 0.43\ 0]$. LMIs (14) are feasible with $\bar{\alpha} = 3.9$, $\mu = 0.027$, output delay $\mu\bar{\xi} = 13\,\text{ms}$, and delay varying rate $d = 0.2$. The LMIs in (20) are feasible with $\delta_1 = 16$, $\delta_2 = 1$, $k_1 = 0.93$, and $k_2 = 0.07$. Hence, the observer (5) with gains in (4) and (6) has been designed. This ensures that the error variable is bounded in the range $|e_{2_i}(t)| \leq 0.53$, according to the estimate (24). To verify the estimation, Fig. 1 is plotted using the sign function. It can be seen that every error variable is stabilized into a bound $|e_{2_i}| \leq 0.18$. The high degree of switching is acceptable for an observer error signal; this is avoided in the reconstruction of the fault signals. For illustration, an input fault $f_i(t) = 0.6sin(5t)$ is reconstructed in Fig. 2, where the smoothing parameter $r = 0.5$ in (43) was chosen.

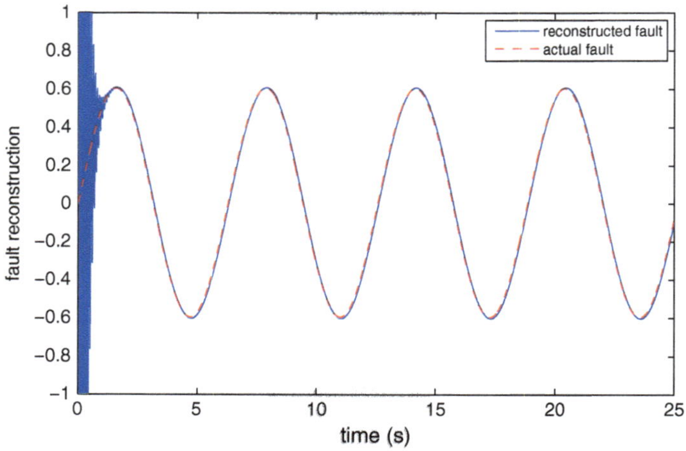

Fig. 3 Output fault reconstructions under uncertainties

6.2 Reconstruction of Sensor Fault

By choosing tuning parameters $\varepsilon = 0.13$, $h = 20\,\text{ms}$, $\gamma = 0.45$, $W_1 = \begin{bmatrix} 0.1 & 0.3 \end{bmatrix}$, and $A_f = 0.1 I_p$ in (42) and taking into account that the augmented system (33) is already in the required partitioned form (2), the observer gain is obtained as

$$L = \begin{bmatrix} -5.16 & -6.87 & 0 \\ 1.59 & -40.96 & 0 \\ -42.84 & 16.24 & 0 \\ -44.46 & -511.02 & 0 \end{bmatrix},$$

and LMIs (39) and (20) are feasible with $\mu = 0.019$, $\mu\bar{\xi} = 20\text{ms}$, $d = 0.1$, $\bar{\alpha} = 2.2$, $\bar{b} = 6$, $k_1 = 0.03$, $k_2 = 0.97$, $\delta_1 = 50$, and $\delta_2 = 40$. The observer gives the bound $|e_{2_i}| \leq 3.4$ on the errors. Figure 3 shows the fault reconstruction under the fault $f_o = 0.6 sin(t)$, while uncertainties remain as before and a smoothing factor $r = 3$ is used.

6.3 Comparison with a Classical Fault Reconstruction Scheme

We now use the classical observer design in [8], which uses the equivalent injection term (43) only to reconstruct the fault, but does not consider the effect of output measurement delay. This will benchmark the proposed design. Comparing the fault reconstruction in Fig. 4 to the previous results, the proposed method produces much higher precision. Decreasing r below its given values in Fig. 4 will not filter out the high switching terms in simulations. The precision of the fault reconstruction in (43)

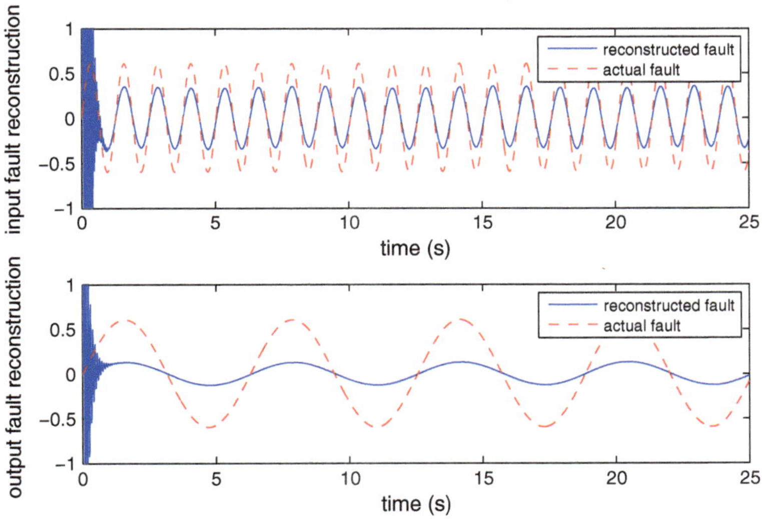

Fig. 4 Fault reconstructions using only the equivalent injection term

depends heavily on the value of the smoothing term r in the equivalent injection. Larger r causes a larger reconstruction error. This error is compensated in our result, using the linear term related to singular perturbation parameter μ, despite using larger value of r. Hence, the proposed observer design method has significant advantages compared with the classical approach, if the output is delayed. The $\bar{e}_2$ term in (28) is pertinent to the reconstruction accuracy. For the observer in [8], the equivalent term is assumed to be zero in an ideal sliding case.

7 Conclusion

This chapter developed an observer design framework for systems with multiple outputs, where the outputs are delayed and thus the output error signals used to drive the observer are subject to delay. A singular perturbation approach is employed for the analysis and ensures an ultimate bound on the error dynamics that is proportional to the time delay and the switching gain. A corresponding robust fault reconstruction technique has been proposed using a robust optimization technique and finds a sufficiently small value of the singular perturbation parameter. It is demonstrated that the faults can be reconstructed reliably, even if the measured outputs are subject to delay and system uncertainties are present.

8 Proof of Theorem 1

The i-th component of (8) with the initial condition (9) can be represented in the form of an integral equation [21]

$$\bar{e}_{2_i}(t) = \varphi(t, t_0, \mu)\bar{e}_{2_i}(t_0) + \int_{t_0}^{t} \varphi(t, s, \mu)\Big[[\bar{A}_{21_i}\ \bar{A}_{22_i}]\bar{e}(s) + \bar{A}_{d22_i}\bar{e}_2(s - \mu\xi(t))$$
$$+ (TD_1)_i \cdot f_i(s) + (TM_2)_i\zeta(s, y, u) - (\|TD_1\|\Delta + \delta_1\Delta + \delta_2\beta$$
$$+ \|TM_2\|\beta)\ \mathrm{sign}\ \bar{e}_{2_i}(s - \mu\xi(s))\Big]ds. \tag{44}$$

The feasibility of (20) implies the bound (17), and then the following inequality holds for large t:

$$\begin{aligned} &|[\bar{A}_{21_i}\ \bar{A}_{22_i}]\bar{e}(s) + \bar{A}_{d22_i}\bar{e}_2(s - \mu\xi(t)) + (TD_1)_i f_i(s) + (TM_2)_i\zeta(s, y, u) \\ &-(\|TD_1\|\Delta + \delta_1\Delta + \delta_2\beta + \|TM_2\|\beta)\ \mathrm{sign}\ \bar{e}_{2_i}(s - \mu\xi(s))| < M_0 \end{aligned} \tag{45}$$

Taking (23) and (45) into account, it is established from (44) that for large t, we have

$$\begin{aligned} &|\bar{e}_{2_i}(t+\theta) - \bar{e}_{2_i}(t)| \le \Big| \int_{t+\theta}^{t} \varphi(t, s, \mu)\Big([\bar{A}_{21_i}\ \bar{A}_{22_i}]\bar{e}(s) + \bar{A}_{d22_i}\bar{e}_2(s - \mu\xi(t)) + (TD_1)_i \\ &\cdot f_i(s) + (TM_2)_i\zeta(s, y, u) - (\|TD_1\|\Delta + \delta_1\Delta + \delta_2\beta + \|TM_2\|\beta)\ \mathrm{sign}\ \bar{e}_{2_i}\big(s - \mu\xi(s)\big)\Big)ds\Big| \\ &< M_0 \int_{t+\theta}^{t} e^{\frac{\alpha_2(t-s)}{\mu}} ds < \mu M_0 \frac{1-e^{2\alpha_2 h}}{\alpha_2} \le 2M_0\mu\bar{\xi} \end{aligned} \tag{46}$$

where $\theta \in [-2\mu\bar{\xi},\ 0]$. Therefore,

$$\bar{e}_{2_i}(t) - 2M_0\mu\bar{\xi} < \bar{e}_{2_i}(t+\theta) < \bar{e}_{2_i}(t) + 2M_0\mu\bar{\xi} \tag{47}$$

holds for large t, and the following implication holds:

$$|\bar{e}_{2_i}(t)| \ge 2M_0\mu\bar{\xi} \Rightarrow \mathrm{sign}\ \bar{e}_{2_i}(t+\theta) = \mathrm{sign}\ \bar{e}_{2_i}(t) \tag{48}$$

for large enough t. Thus, from (17), (45) and (48), the following holds for sufficiently large t:

$$\begin{aligned} &|\bar{e}_{2_i}(t)| \ge 2M_0\mu\bar{\xi} \Rightarrow \\ &\bar{e}_{2_i}^T(t)\big[[\bar{A}_{21_i}\ \ \bar{A}_{22_i}]\bar{e}(t+\theta) + \bar{A}_{d22_i}\bar{e}_2(t - \mu\xi(t) + \theta) + (TD_1)_i f_i(t+\theta) \\ &+ (TM_2)_i\zeta(s+\theta, y, u) - (\|TD_1\|\Delta + \delta_1\Delta + \delta_2\beta + \|TM_2\|\beta)\ \mathrm{sign}\ \bar{e}_{2_i}(t+\theta)\big] \\ &< |\bar{e}_{2_i}(t)|\big(|[\bar{A}_{21_i}\ \ \bar{A}_{22_i}]\bar{e}(t+\theta)| + |[0\ \ \bar{A}_{d22_i}]\bar{e}(t - \mu\xi(t) + \theta)| \\ &+ \|TD_1\|\Delta + \|TM_2\|\beta\big) - (\|TD_1\|\Delta + \delta_1\Delta + \delta_2\beta + \|TM_2\|\beta)|\bar{e}_{2_i}(t)| < 0 \end{aligned} \tag{49}$$

It will be shown next that the $\bar{e}_{2_i}$-component of the solutions of (8) exponentially converges to the ball (24). Moreover, for sufficiently large t, whenever $\bar{e}_{2_i}(t)$ reaches the ball (24), it will never leave it. Taking into account (49), it follows that for sufficiently large t, one has

$$\begin{aligned}
&|\bar{e}_{2_i}(t)| \geq 2M_0\mu\bar{\xi} \Rightarrow \\
&\tfrac{d}{dt}\mu\bar{e}_{2_i}^2(t) = 2\mu\bar{e}_{2_i}(t)\dot{\bar{e}}_{2_i}(t) = 2\bar{e}_{2_i}(t)\big[-\bar{e}_{2_i}(t-\mu\xi(t)) + \mu\big([\bar{A}_{21_i}\;\; \bar{A}_{22_i}]\bar{e}(t) \\
&+[0\;\; \bar{A}_{d22_i}]\bar{e}(t-\mu\xi(t)) + (TD_1)_i f_i(t) + (TM_2)_i\zeta(t,y,u) \\
&-(\|TD_1\|\Delta + \delta_1\Delta + \delta_2\beta + \|TM_2\|\beta)\,\mathrm{sign}\;\bar{e}_{2_i}(t)\big)\big] \\
&\leq -2\bar{e}_{2_i}(t)\big(\bar{e}_{2_i}(t) - \textstyle\int_{t-\mu\xi(t)}^{t}\dot{\bar{e}}_{2_i}(s)ds\big) \\
&= -2\bar{e}_{2_i}^2(t) + 2\bar{e}_{2_i}(t)\textstyle\int_{t-\mu\xi(t)}^{t}\Big[-\tfrac{\bar{e}_{2_i}(s-\mu\xi(t))}{\mu} + [\bar{A}_{21_i}\;\; \bar{A}_{22_i}]\bar{e}(s) \\
&+[0\;\; \bar{A}_{d22_i}]\bar{e}(s-\mu\xi(t)) + (TD_1)_i f_i(s) + (TM_2)_i\zeta(t,y,u) \\
&-(\|TD_1\|\Delta + \delta_1\Delta + \delta_2\beta + \|TM_2\|\beta)\,\mathrm{sign}\;\bar{e}_{2_i}(s)\Big]ds \\
&\leq -2\bar{e}_{2_i}^2(t) - 2\tfrac{\bar{e}_{2_i}(t)}{\mu}\textstyle\int_{t-\mu\xi(t)}^{t}\bar{e}_{2_i}(s-\mu\xi(t))ds.
\end{aligned}$$

Therefore, given that (48) holds for large enough t, it follows that

$$-\int_{t-\mu\xi(t)}^{t}\bar{e}_{2_i}(t)\bar{e}_{2_i}(s-\mu\xi(t))ds \leq 0.$$

Hence

$$|\bar{e}_{2_i}(t)| \geq 2M_0\mu\bar{\xi} \Rightarrow \frac{d}{dt}\mu\bar{e}_{2_i}^2(t) \leq -2\bar{e}_{2_i}^2(t). \tag{50}$$

Assume now that for large enough t_1, the $\bar{e}_{2_i}$ component of the solution to (1) is outside the ball (24). It then follows from (50) that for all $t \geq t_1$ such that

$$|\bar{e}_{2_i}(t)| \geq 2M_0\mu\bar{\xi},$$

one has

$$\bar{e}_{2_i}^2(t) \leq e^{-\frac{2}{\mu}(t-t_1)}\bar{e}_{2_i}^2(t_1),$$

i.e., $\bar{e}_{2_i}$ exponentially converges to the ball (24). Let $t_2 > t_1$ be the time when

$$|\bar{e}_{2_i}(t_2)| = 2M_0\mu\bar{\xi}.$$

Then, (50) implies that

$$\bar{e}_{2_i}^2(t_2^+) < \bar{e}_{2_i}^2(t_2).$$

Therefore, whenever $\bar{e}_{2_i}(t)$ attains the ball (24), it will never leave it.

Acknowledgments This work was partially supported by Israel Science Foundation Grant Number 754/10.

References

1. Ahmed-Ali, T., Cherrier, E., Lamnabhi-Lagarrigue, F.: Cascade high gain predictors for a class of nonlinear systems. IEEE Trans. Autom. Control **57**(1), 224–229 (2012)
2. Ahmed-Ali, T., Karafyllis, I., Lamnabhi-Lagarrigue, F.: Global exponential sampled-data observers for nonlinear systems with delayed measurements. Syst. Control Lett. **62**(7), 539–549 (2013)
3. Chen, J., Patton, R., Zhang, H.: Design of unknown input observers and robust fault detection filters. Intern. J. Control **63**(1), 85–105 (1996)
4. Choi, H., Chung, M.: Observer-based H_∞ controller design for state delayed linear systems. Automatica **32**(7), 1073–1075 (1996)
5. Darouach, M.: Linear functional observers for systems with delays in the state variables. IEEE Trans. Autom. Control **46**(3), 491–497 (2001)
6. Edwards, C., Spurgeon, S.: On the development of discontinuous observers. Intern. J. Control **59**(5), 1211–1229 (1994)
7. Edwards, C., Spurgeon, S.: Sliding Mode Control: Theory and Applications. Taylor and Francis, Bristol, PA (1998)
8. Edwards, C., Spurgeon, S., Patton, P.: Sliding mode observers for fault detection and isolation. Automatica **36**(4), 541–553 (2000)
9. Fridman, E.: A refined input delay approach to sampled-data control. Automatica **46**(2), 421–427 (2010)
10. Fridman, E., Dambrine, M.: Control under quantization, saturation and delay: a LMI approach. Automatica **45**(10), 2258–2264 (2009)
11. Fridman, E., Seuret, A., Richard, J.-P.: Robust sampled-data stabilization of linear systems: an input delay approach. Automatica **40**(8), 1441–1446 (2004)
12. Floquet, T., Barbot, J., Perruquetti, W., Djemai, M.: On the robust fault detection via a sliding mode disturbance observer. Intern. J. Control **77**(7), 622–629 (2004)
13. Germani, A., Manes, C., Pepe, P.: A new approach to state observation of nonlinear systems with delyed output. IEEE Trans. Autom. Control **47**(1), 96–101 (2002)
14. Germani, A., Pepe, P.: A State observer for a class of nonlinear systems with multiple discrete and distributed time delays. Eur. J. Control **11**(3), 196–205 (2005)
15. Han, X., Fridman, E., Spurgeon, S., Edwards, C.: On the design of sliding mode static output feedback controllers for systems with state delay. IEEE Trans. Ind. Electron. **56**(9), 3656–3664 (2009)
16. Han, X., Fridman, E., Spurgeon, S.: Sliding mode control in the presence of input delay: a singular perturbation approach. Automatica **48**(8), 1904–1912 (2012)
17. Han, X., Fridman, E., Spurgeon, S.: Sampled data sliding mode observer for robust fault reconstruction: a time delay approach. J. Frankl. Inst. **351**(4), 2125–2142 (2013)
18. Ibrir, S.: Observer-based control of a class of time-delay nonlinear systems having triangular structure. Automatica **47**(2), 388–394 (2011)
19. Jafarov, E.: Design modification of sliding mode observers for uncertain MIMO systems without and with time-delay. Asian J. Control **7**(4), 380–392 (2005)
20. Kokotovic, P., Khalil, H., O'Reilly, J.: Singul. Perturbat. Methods Control: Anal. Des. Academic Press, New York (1986)
21. Kolmanovskii, V., Myshkis, A.: Applied Theory of Functional Differential Equations. Series Mathematics and Its Applications (Soviet Series), vol. 85. Kluwer, Dordrecht (1992)
22. Mikheev, Y., Sobolev, V., Fridman, E.: Asymptotic analysis of digital control systems. Autom. Remote Control **49**(9), 1175–1180 (1988)

23. Niu, Y., Lam, J., Wang, X., Ho, D.: Observer Based sliding mode control for nonlinear state-delayed systems. Intern. J. Syst. Sci. **35**(2), 139–150 (2004)
24. Nounou, M., Nounou, H., Mahmoud, M.: Robust adaptive sliding mode control for continuous time delay systems. IMA J. Math. Control Inf. **24**(3), 299–313 (2007)
25. Spurgeon, S.: Sliding mode observers—a survey. Intern. J. Syst. Sci. **39**(8), 751–764 (2008)
26. Subbarao, K., Muralidhar, P.: State observer for linear systems with piece-wise constant output delays. IET Control Theory Appl. **3**(8), 1017–1022 (2009)
27. Suplin, V., Fridman, E., Shaked, U.: Sampled-data H_∞ control and filtering: nonuniform uncertain sampling. Automatica **43**(6), 1072–1083 (2007)
28. Tan, C., Edwards, C.: An LMI approach for design sliding mode observers. Intern. J. Control **74**(16), 1559–1568 (2001)
29. Tan, C., Edwards, C.: Sliding mode observers for detection and reconstruction of sensor faults. Automatica **38**(10), 1815–1821 (2002)
30. Tan, C., Edwards, C.: Sliding mode observers for robust detection and reconstruction of actuator and sensor faults. Intern. J. Robust Nonlinear Control **13**(5), 443–463 (2003)

A Lyapunov-Krasovskii Methodology for a Class of Large-Scale Systems with Neutral-type Delays in an iISS Framework

Hiroshi Ito and Frédéric Mazenc

Abstract This chapter presents a framework for verifying stability and robustness of dynamical networks consisting of neutral subsystems subject to disturbances in the spirit of integral input-to-state stability. In addition to neutral-type delays in subsystems, time-delays are allowed to reside in both subsystems and interconnection channels. A small-gain condition is proposed for constructing a Lyapunov-Krasovskii functional to establish stability and robustness of the network. No assumption is made on the network topology.

1 Introduction

Large-scale dynamical systems in the real world are often subject to time delay arising from both communication between subsystems and the mechanism of internal processes. Typically, an internal delay can appear in a system equation in two ways. One is in the form of delayed state variables, which is referred to as a retarded-type delay in the literature. The other appears as the form of delayed derivative of state variables. Time-delay systems of neutral type are the most well-known class of dynamical systems depicting such features [8, 22]. Neutral-type delays often arise or are utilized in practical examples such as population dynamics [23], circuits of transmission lines [2, 3], and Hopfield neural networks, cellular neural networks, Cohen-Grossberg neural networks and projection networks [4, 5, 27].

The objective of this chapter is to propose a framework for verifying robustness of large-scale systems involving neutral, retarded and communication delays in the

H. Ito (✉)
Department of Systems Design and Informatics, Kyushu Institute of Technology, Iizuka, Fukuoka 820-8502, Japan
e-mail: hiroshi@ces.kyutech.ac.jp

F. Mazenc
EPI Inria DISCO, Laboratory of Signals and Systems (L2S, UMR CNRS 8506), CNRS-CentraleSupélec-Université Paris Sud, Gif-sur-Yvette, France
e-mail: Frederic.MAZENC@lss.supelec.fr

I. Karafyllis et al. (eds.), *Recent Results on Nonlinear Delay Control Systems*, Advances in Delays and Dynamics 4, DOI 10.1007/978-3-319-18072-4_10

sense of integral input-to-state stability (iISS). For systems without delays, iISS is a widely accepted notion that allows a system to be blown up by an input of finite magnitude even if autonomous dynamics of the system is stable [31]. Another popular notion is input-to-state stability (ISS) that guarantees the state of a system to be bounded for any bounded input [30]. ISS systems are always iISS, but the converse does not hold true. In contrast to ISS, the notion of iISS allows us to cope with saturation mechanisms and limitations of contractibility which may often arise in components of systems of large size.

To the best of the authors' knowledge, virtually no attempts to globally analyze nonlinear networks of iISS neutral systems have been made until very recently in the literature although linear systems and local analysis of nonlinear systems have been studied extensively, In fact, even its partial problems have been considered to be too hard. As for ISS-type analysis which is a partial problem, one important progress has been made for nonlinear neutral systems based on coupled delay differential and difference equations [21, 29]. Very recently in [25], an approach to iISS-type analysis has been developed for nonlinear neutral systems without transforming the neutral systems into coupled delay differential and difference equations. The study focuses on only neutral-type delays without addressing other types of delays. The purpose of this chapter is to show how to make use of the approach for dealing with interconnections of systems subject to communication delays as well as retarded-type delays. A method for analyzing iISS of network of arbitrary size and arbitrary network topology is proposed. The method provides us with stability and robustness criteria as a unified small-gain-type condition. A network remains stable or robust irrespective of the delay amount if subsystems exhibit delay-independent properties satisfying the small-gain condition. The proposed method is based on construction of Lyapunov-Krasovskii functionals in an iISS setting. Dissipative inequalities of subsystems are utilized for the construction. In the presence of neutral-type delays, the construction requires the information about nonlinear bounds of state variation at subsystems in addition to the dissipative inequalities. A preliminary version of the results in this chapter has been presented in [17] without any proofs. The main theorem in this chapter admits delays in communication channels as well as retarded-type delays. The formula and guidelines for constructing Lyapunov-Krasovskii functionals are restructured in this chapter.

In this chapter, the symbol $|\cdot|$ denotes the Euclidean norm of a real vector in $\mathbb{R}^n$ of a compatible dimension n. The set of non-negative real numbers $[0, +\infty)$ is denoted by $\mathbb{R}_+$. We also use the extended space $\overline{\mathbb{R}}_+ = [0, \infty]$. For a measurable and essentially bounded $\mathbb{R}^n$-valued function u defined on an interval $\mathscr{S} \subseteq \mathbb{R}$, we use $\|u\|_\infty = \text{ess sup}_{t\in\mathscr{S}} |u(t)|$. By $u_{\mathscr{S}}$ it is meant that $u_{\mathscr{S}}(t) = u(t)$ for all $t \in \mathscr{S}$ and $=0$ elsewhere. The space of measurable and essentially bounded $\mathbb{R}^n$-valued function on $\mathscr{S}$ is denoted by $\mathscr{L}^n_\infty\mathscr{S}$. A function $u : \mathscr{S} \to \mathbb{R}^n$ defined for a bounded interval $\mathscr{S} \subseteq \mathbb{R}$ is said to be piecewise continuously differentiable if it is continuous on $\mathscr{S}$ and, for any real number $c \in \mathscr{S}$, the derivative $\dot{u}$ is continuous over $\{t \in \mathscr{S} : t \leq c\}$, except at a finite number of points. Let $\mathscr{PC}^n_1\mathscr{S}$ denote the set of all $\mathbb{R}^n$-valued functions defined on a given bounded interval $\mathscr{S}$ that are continuous, piecewise continuously differentiable and with an essentially bounded first derivative. Given

$T > 0$ or $T = \infty$ and $\Delta > 0$, for each $t \in [0, T)$, x_t is defined for $x \in \mathscr{PC}_1^n[-\Delta, T)$ by $x_t = x(t+\theta)$ for all $\theta \in [-\Delta, 0]$. For a function $\zeta : \mathbb{R}_+ \to \mathbb{R}_+$, we write $\zeta \in \mathscr{P}$ if it is continuous and satisfies $\zeta(0) = 0$, and $\zeta(s) > 0$ for all $s > 0$. For a function $\zeta \in \mathscr{P}$, we write $\zeta \in \mathscr{J}$ if it is non-decreasing. A function is said to be of class $\mathscr{K}$ if it belongs to $\mathscr{J}$ and is strictly increasing; of class $\mathscr{K}_\infty$ if it is of class $\mathscr{K}$ and is unbounded. A function $\beta : \mathbb{R}_+^2 \to \mathbb{R}_+$ is of class $\mathscr{KL}$ if for each fixed t the function $s \mapsto \beta(s, t)$ is of class $\mathscr{K}$ and for each fixed s the function $t \mapsto \beta(s, t)$ is non-increasing and goes to zero as $t \to +\infty$. For $\zeta \in \mathscr{K}$, the operator $\zeta^\ominus\colon \overline{\mathbb{R}}_+ \to \overline{\mathbb{R}}_+$ is defined as $\zeta^\ominus(s) = \sup\{z \in \mathbb{R}_+ : \zeta(z) \le s\}$. By definition, $\zeta_i^\ominus(s) = \zeta^{-1}(s)$ for $0 \le s < \lim_{\tau\to\infty} \zeta(\tau)$, and $\zeta^\ominus(s) = \infty$ elsewhere. A function $\zeta \in \mathscr{K}$ is extended to $\zeta\colon \overline{\mathbb{R}}_+ \to \overline{\mathbb{R}}_+$ as $\zeta(s) := \sup_{z\in\{y\in\mathbb{R}_+ : y\le s\}} \zeta(z)$. Composition of two functions $\zeta_1, \zeta_2 : \overline{\mathbb{R}}_+ \to \overline{\mathbb{R}}_+$ is denoted as $\zeta_1 \circ \zeta_2$. For brevity, a nonstandard symbol for repeated composition is employed in this chapter as $\bigodot_{i=1}^n \zeta_i = \zeta_1 \circ \zeta_2 \circ \cdots \circ \zeta_n$. The symbols $\vee$ and $\wedge$ represent the logical sum and the logical product, respectively.

2 Network Description and Problem Formulation

Consider the network Σ consisting of subsystems described by

$$\begin{aligned}\Sigma_i : \dot{x}_i(t) = f_i(x_i(t), \dot{x}_i(t-\Delta_i), x_1(t-\Delta_{i,1}), x_2(t-\Delta_{i,2}),\\ \ldots, x_n(t-\Delta_{i,n}), r_i(t)) \quad (1)\\ x_{i,0} = \xi_{i,0},\ \dot{x}_{i,0} = \dot{\xi}_{i,0}, \quad i = 1, 2, \ldots, n,\end{aligned}$$

where $x_i(t) \in \mathbb{R}^{N_i}$, $r_i(t) \in \mathbb{R}^{M_i}$ and $t \in \mathbb{R}_+$. Equation (1) is understood to hold almost everywhere in t. The number of subsystems Σ_i in the overall network Σ is $n \ge 2$. The real numbers $\Delta_i \in \mathbb{R}_+$ are neutral-type delays at subsystem Σ_i, while $\Delta_{i,j} \in \mathbb{R}_+$ $(i \ne j)$ are discrete delays in the commutation channel connecting Σ_j to Σ_i. The real numbers $\Delta_{i,i} \in \mathbb{R}_+$ are discrete delays at Σ_i, i.e., delays of retarded type at Σ_i. We define $\Delta := \max_i\{\Delta_i, \max_j\{\Delta_{i,j}\}\} \ge 0$. Suppose that the functions $f_i\colon \mathbb{R}^{N_i} \times \mathbb{R}^{N_i} \times \mathbb{R}^{N_1} \times \ldots \times \mathbb{R}^{N_n} \times \mathbb{R}^{M_i} \to \mathbb{R}^{N_i}$ are locally Lipschitz and satisfy $f_i(0, \ldots, 0, 0) = 0$. The external signal $r_i(t)$ is assumed to be piecewise continuous. This chapter also assumes that $\xi_{i,0} \in \mathscr{PC}_1^{N_i}[-\Delta, 0]$, $i = 1, 2, \ldots, n$. The state and the disturbance of the network Σ is compactly denoted by $x(t) = [x_1(t)^T, \ldots, x_n(t)^T]^T \in \mathbb{R}^N$ and $r(t) = [r_1(t)^T, \ldots, r_n(t)^T]^T \in \mathbb{R}^M$, respectively. We use the following terminologies:

Definition 1 The network Σ is called iISS with respect to input r and state x if there exist $\chi \in \mathscr{K}_\infty$, $\beta \in \mathscr{KL}$ and $\eta \in \mathscr{K}$ such that, for any initial condition ξ_0 and $\dot{\xi}_0$ and for any input r, the corresponding solution exists for all $t \ge 0$ and satisfies

$$\chi(|x(t)|) \le \beta(\|\xi_0\|_\infty + \|\dot{\xi}_0\|_\infty, t) + \int_0^t \eta(|r(\tau)|)d\tau \quad (2)$$

for all $t \in \mathbb{R}_+$.

Definition 2 The network Σ is called ISS with respect to input r and state x if there exist $\beta \in \mathscr{KL}$ and $\gamma \in \mathscr{K}$ such that for any initial conditions ξ_0 and $\dot{\xi}_0$ and for any input r, the corresponding solution exists for all $t \geq 0$ and satisfies

$$|x(t)| \leq \beta(\|\xi_0\|_\infty + \|\dot{\xi}_0\|_\infty, t) + \gamma(\|r_{[0,t]}\|_\infty),\ \forall t \in \mathbb{R}_+. \tag{3}$$

The above two properties coincide with iISS and the ISS, respectively, as defined in [28], if we remove $\dot{\xi}_0$. As in [25], the derivative is introduced in each definition since the response of (1) defined with $\dot{x}_i(t-\Delta_i)$ is affected by $\dot{\xi}_0$. A system is iISS (resp. ISS) without $\dot{\xi}_0$ only if it is iISS (resp. ISS) with $\dot{\xi}_0$.

Remark 1 Property (3) of ISS is different from the one in [29] since it involves the derivative of the initial condition. The approach in [21, 29] defines ISS for coupled delay differential and difference equations into which neutral systems can be transformed with an auxiliary variable, under a sewing initial condition [8, 22, 26]. In such a formulation, absolutely continuous solutions are addressed with respect to less regular spaces of initial conditions and the external input of the transformed system in [21, 29]. For such coupled delay differential and difference equations, ISS is defined in terms of both states of differential and difference equations.

The goal of this chapter is to derive a condition under which the network Σ is iISS or ISS with respect to the entire disturbance r. In this chapter, we do not assume that f is known precisely. Instead, we assume the following:

Assumption 1 There exist continuously differentiable functions $V_i : \mathbb{R}^{N_i} \to \mathbb{R}_+$ and continuous functions $\underline{\alpha}_i, \overline{\alpha}_i \in \mathscr{K}_\infty$, $\alpha_i \in \mathscr{K}$, $\psi_i, \sigma_{i,j}, \kappa_i \in \mathscr{K} \cup \{0\}$ for $i, j = 1, 2, \ldots, n$ such that

$$\underline{\alpha}_i(|a|) \leq V_i(a) \leq \overline{\alpha}_i(|a|), \quad \forall a \in \mathbb{R}^{N_i} \tag{4}$$

holds and

$$\frac{\partial V_i}{\partial a} f_i(a, b, c, d) \leq -\alpha_i(|a|) + \psi_i(|b|) + \sum_{j=1}^{n} \sigma_{i,j}(|c_j|) + \kappa_i(|d|) \tag{5}$$

holds for all $(a, b, c, d) \in \mathbb{R}^{N_i} \times \mathbb{R}^{N_i} \times \mathbb{R}^N \times \mathbb{R}^{M_i}$, where $c = [c_1, c_2, \ldots, c_n]^T$.

Assumption 2 There exist continuous functions $\bar{g}_i, \bar{f}_i, \bar{g}_{i,j}, \bar{\kappa}_i \in \mathscr{K} \cup \{0\}$ for $i, j = 1, 2, \ldots, n$ such that

$$|f_i(a, b, c, d)| \leq \bar{g}_i(|a|) + \bar{f}_i(|b|) + \sum_{j=1}^{n} \bar{g}_{i,j}(|c_j|) + \bar{\kappa}_i(|d|) \tag{6}$$

holds for all $(a, b, c, d) \in \mathbb{R}^{N_i} \times \mathbb{R}^{N_i} \times \mathbb{R}^N \times \mathbb{R}^{M_i}$, where $c = [c_1, c_2, \ldots, c_n]^T$.

Note that $\sigma_{i,i}$ and $\bar{g}_{i,i}$ are redundant in (5) and (6) if $\Delta_{i,i} = 0$. Due to the presence of b representing the delayed $\dot{x}$, Assumption 1 does not precisely guarantee each subsystem Σ_i to be iISS. If $\psi_i = 0$ and $\sigma_{i,i} = 0$ hold, Assumption 1 is nothing but requiring each subsystem Σ_i to be iISS with respect to signals x_j $(j \neq i)$ from the other systems and the external signal r_i. In the case where $\psi_i = 0$ is satisfied for all i in Assumption 1, a condition under which the network Σ is iISS or ISS has been proposed in [16]. This article extends that result to networks allowing for the neutral terms $\psi_i \neq 0$ by making use of the additional information Assumption 2 provides.

Remark 2 Throughout this chapter, (1) can be replaced by a more general representation

$$\Sigma_i : \dot{x}_i(t) = f_i(x_i(t), \dot{x}_i(t - \Delta_i), x_{1,t}, x_{2,t}, \ldots, x_{n,t}, r_i(t)) \tag{7}$$

as long as Assumptions 1–2 hold. For example, the one dimensional system

$$\dot{x}_i(t) = -x_i(t) + \frac{\varepsilon_i}{1 + h_i(t)} \dot{x}_i(t - \Delta_i) + r_i(t)$$

with $h_i(t) = \int_{t-\delta_i}^{t} x_i(s)^2 ds$, any $\delta_i \in \mathbb{R}_+$ and any $\varepsilon_i \in \mathbb{R}$ satisfy Assumptions 1–2.

3 Translation into a Lyapunov-Krasovskii Functional

The iISS and ISS properties of the network Σ are characterized in terms of a Lyapunov-Krasovskii functional in this section. Clarifying a condition under which such a functional can be constructed leads us to a criterion establishing iISS and ISS. For this end, we first define a directed graph G associated with the network Σ. External inputs r_i are not used for defining the graph. In this chapter, a directed graph is allowed to have loops. A loop is an arc that connects a vertex to itself. Let the vertex set and the arc set be denoted by $\mathscr{V}(G)$ and $\mathscr{A}(G)$. Elements of $\mathscr{V}(G)$ are subsystems Σ_i, $i = 1, 2, \ldots, n$, and we write merely i instead of Σ_i for brevity:

$$\mathscr{V}(G) = \{1, 2, \ldots, n\}. \tag{8}$$

Each element of $\mathscr{A}(G)$ is an ordered pair (i, j) which is directed away from the jth vertex and directed toward the ith vertex. It is defined in the following way:

$$\{(i, j) \in \mathscr{A}(G) \Leftrightarrow \sigma_{i,j} + \psi_i \circ \bar{g}_{i,j} \neq 0\}, \quad i \neq j \tag{9}$$

$$\{(i, i) \in \mathscr{A}(G) \Leftrightarrow \sigma_{i,i} + \psi_i \circ \bar{g}_{i,i} + \psi_i \circ \bar{g}_i \neq 0\}. \tag{10}$$

Let $\mathscr{C}(G)$ and $\mathscr{P}(G)$ denote the set of all directed cycle graphs and directed path graphs, respectively, contained in the directed graph G. Given a directed cycle or a directed path U of length k, we employ the notation

$$|U| = k, \quad U = (u(1), u(2), \ldots, u(k), u(k+1)),$$

where $u(i)$'s are all vertices comprising U. If U is a directed cycle, we have $u(1) = u(k+1)$. The starting vertex of the directed path U is $u(k+1)$, and the ending vertex is $u(1)$. Let $\mathscr{L}(G)$ denote the set of all loops contained in G, and we write $|U| = 1$ and $u(1) = u(2)$ for $U \in \mathscr{L}(G)$. In the same way, we write $|U| = 0$ for $U \in \mathscr{V}(G)$. In this chapter, the cycle subgraph (resp., the path subgraph, the loop subgraph) consisting of the cycle (resp., the path, the loop) is also denoted by U. In the rest of this chapter, the term "directed" is omitted in referring to graphs.

Consider J_U, d_i, and $d_{i,j}$ in $\mathbb{R}_+$ satisfying

$$1 = d_i \sum_{U \in \{W \in \mathscr{C}(G) \cup \mathscr{P}(G) \cup \mathscr{L}(G) \cup \mathscr{V}(G) : \mathscr{V}(W) \ni i\}} J_U, \qquad \forall i \in \mathscr{V}(G) \tag{11}$$

$$1 = d_{i,j} \sum_{U \in \{W \in \mathscr{C}(G) \cup \mathscr{P}(G) \cup \mathscr{L}(G) : \mathscr{A}(W) \ni (i,j)\}} J_U, \qquad \forall (i,j) \in \mathscr{A}(G) \tag{12}$$

$$0 = d_{i,j}, \qquad \forall (i,j) \notin \mathscr{A}(G). \tag{13}$$

For each loop $(i, i) \in \mathscr{A}(G)$ the set of all U's in (12) is a singleton. Thus, $1 = d_{i,i} J_U$ holds for the loop subgraph U at vertex i. The set of non-zero J_U's fulfilling (11) and (12) defines a covering of the graph G by cycles, loops, paths and vertices. A subgraph U is adopted to cover a part of G if and only if $J_U \neq 0$. Multiple subgraphs U defining a covering can overlap. The set of subgraphs U covering G is not unique, but there is always such a set of subgraphs. For any real numbers $J_U > 0$ chosen for each covering, we can always compute $d_i > 0$ and $d_{i,j} > 0$ satisfying (11)–(13).

For each $i = 1, 2, \ldots, n$, let the integer $H_i \in \{0, 1, 2, \ldots, n+2\}$ be the number of non-zero functions in $\bar{g}_i$, $\bar{f}_i$, $\bar{g}_{i,1}$, ..., $\bar{g}_{i,n}$ and $\bar{\kappa}_i$. For each $i = 1, 2, \ldots, n$, let $h_i \in \{0, 1, 2\}$ be the number of non-zero functions in $\sigma_{i,i}$ and ψ_i. The next theorem transforms iISS and ISS of the network Σ into a set of inequalities (in terms of $\bar{\zeta}_i$ and $\bar{\xi}_{ij}$) which we solve for λ_i. The iISS and ISS of the network Σ is established via a Lyapunov-Krasovskii functional constructed with the functions λ_i, $i = 1, 2, \ldots, n$ as follows:

Theorem 1 *Suppose that the network Σ satisfies Assumptions* 1 *and* 2. *Assume that, for each $i = 1, 2, \ldots, n$, there exist real numbers $\mu_i > 0$ and $p_i > 1$ such that*

$$\psi_i(H_i \bar{f}_i(s)) \leq \frac{p_i - 1}{p_i e^{\mu_i \Delta_i}} \psi_i(s), \quad \forall s \in \mathbb{R}_+ \tag{14}$$

$$\psi_i \neq 0 \Rightarrow \left\{ d_{i,i} \leq d_{i,j} p_i e^{\mu_i \Delta_i}, \quad \forall (i,j) \in \mathscr{A}(G) \right\} \tag{15}$$

hold. Suppose that there exist continuous functions $\lambda_i : \mathbb{R}_+ \to \mathbb{R}_+$, $i = 1, 2, \ldots, n$, *and constants* $\mu_{i,j} > 0$, $i, j = 1, 2, \ldots, n$, *such that*

$$\lambda_i \text{ is nondecreasing and } \lambda_i(s) > 0, \quad \forall s \in (0, \infty), \quad i = 1, 2, \ldots, n, \tag{16}$$

$$\{\lim_{s\to\infty} \alpha_i(s) = \infty \vee \lim_{s\to\infty} \lambda_i(s) < \infty\}, \quad i = 1, 2, \ldots, n \tag{17}$$

$$\sum_{U\in\mathscr{C}(G)\cup\mathscr{P}(G)\cup\mathscr{L}(G)} \sum_{i=1}^{|U|} \{-\bar{\zeta}_{u(i)}(s_{u(i)}) + \bar{\xi}_{u(i),u(i+1)}(s_{u(i+1)})\} \le 0, \qquad \forall s_1, \ldots, s_n \in \mathbb{R}_+ \tag{18}$$

hold, where, for $i, j = 1, 2, \ldots, n$,

$$1 < \tau_i < c_i \tag{19}$$

$$\hat{\alpha}_i(s) = d_i \alpha_i(s), \quad \hat{\sigma}_i(s) = d_{i,i}\tilde{\sigma}_i(s) \tag{20}$$

$$\tilde{\sigma}_{i,j}(s) = \begin{cases} e^{\mu_{i,j}\Delta_{i,j}}[\sigma_{i,j}(s) + h_i p_i e^{\mu_i \Delta_i}\psi_i(H_i \bar{g}_{i,j}(s))], & i \ne j \\ e^{\mu_{i,j}\Delta_{i,j}}[h_i\sigma_{i,j}(s) + h_i p_i e^{\mu_i \Delta_i}\psi_i(H_i \bar{g}_{i,j}(s))] \\ \qquad\qquad +h_i p_i e^{\mu_i \Delta_i}\psi_i(H_i \bar{g}_i(s)), & i = j \end{cases} \tag{21}$$

$$\bar{\zeta}_i = \frac{\tau_i}{c_i}\left(1 - \frac{1}{\tau_i}\right)\lambda_i(\underline{\alpha}_i(s))\hat{\alpha}_i(\overline{\alpha}_i^{-1}(\underline{\alpha}_i(s))) \tag{22}$$

$$\bar{\xi}_{i,j}(s) = \lambda_i(\overline{\alpha}_i(\hat{\alpha}_i^{\ominus}(\tau_i\hat{\sigma}_{i,j}(s))))\,\hat{\sigma}_{i,j}(s). \tag{23}$$

Then the functional $V : \mathscr{L}_\infty^N[-\Delta, 0] \times \mathscr{L}_\infty^N[-\Delta, 0] \to \mathbb{R}_+$ *given by*

$$V(\varphi, \dot{\varphi}) = \sum_{i=1}^{n} \int_0^{V_i(\varphi_i(0))} \lambda_i(s)ds + \sum_{(i,i)\in\mathscr{A}(G)} X_i(\dot{\varphi}_i) + \sum_{(i,j)\in\mathscr{A}(G)} Y_{i,j}(\varphi_j) \tag{24}$$

achieves

$$\dot{V}(t) \le -\delta\sum_{i=1}^{n}\bar{\zeta}_i(|x_i(t)|) - \sum_{(i,i)\in\mathscr{A}(G)} p_i\mu_i \int_{-\Delta_i}^{0} E_i(s)\xi_i(|\dot{x}_i(t+s)|)ds - \sum_{(i,j)\in\mathscr{A}(G)} \mu_{i,j}\int_{-\Delta_{i,j}}^{0} E_{i,j}(s)\beta_{i,j}(|x_j(t+s)|)ds \; + \; e(|r(t)|) \tag{25}$$

almost everywhere in $t \in \mathbb{R}_+$ *along the trajectories of* Σ *with a constant* $\delta > 0$ *and a function* $e \in \mathscr{K} \cup \{0\}$, *where the functionals*

$$X_j, Y_{i,j} : \mathscr{L}_\infty^{N_j}[-\Delta, 0] \to \mathbb{R}_+$$

are given by

$$X_i(\dot{\varphi}_i) = p_i \int_{-\Delta_i}^{0} E_i(s)\xi_i(|\dot{\varphi}_i(s)|)ds, \tag{26}$$

$$E_i(s) = e^{\mu_i(s+\Delta_i)}, \quad \xi_i(s) = \lambda_i(\overline{\alpha}_i(\hat{\alpha}_i^{\ominus}(\tau_i h_i d_{i,i}\psi_i(s))))\, h_i\psi_i(s), \tag{27}$$

$$Y_{i,j}(\varphi_j) = \int_{-\Delta_{i,j}}^{0} E_{i,j}(s)\beta_{i,j}(|\varphi_j(s)|)ds, \quad E_{i,j}(s) = e^{\mu_{i,j}(s+\Delta_{i,j})}, \tag{28}$$

$$\begin{aligned}\beta_{i,j}(s) = {} & \lambda_i(\overline{\alpha}_i(\hat{\alpha}_i^{\ominus}(\tau_i d_{i,j}\omega_{i,j}(s))))\, \omega_{i,j}(s) \\ & + p_i e^{\mu_i\Delta_i}\lambda_i(\overline{\alpha}_i(\hat{\alpha}_i^{\ominus}(\tau_i h_i d_{i,i}\psi_i(H_i\bar{g}_{i,j}(s)))))\, h_i\psi_i(H_i\bar{g}_{i,j}(s)),\end{aligned} \tag{29}$$

$$\omega_{i,j}(s) = \begin{cases} h_i\sigma_{i,i}(s), & \textit{if} \ \ i = j \\ \sigma_{i,j}(s), & \textit{otherwise}. \end{cases} \tag{30}$$

Note that the pair (9)–(10) is equivalent to

$$\{(i,j) \in \mathscr{A}(G) \ \Leftrightarrow \ \hat{\sigma}_{i,j} \neq 0\}. \tag{31}$$

The following theorem states that the functional V achieving (25) is an iISS Lyapunov-Krasovskii functional. It becomes an ISS Lyapunov-Krasovskii functional if $\alpha_i \in \mathscr{K}_\infty$ holds for $i = 1, 2, \ldots, n$.

Theorem 2 *The network Σ is iISS with respect to input r and state x if the functional $V : \mathscr{L}_\infty^N[-\Delta, 0] \times \mathscr{L}_\infty^N[-\Delta, 0] \to \mathbb{R}_+$ given by* (24) *achieves* (25) *almost everywhere in $t \in \mathbb{R}_+$ along the trajectories of Σ with $p_i > 1$, $\mu_i > 0$, $\mu_{i,j} > 0$, $\delta > 0$, $e \in \mathscr{K} \cup \{0\}$,* (16)–(17), (19)–(22) *and* (26)–(30). *Furthermore, if $\alpha_i \in \mathscr{K}_\infty$ is satisfied for $i = 1, 2, \ldots, n$ additionally, then the network Σ is ISS.*

It is stressed that property (17) ensures that $\bar{\xi}_{i,j}$, ξ_i and $\beta_{i,j}$ are class $\mathscr{K}$ functions (or zero) defined on $\mathbb{R}_+$. Recall that the functions $\hat{\alpha}_i$ are not guaranteed to be invertible on the whole $\mathbb{R}_+$. The notation $\hat{\alpha}^{\ominus_i}$ allows us to present all the cases requiring appropriate interpretations of many $\hat{\alpha}_i^{-1}$ in a unified presentation [12]. It is demonstrated in [9, 11, 12, 14] that stability conditions for interconnections of non-ISS subsystems are necessarily asymmetric and it does not admit a single expression unless we employ the notation $\hat{\alpha}_i^{\ominus}$. It is also emphasized that for any given $d_{i,i} > 0$, $d_{i,j} > 0$ and $\mu_i > 0$, property (15) can always be satisfied by large p_i. The right hand side of (14) is an increasing function of p_i. Therefore, properties (14) and (15) can be achieved simultaneously if there exist $\mu_i > 0$ and $p_i > 1$ such that (14) holds.

Remark 3 Since $e^{\mu_i \Delta_i} \geq 1$, the implication condition (14) requires $\bar{f}_i$ to be bounded if ψ_i is bounded. This requirement is not restrictive since we use the pair of (5) and (6). The boundedness of $\bar{f}_i$ is not necessary if ψ_i is unbounded. The larger p_i is, the larger $\hat{\sigma}_{i,j}$ is, which will appear in the stability condition (37).

Remark 4 If ψ_i and λ_i are a linear function and a constant, respectively, then we can use $H_i = 1$ and $h_i = 1$ in Theorem 1. The use of constant λ_i's does not cause any conservativeness if $k_{i,j}\alpha_j = \sigma_{i,j}$, $\bar{k}_{i,j}\alpha_j = \bar{g}_{i,j}$ and $k_i\alpha_i = \bar{g}_i$ hold for some constants $k_{i,j}, \bar{k}_{i,j}, k_i \geq 0$. This can be verified in the same way as in the delay free case [6].

Remark 5 If the neutral-type delays are absent, condition (14) vanishes. Then taking $\psi_i = 0$ for all i reduces the Lyapunov-Krasovskii functional (24) to the functional of a result presented in [16]. The previous studies [16, 18], which do not accommodate neutral-type delays, use $E_{i,j}(s) = \left(1 + \mu_{i,j}\left(1 + s/\Delta_{i,j}\right)\right)$ instead of

$$E_{i,j}(s) = e^{\mu_{i,j}(s+\Delta_{i,j})}. \tag{32}$$

This replacement of $E_{i,j}$ produces a qualitative change in the iISS and ISS estimates for the network. However, we can still use $E_{i,j}(s) = \left(1+\mu_{i,j}\left(1+s/\Delta_{i,j}\right)\right)$ instead of (32) for neutral-type delays. Also, as in [16, 18], we can include distributed delays in this chapter by assuming convexity of ψ and using Jensen's inequality.

4 A Small-Gain Condition

We next present λ_i's satisfying the requirements in Theorem 1 for constructing a functional V to verify iISS and ISS of the network Σ. Deriving a condition under which such λ_i's exist, we propose a sufficient condition for iISS (or ISS) of Σ.

Let $\mathscr{CP}(i, j)$ denote the set of all cycles and paths from vertex j to vertex i of the graph G. The number of elements in $\mathscr{CP}(i, j)$ is finite. Let $\hat{\sigma}_{i,j} \in \mathscr{K} \cup \{0\}$ defined in (20) be the weight of the arc directed away from vertex j and directed toward vertex i of the graph G. Let $\hat{\alpha}_i^{\ominus}$ be the weight of vertex i. Now, we define functions $F_{i,j}$ for all the ordered pairs (i, j) of $\mathscr{V}(G)$ by

$$F_{i,j}(s) = \max\nolimits_{U \in \mathscr{CP}(i,j)} \hat{\sigma}_{u(1),u(2)} \circ \bigodot_{i=2}^{|U|} \hat{\alpha}_{u(i)}^{\ominus} \circ \tau_{u(i)} \hat{\sigma}_{u(i),u(i+1)}(s), \quad i \neq j \tag{33}$$

$$F_{i,i}(s) = \hat{\sigma}_{i,i} \tag{34}$$

with $\tau_i > 1$ to be determined later for $i = 1, 2, \ldots, n$. This function $F_{i,j}$ can be considered as the maximum composite weight of the paths or cycles from vertex j to vertex i of the graph G. The factors $\tau_i > 1$ can be used to increase the composite weight. For all ordered pairs (i, j) in $\mathscr{V}(G)$, we have the function $F_{i,j}$. Thus, a complete graph with loops at some vertices $\{1, 2, \ldots, n\}$ is constructed with the weight $F_{i,j}$ at each arc (i, j) and the weight $\hat{\alpha}_i^{\ominus}$ at each vertex i. We can prove the following property for $F_{i,j}$ easily as done in [15].

Lemma 1 *Consider $\hat{\alpha}_i \in \mathcal{K}$, $\hat{\sigma}_{i,j} \in \mathcal{K} \cup \{0\}$, $i, j = 1, 2, \ldots, n$, satisfying*

$$\left\{\lim_{s\to\infty} \hat{\alpha}_j(s) = \infty \ \vee \ \lim_{s\to\infty} \sum_{i=1}^{n} \hat{\sigma}_{i,j}(s) < \infty\right\}, \quad j = 1, 2, \ldots, n. \tag{35}$$

Then, for arbitrary given $\tau_i > 1$, we have $F_{i,j} \in \mathcal{J}$ for $i, j = 1, 2, \ldots, n$, and

$$\left\{\lim_{s\to\infty} \alpha_j(s) = \infty \ \vee \ \lim_{s\to\infty} \sum_{i=1}^{n} F_{i,j}(s) < \infty\right\}, \quad j = 1, 2, \ldots, n. \tag{36}$$

Using the vertex- and arc-weighted complete graph of $\mathcal{V}(G)$ build with the weights $F_{i,j}$ and $\hat{\alpha}_i^{\ominus}$, we can construct the desired λ_i, $i, j = 1, 2, \ldots, n$.

Theorem 3 *Consider $\hat{\alpha}_i \in \mathcal{K}$, $\hat{\sigma}_{i,j} \in \mathcal{K} \cup \{0\}$, $\underline{\alpha}_i, \overline{\alpha}_i \in \mathcal{K}_\infty$, $i, j = 1, 2, \ldots, n$, satisfying* (35)*. Let the graph G be defined by* (8) *and* (31)*. Assume that there exist $c_i > 1$, $i = 1, 2, \ldots, n$ such that*

$$\bigodot_{i=1}^{|U|} \underline{\alpha}_{u(i)}^{-1} \circ \overline{\alpha}_{u(i)} \circ \hat{\alpha}_{u(i)}^{\ominus} \circ c_{u(i)} \hat{\sigma}_{u(i),u(i+1)}(s) \le s, \quad \forall s \in \mathbb{R}_+ \tag{37}$$

holds for all cycle subgraphs and all loop subgraphs $U \in \mathcal{C}(G) \cup \mathcal{L}(G)$. Let τ_i and $\rho \ge 0$ be such that (19) *and*

$$\left(\frac{\tau_i}{c_i}\right)^{\rho} \le \tau_i - 1, \quad i = 1, 2, \ldots, n \tag{38}$$

are satisfied. Furthermore, define $\overline{\lambda}_i \in \mathcal{J}$, $i = 1, 2, \ldots, n$, by

$$\overline{\lambda}_i(s) = \left[\frac{1}{\tau_i} \hat{\alpha}_i(\overline{\alpha}_i^{-1}(s))\right]^{\rho} \prod_{j \in \mathcal{V}(G) - \{i\}} \left[F_{j,i}(\underline{\alpha}_i^{-1}(s))\right]^{\rho+1}, \tag{39}$$

and let $\nu_i : (0, \infty) \to \mathbb{R}_+$, $i = 1, 2, \ldots, n$, be continuous functions fulfilling

$$0 < \nu_i(s) < \infty, \ s \in (0, \infty), \ i = 1, 2, \ldots, n \tag{40}$$

$$\lim_{s\to\infty} \alpha_i(s) = \infty \ \vee \ \lim_{s\to\infty} \nu_i(s) < \infty \tag{41}$$

$$\overline{\lambda}_i(s)\nu_i(s) : \text{non-decreasing continuous for } s \in (0, \infty) \tag{42}$$

and

$$\nu_{u(j)} \circ \overline{\alpha}_{u(j)} \circ \hat{\alpha}_{u(j)}^{\ominus} \circ \tau_{u(j)} \hat{\sigma}_{u(j),u(j+1)}(s) \\ \leq \left(\frac{c_{u(j+1)}}{\tau_{u(j+1)}} \right)^{\rho} (\tau_{u(j+1)} - 1) \nu_{u(j+1)} \circ \underline{\alpha}_{u(j+1)}(s) \tag{43}$$

for all $s \in (0, \infty)$*, all* $j = 1, 2, \ldots, |U|$*, and all cycle subgraphs* $U \in \mathscr{C}(G)$*. Then non-decreasing continuous functions* $\lambda_i : \mathbb{R}_+ \to \mathbb{R}_+$*,* $i = 1, 2, \ldots, n$*, defined by*

$$\lambda_i(s) = \begin{cases} \overline{\lambda}_i(s)\nu_i(s), & s \in (0, \infty) \\ \lim_{s \to 0^+} \overline{\lambda}_i(s)\nu_i(s), & s = 0 \end{cases} \tag{44}$$

achieve (16)–(18) *for* $\bar{\zeta}_i$ *and* $\bar{\xi}_{i,j}$ *defined in* (22) *and* (23).

With the help of Lemma 1, a technical result in [15] proves Theorem 3. Note that, for a loop $U \in \mathscr{L}(G)$, condition (37) indicates

$$\underline{\alpha}_{u(1)}^{-1} \circ \overline{\alpha}_{u(1)} \circ \hat{\alpha}_{u(1)}^{\ominus} \circ c_{u(1)} \hat{\sigma}_{u(1),u(1)}(s) \leq s, \ \forall s \in \mathbb{R}_+. \tag{45}$$

Note that $\hat{\alpha}_1, \ldots, \hat{\alpha}_n \in \mathscr{K}_\infty$ or equivalently $\alpha_1, \ldots, \alpha_n \in \mathscr{K}_\infty$ fulfills (35). Indeed, property (35) is identical with

$$\left\{ \lim_{s \to \infty} \alpha_j(s) = \infty \vee \lim_{s \to \infty} \psi_j(\bar{g}_j(s)) + \sum_{i=1}^{n} \sigma_{i,j}(s) + \psi_i(\bar{g}_{i,j}(s)) < \infty \right\}, \ j = 1, 2, \ldots, n. \tag{46}$$

Since there always exist τ_i and $\rho \geq 0$ fulfilling (19) and (38), and since continuous functions ν_i satisfying (40)–(43) also always exist (see also Remark 6), Theorems 1, 2 and 3 lead us to the following main result:

Proposition 1 *Consider the network* Σ *satisfying Assumptions* 1 *and* 2 *and* (46)*. Assume that, for each* i*, there exist real numbers* $\mu_i > 0$ *and* $p_i > 1$ *such that* (14) *holds. If there exist* $d_i, d_{i,j} > 0$ *and* $c_i > 1$*,* $i, j = 1, 2, \ldots, n$ *such that* (37) *is satisfied for all* $U \in \mathscr{C}(G) \cup \mathscr{L}(G)$*, then the network* Σ *is iISS with respect to input* r *and state* x*. If* $\alpha_i \in \mathscr{K}_\infty$ *holds for* $i = 1, 2, \ldots, n$ *additionally, the network* Σ *is ISS.*

Condition (37) is invariant under cyclic shifting of vertices, by the definition of $\hat{\alpha}_i^{\ominus}$ [12]. The collection (37) of inequalities implies that all cycles and loops in the graph G have a small-gain property. When $\Delta = 0$ and $\sigma_{i,i} = \psi_i = 0$, $i = 1, 2, \ldots, n$, Proposition 1 reduces to the iISS small-gain result for delay-free networks in [15].

Remark 6 The flexibility of ν_i for constructing λ_i in (44) is useful for modifying V when we expand the network afterwards [11]. The flexibility can also be used for simplifying λ_i by canceling some complexity arising in $\overline{\lambda}_i$. The reader may consult the delay-free case [15] for some examples of ν_i. The simplest choice is

$$\nu_1(s) = \nu_2(s) = \ldots = \nu_n(s) = \text{constant} > 0$$

that always fulfills (40)–(43).

Remark 7 The small-gain property (37) is required for only cycles and loops. If neither cycles nor loops are involved in G and (14) is met, iISS of the system Σ is always guaranteed and a functional $V(\varphi, \dot{\varphi})$ verifying the iISS is computed as in (24) with (44). This consequence is due to $\alpha_i \in \mathcal{K}$, $i = 1, 2, \ldots, n$, and (46). This fact conforms to the observation given for delay-free systems in [10].

5 Proofs of Theorems

This section sketch proofs of the key theorems. For Theorem 3, see [15]. Proposition 1 follows directly from Theorems 1, 2 and 3.

5.1 Proof of Theorem 1

We begin with the proof of Theorem 1. Consider the functional in (24) with

$$\varphi_i(s) = x_i(t+s) \quad \text{and} \quad \dot{\varphi}_i(s) = \dot{x}_i(t+s)$$

for $s \in [-\Delta, 0]$. Let $t_h \in \mathbb{R}_+$ be such that $[-\Delta, t_h)$ is the maximal interval over which system Σ admits a unique piecewise continuously differentiable solution. By (11), (12) and (20), the functional V satisfies

$$\dot{V}(t) \leq D(t) + \sum_{i=1}^{n} \lambda_i(V_i(x_i(t)))\kappa_i(|r_i(t)|) \tag{47}$$

almost everywhere in $t \in [0, t_h)$ along the trajectories $x(t)$ of Σ, where

$$\begin{aligned}
D(t) &= \sum_{i=1}^{n} \lambda_i(V_i(x_i(t)))\Bigg[-\alpha_i(|x_i(t)|) + \psi_i(|\dot{x}_i(t-\Delta_i)|) \\
&\quad + \sum_{j=1}^{n} \sigma_{i,j}(|x_j(t-\Delta_{i,j})|)\Bigg] + \sum_{(i,i)\in\mathcal{A}(G)} \frac{d}{dt}\{X(\dot{x}_i(t))\} + \sum_{(i,j)\in\mathcal{A}(G)} \frac{d}{dt}\{Y_{i,j}(x_j(t))\} \\
&= \sum_{U\in\mathcal{C}(G)} J_U \sum_{i=1}^{|U|} \lambda_{u(i)}(V_{u(i)}(x_{u(i)}(t)))\{-\hat{\alpha}_{u(i)}(|x_{u(i)}(t)|) \\
&\qquad + d_{u(i),u(i+1)}\sigma_{u(i),u(i+1)}(|x_{u(i+1)}(t-\Delta_{u(i),u(i+1)})|)\}
\end{aligned}$$

$$
\begin{aligned}
&+ \sum_{U\in\mathscr{P}(G)} J_U \sum_{i=1}^{|U|} \lambda_{u(i)}(V_{u(i)}(x_{u(i)}(t)))\{-\hat{\alpha}_{u(i)}(|x_{u(i)}(t)|) \\
&\qquad\qquad + d_{u(i),u(i+1)}\sigma_{u(i),u(i+1)}(|x_{u(i+1)}(t-\Delta_{u(i),u(i+1)})|)\} \\
&- \sum_{U\in\mathscr{V}(G)} J_U \sum_{i=1}^{|U|} \lambda_{u(1)}(V_{u(1)}(x_{u(1)}(t)))\hat{\alpha}_{u(1)}(|x_{u(1)}(t)|) \\
&+ \sum_{U\in\mathscr{L}(G)} J_U \lambda_{u(1)}(V_{u(1)}(x_{u(1)}(t)))\{-\hat{\alpha}_{u(1)}(|x_{u(1)}(t)|) \\
&\qquad\qquad + d_{u(i),u(i)}\sigma_{u(i),u(i)}(|x_{u(i)}(t-\Delta_{u(i),u(i)})|)\} \\
&+ \sum_{U\in\mathscr{L}(G)} \left\{\lambda_{u(1)}(V_{u(1)}(x_{u(1)}(t)))\psi_{u(1)}(|\dot{x}_{u(1)}(t-\Delta_{u(1)})|) + \frac{d}{dt}\{X(\dot{x}_{u(1)}(t))\}\right\} \\
&+ \sum_{(i,j)\in\mathscr{A}(G)} \frac{d}{dt}\{Y_{i,j}(x_j(t))\}. \qquad (48)
\end{aligned}
$$

Using the definition of E_i in (27) yields

$$
\begin{aligned}
&\frac{d}{dt}\int_{-\Delta_i}^{0} E_i(s)\xi_i(|\dot{\varphi}_i(s)|)ds = \frac{d}{dt}\int_{t-\Delta_i}^{t} E_i(s-t)\xi_i(|\dot{\varphi}_i(s-t)|)ds \\
&= -\mu_i \int_{-\Delta_i}^{0} E_i(s)\xi_i(|\dot{x}_i(t+s)|)ds + e^{\mu_i\Delta_i}\xi_i(|\dot{x}_i(t)|) - \xi_i(|\dot{x}_i(t-\Delta_i)|).
\end{aligned}
$$

From (26), (6), definition of H_i and $\psi_i \in \mathscr{K}\cup\{0\}$, we deduce that

$$
\begin{aligned}
\frac{d}{dt}\{X(\dot{x}_i(t))\} \leq{}& -p_i\mu_i \int_{-\Delta_i}^{0} E_i(s)\xi_i(|\dot{x}_i(t+s)|)ds + p_i e^{\mu_i\Delta_i}\xi_i(H_i\bar{f}_i(|\dot{x}_i(t-\Delta_i)|)) \\
&- p_i\xi_i(|\dot{x}_i(t-\Delta_i)|) + p_i e^{\mu_i\Delta_i}\xi_i(H_i\bar{g}_i(|x_i(t)|)) \\
&+ p_i e^{\mu_i\Delta_i}\sum_{j=1}^{n}\xi_i(H_i\bar{g}_{i,j}(|x_j(t-\Delta_{i,j})|)) + p_i e^{\mu_i\Delta_i}\xi_i(H_i\bar{\kappa}_i(|r_i(t)|)). \qquad (49)
\end{aligned}
$$

From ξ_i in (27) it follows that

$$
\begin{aligned}
&p_i e^{\mu_i\Delta_i}\xi_i(H_i\bar{g}_{i,j}(s)) = \\
&\qquad p_i e^{\mu_i\Delta_i}\lambda_i(\overline{\alpha}_i(\hat{\alpha}_i^{\ominus}(\tau_i h_i d_{i,i}\psi_i(H_i\bar{g}_{i,j}(s)))))\, h_i\psi_i(H_i\bar{g}_{i,j}(s)) \qquad (50) \\
&p_i e^{\mu_i\Delta_i}\xi_i(H_i\bar{g}_i(s)) = \\
&\qquad p_i e^{\mu_i\Delta_i}\lambda_i(\overline{\alpha}_i(\hat{\alpha}_i^{\ominus}(\tau_i h_i d_{i,i}\psi_i(H_i\bar{g}_i(s)))))\, h_i\psi_i(H_i\bar{g}_i(s)). \qquad (51)
\end{aligned}
$$

Property (14) and

$$(p_i-1)/(p_i e^{\mu_i\Delta_i}) < 1$$

imply that

$$H_i \bar{f}_i(s) < s, \quad \forall\, s \in \mathbb{R}_+.$$

Since λ_i is a non-decreasing, the definition ξ_i in (27) together with (14) yields

$$p_i e^{\mu_i \Delta_i} \xi_i(H_i \bar{f}_i(s)) \leq (p_i - 1)\xi_i(s). \tag{52}$$

Due to (16)–(17), (19) and the definition of h_i, it can be verified that

$$\begin{aligned}
&\lambda_i(V_i(x_i(t)))\{-\hat{\alpha}_i(|x_i(t)|) + d_{i,i}\sigma_{i,i}(x_i(t-\Delta_{i,i})) + d_{i,i}\psi_i(\dot{x}_i(t-\Delta_i))\} \\
&\quad \leq -\left(1 - \frac{1}{\tau_i}\right)\lambda_i(\underline{\alpha}_i(|x_i(t)|))\hat{\alpha}_i(\overline{\alpha}_i^{-1}(\underline{\alpha}_i(|x_i(t)|))) \\
&\qquad + \lambda_i(\overline{\alpha}_i(\hat{\alpha}_i^{\ominus}(\tau_i h_i d_{i,i}\sigma_{i,i}(x_i(t-\Delta_{i,i})))))h_i d_{i,i}\sigma_{i,i}(x_i(t-\Delta_{i,i})) \\
&\qquad + \lambda_i(\overline{\alpha}_i(\hat{\alpha}_i^{\ominus}(\tau_i h_i d_{i,i}\psi_i(\dot{x}_i(t-\Delta_i)))))h_i d_{i,i}\psi_i(\dot{x}_i(t-\Delta_i)).
\end{aligned} \tag{53}$$

Properties (49), (50), (51), (52) (53), (12) and definitions (19), (22) and (27) give

$$\begin{aligned}
&\sum_{U\in\mathscr{L}(G)} J_U \lambda_{u(1)}(V_{u(1)}(x_{u(1)}(t)))\Big\{-\hat{\alpha}_{u(1)}(|x_{u(1)}(t)|) \\
&\qquad\qquad + d_{u(i),u(i)}\sigma_{u(i),u(i)}(|x_{u(i)}(t-\Delta_{u(i),u(i)})|)\Big\} \\
&\quad + \sum_{U\in\mathscr{L}(G)} \Big\{\lambda_{u(1)}(V_{u(1)}(x_{u(1)}(t)))\psi_{u(1)}(|\dot{x}_{u(1)}(t-\Delta_{u(1)})|) + \frac{d}{dt}\{X(\dot{x}_{u(1)}(t))\}\Big\} \\
&\quad \leq -\sum_{U\in\mathscr{L}(G)} J_U \Big\{\frac{c_{u(1)}}{\tau_{u(1)}}\bar{\zeta}_{u(1)}(|x_{u(1)}(t)|) \\
&\quad + \lambda_i(\overline{\alpha}_i(\hat{\alpha}_i^{\ominus}(\tau_i h_{u(1)} d_{u(1),u(1)}\sigma_{u(1),u(1)}(x_i(t-\Delta_{u(1),u(1)}))))) \\
&\qquad\qquad \cdot h_{u(1)} d_{u(1),u(1)}\sigma_{u(1),u(1)}(x_i(t-\Delta_{u(1),u(1)}))\Big\} \\
&\quad - p_{u(1)}\mu_{u(1)}\int_{-\Delta_{u(1)}}^{0} E_{u(1)}(s)\xi_{u(1)}(|\dot{x}_{u(1)}(t+s)|)ds \\
&\quad + p_{u(1)}e^{\mu_{u(1)}\Delta_{u(1)}} \\
&\qquad \cdot \lambda_{u(1)}(\overline{\alpha}_{u(1)}(\hat{\alpha}_{u(1)}^{\ominus}(\tau_{u(1)} h_{u(1)} d_{u(1),u(1)}\psi_{u(1)}(H_i \bar{g}_{u(1)}(x_{u(1)}(t)))))) \\
&\qquad\qquad \cdot h_{u(1)}\psi_i(H_i \bar{g}_{u(1)}(|x_{u(1)}(t)|)) \\
&\quad + \sum_{j=1}^{n} p_{u(1)}e^{\mu_{u(1)}\Delta_{u(1)}} \\
&\qquad \cdot \lambda_{u(1)}(\overline{\alpha}_{u(1)}(\hat{\alpha}_{u(1)}^{\ominus}(\tau_{u(1)} h_{u(1)} d_{u(1),u(1)}\psi_{u(1)}(H_i \bar{g}_{u(1),j}(|x_j(t-\Delta_{i,j})|))))) \\
&\qquad\qquad \cdot h_{u(1)}\psi_{u(1)}(H_i \bar{g}_{u(1),j}(|x_j(t-\Delta_{i,j})|)) \\
&\quad + p_{u(1)}e^{\mu_{u(1)}\Delta_{u(1)}}\xi_{u(1)}(H_{u(1)}\bar{\kappa}_{u(1)}(|r_{u(1)}(t)|)).
\end{aligned} \tag{54}$$

Define

$$\mathscr{E}(i,j)=\{W\in\mathscr{C}(G)\cup\mathscr{P}(G)\cup\mathscr{L}(G):\mathscr{A}(W)\ni(i,j)\},\quad (i,j)\in\mathscr{A}(G). \tag{55}$$

Definition (28) gives

$$\frac{d}{dt}\int_{t-\Delta_{i,j}}^{t}E_i(s-t)\beta_{i,j}(|\varphi_j(s-t)|)ds=-\mu_{i,j}\int_{-\Delta_{i,j}}^{0}E_i(s)\beta_{i,j}(|x_j(t+s)|)ds \\ +e^{\mu_{i,j}\Delta_{i,j}}\beta_{i,j}(|x_j(t)|)-\beta_{i,j}(|x_j(t-\Delta_{i,j})|). \tag{56}$$

Thus, by virtue of (29) and (12), it holds that

$$\sum_{U\in\mathscr{E}(i,j)}J_U\lambda_i(\overline{\alpha}_i(\hat{\alpha}_i^{\ominus}(\tau_i d_{i,j}\omega_{i,j}(x_j(t-\Delta_{i,j})))))d_{i,j}\omega_{i,i}(x_j(t-\Delta_{i,j})) \\ +p_ie^{\mu_i\Delta_i}\lambda_i(\overline{\alpha}_i(\hat{\alpha}_i^{\ominus}(\tau_ih_id_{i,i}\psi_i(H_i\bar{g}_{i,j}(|x_j(t-\Delta_{i,j})|))))) \\ \cdot h_i\psi_i(H_i\bar{g}_{i,j}(|x_j(t-\Delta_{i,j})|)) \\ +\frac{d}{dt}\{Y_{i,j}(x_j(t))\} \\ \le-\mu_{i,j}\int_{-\Delta_{i,j}}^{0}E_i(s)\beta_{i,j}(|x_j(t+s)|)ds \\ +e^{\mu_{i,j}\Delta_{i,j}}\lambda_i(\overline{\alpha}_i(\hat{\alpha}_i^{\ominus}(\tau_id_{i,j}\omega_{i,j}(|x_j(t)|))))\,\omega_{i,j}(|x_j(t)|) \\ +e^{\mu_{i,j}\Delta_{i,j}}p_ie^{\mu_i\Delta_i}\lambda_i(\overline{\alpha}_i(\hat{\alpha}_i^{\ominus}(\tau_ih_id_{i,i}\psi_i(H_i\bar{g}_{i,j}(|x_j(t)|))))) \\ \cdot h_i\psi_i(H_i\bar{g}_{i,j}(|x_j(t)|)) \tag{57}$$

for $(i,j)\in\mathscr{A}(G)$. By (12), the non-decreasingness of λ_i and $p_ie^{\mu_i\Delta_i}\ge 1$, we have

$$p_ie^{\mu_i\Delta_i}\lambda_i(\overline{\alpha}_i(\hat{\alpha}_i^{\ominus}(\tau_ih_id_{i,i}\psi_i(H_i\bar{g}_{i,j}(s)))))\,h_i\psi_i(H_i\bar{g}_{i,j}(s))\le \\ \sum_{U\in\mathscr{E}(i,j)}J_U\lambda_i(\overline{\alpha}_i(\hat{\alpha}_i^{\ominus}(\tau_i p_ie^{\mu_i\Delta_i}h_id_{i,i}\psi_i(H_i\bar{g}_{i,j}(s)))))\,p_ie^{\mu_i\Delta_i}h_id_{i,j}\psi_i(H_i\bar{g}_{i,j}(s)) \tag{58}$$

for $(i,j)\in\mathscr{A}(G)$, and

$$p_ie^{\mu_i\Delta_i}\lambda_i(\overline{\alpha}_i(\hat{\alpha}_i^{\ominus}(\tau_ih_id_{i,i}\psi_i(H_i\bar{g}_i(s)))))\,h_i\psi_i(H_i\bar{g}_i(s)) \\ \le\lambda_i(\overline{\alpha}_i(\hat{\alpha}_i^{\ominus}(\tau_ip_ie^{\mu_i\Delta_i}h_id_{i,i}\psi_i(H_i\bar{g}_i(s)))))\,p_ie^{\mu_i\Delta_i}h_id_{i,i}\psi_i(H_i\bar{g}_i(s)) \tag{59}$$

for $(i, i) \in \mathscr{A}(G)$. Now, it is verified that for an arbitrary non-decreasing continuous function $q : \mathbb{R}_+ \to \mathbb{R}_+$, we have

$$q(a_1 c)b_1 + q(a_1 c)b_2 \leq q(a_1 cd + a_2 cd)(b_1 + b_2) \tag{60}$$

for all a_1, a_2, b_1, b_2, c, and d in $\mathbb{R}_+$ satisfying $d \geq 1$ and $a_2 d \geq a_1$. Using definitions (20), (21) and property (15) in the application of (60), we can obtain

$$\begin{aligned}
&e^{\mu_{i,j}\Delta_{i,j}}\lambda_i(\overline{\alpha}_i(\hat{\alpha}_i^{\ominus}(\tau_i d_{i,j}\sigma_{i,j}(s))))\, d_{i,j}\sigma_{i,j}(s)\\
&+e^{\mu_{i,j}\Delta_{i,j}} p_i e^{\mu_i \Delta_i}\lambda_i(\overline{\alpha}_i(\hat{\alpha}_i^{\ominus}(\tau_i h_i d_{i,i}\psi_i(H_i\bar{g}_{i,j}(s)))))\, h_i d_{i,j}\psi_i(H_i\bar{g}_{i,j}(s))\\
&\leq \lambda_i(\overline{\alpha}_i(\hat{\alpha}_i^{\ominus}(\tau_i\hat{\sigma}_{i,j}(s))))\, \hat{\sigma}_{i,j}(s)
\end{aligned} \tag{61}$$

for $(i, j) \in \mathscr{A}(G)$ and $i \neq j$. In the case of $(i, i) \in \mathscr{A}(G)$,

$$\begin{aligned}
&e^{\mu_{i,i}\Delta_{i,i}}\lambda_i(\overline{\alpha}_i(\hat{\alpha}_i^{\ominus}(\tau_i h_i d_{i,i}\sigma_{i,i}(s))))\, h_i d_{i,i}\sigma_{i,i}(s)\\
&+e^{\mu_{i,i}\Delta_{i,i}} p_i e^{\mu_i \Delta_i}\lambda_i(\overline{\alpha}_i(\hat{\alpha}_i^{\ominus}(\tau_i h_i d_{i,i}\psi_i(H_i\bar{g}_{i,i}(s)))))\, h_i d_{i,i}\psi_i(H_i\bar{g}_{i,i}(s))\\
&+p_i e^{\mu_i \Delta_i}\lambda_i(\overline{\alpha}_i(\hat{\alpha}_i^{\ominus}(\tau_i p_i e^{\mu_i \Delta_i} h_i d_{i,i}\psi_i(H_i\bar{g}_i(s)))))\, h_i d_{i,i}\psi_i(H_i\bar{g}_i(s))\\
&\leq \lambda_i(\overline{\alpha}_i(\hat{\alpha}_i^{\ominus}(\tau_i\hat{\sigma}_{i,j}(s))))\, \hat{\sigma}_{i,j}(s).
\end{aligned} \tag{62}$$

Hence, substituting (18) into D with $\bar{\zeta}_i$ and $\bar{\xi}_{i,j}$ defined in (22) and (23) leads to

$$\begin{aligned}
D(t) \leq &-\sum_{U\in\mathscr{C}(G)\cup\mathscr{P}(G)\cup\mathscr{L}(G)} \sum_{i=1}^{|U|}\left(\frac{c_{u(i)}}{t_{u(i)}} - 1\right)\bar{\zeta}_{u(i)}(|x_{u(1)}(t)|)\\
&-\sum_{U\in\mathscr{V}(G)} J_U \sum_{i=1}^{|U|}\lambda_{u(1)}(V_{u(1)}(x_{u(1)}(t)))\hat{\alpha}_{u(1)}(|x_{u(1)}(t)|)\\
&-\sum_{(i,i)\in\mathscr{A}(G)} p_i\mu_i\int_{-\Delta_i}^{0} E_i(s)\xi_i(|\dot{x}_i(t+s)|)ds + p_i e^{\mu_i\Delta_i}\xi_i(H_i\bar{\kappa}_i(|r_i(t)|))\\
&-\sum_{(i,j)\in\mathscr{A}(G)} \mu_{i,j}\int_{-\Delta_{i,j}}^{0} E_i(s)\beta_{i,j}(|x_j(t+s)|)ds.
\end{aligned} \tag{63}$$

Applying (19) and (16)–(17) to (47) provides appropriate $\delta > 0$ and $e \in \mathscr{K} \cup \{0\}$ such that all trajectories of Σ satisfy (25) almost everywhere in $t \in [0, t_h)$. Although t_h was assumed to be finite, the argument in [25] can verify that (25) and (6) imply essential boundedness of $\dot{x}$ over $[0, t_h)$. Hence, t_h cannot be finite, so (25) for $t \in \mathbb{R}_+$.

5.2 Proof of Theorem 2

We next provide Theorem 2. By (4) and (24), there are $\underline{\gamma}_a$ and $\overline{\gamma}_a$ in $\mathscr{K}_\infty$ such that

$$\underline{\gamma}_a(|\varphi(0)|) \leq V(\varphi) \leq \overline{\gamma}_a(\|\varphi\|_\infty + \|\dot{\varphi}\|_\infty).$$

The application of the comparison principle used in [1, 25] to (25) proves (2). A standard technique to establish ISS [25, 32] can also lead us to (3) for the delay network Σ in the case of $\alpha_i \in \mathcal{K}_\infty$, $i = 1, 2, \ldots, n$.

6 Conclusion

This chapter provided a solution to the problem of verifying stability and robustness of dynamical networks involving neutral, retarded and communication time delays. It treated networks including neutral-type delays in the framework of iISS (which is beyond ISS) by constructing Lyapunov-Krasovskii functionals to conclude iISS of the overall network from iISS-type information of subsystems. When a network has no delays, the proposed Lyapunov-Krasovskii functional and the stability criterion provided here reduce to the ones of the delay-free result in [15]. When the network has no neutral-type delays, the functional and criterion become identical with the ones derived in [16]. Nevertheless, this chapter used a slightly different technique, i.e., the choice of $E_{i,j}$ which is proposed in [25].

For brevity, this chapter does not explicitly show formulas for distributed delays. For ISS networks without neutral-type delays, this chapter provides a Lyapunov-type constructive alternative to the trajectory-based result in [33].

It is demonstrated in [13] that the maximization technique that is popular for ISS systems [7, 19, 20, 24] is incapable of producing Lyapunov functions or functionals guaranteeing global stability for interconnections involving non-ISS subsystems. To overcome this difficulty, this chapter extended the sum-type construction of Lyapunov functions pursued in [9, 14, 15] further to neutral systems. The proposed method deals with local delays and communication delays at the same time, in the iISS framework. The idea of this simultaneous treatment is employed in the ISS framework by a recent result [20].

For the sake of succinctly conveying the main idea, other than neutral-type delays, the formula presented in this chapter only allows for a single discrete delay in each subsystem and a single discrete delay at each coupling channel. Combining this chapter with the previous work [16], the reader can obtain stability criteria and procedures for constructing Lyapunov-Krasovskii functionals in the presence of multiple and various types of delays at each subsystem and channel. In the presence of multiple delays, pursuing optimality to reduce possible conservativeness in applying the stability criterion and the assumptions with non-unique parameters to a network is an important topic of feature research.

Acknowledgments A part of this work was inspired by discussions with Z.P. Jiang and P. Pepe the first author had concerning iISS of delay systems and the ISS small-gain technique for delay systems.

References

1. Angeli, D., Sontag, E., Wang, Y.: A characterization of integral input-to-state stability. IEEE Trans. Autom. Control **45**(6), 1082–1097 (2000)
2. Bellen, A., Guglielmi, N., Ruehli, A.: Methods for linear systems of circuit delay differential equations of neutral type. IEEE Trans. Circuits Syst.: Fundam. Theory Appl. **46**(1), 212–216 (1999)
3. Brayton, R.: Bifurcation of periodic solutions in a nonlinear difference-differential equation of neutral type. Q. J. Appl. Math. **24**(3), 215–224 (1966)
4. Cheng, C.-J., Liao, T.-L., Yan, J.-J., Hwang, C.-C.: Globally asymptotic stability of a class of neutral-type neural networks with delays. IEEE Trans. Syst. Man Cybern. Part B: Cybern. **36**(5), 1191–1195 (2006)
5. Cheng, L., Hou, Z.-G., Tan, M.: A neutral-type delayed projection neural network for solving nonlinear variational inequalities. IEEE Trans. Circuits Syst. II Express Briefs **55**(8), 806–810 (2008)
6. Dashkovskiy, S., Ito, H., Wirth, F.: On a small gain theorem for ISS networks in dissipative Lyapunov form. Eur. J. Control **17**(4), 357–369 (2011)
7. Dashkovskiy, S., Ruffer, B., Wirth, F.: Small gain theorems for large scale systems and construction of ISS Lyapunov functions. SIAM J. Control Optim. **48**(6), 4089–4118 (2010)
8. Hale, J., Verduyn Lunel, S.: Introduction to Functional Differential Equations. Springer, New York (1993)
9. Ito, H.: State-dependent scaling problems and stability of interconnected iISS and ISS systems. IEEE Trans. Autom. Control **51**(10), 1626–1643 (2006)
10. Ito, H.: A Lyapunov approach to cascade interconnection of integral input-to-state stable systems. IEEE Trans. Autom. Control **55**(3), 702–708 (2010)
11. Ito, H.: Construction of Lyapunov functions for networks of iISS systems: an explicit solution for a cyclic structure. In: Proceedings of the American Control Conference, pp. 196–201 (2010)
12. Ito, H.: Necessary conditions for global asymptotic stability of networks of iISS systems. Math. Control Signals Systems **24**(1), 55–74 (2012)
13. Ito, H., Dashkovskiy, S., Wirth, F.: Capability and limitation of max- and sum-type construction of Lyapunov functions for networks of iISS systems. Automatica **48**(6), 1197–1204 (2012)
14. Ito, H., Jiang, Z.-P.: Necessary and sufficient small gain conditions for integral input-to-state stable systems: a Lyapunov perspective. IEEE Trans. Autom. Control **54**(10), 2389–2404 (2009)
15. Ito, H., Jiang, Z.-P., Dashkovskiy, S., Ruffer, B.: Robust stability of networks of iISS systems: Construction of sum-type Lyapunov functions. IEEE Trans. Autom. Control **58**(5), 1192–1207 (2013)
16. Ito, H., Jiang, Z.-P., Pepe, P.: Construction of Lyapunov-Krasovskii functionals for networks of iISS retarded systems in small-gain formulation. Automatica **49**(11), 3246–3257 (2013)
17. Ito, H., Mazenc, F.: An iISS formulation for establishing robust stability of dynamical networks with neutral, retarded and communication delay. In: Proceedings of the American Control Conference, pp. 6394–6399 (2012)
18. Ito, H., Pepe, P., Jiang, Z.-P.: A small-gain condition for iISS of interconnected retarded systems based on Lyapunov-Krasovskii functionals. Automatica **46**(10), 1646–1656 (2010)
19. Jiang, Z.-P., Mareels, I., Wang, Y.: A Lyapunov formulation of the nonlinear small-gain theorem for interconnected ISS systems. Automatica **32**(8), 1211–1215 (1996)
20. Karafyllis, I., Jiang, Z.-P.: A vector small-gain theorem for general non-linear control systems. IMA J. Math. Control Inf. **28**(3), 309–344 (2011)
21. Karafyllis, I., Pepe, P., Jiang, Z-P.: Stability results for systems described by coupled retarded functional differential equations and functional difference equations. Nonlinear Anal. Theory Methods Appl. **71**(7–8), 3339–3362 (2009)
22. Kolmanovskii, V., Myshkis, A.: Introduction to the Theory and Applications of Functional Differential Equations. Kluwer Academic Publishers, Dordrecht (1999)
23. Kuang, Y.: Delay Differential Equations with Applications in Population Dynamics. Academic Press, San Diego (1993)

24. Liu, T., Hill, D., Jiang, Z.-P.: Lyapunov formulation of ISS small-gain in continuous-time dynamical networks. Automatica **47**(9), 2088–2093 (2011)
25. Mazenc, F., Ito, H.: Lyapunov technique and backstepping for nonlinear neutral systems. IEEE Trans. Autom. Control **58**(2), 512–517 (2013)
26. Niculescu, S.-I.: Delay Effects on Stability: A Robust Control Approach. Springer, Berlin (2001)
27. Park, J.: Synchronization of cellular neural networks of neutral type via dynamic feedback controller. Chaos, Solitons Fractals **42**(3), 1299–1304 (2009)
28. Pepe, P., Jiang, Z.-P.: A Lyapunov-Krasovskii methodology for ISS and iISS of time-delay systems. Syst. Control Lett. **55**(12), 1006–1014 (2006)
29. Pepe, P., Karafyllis, I., Jiang, Z.-P.: On the Liapunov-Krasovskii methodology for the ISS of systems described by coupled delay differential and difference equations. Automatica **44**(9), 2266–2273 (2008)
30. Sontag, E.: Smooth stabilization implies coprime factorization. IEEE Trans. Autom. Control **34**(4), 435–443 (1989)
31. Sontag, E.: Comments on integral variants of ISS. Syst. Control Lett. **34**(1–2), 93–100 (1998)
32. Sontag, E., Wang, Y.: On characterizations of input-to-state stability property. Syst. Control Lett. **24**(5), 351–359 (1995)
33. Tiwari, S., Wang, Y., Jiang, Z.-P.: Nonlinear small-gain theorems for large-scale time-delay systems. Dyn. Continuous Discrete Impulsive Syst. Series A: Math. Anal. **19**(1), 27–63 (2012)

Input-to-State Stability of Switched Systems with Time Delays

Zhong-Ping Jiang, Yuandan Lin and Yuan Wang

Abstract This chapter deals with the input-to-state stability of switched nonlinear systems with time delays. The proposed results demonstrate a connection between small-gain arguments in the context of input-to-state stability and the traditional Lyapunov-Razumikhin method for switched systems. By using the notion of average dwell time, it is shown that a switching among ISS systems with compatible Lyapunov-Razumikhin functions will not destroy the stability property if the switching is not too fast on average. Particularly, it is shown that the existence of a common Razumikhin function is sufficient to guarantee input-to-state stability for time-delayed systems under arbitrary switching.

1 Introduction

This chapter studies the input-to-state stability of switched systems consisting of a set of delay differential equations, called subsystems, and a discrete logic rule that governs the switchings among the subsystems. Such systems commonly arise in applications ranging from manufacturing, communication networks, and computer synchronization, to chemical processes; see [9] and references therein. Due to their wide applications, stability and stabilization of switched systems have been intensively studied. The main tools used in the area include common Lyapunov functions, compatible Lyapunov functions, and average dwell time; see for instance the monograph [9] and the survey [10].

Z.-P. Jiang—Work partially supported by the NSF grants DMS-0906659 and ECCS-1230040.
Y. Wang—Work partially supported by the NSF grant DMS-0906918.

Z.-P. Jiang (✉)
New York University, Brooklyn, NY 11201, USA
e-mail: zjiang@nyu.edu

Y. Lin · Y. Wang
Florida Atlantic University, Boca Raton, FL 33431, USA
e-mail: lin@math.fau.edu

Y. Wang
e-mail: ywang@math.fau.edu

I. Karafyllis et al. (eds.), *Recent Results on Nonlinear Delay Control Systems*,
Advances in Delays and Dynamics 4, DOI 10.1007/978-3-319-18072-4_11

In practice, a system is often affected by time delays. In such a system, the evolution of a process depends not only on the current values of the state variables, but also on the values of state variables in a past time interval. Mathematically, such a system is modeled by differential equations with time delays; see e.g., [6, 8]. The effect of delays adds more complexity to the analysis and synthesis of switched systems. There has been rapid progress for linear switched systems with delays. For instance, sufficient criteria were obtained by the method of common Lyapunov-Krasovskii functionals in [7]; L_2 stability was studied by compatible Lyapunov-Krasovskii functionals in [17]; and the stability problem was addressed based on constructions of piecewise time-varying Lyapunov-Krasovskii functionals in [1]. However, most past work, including the papers mentioned above, has mainly focused on linear systems or is based on linear approaches. Fundamentally nonlinear approaches for switched systems with time delays still remain largely open. The recent work [12] on several stability properties was based on Lyapunov-Krasovskii functionals. On the other hand, the task of obtaining a Lyapunov-Krasovskii functional is much more challenging than its delay-free analogue. Another approach was used in [6], where input-to-state stability was studied for dynamical systems which may not satisfy the semigroup property. Thus, switched systems, or more generally, hybrid systems, with or without delays can be studied from the point view of [6] in a different angle.

Our main objective is to develop stability criteria by the Lyapunov-Razumikhin approach; see [2]. The Razumikhin theorem provides a convenient tool for determining asymptotic stability without invoking a Lyapunov-Krasovskii functional defined on an infinite dimensional space. In the 1990s, Teel's work [18] identified an interesting connection between the Razumikhin theorem and the small-gain theorem, which was an effective tool in robust stability analysis. The idea in [18] was developed further in [4] for control Lyapunov functions. In the more recent works [19, 20], the Razumikhin approach was used to develop small-gain theorems and several stability criteria for systems affected by time delays. The advantage of this approach is that it allows one to convert stability analysis for a delay system into a more extensively studied problem of robust stability for a delay-free system. It is well understood that even if each individual subsystem has a certain stability property, the switching may destroy the stability; see [9]. It is thus desirable for subsystems of a switched system to possess a common Lyapunov function or Lyapunov-Razumikhin function. However, this condition can often be too strong. Frequently in practice, the subsystems will each have an individual Lyapunov-Razumikhin function. In such a situation, the average dwell time (which is a notion introduced in [3]) provides an efficient tool to estimate how much the effect of the switching can be tolerated based on how frequent the switchings occur. By applying the notion of average dwell time with the Lyapunov-Razumikhin approach, we will extend the results in [3] to nonlinear switched systems with time delays. As a special case, we show that the existence of a common Lyapunov-Razumikhin function among all of the individual subsystems is sufficient for the switched system to be uniformly globally asymptotically stable.

This chapter is organized as follows. In Sect. 2, we briefly review some preliminary results on the Razumikhin approach for nonlinear systems with delays. In Sect. 3, we present our main results, including the consequences in a few special cases, such as common Lyapunov-Razumikhin functions and the force-free case. In Sect. 4, we provide the proofs of the main results, followed by some brief concluding remarks in Sect. 5. To make the work more self contained, we give some discussions on Lyapunov results for systems with inputs taking values in a function space (which is infinite dimensional).

Notation. In this chapter, we use $|\cdot|$ to denote the Euclidean norm of vectors, and $\|\cdot\|_I$ denotes the essential supremum norm of a measurable and locally essentially bounded function on any interval $I \subseteq \mathbb{R}$. When $I = [0, \infty)$, $||\cdot||$ will be used to denote $\|\cdot\|_I$. For $\varphi = (\varphi_1, \cdots, \varphi_k)^T$ defined on I, $||\varphi||_I = \max_{1\le i\le k}\{||\varphi_i||_I\}$. A continuous function $\alpha : \mathbb{R}_{\ge 0} \to \mathbb{R}_{\ge 0}$ is of *class* $\mathscr{K}$ provided it is positive definite and strictly increasing; and is of *class* $\mathscr{K}_\infty$ provided it is of class $\mathscr{K}$ and also unbounded. A function $\beta : \mathbb{R}_{\ge 0} \times \mathbb{R}_{\ge 0} \to \mathbb{R}_{\ge 0}$ is of *class* $\mathscr{KL}$ provided (i) for each fixed $t \ge 0$, the function $\beta(\cdot, t)$ is of class $\mathscr{K}$ and (ii) for each fixed $s \ge 0$, the function $\beta(s, t)$ decreases to 0 as $t \to \infty$. For a positive semi-definite function $\kappa : \mathbb{R}_+ \to \mathbb{R}_+$, we say that $\kappa < \text{id}$ if $\kappa(s) < s$ for all $s > 0$.

2 Preliminaries

Given $\theta > 0$, let $\mathscr{X} = C([-\theta, 0])$ be the space of all continuous functions from $[-\theta, 0]$ to $\mathbb{R}$, equipped with the norm $||\cdot||_{[-\theta,0]}$. For a function φ defined on $[-\theta, b)$, where $b > 0$ is a given constant, let φ_t denote the function defined by $\varphi_t(s) = \varphi(t + s)$ for all $s \in [-\theta, 0]$.

Consider a *control system with time-delay* as follows:

$$\Sigma^o : \quad \begin{aligned} \dot{x}(t) &= f(x(t), x_t, u(t)), \\ x(s) &= \xi(s), \qquad s \in [-\theta, 0], \end{aligned} \tag{1}$$

where for each $t \ge 0$, $x(t) \in \mathbb{R}^n$ and $u(t) \in \mathbb{R}^m$. Assume that each input $u(\cdot)$ is a measurable and locally essentially bounded function from $\mathbb{R}_{\ge 0}$ to $\mathbb{R}^m$; and that the map $f : \mathbb{R}^n \times \mathscr{X}^n \times \mathbb{R}^m \to \mathbb{R}^n$ satisfies the following properties:

- f is completely continuous (that is, f is continuous, and the closure of $f(B)$ is compact for each bounded set B), and
- f is Lipschitz on every compact subset of $\mathbb{R}^n \times \mathscr{X}^n \times \mathbb{R}^m$.

Under the assumptions stated above, the existence, uniqueness, and the maximum continuation of solutions hold for (1), that is, for each input function $u(\cdot)$ and each initial state $\xi \in \mathscr{X}$, there is a unique solution $x(\cdot)$ defined on a maximum interval $[-\theta, T)$ with $0 < T \le \infty$; see [2, 19].

A system Σ^o as in (1) is said to be *input-to-state stable* (ISS) provided there exist functions $\beta \in \mathscr{KL}$ and $\gamma \in \mathscr{K}$ such that the following holds along each trajectory:

$$|x(t)| \le \max\left\{\beta\left(||\xi||_{[-\theta,0]}, t\right), \gamma(||u||)\right\} \qquad \forall t \ge 0.$$

To utilize the Lyapunov-Razumikhin approach, we associate the following system to the system Σ^o from (1):

$$\Sigma^w : \quad \dot{x}(t) = f(x(t), w(t), u(t)), \quad x(0) = x_o, \tag{2}$$

where $x_o \in \mathbb{R}^n$, the inputs represented by w are continuous functions from $[0, \infty)$ to $C[-\theta, 0]$, and the inputs represented by u are measurable and locally essentially bounded functions from $[0, \infty)$ to $\mathbb{R}^m$. Notice that one may treat system (1) as an interconnected system composed of subsystem (2) and the delay operator $w(t) = x_t$.

Though system (2) is slightly different from a usual control system where both the input functions u and w take values in Euclidean spaces, many stability notions (including input-to-state stability) apply to a system as in Σ^w, where the input w takes values in a space of functions; see some discussions in the appendix. Specifically, a system Σ^w as in (2) is ISS provided there exist functions $\beta \in \mathscr{KL}$, $\kappa \in \mathscr{K}$, and $\gamma \in \mathscr{K}$ such that

$$|x(t)| \leq \max\left\{\beta(|x_o|, t),\ \kappa\left(||w||_{[0,\infty)}\right),\ \gamma(||u||)\right\} \tag{3}$$

holds for all $t \geq 0$, all w, and all u. It can be seen that if (3) holds for Σ^w, then the following holds for each trajectory of Σ^o:

$$|x(t)| \leq \max\left\{\beta(|x(0)|, t),\ \kappa\left(||x||_{[-\theta,t)}\right),\ \gamma(||u||)\right\}$$

on its maximum interval. The following small-gain result provides a convenient tool for stability analysis of systems with delays as in (1):

Proposition 1 ([19]) *Consider a system Σ^o as defined in* (1). *Assume that its corresponding system Σ^w is ISS with gain functions κ in w and γ in u as in* (3). *If the small-gain condition $\kappa < \mathrm{id}$ holds, then the system Σ^o is ISS.*

In the Lyapunov context, Proposition 1 is related to the Razumikhin theorem. See [2, 18, 19] for the Razumikhin theorem, which we state next.

Proposition 2 *Consider a system Σ^o as in* (1). *Assume that there exists some C^1 function $V : \mathbb{R}^n \to \mathbb{R}_{\geq 0}$ such that the following two conditions hold:*

- *there exist $\overline{\alpha}$ $\underline{\alpha} \in \mathscr{K}_\infty$ such that*

$$\underline{\alpha}(|x|) \leq V(x) \leq \overline{\alpha}(|x|) \quad \forall x \in \mathbb{R}^n; \tag{4}$$

- *there exist a continuous positive definite function α, some $\chi \in \mathscr{K}_\infty$ satisfying $\chi < \mathrm{id}$, and some $\gamma \in \mathscr{K}$ such that the following holds for all $x \in \mathbb{R}^n$, all $\mathrm{w} \in C[-\theta, 0]$, and all $u \in \mathbb{R}^m$:*

$$V(x) \geq max\left\{\chi\left(||V(\mathrm{w})||_{[-\theta,0]}\right),\ \gamma(|u|)\right\} \\ \Rightarrow \quad DV(x)f(x, \mathrm{w}, u) \leq -\alpha(|x|) \tag{5}$$

Then the system Σ^o is ISS.

Remark 1 For a system Σ^w as in (2) where both u and w take values in Euclidean spaces, it is known that the system is ISS if and only if it admits a C^1 ISS-Lyapunov function satisfying (4)–(5) with the term $\chi(||V(\mathrm{w})||_{[-\theta,0]})$ replaced by $\chi(V(\mathrm{w}))$; see [16]. For the more general case where w takes values in a function space, the Lyapunov sufficiency implication still holds; see the appendix. Assume that such a Lyapunov function exists for Σ^w. By [13, Lemma 11], it follows that for each $\lambda > 0$, there exists a function $\rho \in \mathscr{K}_\infty$ such that for the function W given by $W = \rho \circ V$, property (5) on V becomes:

$$W(x) \geq \max\left\{\tilde{\chi}\left(||W(\mathrm{w})||_{[-\theta,0]}\right),\ \tilde{\gamma}(|u|)\right\}$$
$$\Rightarrow\ DW(x)f(x,\mathrm{w},u) \leq -\lambda W(x), \tag{6}$$

where $\tilde{\chi} = \rho \circ \chi \circ \rho^{-1}$, and $\tilde{\gamma} = \rho \circ \gamma$. Note that if $\chi < \mathrm{id}$, it also holds that $\tilde{\chi} < \mathrm{id}$.

If there is a W satisfying (6) for Σ^w, then

$$W(x(t)) \leq \max\left\{W(x(0))e^{-\lambda t},\ \tilde{\chi}(||W(w(\cdot))||_{[0,t]}),\ \tilde{\gamma}(||u||)\right\}$$

along all trajectories of (2). Furthermore, if $\tilde{\chi} < \mathrm{id}$, then the system Σ^o is ISS.

3 Main Results

Consider a *switched system* described by

$$\Sigma_\sigma^o:\quad \dot{x}(t) = f_{\sigma(t)}(x(t), x_t, u(t)), \tag{7}$$

where the switching signal $\sigma : [0,\infty) \to \Gamma$ is piecewise constant and continuous from the right, and where Γ is an index set; the state variable $x(\cdot)$ takes values in $\mathbb{R}^n$, and the input function $u(\cdot)$ takes values in $\mathbb{R}^m$. Assume that for each $p \in \Gamma$, the function $f_p : \mathbb{R}^n \times \mathscr{X}^n \times \mathbb{R}^m \to \mathbb{R}^n$ is completely continuous Lipschitz on compact sets. The inputs are measurable and locally essentially bounded functions.

Under the assumption on f_p stated above, the existence, uniqueness, and the maximum continuation of solutions hold for (7), that is, for each $\xi \in \mathscr{X}$, each switching signal $\sigma(\cdot)$, and each input $u(\cdot)$, there is a unique trajectory of the system (7), satisfying the initial condition $x_0 = \xi$ (i.e., $x(s) = \xi(s)$ for all $s \in [-\theta, 0]$), defined on a maximum interval $[-\theta, T)$; see [2, 19]. Such a trajectory will be denoted by $x_\sigma(\cdot, \xi, u)$.

Let $\mathscr{F}$ be a set of switching functions. A system as in (7) is said to be *uniformly input-to-state stable* (or uniformly ISS) over $\mathscr{F}$ provided there exist $\beta \in \mathscr{KL}$ and $\gamma \in \mathscr{K}$ such that for all input functions u and all switching signals $\sigma \in \mathscr{F}$, the following holds:

$$|x_\sigma(t,\xi,u)| \leq \max\left\{\beta\left(||\xi||_{[-\theta,0]}, t\right),\ \gamma(||u||)\right\}\quad \forall t \geq 0.$$

Corresponding to a switched system Σ_σ^o as in (7), we associate with it the delay-free system

$$\Sigma_\sigma^w: \quad \dot{x}(t) = f_{\sigma(t)}(x(t), w(t), u(t)), \tag{8}$$

where w takes values in $C([-\theta, 0])$. For each individual system

$$\Sigma_p^o: \quad \dot{x}(t) = f_p(x(t), x_t, u(t)), \tag{9}$$

we associate with it the delay-free system

$$\Sigma_p^w: \quad \dot{x}(t) = f_p(x(t), w(t), u(t)). \tag{10}$$

Assume that for each $p \in \Gamma$, the system Σ_p^w admits an ISS-Lyapunov function V_p satisfying the following:

$$\underline{\alpha}(|x|) \leq V_p(x) \leq \overline{\alpha}(|x|) \quad \forall x \in \mathbb{R}^n, \tag{11}$$

where $\underline{\alpha}$ and $\overline{\alpha}$ are in $\mathscr{K}_\infty$; and

$$\begin{aligned} V_p(x) \geq \max\left\{\kappa\left(||V_p(\mathrm{w}(\cdot))||_{[-\theta,0]}\right),\ \rho(|u|)\right\} \\ \Rightarrow DV_p(x) f_p(x, \mathrm{w}, u) \leq -\lambda V_p(x), \end{aligned} \tag{12}$$

and where κ and ρ are in $\mathscr{K}$ and $\lambda > 0$. Then the following holds along all trajectories of Σ_p^w:

$$V_p(x(t)) \leq \max\left\{V_p(x(0))e^{-\lambda t},\ \kappa(||V_p(w(\cdot))||_{[0,t]}),\ \rho(||u||)\right\}$$

for all $t \geq 0$. Under the additional small-gain condition $\kappa < \mathrm{id}$, the delay system Σ_p^o is ISS.

To consider the uniform-ISS property for switched systems with time-delays, we consider the *average dwell-time*, a notion introduced in [3]. For each switching signal σ, let $N_\sigma(t_0, t_1)$ denote the number of switchings occurring in the interval (t_0, t_1), (i.e., the number of discontinuous points of $\sigma(\cdot)$ on (t_0, t_1)). For any constant $\tau > 0$ and any positive integer N_0, let $\mathscr{S}[\tau, N_0]$ denote the set of all switching signals for which the following holds for all $0 \leq t_0 < t_1$:

$$N_\sigma(t_0, t_1) \leq N_0 + \frac{t_1 - t_0}{\tau}. \tag{13}$$

For a switching signal, $\sigma \in \mathscr{S}[\tau, N_0]$, N_0 is called a chatter bound of σ, and τ is called the average dwell time of σ; see [3]. Below we state our main result. We will provide the proof in Sect. 4.

Theorem 1 *Consider a switched system with time delays as in* (7). *Assume that for each* $p \in \Gamma$, *there is a* C^1 *function* $V_p : \mathbb{R}^n \to \mathbb{R}_{\geq 0}$ *such that the following holds:*

(a) for some $\underline{\alpha}$ *and* $\overline{\alpha}$ *in* $\mathcal{K}_\infty$, (11) *holds;*
(b) for some κ *and* ρ *in* $\mathcal{K}$ *and* $\lambda > 0$, (12) *holds;*
(c) there is some $\mu \geq 1$ *such that*

$$V_p(x) \leq \mu V_q(x) \quad \forall x \in \mathbb{R}^n, \ \forall p, q \in \Gamma;$$

(d) for some integer $N_0^* \geq 0$, *we have* $\mu^{N_0^*+2}\kappa(s) < s$ *for all* $s > 0$.

Set

$$\tau^* = \frac{2 \ln \mu}{\lambda}. \tag{14}$$

Then the system (7) *is uniformly-ISS over* $\mathcal{S}[\tau, N_0]$ *for all* $\tau \geq \tau^*$ *and all* $N_0 \leq N_0^*$.

Observe that if one allows the chatter bound N_0 to be zero, then condition (d) in the above theorem can be reduced to $\mu\kappa(\mu s) < s$ for all $s > 0$. For a system without switching, the usual Razumikhin condition is $\kappa <$ id. It should be seen that the factor μ in the condition $\mu\kappa(\mu s) < s$ provides a stability margin to absorb the effect of switchings.

For an unforced system

$$\dot{x}(t) = f(x(t), x_t), \tag{15}$$

the ISS property reduces to global asymptotic stability (or GAS). Correspondingly, a switched system

$$\dot{x}(t) = f_{\sigma(t)}(x(t), x_t) \tag{16}$$

without input is said to be uniformly-GAS over a collection $\mathcal{F}$ of switching signals provided for some $\beta \in \mathcal{KL}$, the following holds along every trajectory of (16) for which $\sigma \in \mathcal{F}$:

$$|x_\sigma(t)| \leq \beta(||x_\sigma||_{[-\theta,0]}, t) \qquad \forall t \geq 0.$$

As a consequence of Theorem 1, we have the following:

Corollary 1 *Consider a switched system without input as in* (16). *Assume that for each* $p \in \Gamma$, *there is a* C^1 *function* $V_p : \mathbb{R}^n \to \mathbb{R}_{\geq 0}$ *such that the following holds:*

(a) for some $\underline{\alpha}$ *and* $\overline{\alpha}$ *in* $\mathcal{K}_\infty$, (11) *holds;*
(b) for some κ *and* $\lambda > 0$, *the following holds for all* $x \in \mathbb{R}^n$ *and all* $\mathrm{w} \in C[-\theta, 0]$:

$$V_p(x) \geq \kappa(||V_p(\mathrm{w}(t))||_{[-\theta,0]}) \Rightarrow DV_p(x) f_p(x, \mathrm{w}) \leq -\lambda V_p(x);$$

(c) there is some $\mu \geq 1$ *such that*

$$V_p(x) \leq \mu V_q(x) \quad \forall x \in \mathbb{R}^n, \ \forall p, q \in \Gamma;$$

(d) for some integer $N_0^* \geq 0$*, we have* $\mu^{N_0^*+2}\kappa(s) < s$ *for all* $s > 0$.

Then the system (7) *is uniformly-GAS over* $\mathscr{S}[\tau, N_0]$ *for all* $\tau \geq \tau^*$ *and all* $N_0 \leq N_0^*$, *where* τ^* *is as in* (14).

Example 1 Consider a system that switches between

$$\Sigma_1^o : \begin{array}{l} \dot{x}_1(t) = -2x_1(t) + \displaystyle\int_{t-\theta}^{t} x_1(s)\,ds, \\ \dot{x}_2(t) = 2x_1(t) - 2x_2(t), \end{array} \tag{17}$$

and

$$\Sigma_2^o : \begin{array}{l} \dot{x}_1(t) = -x_1(t) + x_1(t)x_2(t) - 2x_1^3(t) \\ \dot{x}_2(t) = x_1(t) - x_2(t). \end{array} \tag{18}$$

Note that the x_1-subsystem of Σ_1^o can be rewritten in the form $\dot{x}_1(t) = -2x_1(t) + \int_{-\theta}^{0}(x_1)_t(s)\,ds$, and thus, the corresponding system Σ_1^w is given by

$$\Sigma_1^w : \begin{array}{l} \dot{x}_1(t) = -2x_1(t) + \displaystyle\int_{-\theta}^{0} [w_1(t)](s)\,ds, \\ \dot{x}_2(t) = 2x_1(t) - 2x_2(t), \end{array}$$

where w_1 is treated as the disturbance input for Σ_1^w. Let $V_1(x) = (x_1^2 + x_2^2)/2$. Then for Σ_1^w, one has

$$\begin{aligned} DV_1(x) f_1(x, \mathrm{w}) &= -2x_1^2 + x_1 \int_{-\theta}^{0} \mathrm{w}_1(s)\,ds + 2x_1x_2 - 2x_2^2 \\ &\leq -x_1^2 - x_2^2 + \theta |x_1|\,||\mathrm{w}_1|| \end{aligned}$$

Hence, whenever $\theta^2 ||\mathrm{w}||^2_{[-\theta,0]} \leq \frac{|x|^2}{2}$, it holds that

$$-x_1^2 - x_2^2 + \theta |x_1|\,||\mathrm{w}_1||_{[-\theta,0]} \leq -\frac{1}{4}(x_1^2 + x_2^2).$$

Consequently, one has the following:

$$V_1(x) \geq 2\theta^2 ||V_1(\mathrm{w}(\cdot))||_{[-\theta,0]} \;\Rightarrow\; DV_1 f_1(x, \mathrm{w}) \leq -\frac{1}{2}V(x).$$

One concludes that for the Σ_1^o-subsystem, we have $\lambda = -1/2$ and $\kappa(s) = 2\theta^2 s$.

The system Σ_2^o is delay-free, and a Lyapunov function can be chosen as $V_2(x) = x_1^2 + \frac{x_2^2}{2}$, so that

$$\begin{aligned} DV_2(x) f_2(x) &= -2x_1^2 + 2x_1^2 x_2 - 4x_1^4 + x_1 x_2 - x_2^2 \\ &\le -x_1^2 + 2x_1^2 x_2 - 4x_1^4 - \frac{1}{2}x_2^2 \\ &\le -x_1^2 - \frac{1}{4}x_2^2 \le -\frac{V_2(x)}{2}. \end{aligned}$$

Thus, for the collection $\{\Sigma_p : p = 1, 2\}$, we take $\mu = 2, \lambda = 1/2$, and $\kappa(s) = 2\theta^2 s$. To apply Theorem 1, let

$$\tau^* = \frac{2\ln 2}{1/2} = 4\ln 2.$$

For any given $N_0 \ge 0$, let

$$\theta^* = \frac{1}{\sqrt{2^{N_0+3}}} = \frac{1}{2^{(N_0+3)/2}}.$$

Then, by Theorem 1, it follows that whenever θ is small enough (i.e., $\theta < \theta^*$), the system Σ_σ^o is uniformly-GAS over $\mathscr{S}[\tau, N_0]$ *for all* $\tau \ge \tau^*$.

Note that in this example, the estimates on parameters to guarantee the stability can be conservative and are not necessarily the best. Rather, the estimates only provide sufficient conditions for the stability property as stated in Theorem 1.

In the remainder of this section, we discuss two special yet interesting cases.

3.1 Discrete Delays

When a system is only affected by discrete delays, in the sense that it has the form

$$\Sigma_\sigma^o : \dot{x}(t) = f_{\sigma(t)}(x(t), x(\theta_1(t)), \ldots, x(\theta_k(t)), u(t)), \tag{19}$$

where $\theta_i : [0, \infty) \to [-\theta, \infty)$ is measurable and satisfying $t - \theta \le \theta_i(t) \le t$ for all $1 \le i \le k$, the corresponding system Σ_σ^w can be simplified to

$$\Sigma_\sigma^w : \quad \dot{x}(t) = f_{\sigma(t)}(x(t), w_1(t) \ldots, w_k(t), u(t)), \tag{20}$$

where $w_1, \ldots, w_k$ and u are input signals for Σ_σ^w. In this case, Σ_σ^w becomes a usual system affected by external inputs and taking values in finite dimensional spaces. Condition (12) becomes:

$$V_p(x) \geq \max\left\{\kappa(V_p(\mathrm{w}_1)), \ldots, \kappa(V_p(\mathrm{w}_k)), \rho(|u|)\right\} \\ \Rightarrow DV_p(x) f_p(x, \mathrm{w}_1, \ldots, \mathrm{w}_k, u) \leq -\lambda V_p(x) \qquad (21)$$

for all $x \in \mathbb{R}^n$, all $(\mathrm{w}_1, \ldots, \mathrm{w}_k) \in \mathbb{R}^{nk}$, and all $u \in \mathbb{R}^m$. For a system affected by discrete delays, the statement of Theorem 1 holds with property (12) replaced by (21).

3.2 Common Lyapunov-Razumikhin Functions

Assume that the individual systems Σ_p^o admit a common Lyapunov-Razumikhin function V, that is, for all $p \in \Gamma$, $V_p = V$, where V satisfies conditions (11)–(12). Then $\mu = 1$. In this case, Theorem 1 leads to the following; see Remark 2 at the end of Sect. 4.

Corollary 2 *For a switched system Σ_σ^o with time delays as in* (7)*, assume that there is some C^1 function $V : \mathbb{R}^n \to \mathbb{R}_{\geq 0}$ such that*

(a) for some $\underline{\alpha}$ and $\overline{\alpha}$ in $\mathscr{K}_\infty$, we have

$$\underline{\alpha}(|x|) \leq V(x) \leq \overline{\alpha}(|x|) \quad \forall x \in \mathbb{R}^n;$$

(b) for some κ and ρ in $\mathscr{K}_\infty$ and some $\lambda > 0$, we have

$$V(x) \geq max\left\{\kappa\left(||V(\mathrm{w}(\cdot))||_{[-\theta,0]}\right),\ \rho(|u|)\right\} \\ \Rightarrow DV(x) f_p(x, \mathrm{w}, u) \leq -\lambda V(x)$$

for all $p \in \Gamma$; and

(c) $\kappa <$ id.

Then the system Σ_σ^o is uniformly-ISS under arbitrary switching.

4 Proof of Main Theorem

In this section, we provide the proof of Theorem 1. Throughout this section, we assume that conditions (a)–(d) of Theorem 1 hold.

Let $\sigma \in \mathscr{S}[\tau, N_0]$, where $\tau \geq \tau^*$ and $N_0 \leq N_0^*$, and let $0 < t_1 < t_2 < \cdots$ be the discontinuous points of σ. Assume that $\sigma(t) = p_i$ on $[t_{i-1}, t_i)$, where $t_0 = 0$. For a given initial state x_0 and a given pair (w, u) of inputs, consider the corresponding trajectory $x_\sigma(t)$ of the system Σ_σ^w.

On $[t_0, t_1)$, it holds that

$$V_{p_1}(x_\sigma(t)) \le \max\left\{V_{p_1}(x_\sigma(0))e^{-\lambda t},\ \kappa(||V_{p_1}(w)||_{[0,t)}),\ \rho(||u||)\right\},$$

and in particular,

$$V_{p_1}(x_\sigma(t_1)) \le \max\left\{V_{p_1}(x_\sigma(0))e^{-\lambda t_1}, \kappa(||V_{p_1}(w)||_{[0,t_1)}),\ \rho(||u||)\right\}.$$

On $[t_1, t_2)$,

$$\begin{aligned}
V_{p_2}(x_\sigma(t)) &\le \max\left\{V_{p_2}(x_\sigma(t_1))e^{-\lambda(t-t_1)},\ \kappa(||V_{p_2}(w)||_{[t_1,t)}),\ \rho(||u||)\right\}\\
&\le \max\left\{\mu V_{p_1}(x_\sigma(t_1))e^{-\lambda(t-t_1)},\ \kappa(||V_{p_2}(w)||_{[t_1,t)}),\ \rho(||u||)\right\}\\
&\le \max\Big\{\mu V_{p_1}(x_\sigma(0))e^{-\lambda t}, \kappa(||V_{p_2}(w)||_{[t_1,t)}),\ \rho(||u||),\\
&\qquad \mu e^{-\lambda(t-t_1)}\kappa(||V_{p_1}(w)||_{[0,t_1)}),\ \mu e^{-\lambda(t-t_1)}\rho(||u||)\Big\}.
\end{aligned}$$

Inductively, one concludes that on $[t_k, t_{k+1})$, we have

$$\begin{aligned}
V_{p_{k+1}}(x_\sigma(t)) \le \max\Big\{&\mu^k V_{p_1}(x_\sigma(0))e^{-\lambda t},\ \ \mu^k e^{-\lambda(t-t_1)}\rho(||u||),\\
&\mu^k e^{-\lambda(t-t_1)}\kappa(||V_{p_1}(w)||_{[0,t_1)}),\ \ldots,\\
&\mu e^{-\lambda(t-t_k)}\kappa(||V_{p_k}(w)||_{[t_{k-1},t_k)}),\ \kappa(||V_{p_{k+1}}(w)||_{[t_k,t)})\Big\},
\end{aligned} \tag{22}$$

where the terms represented by "..." are of the form

$$\mu^{k-j}e^{-\lambda(t-t_{j+1})}\kappa(||V_{p_{j+1}}(w)||_{[t_j,t_{j+1})}),\ \mu^{k-j}e^{-\lambda(t-t_{j+1})}\rho(||u||), \tag{23}$$

where $j = 1, \ldots, k-1$. It was shown in [3] that for all $t > t_k$, we have

$$\mu^k e^{-\frac{\lambda t}{2}} \le \mu^{N_0}.$$

For all $t > t_k$ and $j \le k$, we have $k - j \le N(t_{j+1}, t) + 1$, and by the arguments used in [3], one can see that

$$\begin{aligned}
\mu^{k-j}e^{-\frac{\lambda(t-t_{j+1})}{2}} &\le \mu^{N(t_{j+1},t)+1}e^{-\frac{\lambda(t-t_{j+1})}{2}}\\
&\le \mu e^{(\ln\mu)N(t_{j+1},t)-\frac{\lambda(t-t_{j+1})}{2}}\\
&\le \ \mu e^{(\ln\mu)(N_0+\frac{t-t_{j+1}}{\tau^*})-\frac{\lambda(t-t_{j+1})}{2}}\\
&\le \mu^{N_0+1}e^{(\ln\mu)\frac{t-t_{j+1}}{\tau^*}-\frac{\lambda(t-t_{j+1})}{2}}\\
&\le \mu^{N_0+1},
\end{aligned}$$

where we have used the fact that $\frac{\ln \mu}{\tau^*} \le \frac{\lambda}{2}$. Hence, for the terms as in (23), we have the following estimates:

$$\mu^{k-j} e^{-\lambda(t-t_{j+1})} \rho(||u||) \le \mu^{N_0+1} \rho(||u||),$$

and

$$\mu^{k-j} e^{-\lambda(t-t_{j+1})} \kappa(r_j) \le \mu^{N_0+1} \kappa(r_j),$$

where $r_j = ||V_{p_{j+1}}(w)||_{[t_j-\theta, t_{j+1})}$. It then follows from (22) that for all $t \in [t_k, t_{k+1})$, we have

$$\begin{aligned} V_{p_{k+1}}(x_\sigma(t)) \le \max\Big\{ & V_{p_1}(x_\sigma(0))\mu^{N_0} e^{-\frac{\lambda t}{2}},\ \mu^{N_0+1}\kappa(||V_{p_1}(w)||_{[0,t_1)}), \ldots, \\ & \mu^{N_0+1}\kappa(||V_{p_2}(w)||_{[t_1,t_2)}),\ \cdots, \\ & \mu^{N_0+1}\kappa(||V_{p_k}(w)||_{[t_{k-1},t_k)}), \\ & \kappa(||V_{p_{k+1}}(w)||_{[t_k,t)}),\ \mu^{N_0+1}\rho(||u||)\Big\} \\ \le \max\Big\{ & V_{p_1}(x_\sigma(0))\mu^{N_0} e^{-\frac{\lambda t}{2}}, \\ & \mu^{N_0+1}\kappa(\mu||V_{\sigma(t)}(w)||_{[0,t)}),\ \mu^{N_0+1}\rho(||u||)\Big\}. \end{aligned}$$

Since k can be arbitrarily chosen, it holds that

$$\begin{aligned} V_{\sigma(t)}(x_\sigma(t)) \le \max\Big\{ & V_{\sigma(0)}(x_\sigma(0))\mu^{N_0} e^{-\frac{\lambda t}{2}}, \\ & \mu^{N_0+1}\kappa(\mu||V_\sigma(w)||_{[0,t)}),\ \mu^{N_0+1}\rho(||u||)\Big\}, \end{aligned} \tag{24}$$

where $\sigma(t) = p_1$ on $[-\theta, 0]$. We have thus shown the following:

Lemma 1 *For each $\tau \ge \tau^*$ and each $\sigma \in \mathscr{S}[\tau, N_0]$, the system Σ_σ^w is forward complete, and property* (24) *holds for every trajectory of Σ_σ^w.*

Let ψ_σ be any trajectory of Σ_σ^o, where $\sigma \in \mathscr{S}[\tau, N_0]$ and $\tau \ge \tau^*$. Then, ψ_σ is a trajectory of the closed-loop system of Σ_σ^w under the feedback $w(t) = (\psi_\sigma)_t$, and hence, the following holds on the maximum interval of ψ_σ:

$$\begin{aligned} V_{\sigma(t)}(\psi_\sigma(t)) \le \max\Big\{ & V_{\sigma(0)}(\psi_\sigma(0))\mu^{N_0} e^{-\frac{\lambda t}{2}}, \\ & \mu^{N_0+1}\kappa(\mu \cdot ||V_\sigma(\psi_\sigma(\cdot))||_{[-\theta,t)}),\ \mu^{N_0+1}\rho(||u||)\Big\}. \end{aligned} \tag{25}$$

Let $v(t) = V_{\sigma(t)}(\psi_\sigma(t))$, and rewrite (25) as

$$v(t) \le \max\left\{v(0)\mu^{N_0}e^{-\frac{\lambda t}{2}},\ \tilde{\kappa}(||v||_{[-\theta,t)}),\ \mu^{N_0+1}\rho(||u||)\right\}$$

on the maximum interval of ψ_σ, where $\tilde{\kappa}(s) = \mu^{N_0+1}\kappa(\mu s)$. By [5, Remark 3.1], we have $\mu^{N_0+1}\kappa(\mu s) < s$ for all $s > 0$.

Fix $b > 0$ such that ψ_σ is defined on $[0, b)$. Then, for any $t_1 \in (0, b)$, we have

$$\begin{aligned}||v(t)||_{[0,t_1)} &\le \max\left\{v(0)\mu^{N_0},\ \tilde{\kappa}(||v||_{[-\theta,t_1)}),\ \mu^{N_0+1}\rho(||u||)\right\}\\ &\le \max\left\{v(0)\mu^{N_0},\ \tilde{\kappa}(||v||_{[-\theta,0]}),\ \tilde{\kappa}(||v||_{[0,t_1)}),\ \mu^{N_0+1}\rho(||u||)\right\}\end{aligned}$$

Since $\tilde{\kappa}(s) < s$ for all $s > 0$ and $||v(t)||_{[0,t_1)} < \infty$, it follows that

$$||v(t)||_{[0,t_1)} \le \max\left\{v(0)\mu^{N_0},\ \tilde{\kappa}(||v||_{[-\theta,0]}),\ \mu^{N_0+1}\rho(||u||)\right\},$$

and by continuity,

$$v(t_1) \le \max\left\{v(0)\mu^{N_0},\ \tilde{\kappa}(||v||_{[-\theta,0]}),\ \mu^{N_0+1}\rho(||u||)\right\}. \tag{26}$$

Note that (26) holds on every interval $[0, b)$ on which ψ_σ is defined. It follows that Σ_σ^o is forward complete provided that $\sigma \in \mathscr{S}[\tau, N_0]$. As a consequence, (26) holds for all $t_1 \ge 0$. Also note that $\tilde{\kappa}(s) \le s \le \mu_0^N s$ for $s \ge 0$. Hence, one concludes the following:

Lemma 2 *For each $\tau \ge \tau^*$ and each $\sigma \in \mathscr{S}[\tau, N_0]$, the system Σ_σ^o is forward complete, and*

$$V_{\sigma(t)}(\psi_\sigma(t)) \le max\left\{\mu^{N_0}\,||V_\sigma(\psi_\sigma)||_{[-\theta,0]},\ \mu^{N_0+1}\rho(||u||)\right\}\quad \forall t \ge 0 \tag{27}$$

holds for each trajectory of Σ_σ^o.

To complete the proof of Theorem 1, note that (25) holds for all $t \ge 0$ along every trajectory of Σ_σ^o. For any $c > 0$ and $t \ge c$, it holds that

$$V_{\sigma(t)}(\psi_\sigma(t)) = V_{\tilde{\sigma}(t-c)}(\tilde{\psi}_{\tilde{\sigma}}(t-c)),$$

where $\tilde{\sigma}$ is the switching signal defined by $\tilde{\sigma}(t) = \sigma(t + c)$, $\tilde{\psi}_{\tilde{\sigma}}$ is the trajectory of $\Sigma_{\tilde{\sigma}}$ corresponding to the initial value $\tilde{\psi}_{\tilde{\sigma}}(s) = \psi_\sigma(c + s)$ for all $s \in [-\theta, 0]$, and the input $\tilde{u}$ given by $\tilde{u}(t) = u(t + c)$. Observe that $\tilde{\sigma} \in \mathscr{S}[\tau, N_0]$. Applying Lemma 2, we get the following for any $t \ge c$:

$$\begin{aligned} V_{\sigma(t)}(\psi_\sigma(t)) &= V_{\tilde{\sigma}(t-c)}(\tilde{\psi}_{\tilde{\sigma}}(t-c)) \\ &\le \max\Big\{ V_{\tilde{\sigma}(0)}(\tilde{\psi}_{\tilde{\sigma}}(0))\mu^{N_0} e^{-\frac{\lambda(t-c)}{2}}, \\ &\qquad \tilde{\kappa}(\|V_{\tilde{\sigma}}(\tilde{\psi}_{\tilde{\sigma}})\|_{[-\theta,t-c)}),\ \mu^{N_0+1}\rho(\|\tilde{u}\|)\Big\} \\ &\le \max\Big\{ V_{\sigma(c)}(\psi_\sigma(c))\mu^{N_0} e^{-\frac{\lambda(t-c)}{2}}, \\ &\qquad \tilde{\kappa}(\|V_{\sigma}(\psi_{\sigma})\|_{[c-\theta,t)}),\ \mu^{N_0+1}\rho(\|u\|)\Big\}. \end{aligned} \tag{28}$$

By (27), we have

$$V_{\sigma(c)}(\psi_\sigma(c)) \le \max\left\{\mu^{N_0}\|V_\sigma(\psi_\sigma)\|_{[-\theta,0]},\ \mu^{N_0+1}\rho(\|u\|)\right\}$$

for all $c \ge 0$. It thus follows from (28) that

$$\begin{aligned} V_{\sigma(t)}(\psi_\sigma(t)) \le \max\Big\{&\mu^{2N_0}\|V_\sigma(\psi_\sigma)\|_{[-\theta,0]}e^{-\frac{\lambda(t-c)}{2}}, \\ &\tilde{\kappa}(\|V_\sigma(\psi_\sigma)\|_{[c-\theta,t)}),\ \mu^{N_0+1}\rho(\|u\|)\Big\} \end{aligned}$$

for all $0 \le c \le t$. In particular, with $c = t/2$, we have

$$\begin{aligned} V_{\sigma(t)}(\psi_\sigma(t)) \le \max\Big\{&\mu^{2N_0}\|V_\sigma(\psi_\sigma)\|_{[-\theta,0]}\, e^{-\frac{\lambda t}{4}}, \\ &\tilde{\kappa}(\|V_\sigma(\psi_\sigma)\|_{[(t/2)-\theta,\ t)}),\ \mu^{N_0+1}\rho(\|u\|)\Big\} \end{aligned}$$

for all $t \ge 0$. By [19, Lemma 5.4] and the proof of [19, Theorem 1], one concludes that under the small-gain condition $\tilde{\kappa} < \mathrm{id}$, there exist some $\beta \in \mathcal{KL}$ and $\hat{\rho} \in \mathcal{K}$ such that

$$V_{\sigma(t)}(\psi_\sigma(t)) \le \max\left\{\beta(\|V_\sigma(\psi_\sigma)\|_{[-\theta,0]},\ t),\ \hat{\rho}(\|u\|)\right\} \quad \forall t \ge 0.$$

It follows that

$$|\psi_\sigma(t)| \le \max\left\{\beta_1(\|\psi_\sigma\|_{[-\theta,0]},\ t),\ \rho_1(\|u\|)\right\} \quad \forall t \ge 0 \tag{29}$$

for all trajectories of Σ_σ^o for which $\sigma \in \mathcal{S}[\tau, N_0]$, $\tau \ge \tau^*$, and $N_0 \le N_0^*$, where $\beta_1(r,t) = \underline{\alpha}^{-1} \circ \beta(\overline{\alpha}(r), t)$ and $\rho_1 = \underline{\alpha}^{-1} \circ \hat{\rho}$. We conclude that the system Σ_σ^o is uniformly-ISS over $\mathcal{S}[\tau, N_0]$ with $\tau \ge \tau^*$ and $N_0 \le N_0^*$.

Remark 2 Assume that the family of systems $\{\Sigma_p^o : p \in \Gamma\}$ admits a common Lyapunov-Razumikhin function. Then $\mu = 1$ and (25) holds uniformly for all switching signals σ (i.e., all piecewise constant functions). Consequently, (29) holds uniformly over the set of all switchings.

5 Conclusions

The notion of average dwell time is applied to nonlinear switched systems with time delays. It was shown that a switching among ISS systems with compatible Lyapunov-Razumikhin functions will not destroy the stability property if the switching is not too fast on average. The main approach is the Lyapunov-Razumikhin method, in conjunction with small-gain theory. Compared with the non-switching case, a more strenuous small-gain condition on the gain function in the delayed state variables is needed to absorb the effect of chatters.

Appendix

Although it is common in the study of the ISS property to consider inputs that take values in an Euclidean space $\mathbb{R}^m$ or a subspace in $\mathbb{R}^m$, the concept of input-to-state stability may also apply to the more general context when the inputs take values in a normed space which may be infinite dimensional. This is in particular the case in the contexts when Lyapunov-Razumikhin approach is used to deal with the delays on state variables (as in this work). Below we consider such a case when the input takes values in $C[-\theta, 0]$.

Consider a system as follows:

$$\dot{x}(t) = f(x(t), v(t), u(t)), \tag{30}$$

where for each $t \geq 0$, we have $x(t) \in \mathbb{R}^n$, $v(t) \in C[-\theta, 0]$, and $u(t) \in \mathbb{R}^m$. Assume that $f : \mathbb{R}^n \times C[-\theta, 0] \times \mathbb{R}^m \to \mathbb{R}^n$ is completely continuous, and Lipschitz on each compact set. The input v is a continuous function from $\mathbb{R}_{\geq 0}$ to $C[-\theta, 0]$, and the input u is a measurable, locally essentially bounded function from $\mathbb{R}_{\geq 0}$ to $\mathbb{R}^m$.

For any given interval I, we define $||v||_I = \sup\{||v(t)|| : t \in I\}$. Note that since $v : \mathbb{R}_{\geq 0} \to C[-\theta, 0]$ is continuous, we know that $v(I)$ is compact if I is compact. Consequently, for any bounded interval I, we know that $||v||_I$ is finite. In the case when $I = [0, \infty)$, $||v||_I$ will be simply denoted by $||v||$.

For each pair (v, u), the map $F(x, t) = f(x, v(t), u(t))$ satisfies the following:

- for each fixed x, $F(x, \cdot)$ is measurable and locally essentially bounded;
- F is locally Lipschitz on x in the sense that for any compact subset $K \subseteq \mathbb{R}^n$ and any interval $[a, b] \subseteq [0, \infty)$, there exists a locally essentially bounded function $\alpha(\cdot)$ such that

$$|F(x_1, t) - F(x_2, t)| \leq \alpha(t)|x_1 - x_2|$$

  for almost all $t \in [a, b]$ and all $x_1, x_2 \in K$.

As a consequence, for each input pair (v, u) and each initial value condition $x(0) = x_0$, there exists a unique maximum solution of (30) defined on some interval $[0, T)$; see e.g., [15]. Such a solution will be denoted by $x(t, x_0, v, u)$.

As in the case when the input functions take values in finite dimensional spaces, the ISS property for (30) can also be defined for a system as in (30):

Definition 1 ([14]) A system as in (30) is input-to-state stable (ISS) provided there exist functions $\beta \in \mathcal{KL}, \kappa \in \mathcal{K}$ and $\gamma \in \mathcal{K}$ such that

$$|x(t, x_0, v, u)| \leq \max\{\beta(|x_0|, t),\ \kappa(||v||),\ \gamma(||u||)\} \quad \forall t \geq 0$$

holds for all inputs v and u.

Observe that by causality, the above inequality is equivalent to:

$$|x(t, x_0, v, u)| \leq \max\left\{\beta(|x_0|, t),\ \kappa(||v||_{[0,t)}),\ \gamma(||u||_{[0,t)})\right\} \quad \forall t \geq 0.$$

Also note that if $v(t) = x_t$, then $||v||_{[0,t]} = ||x||_{[-\theta,t]}$. (Recall that for each $t \geq 0$, the function $x_t : C[-\theta, 0] \to \mathbb{R}^n$ is defined by $x_t(s) = x(t + s)$.)

A C^1 function $V : \mathbb{R}^n \to \mathbb{R}_{\geq 0}$ is an ISS-Lyapunov function for the system (30) provided

- for some $\underline{\alpha}$ and $\overline{\alpha}$ in $\mathcal{K}_\infty$, we have

$$\underline{\alpha}(|x|) \leq V(x) \leq \overline{\alpha}(|x|) \qquad \forall x;$$

- for some $\chi \in \mathcal{K}$ and ρ in $\mathcal{K}$ and some continuous positive definite function α, the following holds for all $x \in \mathbb{R}^n$, $\mathrm{v} \in C[-\theta, 0]$, and $u \in \mathbb{R}^m$:

$$V(x) \geq \max\{\chi(||\mathrm{v}||_{[-\theta,0]}),\ \rho(|u|)\} \Rightarrow DV(x)f(x, \mathrm{v}, u) \leq -\alpha(|x|). \quad (31)$$

As in the case when the inputs only take values in finite dimensional spaces, the following result can be proven, by following exactly the same proof as given in [14]:

Proposition 3 *A system as in* (30) *is ISS if it admits an ISS-Lyapunov function.*

More precisely, if there is an ISS-Lyapunov function for (30) satisfying (31), then for some $\beta \in \mathcal{KL}$, we have

$$|V(x(t, x_0, v, u))| \leq \max\left\{\beta(V(x_0), t),\ \chi(||v||_{[0,t)}),\ \rho(||u||_{[0,t)})\right\} \quad \forall t \geq 0.$$

We conjecture that the converse of Proposition 3 also holds. Many of the proofs of converse Lyapunov theorems in the literature (such as [11, 16]) may still work in the more general case when v takes values in a normed space (or Banach space), yet one cannot claim that the converse Lyapunov theorem holds for such systems before all details are validated.

References

1. Chen, W., Zheng, W.-X.: Stability analysis for switched systems with time-varying delay and parametric uncertainties. In: Proceedings of the 49th Conference on Decision and Control, pp. 426–431 (2010)
2. Hale, J., Verduyn Lunel, S.: Introduction to Functional Differential Equations. Springer Verlag, New York (1993)
3. Hespanha, J., Morse, A.: Stability of switched systems with average dwell-time. In: Proceedings of the 38th Conference on Decision and Control, pp. 2655–2660 (1999)
4. Jankovic, M.: Control Lyapunov-Razumikhin functions and robust stabilization of time delay systems. IEEE Trans. Autom. Control **46**(7), 1048–1060 (2001)
5. Jiang, Z.-P., Mareels, I., Wang, Y.: A Lyapunov formulation of the nonlinear small-gain theorem for interconnected ISS systems. Automatica **32**(8), 1211–1215 (1996)
6. Karafyllis, I., Jiang, Z.-P.: Stability and Stabilization of Nonlinear Systems. Springer, London (2011)
7. Kim, S., Campell, S., Liu, X.: Stability of a class of linear switching systems with time delays. IEEE Trans. Circ. Syst. **53**(2), 384–393 (2006)
8. Kolmanovskii, V., Myshkis, A.: Introduction to the Theory and Applications of Functional Differential Equations. Kluwer Academic Publishers, Dordrecht, The Netherlands (1999)
9. Liberzon, D.: Switching in Systems and Control. Birkhauser, Boston (2003)
10. Lin, H., Antsaklis, P.: Stability and stabilizability of switched linear systems: a survey of recent results. IEEE Trans. Autom. Control **54**(2), 308–322 (2009)
11. Lin, Y., Sontag, E., Wang, Y.: A smooth converse Lyapunov theorem for robust stability. SIAM J. Control Optim. **34**(1), 124–160 (1996)
12. Liu, J., Liu, X., Xie, W.-C.: Input-to-state stability of impulsive and switching hybrid systems with time-delay. Automatica **47**(5), 899–908 (2011)
13. Praly, L., Wang, Y.: Stabilization in spite of matched unmodeled dynamics and an equivalent definition of input-to-state stability. Math. Control Signals Syst. **9**(1), 1–33 (1996)
14. Sontag, E.: Smooth stabilization implies coprime factorization. IEEE Trans. Autom. Control **34**(4), 435–443 (1989)
15. Sontag, E.: Mathematical Control Theory: Deterministic Finite Dimensional Systems, 2nd edn. Springer, New York (1998)
16. Sontag, E., Wang, Y.: On characterizations of input-to-state stability property. Syst. Control Lett. **24**(5), 351–359 (1995)
17. Sun, X.-M., Zhao, J., Hill, D.: Stability and L_2-gain analysis for switched delay systems: a delay-dependent method. Automatica **42**(10), 1769–1774 (2006)
18. Teel, A.: Connections between Razumikhin-type theorems and the ISS nonlinear small gain theorem. IEEE Trans. Autom. Control **43**(7), 960–964 (1998)
19. Tiwari, S., Wang, Y., Jiang, Z.-P.: Nonlinear small-gain theorems for large-scale time-delay systems. Dyn. Continuous Discrete Impulsive Syst. Ser. A: Math. Anal. **19**(1), 27–63 (2012)
20. Tiwari, S., Wang, Y., Jiang, Z.-P.: Remarks on integral-ISS for systems with delays. In: Proceedings of the 10th World Congress on Intelligent Control and Automation, pp. 2227–2232 (2012)

A Note on Converse Lyapunov Theorems for Neutral Systems

Iasson Karafyllis and Pierdomenico Pepe

Abstract We provide various extensions of recent results on the existence of Lyapunov–Krasovskii functionals for uncertain systems described by neutral functional differential equations. We consider nonlinear neutral delay systems for which the difference operator and the right-hand side of the differential equations are Lipschitz on bounded sets.

1 Introduction

Converse Lyapunov theorems for neutral functional differential equations recently appeared in [9]. The results provided in [9] assumed:

- The difference operator was assumed to be linear and strongly stable in the sense described in [2].
- The system was disturbance-free, i.e., no disturbances were involved in the right hand side of the differential equations.

This work is devoted to the relaxation of the above assumptions. More specifically, we provide converse Lyapunov results for uncertain neutral functional differential equations in Hale's form for which the difference operator can be a nonlinear mapping that is Lipschitz on bounded sets. The construction of the Lyapunov functional is based on recent results in [3–5].

The theory of converse Lyapunov results for neutral functional differential equations is rich; see [9] and the references therein. However, no results seem to cover the case of uncertain neutral functional differential equations or neutral functional

I. Karafyllis (✉)
Department of Mathematics, National Technical University of Athens,
Heroon Polytechniou 9, 15780 Athens, Greece
e-mail: iasonkar@central.ntua.gr

P. Pepe
Department of Information Engineering, Computer Science and Mathematics,
University of L'Aquila, 67100 L'Aquila, Italy
e-mail: pierdomenico.pepe@univaq.it

I. Karafyllis et al. (eds.), *Recent Results on Nonlinear Delay Control Systems*,
Advances in Delays and Dynamics 4, DOI 10.1007/978-3-319-18072-4_12

differential equations with vanishing perturbations. The development of converse Lyapunov results for uncertain finite-dimensional differential equations has opened the way for (i) the development of input-to-state stabiliy theory (by means of what is now called an ISS-Lyapunov function), and (ii) the development of robust feedback stabilizers. It is expected that the same development will appear in the area of neutral functional differential equations.

The rest of this chapter is organized as follows. Section 3 provides existence, uniqueness, and continuation results which are necessary to develop the theory of neutral functional differential equations in Hale's form with inputs. Specific results are provided for the Lipschitz continuity of the solutions with respect to initial conditions. Section 4 constructs a Lyapunov functional for uncertain neutral functional differential equations with a robust global asymptotically stable (RGAS) equilibrium point. Various characterizations of the RGAS equilibrium are provided and emphasis is placed on the evaluation of the derivative of the Lyapunov functional without knowledge of the solution. The method used in [7, 8] is exploited.

2 Notation

Throughout this chapter, we adopt the following notation:

- $\mathbb{R}_+ := [0, +\infty)$.
- By $C^0(A\ ;\ \Omega)$, we denote the class of all continuous functions on $A \subseteq \mathbb{R}^n$, which take values in $\Omega \subseteq \mathbb{R}^m$. For each $x \in C^0([-r, 0]; \mathbb{R}^n)$, we define $\|x\| := \sup\{|x(\theta)| : \theta \in [-r, 0]\}$, where $r > 0$ is a constant. We also use the convention $x \in C^0([-r, 0]; \mathbb{R}^n) = \mathbb{R}^n$ for $r = 0$.
- We say that a function $\rho : \mathbb{R}_+ \to \mathbb{R}_+$ is positive definite provided $\rho(0) = 0$ and $\rho(s) > 0$ for all $s > 0$. By K, we denote the set of positive definite, increasing and continuous functions. We say that a positive definite, increasing and continuous function $\rho : \mathbb{R}_+ \to \mathbb{R}_+$ is of class K_∞ provided $\lim_{s\to+\infty} \rho(s) = +\infty$. By KL, we denote the set of all continuous functions $\sigma = \sigma(s, t) : \mathbb{R}_+ \times \mathbb{R}_+ \to \mathbb{R}_+$ with the properties: (i) for each $t \geq 0$, the mapping $\sigma(\cdot\,, t)$ is of class K and (ii) for each $s \geq 0$, the mapping $\sigma(s, \cdot)$ is non-increasing with $\lim_{t\to+\infty} \sigma(s, t) = 0$.
- Let $x \in \mathbb{R}^n$. By $|x|$ we denote the Euclidean norm of $x \in \mathbb{R}^n$.
- Let $I \subseteq \mathbb{R}_+ := [0, +\infty)$ be an interval. By $L^\infty(I; U)$, we denote the space of all measurable and essentially bounded functions u defined on I and valued in $U \subseteq \mathbb{R}^m$. When $I \subseteq \mathbb{R}_+$ is an unbounded interval, we use $L^\infty_{loc}(I;U)$ to denote the space of all measurable and locally essentially bounded functions u defined on I and taking values in $U \subseteq \mathbb{R}^m$.

3 System-Theoretic Results for Neutral Systems

Consider the following uncertain neutral delay system:

$$\begin{aligned} &\tfrac{d}{dt}\left(x(t) - G(x_t)\right) = f(d(t), x_t, u(t)) \\ &x(t) \in \mathbb{R}^n,\ d(t) \in D,\ u(t) \in U, \end{aligned} \tag{1}$$

where $D \subseteq \mathbb{R}^l$ is a non-empty set, $U \subseteq \mathbb{R}^m$ is a non-empty set with $0 \in U$, $r > 0$ is the maximum delay (which is a constant), $x_t \in C^0([-r,0];\mathbb{R}^n)$ is defined by $(x_t)(s) = x(t+s)$ for all $s \in [-r,0]$, and $f : D \times C^0([-r,0];\mathbb{R}^n) \times U \to \mathbb{R}^n$ and $G : C^0([-r,0];\mathbb{R}^n) \to \mathbb{R}^n$ are continuous mappings, under the following assumptions:

(H1) There exists a constant $\tau \in (0,r]$ and a mapping $\tilde{G} : C^0([-r+\tau,0];\mathbb{R}^n) \to \mathbb{R}^n$ such that for every choice of $x \in C^0([-r,0];\mathbb{R}^n)$, we have $G(x) = \tilde{G}(\delta_\tau x)$, where δ_τ is the shift operator defined by $(\delta_\tau x)(s) = x(s-\tau)$ for all $s \in [-r+\tau, 0]$.

(H2) There exists a continuous non-decreasing function $L_G : \mathbb{R}_+ \to \mathbb{R}_+$ such that $\left|\tilde{G}(x) - \tilde{G}(y)\right| \leq L_G(\|x\| + \|y\|)\,\|x-y\|$ holds for all $x, y \in C^0([-r+\tau,0];\mathbb{R}^n)$.

(H3) There exists a continuous non-decreasing function $L_f : \mathbb{R}_+ \to \mathbb{R}_+$ such that $|f(d,x,u) - f(d,y,u)| \leq L_f(\|x\| + \|y\| + |u|)\,\|x-y\|$ holds for all x and y in $C^0([-r,0];\mathbb{R}^n)$ and all $(d,u) \in D \times U$.

(H4) There exists a non-decreasing, continuous function $a : \mathbb{R}_+ \to \mathbb{R}_+$ satisfying $a(0) = 0$ such that the following inequalities hold for all $(d,u,x) \in D \times U \times C^0([-r,0];\mathbb{R}^n)$:

$$|f(d,x,u)| \leq a(\|x\| + |u|),\ |G(x)| \leq a(\|x\|) \tag{2}$$

An explanation of the role of each one of the assumptions is given next:

- Assumption (H1) guarantees that the mapping $G(x)$ appearing in the difference operator of equation (1) is independent of $x(0)$ [2]. This assumption is crucial for the existence of solutions of (1).
- Assumption (H2) guarantees that the mapping $G(x)$ appearing in the difference operator of equation (1) is Lipschitz on bounded sets. This assumption is not needed for the existence and uniqueness of solutions of (1), but it is crucial for the derivation of useful inequalities that show the dependence of the solution of (1) on the initial condition below.
- Assumption (H3) guarantees that the mapping $f(d,x,u)$ on the right hand side of (1) is Lipschitz on bounded sets. This assumption is crucial for the existence and uniqueness of solutions of (1), but also is crucial for the derivation of useful inequalities that show the dependence of the solution of (1) to the initial condition below.
- Assumption (H4) guarantees that $0 \in C^0([-r,0];\mathbb{R}^n)$ is a robust equilibrium point for system (1).

We searched the literature for a result that guarantees the existence and uniqueness and continuation of solutions of (1), but we have not been able to find one. Therefore, we next give such a result:

Theorem 1 *Consider system* (1) *under (H1)–(H4). For all* $d \in L^\infty_{loc}(\mathbb{R}_+; D)$, $u \in L^\infty_{loc}(\mathbb{R}_+; U)$, *and* $x_0 \in C^0([-r, 0]; \mathbb{R}^n)$, *there exists* $t_{\max} \in (0, +\infty]$ *and a unique mapping* $x \in C^0([-r, t_{\max}); \mathbb{R}^n)$ *that satisfies* $x(s) = (x_0)(s)$ *for all* $s \in [-r, 0]$ *and* $\frac{d}{dt}(x(t) - G(x_t)) = f(d(t), x_t, u(t))$ *for* $t \in [0, t_{\max})$ *a.e.. Moreover, if* $t_{\max} < +\infty$ *then* $\sup_{t\to t_{\max}^-} |x(t)| = +\infty$.

Proof Let $d \in L^\infty_{loc}(\mathbb{R}_+; D)$, $u \in L^\infty_{loc}(\mathbb{R}_+; U)$ and $x_0 \in C^0([-r, 0]; \mathbb{R}^n)$ be given. By virtue of Assumption (H1), the existence of $\delta \in (0, \tau]$ and a mapping $x \in C^0([-r, \delta]; \mathbb{R}^n)$ that satisfies $x(s) = (x_0)(s)$ for all $s \in [-r, 0]$ and $\frac{d}{dt}(x(t) - G(x_t)) = f(d(t), x_t, u(t))$ for $t \in [0, \delta]$ a.e., is equivalent to the existence of a fixed point for the mapping $P : S \to C^0([-r, \delta]; \mathbb{R}^n)$, where

$$\begin{aligned}(Px)(t) &= \xi(t) + \int_0^t f(d(s), x_s, u(s))ds, \text{ for all } t \in (0, \delta],\\ (Px)(t) &= (x_0)(t), \text{ for all } t \in [-r, 0]\end{aligned} \tag{3}$$

$$\begin{aligned}\xi(t) &= (x_0)(t) \text{ for all } t \in [-r, 0],\\ \xi(t) &:= x_0(0) + \tilde{G}(B(x_0, t)) - \tilde{G}(\delta_\tau x_0) \text{ for all } t \in (0, \delta]\end{aligned} \tag{4}$$

$$(B(x_0, t))(s) = x_0(t - \tau + s) \text{ for all } s \in [-r + \tau, 0] \tag{5}$$

$$S := \{x \in C^0([-r, \delta]; \mathbb{R}^n) : \|x - \xi\| \le 1,\ x(s) = (x_0)(s) \text{ for } s \in [-r, 0]\} \tag{6}$$

Notice that $\xi \in C^0([-r, \delta]; \mathbb{R}^n)$. The mapping $P : S \to C^0([-r, \delta]; \mathbb{R}^n)$ is well-defined, because for every $x \in C^0([-r, \delta]; \mathbb{R}^n)$, $d \in L^\infty_{loc}(\mathbb{R}_+; D)$, $u \in L^\infty_{loc}(\mathbb{R}_+; U)$:

- The mapping $(d, u) \to f(d, x, u)$ is continuous (by Lemma 2.1 on page 40 in [2]), and consequently, the mapping $t \to f(d(t), x_s, u(t))$ is measurable for each fixed $s \in \mathbb{R}$(as the composition of a continuous mapping with a Lebesgue measurable function).
- The mapping $s \to f(d(t), x_s, u(t))$ is continuous in $s \in \mathbb{R}$ for every fixed $t \ge 0$ (define $f(d(t), x_s, u(t)) = f(d(t), x_0, u(t))$ for $s < 0$ and $f(d(t), x_s, u(t)) = f(d(t), x_\delta, u(t))$ for $s > \delta$).
- The mapping $t \to f(d(t), x_t, u(t))$ is measurable.

See [10, p. 474]. Moreover, (2) and (4) give

$$\|\xi\| \le \tilde{a}(\|x_0\|) \tag{7}$$

where $\tilde{a}(s) := s + 2a(s)$ for all $s \geq 0$ and consequently, if $x \in S$ then Assumption (H4) implies that $|f(d(t), x_t, u(t))| \leq \tilde{a}\left(1 + \tilde{a}\left(\|x_0\|\right) + \|u\|\right)$ for $t \in [0, \delta]$ a.e., where $\|u\| = \sup_{0 \leq t \leq \tau} |u(t)|$. Therefore, the mapping $t \to f(d(t), x_t, u(t))$ is essentially bounded. Furthermore, if

$$\delta \leq \frac{1}{\tilde{a}\left(1 + \tilde{a}\left(\|x_0\|\right) + \|u\|\right)},$$

then it follows from definition (3) that $\|Px - \xi\| \leq 1$. Therefore, for

$$\delta = \min\left(\tau, \frac{1}{\tilde{a}\left(1 + \tilde{a}\left(\|x_0\|\right) + \|u\|\right)}\right),$$

the mapping P is a self mapping, i.e., $P : S \to S$. Finally, Assumption (H3) and definition (3) imply that for

$$\delta = \min\left(\tau, \frac{1}{\tilde{a}\left(1 + \tilde{a}\left(\|x_0\|\right) + \|u\|\right)}, \frac{1}{2 + 2L_f\left(2 + 2\tilde{a}\left(\|x_0\|\right) + \|u\|\right)}\right) \tag{8}$$

the mapping P is a contraction. Applying Banach's fixed point theorem on the closed set $S \subseteq C^0([-r, \delta]; \mathbb{R}^n)$ defined by (4), we obtain a unique mapping $x \in C^0\left([-r, \delta]; \mathbb{R}^n\right)$ that satisfies $x(s) = (x_0)(s)$ for all $s \in [-r, 0]$ and

$$\frac{d}{dt}\left(x(t) - G(x_t)\right) = f(d(t), x_t, u(t)) \tag{9}$$

for $t \in [0, \delta]$ a.e.. Since the unique function $x \in C^0\left([-r, \delta]; \mathbb{R}^n\right)$ satisfying $x(s) = (x_0)(s)$ for all $s \in [-r, 0]$ and (9) for $t \in [0, \delta]$ a.e. is bounded, it follows (by repeating the arguments above) that there exists $\delta' > \delta$ and a unique function $x \in C^0\left([-r, \delta']; \mathbb{R}^n\right)$ satisfying $x(s) = (x_0)(s)$ for all $s \in [-r, 0]$ and $\frac{d}{dt}\left(x(t) - G(x_t)\right) = f(d(t), x_t, u(t))$ for $t \in [0, \delta']$ a.e.. More specifically, by (6), $\delta' > \delta$ satisfies

$$\delta' \geq \delta + \min\left(\tau, \frac{1}{\tilde{a}\left(1+\tilde{a}\left(\sup_{\delta-r \leq s \leq \delta}|x(s)|\right) + \sup_{\delta \leq s \leq \delta+\tau}|u(s)|\right)}, \frac{1}{2+2L_f\left(2+2\tilde{a}\left(\sup_{\delta-r \leq s \leq \delta}|x(s)|\right) + \sup_{\delta \leq s \leq \delta+\tau}|u(s)|\right)}\right) \tag{10}$$

Let $t_{\max} = t_{\max}(x_0, d, u) \in (0, +\infty]$ be the least upper bound of all $\delta > 0$ for which there exists a function $x \in C^0\left([-r, \delta]; \mathbb{R}^n\right)$ satisfying $x(s) = (x_0)(s)$ for all $s \in [-r, 0]$ and $\frac{d}{dt}\left(x(t) - G(x_t)\right) = f(d(t), x_t, u(t))$ for $t \in [0, \delta]$ a.e.. If $t_{\max} < +\infty$ (i.e., is finite) then for every $\varepsilon > 0$ there exists $\delta > t_{\max} - \varepsilon$ and a function $x \in C^0\left([-r, \delta]; \mathbb{R}^n\right)$ satisfying $x(s) = (x_0)(s)$ for all $s \in [-r, 0]$ and

$$\frac{d}{dt}\left(x(t) - G(x_t)\right) = f(d(t), x_t, u(t))$$

for $t \in [0, \delta]$ a.e.. It follows in any case ($t_{\max} < +\infty$ or $t_{\max} = +\infty$) that there is a function $x \in C^0([-r, t_{\max}); \mathbb{R}^n)$ satisfying $x(s) = (x_0)(s)$ for all $s \in [-r, 0]$ and $\frac{d}{dt}(x(t) - G(x_t)) = f(d(t), x_t, u(t))$ for $t \in [0, t_{\max})$ a.e.. When $t_{\max} < +\infty$, notice that (10) implies

$$\varepsilon \geq \min\left(\tau, \frac{1}{\tilde{a}(1+\tilde{a}(\sup_{\delta-r\leq s\leq\delta}|x(s)|) + \sup_{\delta\leq s\leq\delta+\tau}|u(s)|)}, \frac{1}{2+2L_f(2+2\tilde{a}(\sup_{\delta-r\leq s\leq\delta}|x(s)|) + \sup_{\delta\leq s\leq\delta+\tau}|u(s)|)}\right) \tag{11}$$

for all $\varepsilon > 0$ and $\delta > t_{\max} - \varepsilon$. Since $u \in L^\infty_{loc}(\mathbb{R}_+; U)$ and $t_{\max} < +\infty$, it follows that

$$\sup_{\delta\leq s\leq\delta+\tau}|u(s)| \leq \sup_{0\leq s\leq t_{\max}+\tau}|u(s)| < +\infty.$$

Inequality (11) cannot hold for all $\varepsilon > 0$ if $\sup_{-r\leq s<t_{\max}}|x(s)| < +\infty$. Therefore, if $t_{\max} < +\infty$ then $\limsup_{t\to t_{\max}^-}|x(t)| = +\infty$. The proof is complete. ◁

Having proven Theorem 1, we are justified to do the following: for arbitrary $x_0 \in C^0([-r, 0]; \mathbb{R}^n)$, $d \in L^\infty_{loc}(\mathbb{R}_+; D)$, and $u \in L^\infty_{loc}(\mathbb{R}_+; U)$, we name the mapping $x \in C^0([-r, t_{\max}); \mathbb{R}^n)$ satisfying $x(s) = (x_0)(s)$ for $s \in [-r, 0]$ and $\frac{d}{dt}(x(t) - G(x_t)) = f(d(t), x_t, u(t))$ for $t \in [0, t_{\max})$ a.e. for certain $t_{\max} = t_{\max}(x_0, d, u) \in (0, +\infty]$ "*the solution $x(t)$ of* (1) *with initial condition $x(s) = (x_0)(s)$ for all $s \in [-r, 0]$, corresponding to inputs $d \in L^\infty_{loc}(\mathbb{R}_+; D)$* and $u \in L^\infty_{loc}(\mathbb{R}_+; U)$", without explicit reference to the maximal existence time of the solution $t_{\max} = t_{\max}(x_0, d, u) \in (0, +\infty]$.

We next consider the issue of Lipschitz dependence on the initial conditions. Consider two solutions $x(t)$ and $y(t)$ of (1) with initial conditions x_0 and y_0 in $C^0([-r, 0]; \mathbb{R}^n)$, respectively, corresponding to the same inputs $d \in L^\infty_{loc}(\mathbb{R}_+; D)$ and $u \in L^\infty_{loc}(\mathbb{R}_+; U)$. Let $t < \min\{t_{\max}(x_0, d, u),\ t_{\max}(y_0, d, u)\}$ be a positive time and define

$$C := \max\left\{\sup_{-r\leq s\leq t}|x(s)|,\ \sup_{-r\leq s\leq t}|y(s)|,\ \sup_{0\leq s\leq t}|u(s)|\right\}.$$

We have:

$$x(p) = G(x_p) + x(0) - G(x_0) + \int_0^p f(d(w), x_w, u(w))dw, \ \ for\ all\ p \in [0, t] \tag{12}$$

$$y(p) = G(y_p) + y(0) - G(y_0) + \int_0^p f(d(w), y_w, u(w))dw, \ \ for\ all\ p \in [0, t] \tag{13}$$

Let us first assume that $p \in [0, \min(t, \tau)]$, where $\tau \in (0, r]$ is the constant involved in (H1). It follows that conditions $G(x_p) = \tilde{G}(\delta_\tau x_p)$ and $G(y_p) = \tilde{G}(\delta_\tau y_p)$. We obtain from (12) and (13):

$$\begin{aligned} |x(p) - y(p)| \leq & \left|\tilde{G}(\delta_\tau x_p) - \tilde{G}(\delta_\tau y_p)\right| + |x(0) - y(0)| \\ & + \left|\tilde{G}(\delta_\tau x_0) - \tilde{G}(\delta_\tau y_0)\right| \\ & + \int_0^p |f(d(w), x_w, u(w)) - f(d(w), y_w, u(w))| \, dw \end{aligned} \tag{14}$$

Assumption (H2) implies that $\left|\tilde{G}(\delta_\tau x_p) - \tilde{G}(\delta_\tau y_p)\right| \leq L_G(2C) \left\|\delta_\tau x_p - \delta_\tau y_p\right\|$ and $\left|\tilde{G}(\delta_\tau x_0) - \tilde{G}(\delta_\tau y_0)\right| \leq L_G(2C) \|\delta_\tau x_0 - \delta_\tau y_0\|$. However, since $p \in [0, \tau]$, it also follows that $\left\|\delta_\tau x_p - \delta_\tau y_p\right\| \leq \|x_0 - y_0\|$. Using the inequalities $\|\delta_\tau x_0 - \delta_\tau y_0\| \leq \|x_0 - y_0\|$ and $|x(0) - y(0)| \leq \|x_0 - y_0\|$, we conclude from (14) that for all $p \in [0, \min(t, \tau)]$, we have:

$$\begin{aligned} |x(p) - y(p)| \leq & (2L_G(2C) + 1) \|x_0 - y_0\| + \\ & \int_0^p |f(d(w), x_w, u(w)) - f(d(w), y_w, u(w))| \, dw \end{aligned} \tag{15}$$

Using Assumption (H3) and (15) we get for all $p \in [0, \min(t, \tau)]$:

$$|x(p) - y(p)| \leq (2L_G(2C) + 1) \|x_0 - y_0\| + L_f(3C) \int_0^p \|x_w - y_w\| \, dw$$

which directly implies for all $p \in [0, \min(t, \tau)]$:

$$\left\|x_p - y_p\right\| \leq (2L_G(2C) + 1) \|x_0 - y_0\| + L_f(3C) \int_0^p \|x_w - y_w\| \, dw \tag{16}$$

Using the Gronwall-Bellman Lemma and (16), we get the following for all $p \in [0, \min(t, \tau)]$:

$$\left\|x_p - y_p\right\| \leq (2L_G(2C) + 1) \exp\left(p L_f(3C)\right) \|x_0 - y_0\| \tag{17}$$

Inequality (17), in conjunction with the semigroup property for (1), implies the following estimate:

$$\sup_{w \leq s \leq p+w} (\|x_s - y_s\|) \leq (2L_G(2C) + 1) \exp\left(p L_f(3C)\right) \|x_w - y_w\| \tag{18}$$

for all $w \geq 0$, $p \in [0, \tau]$ with $p + w \leq t$. Next, using induction, the fact that $2L_G(2C) + 1 \leq \exp(2L_G(2C))$, and (18), we prove:

$$\begin{aligned} & \sup_{0 \leq s \leq j\tau} (\|x_s - y_s\|) \leq \\ & (2L_G(2C) + 1) \exp(2L_G(2C)\, j) \exp\left(j\tau L_f(3C)\right) \|x_0 - y_0\| \end{aligned} \tag{19}$$

for all integers $j \geq 0$ with $j\tau \leq t$. Therefore, (18) and (19) give:

$$\sup_{0\leq s\leq t}(\|x_s - y_s\|) \leq \\ (2L_G(2C) + 1)\exp(2L_G(2C))\exp\left(t\left(L_f(3C) + 2\tau^{-1}L_G(2C)\right)\right)\|x_0 - y_0\| \tag{20}$$

Thus, we have proved:

Theorem 2 *Consider* (1) *under (H1)–(H4). For every x_0 and y_0 in $C^0([-r,0];\mathbb{R}^n)$, $d \in L^\infty_{loc}(\mathbb{R}_+;D)$, $u \in L^\infty_{loc}(\mathbb{R}_+;U)$ and $t \in [0, \min(t_{\max}(x_0,d,u), t_{\max}(y_0,d,u)))$, the solutions $x(t)$ and $y(t)$ of* (1) *with initial conditions x_0 and y_0 in $C^0([-r,0];\mathbb{R}^n)$, respectively, corresponding to the same inputs $d \in L^\infty_{loc}(\mathbb{R}_+;D)$ and $u \in L^\infty_{loc}(\mathbb{R}_+;U)$ satisfy* (20), *where* $C := \max\left(\sup_{-r\leq s\leq t}|x(s)|, \sup_{-r\leq s\leq t}|y(s)|, \sup_{0\leq s\leq t}|u(s)|\right)$.

Theorem 2 shows a version of Lipschitz continuity of the solution of (1) with respect to the initial condition. However, we also know that $0 \in C^0([-r,0];\mathbb{R}^n)$ is a robust equilibrium point [3].

4 Global Stability Results for Neutral Systems

Using the results in [3], we are in a position to define the notion of (uniform) robust global asymptotic stability for system (1) under Assumptions (H1)–(H4) and the additional assumption $U = \{0\}$. The last assumption requires that the only input $u \in L^\infty_{loc}(\mathbb{R}_+;U)$ that can be applied to the system is the zero input and that is why the input $u \in L^\infty_{loc}(\mathbb{R}_+;U)$ becomes irrelevant.

Definition 1 Consider system (1) with $U = \{0\}$ under Assumptions (H1)–(H4) and assume that for every $x_0 \in C^0([-r,0];\mathbb{R}^n)$ and $d \in L^\infty_{loc}(\mathbb{R}_+;D)$, the solution of (1) with initial condition $x_0 \in C^0([-r,0];\mathbb{R}^n)$ corresponding to input $d \in L^\infty_{loc}(\mathbb{R}_+;D)$ exists for all $t \geq 0$. We say that $0 \in C^0([-r,0];\mathbb{R}^n)$ is (uniformly) robustly globally asymptotically stable (RGAS) for (1) provided the following three properties hold:

Robust Lagrange Stability: For every $\varepsilon > 0$, we have

$$\sup\left\{\|x_t\| : t \geq 0,\ \|x_0\| \leq \varepsilon,\ d \in L^\infty_{loc}(\mathbb{R}_+;D)\right\} < +\infty.$$

Robust Lyapunov Stability: For every $\varepsilon > 0$, there exists $\delta := \delta(\varepsilon) > 0$ such that

$$\sup\left\{\|x_t\| : t \geq 0,\ \|x_0\| \leq \delta,\ d \in L^\infty_{loc}(\mathbb{R}_+;D)\right\} < \varepsilon.$$

Uniform Robust Attractivity: For every $\varepsilon > 0$ and $R \geq 0$, there exists $\tau := \tau(\varepsilon, R) > 0$ such that

$$\sup\left\{\|x_t\| : t \geq \tau,\ \|x_0\| \leq R,\ d \in L^\infty_{loc}(\mathbb{R}_+;D)\right\} < \varepsilon.$$

Theorem 1, Theorem 2, and Lemma 4.2 in [3] and Theorem 2.2 in [3] give us the following result:

Theorem 3 *Consider system* (1) *with* $U = \{0\}$ *under Assumptions (H1)–(H4). Then the following statements are equivalent:*

(a) $0 \in C^0([-r, 0]; \mathbb{R}^n)$ *is RGAS.*
(b) There exists $\sigma \in KL$ *such that for every* $x_0 \in C^0([-r, 0]; \mathbb{R}^n)$ *and* $d \in L^\infty_{loc}(\mathbb{R}_+; D)$*, the solution* $x(t)$ *of* (1) *with initial condition* $x_0 \in C^0([-r, 0]; \mathbb{R}^n)$ *corresponding to the input* $d \in L^\infty_{loc}(\mathbb{R}_+; D)$ *exists for all* $t \geq 0$ *and satisfies* $\|x_t\| \leq \sigma(\|x_0\|, t)$ *for all* $t \geq 0$.
(c) System (1) *is robustly forward complete, i.e., for every* $T \geq 0$ *and* $R \geq 0$*, we have* $\sup\left\{ \|x_t\| : t \in [0, T],\ \|x_0\| \leq R,\ d \in L^\infty_{loc}(\mathbb{R}_+; D) \right\} < +\infty$*, where* $x(t)$ *denotes the solution of* (1) *with initial condition* $x_0 \in C^0([-r, 0]; \mathbb{R}^n)$ *corresponding to the input* $d \in L^\infty_{loc}(\mathbb{R}_+; D)$*, and the uniform robust attractivity property from Definition 1 holds.*

Theorem 3.4 in [3] and the results of Theorems 1 and 2 allow us to obtain a complete Lyapunov characterization for the RGAS property for system (1). Notice that Theorem 2 in conjunction with Theorem 3(b) guarantees that property (REG2) in [3] holds.

Theorem 4 *Consider system* (1) *with* $U = \{0\}$ *under Assumptions (H1)–(H4). The equilibrium point* $0 \in C^0([-r, 0]; \mathbb{R}^n)$ *is RGAS for* (1) *if and only if there exist a functional* $V : C^0([-r, 0]; \mathbb{R}^n) \to \mathbb{R}_+$*, a non-decreasing function* $Q : \mathbb{R}_+ \to \mathbb{R}_+$*, and functions* a_1 *and* a_2 *in* K_∞ *such that the following inequalities hold:*

$$a_1(\|x\|) \leq V(x) \leq a_2(\|x\|),\ for\ all\ x \in C^0([-r, 0]; \mathbb{R}^n) \tag{21}$$

$$\begin{aligned} |V(x) - V(y)| \leq Q(\max(\|x\|, \|y\|)) \|x - y\|, \\ for\ all\ x, y \in C^0([-r, 0]; \mathbb{R}^n) \end{aligned} \tag{22}$$

$$\begin{aligned} &V(x_t) \leq \exp(-t)\, V(x_0), \\ &for\ all\ t \geq 0,\ x_0 \in C^0([-r, 0]; \mathbb{R}^n),\ d \in L^\infty_{loc}(\mathbb{R}_+; D) \end{aligned} \tag{23}$$

where $x_t \in C^0([-r, 0]; \mathbb{R}^n)$ *denotes the solution of* (1) *with initial condition* $x_0 \in C^0([-r, 0]; \mathbb{R}^n)$ *corresponding to the input* $d \in L^\infty_{loc}(\mathbb{R}_+; D)$.

Inequality (22) guarantees that the functional $V : C^0([-r, 0]; \mathbb{R}^n) \to \mathbb{R}_+$ is Lipschitz on bounded sets of the state space $C^0([-r, 0]; \mathbb{R}^n)$. However, inequality (22) does not guarantee Fréchet differentiability of the functional $V : C^0([-r, 0]; \mathbb{R}^n) \to \mathbb{R}_+$ nor that the limit $\lim_{t \to 0^+} t^{-1}(V(x_t) - V(x_0))$ exists for the solution $x(t)$ of (1) with initial condition $x_0 \in C^0([-r, 0]; \mathbb{R}^n)$ corresponding to input $d \in L^\infty_{loc}(\mathbb{R}_+; D)$. Inequality (23) guarantees that for every $x_0 \in C^0([-r, 0]; \mathbb{R}^n)$ and $d \in L^\infty_{loc}(\mathbb{R}_+; D)$, the solution $x(t)$ of (1) with initial condition $x_0 \in C^0([-r, 0]; \mathbb{R}^n)$ corresponding to input $d \in L^\infty_{loc}(\mathbb{R}_+; D)$ satisfies $\limsup_{t \to 0^+} t^{-1}(V(x_t) - V(x_0)) \leq -V(x_0)$.

If we wish to use Dini derivatives that do not depend on the solution of (1), then we have to define the following operator for all $x \in C^0([-r,0];\mathbb{R}^n)$, $v \in \mathbb{R}^n$ and $h \in [0,\tau]$:

$$E_h(x;v) = \begin{cases} x(h+\theta), & \theta \in [-r,-h] \\ x(0) + (\theta+h)v - \tilde{G}(\delta_\tau x) + \tilde{G}(\delta_{\tau-\theta-h}x), & \theta \in (-h,0] \end{cases} \tag{24}$$

where $\tau \in (0,r]$ is the constant involved in (H1) and $\tilde{G} : C^0([-r+\tau,0];\mathbb{R}^n) \to \mathbb{R}^n$ is the mapping involved in (H1). When $V : C^0([-r,0];\mathbb{R}^n) \to \mathbb{R}_+$ is a functional that is Lipschitz on bounded sets, then we can use Assumption (H3) and the triangle inequality to obtain the following for sufficiently small $t \geq 0$:

$$\begin{aligned}
&|V(x_t) - V(E_t(x_0; f(\tilde{d}(0),x_0,0)))| \leq R \left\| x_t - E_t(x_0; f(\tilde{d}(0),x_0,0)) \right\| \\
&\leq R \sup_{0\leq s\leq t} \left| \int_0^s \left(f(\tilde{d}(w),x_w,0) - f(\tilde{d}(0),x_0,0) \right) dw \right| \\
&\leq R \sup_{0\leq s\leq t} \left| \int_0^s \left(f(\tilde{d}(w),x_w,0) - f(\tilde{d}(w),x_0,0) \right) dw \right| + \\
&\qquad R \sup_{0\leq s\leq t} \left| \int_0^s \left(f(\tilde{d}(w),x_0,0) - f(\tilde{d}(0),x_0,0) \right) dw \right| \\
&\leq RLt \sup_{0\leq s\leq t} \|x_s - x_0\| + Rt \sup_{0\leq w\leq t} \left| f(\tilde{d}(w),x_0,0) - f(\tilde{d}(0),x_0,0) \right|
\end{aligned}$$

where $R \geq 0$ and $L \geq 0$ are appropriate constants and $x_t \in C^0([-r,0];\mathbb{R}^n)$ denotes the solution of (1) with initial condition $x_0 \in C^0([-r,0];\mathbb{R}^n)$ corresponding to input

$$\tilde{d} \in L^\infty_{loc}(\mathbb{R}_+;D).$$

The above inequality (in conjunction with Lemma 2.1 on page 40 in [2]) guarantees that for every functional $V : C^0([-r,0];\mathbb{R}^n) \to \mathbb{R}_+$ that is Lipschitz on bounded sets, and for every $\tilde{d} \in C^0(\mathbb{R}_+;D)$ and $x_0 \in C^0([-r,0];\mathbb{R}^n)$, the solution $x_t \in C^0([-r,0];\mathbb{R}^n)$ of (1) with initial condition $x_0 \in C^0([-r,0];\mathbb{R}^n)$ corresponding to input $\tilde{d} \in C^0(\mathbb{R}_+;D)$ satisfies this equality:

$$\begin{aligned}
&\limsup_{t\to 0^+} t^{-1}(V(x_t) - V(x_0)) = \\
&\qquad \limsup_{t\to 0^+} t^{-1} \left(V(E_t(x_0; f(\tilde{d}(0),x_0,0))) - V(x_0) \right)
\end{aligned} \tag{25}$$

To obtain a complete Lyapunov characterization, we need to generalize the notion of the stable operator given on page 274 in [2]; see also page 125 in [6].

Definition 2 Let $r > 0$ be a constant and let $K : C^0([-r,0];\mathbb{R}^n) \to \mathbb{R}^n$ be a continuous mapping such that $K(0) = 0$. The mapping $K : C^0([-r,0];\mathbb{R}^n) \to \mathbb{R}^n$ is called strongly stable provided the system

$$x(t) = K(x_t) + u(t) \tag{26}$$

is input-to-state stable (or ISS) [3] with inputs $u \in C^0(\mathbb{R}_+;\mathbb{R}^n)$, i.e., there exist $\sigma \in KL$ and $\gamma \in K$ such that for every $x_0 \in C^0([-r,0];\mathbb{R}^n)$ and $u \in C^0(\mathbb{R}_+;\mathbb{R}^n)$ satisfying $x_0(0) = K(x_0) + u(0)$, the solution $x(t)$ of (26) exists for all $t \geq 0$, is unique, is continuous at all $t \geq 0$, and satisfies

$$\|x_t\| \leq \sigma(\|x_0\|, t) + \gamma\left(\sup_{0\leq s\leq t} |u(s)|\right) \tag{27}$$

for all $t \geq 0$.

We are now ready to state our main result.

Theorem 5 *Consider system* (1) *with* $U = \{0\}$ *under Assumptions (H1)–(H4). The following statements are equivalent:*

(a) The equilibrium point $0 \in C^0([-r,0];\mathbb{R}^n)$ *is RGAS for* (1).
(b) There exists $\sigma \in KL$ *such that for all choices of* $x_0 \in C^0([-r,0];\mathbb{R}^n)$ *and* $d \in L^\infty_{loc}(\mathbb{R}_+;D)$*, the solution* $x(t)$ *of* (1) *with initial condition* $x_0 \in C^0([-r,0];\mathbb{R}^n)$ *corresponding to input* $d \in L^\infty_{loc}(\mathbb{R}_+;D)$ *exists for all* $t \geq 0$ *and satisfies* $\|x_t\| \leq \sigma(\|x_0\|, t)$ *for all* $t \geq 0$.
(c) There exist a functional $V : C^0([-r,0];\mathbb{R}^n) \to \mathbb{R}_+$*, a non-decreasing function* $Q : \mathbb{R}_+ \to \mathbb{R}_+$*, and functions* a_1 *and* a_2 *in* K_∞ *such that inequalities* (21) *and* (22) *hold and the following holds:*

$$\limsup_{h\to 0^+} h^{-1}\left(V(E_h(x; f(d,x,0))) - V(x)\right) \leq -V(x), \quad \text{for all } x \in C^0([-r,0];\mathbb{R}^n), d \in D \tag{28}$$

(d) There exist a functional $V : C^0([-r,0];\mathbb{R}^n) \to \mathbb{R}_+$*, a strongly stable operator* $K : C^0([-r,0];\mathbb{R}^n) \to \mathbb{R}^n$*, a non-decreasing function* $Q : \mathbb{R}_+ \to \mathbb{R}_+$*, a positive definite function* $\rho \in C^0(\mathbb{R}_+;\mathbb{R}_+)$ *and functions* a_1 *and* a_2 *in* K_∞ *such that inequality* (22) *and the following inequalities hold:*

$$a_1(|x(0) - K(x)|) \leq V(x) \leq a_2(\|x\|), \text{ for all } x \in C^0([-r,0];\mathbb{R}^n) \tag{29}$$

$$\limsup_{h\to 0^+} h^{-1}(V(E_h(x; f(d,x,0))) - V(x)) \leq -\rho(V(x)), \text{ for all } x \in C^0([-r,0];\mathbb{R}^n), d \in D \tag{30}$$

Proof Equivalences (a)⇔(b) and (b)⇒(c) are direct consequences of the preceding theorems, and the fact that (25) holds for all $x_0 \in C^0([-r,0];\mathbb{R}^n)$ and $\tilde{d} \in C^0(\mathbb{R}_+;D)$. Indeed, for every $d \in D$, we may consider the continuous input $\tilde{d}(t) \equiv d$ (where $\tilde{d} \in C^0(\mathbb{R}_+;D)$). Inequality (28) is a direct consequence of (25) (for the continuous input $\tilde{d}(t) \equiv d$) and inequality (23). The implication (c)⇒(d) is obvious.

Thus, we are left with the proof of implication (d)⇒(b). Inequality (20) and density of $C^1([-r,0];\mathbb{R}^n)$ in $C^0([-r,0];\mathbb{R}^n)$ implies that it suffices to show that there exists $\hat{\sigma} \in KL$ such that

$$\|x_t\| \leq \hat{\sigma}\left(\|x_0\|, t\right), \ for\ all\ t \geq 0,\ x_0 \in C^1\left([-r,0];\mathbb{R}^n\right),\ d \in L^{\infty}_{loc}(\mathbb{R}_+;D) \tag{31}$$

where $x_t \in C^0([-r,0];\mathbb{R}^n)$ denotes the solution of (1) with (arbitrary) initial condition $x_0 \in C^1([-r,0];\mathbb{R}^n)$ corresponding to (arbitrary) input $d \in L^{\infty}_{loc}(\mathbb{R}_+;D)$. On the other hand, if the initial condition $x_0 \in C^0([-r,0];\mathbb{R}^n)$ is a continuously differentiable mapping, then the solution $x(t)$ of (1) is locally absolutely continuous on $[0, t_{\max})$, where $t_{\max} > 0$ is the maximal existence time. Indeed, using the integral formula

$$x(t) - x(t_0) = G(x_t) - G(x_{t_0}) + \int_{t_0}^{t} f(d(s), x_s, 0)ds \tag{32}$$

for all $t \in [0, t_{\max})$, $t_0 \in [0, t_{\max})$ with $t \geq t_0$ and $t \leq \tau$, We can use Assumptions (H1), (H2), and (H4) to get

$$\begin{aligned}|x(t) - x(t_0)| \leq \big(&L_G\left(2\sup\nolimits_{0\leq s\leq t}(\|x_s\|)\right)\sup\nolimits_{-r\leq s\leq t-\tau}(|\dot{x}_0(s)|) + \\ &a\left(\sup\nolimits_{0\leq s\leq t}(\|x_s\|)\right)\big)(t - t_0)\end{aligned} \tag{33}$$

Inequality (33) shows that $x(t)$ is locally absolutely continuous on $[0, t_{\max}) \bigcap [0, \tau]$. If $t_{\max} > \tau$ then we can repeat the same argument on $[\tau, t_{\max}) \bigcap [\tau, 2\tau]$ and prove that $x(t)$ is locally absolutely continuous on $[0, t_{\max}) \bigcap [0, 2\tau]$. The same argument, repeated as many times as needed, shows that the solution $x(t)$ of (1) is locally absolutely continuous on $[0, t_{\max})$. Using (22), we can establish that the mapping $t \to V(x_t)$ is locally absolutely continuous on $[0, t_{\max})$. Therefore, the limit

$$\lim_{h\to 0^+} h^{-1}\left(V(x_{t+h}) - V(x_t)\right)$$

exists for $t \in [0, t_{\max})$ a.e.. Finally, inequality (22) and definition (24) imply that the following inequality holds for all $t \in [0, t_{\max})$ and $v \in \mathbb{R}^n$:

$$\begin{aligned}V(x_{t+h}) - V(E_h(x_t;v)) \leq R\,\|x_{t+h} - E_h(x_t;v)\| \leq \\ Rh\sup\nolimits_{0<s\leq h}\left|s^{-1}\int_t^{t+s} f(d(w), x_w, 0)dw - v\right|\end{aligned}$$

for appropriate constant $R \geq 0$ and for sufficiently small $h > 0$. Since

$$\lim_{h\to 0^+} h^{-1}\int_t^{t+h} f(d(w), x_w, 0)dw = f(d(t), x_t, 0)\ for\ t \in [0, t_{\max})\ a.e.$$

we get:

$$\begin{aligned}&\lim_{h \to 0^+} \ h^{-1} \left(V(x_{t+h}) - V(x_t) \right) = \\ &\lim_{h \to 0^+} \ h^{-1} \left(V(E_h(x_t;\, f(d(t), x_t, 0))) - V(x_t) \right), \ \ for\ t \in [0, t_{\max})\ a.e.\end{aligned} \tag{34}$$

Equation (34) and inequality (30) imply that:

$$\lim_{h \to 0^+} \ h^{-1} \left(V(x_{t+h}) - V(x_t) \right) \le -\rho \left(V(x_t) \right), \ \ for\ t \in [0, t_{\max})\ a.e. \tag{35}$$

The differential inequality (35) in conjunction with (29) and Lemma 2.14 on page 82 in [3] implies that there exists $\tilde{\sigma} \in KL$ (which depends only on ρ) such that:

$$|x(t) - K(x_t)| \le a_1^{-1} \left(\tilde{\sigma} \left(a_2 \left(\|x_0\| \right),\ t \right) \right), \ \ for\ t \in [0, t_{\max}) \tag{36}$$

Defining $u(t) = x(t) - K(x_t)$ for all $t \in [0, t_{\max})$, it follows from (27) and (36) that

$$\|x_t\| \le \sigma \left(\|x_0\|,\ 0 \right) + \gamma \left(a_1^{-1} \left(\tilde{\sigma} \left(a_2 \left(\|x_0\| \right), 0 \right) \right) \right), \ \ for\ t \in [0, t_{\max}) \tag{37}$$

Inequality (37) and a standard contradiction argument (which uses the conclusion of a previous theorem that if $t_{\max} < +\infty$ then $\limsup_{t \to t_{\max}^-} |x(t)| = +\infty$) implies that $t_{\max} = +\infty$ and that estimates (36) and (37) hold for all $t \ge 0$. Using (27), we obtain the following for all $t \ge 0$:

$$\|x_t\| \le \sigma \left(\left\| x_{t/2} \right\|,\ t/2 \right) + \gamma \left(\sup_{t/2 \le s \le t} |u(s)| \right) \tag{38}$$

Then (36), (37) and (38) give the following for all $t \ge 0$:

$$\begin{aligned}\|x_t\| \le \sigma &\left(\sigma \left(\|x_0\|, 0 \right) + \gamma \left(a_1^{-1} \left(\tilde{\sigma} \left(a_2 \left(\|x_0\| \right), 0 \right) \right) \right),\ t/2 \right) \\ &+ \gamma \left(a_1^{-1} \left(\tilde{\sigma} \left(a_2 \left(\|x_0\| \right), t/2 \right) \right) \right)\end{aligned} \tag{39}$$

It follows that (31) holds with

$$\hat{\sigma}(s,t) := \sigma \left(\sigma(s, 0) + \gamma \left(a_1^{-1} \left(\tilde{\sigma} \left(a_2(s), 0 \right) \right) \right),\ t/2 \right) + \gamma \left(a_1^{-1} \left(\tilde{\sigma} \left(a_2(s), t/2 \right) \right) \right)$$

for all $s \ge 0$ and $t \ge 0$.
The proof is complete. ◁

Remark 1 (a) Differential inequalities like (28) and (30) have been used in the past for neutral systems for the proof of various stability properties [1, 5, 6].
(b) The stability theory provided in the literature (such as [2, 4]) requires the strongly stable operator $K : C^0([-r, 0]; \mathbb{R}^n) \to \mathbb{R}^n$ to coincide with the difference

operator $G : C^0([-r, 0]; \mathbb{R}^n) \to \mathbb{R}^n$. The following example illustrates that this coincidence is not necessary.

Example 1 Consider the linear control system:

$$\begin{aligned} &\dot{x}_1(t) = ax_1(t) + x_2(t-r) \\ &\tfrac{d}{dt}\left(x_2(t) + cx_2(t-\tau)\right) = u(t) \\ &x(t) = (x_1(t), x_2(t)),\ u(t) \in \mathbb{R} \end{aligned} \tag{40}$$

where $a > 0, r \geq \tau > 0$ and $c \in (0, 1)$ are constants. Using the Lyapunov functional

$$\begin{aligned} V(x) = &\tfrac{1}{2}\left(\exp(ar)x_1(0) + \textstyle\int_{-r}^{0} \exp(-as)x_2(s)ds\right)^2 + \gamma \textstyle\int_{-\tau}^{0} \exp(ps)\, x_2^2(s)ds \\ &+\tfrac{M}{2}\left(x_2(0) + cx_2(-\tau) + k\exp(ar)x_1(0) + k\textstyle\int_{-r}^{0} \exp(-as)x_2(s)ds\right)^2 \end{aligned} \tag{41}$$

for all $x \in C^0([-r, 0]; \mathbb{R}^2)$, where $k > 0$, $M > 0$, $p > 0$ and $\gamma > 0$ are constants, we will show that for every $L \geq 1$ and

$$k \in \left(2a + c + 1,\ \frac{\exp(-p\tau)}{2c} - 1 - c\right),$$

the linear feedback law

$$\begin{aligned} u(t) = &-k(a+L)\exp(ar)x_1(t) \\ &-k(a+L)\textstyle\int_{t-r}^{t} \exp(a(t-s))x_2(s)ds - (k+L)x_2(t) - Lcx_2(t-\tau) \end{aligned} \tag{42}$$

globally asymptotically stabilizes $0 \in C^0([-r, 0]; \mathbb{R}^2)$ for the closed-loop system given by (40) and (42), provided that

$$4c^2 + 4(a+1)c - \exp(-p\tau) < 0\,. \tag{43}$$

Indeed, notice that (43) implies the existence of

$$k \in \left(2a + c + 1,\ \frac{\exp(-p\tau)}{2c} - 1 - c\right).$$

Notice that for every choice of $k > 0$, $M > 0$, $p > 0$ and $\gamma > 0$, the functional defined by (41) satisfies (29) for the operator:

$$K(x) := \begin{bmatrix} K_1(x) \\ K_2(x) \end{bmatrix} = \begin{bmatrix} -\int_{-r}^{0} \exp(-a(r+s))x_2(s)ds \\ -cx_2(-\tau) \end{bmatrix} \tag{44}$$

Indeed, notice that the functional defined by (41) satisfies the following inequalities for all $x \in C^0([-r, 0]; \mathbb{R}^2)$:

$$
\begin{aligned}
&A\,(x_1(0)-K_1(x))^2+\tfrac{M}{2}\,(x_2(0)-K_2(x)+k\exp(ar)\,(x_1(0)-K_1(x)))^2\\
&\leq V(x)\leq \gamma\tau\,\|x\|^2+A\left(1+2Mk^2\right)(x_1(0)-K_1(x))^2+M\,(x_2(0)-K_2(x))^2
\end{aligned}
\tag{45}
$$

where $A=\exp(2ar)/2$ and consequently there exist constants $G_2\geq G_1>0$ such that the following inequality holds for all $x\in C^0\left([-r,0];\mathbb{R}^2\right)$:

$$
G_1\,|x(0)-K(x)|^2\leq V(x)\leq(\gamma\tau+G_2)\,\|x\|^2 \tag{46}
$$

Moreover, the operator $K:C^0([-r,0];\mathbb{R}^2)\to\mathbb{R}^2$ defined by (44) is strongly stable for $c\in(0,1)$. Indeed, we notice that for every $x_0\in C^0\left([-r,0];\mathbb{R}^2\right)$ and $u\in C^0(\mathbb{R}_+;\mathbb{R}^2)$ with $x_0(0)=K(x_0)+u(0)$ the solution $x(t)\in\mathbb{R}^2$ of

$$
\begin{aligned}
x_1(t)&=u_1(t)-\textstyle\int_{-r}^{0}\exp(-a(r+s))x_2(t+s)ds\\
&=u_1(t)-\textstyle\int_{t-r}^{t}\exp(a(t-r-w))x_2(w)dw\\
x_2(t)&=u_2(t)-cx_2(t-\tau)
\end{aligned}
\tag{47}
$$

exists for all $t\geq 0$ and is unique and continuous for $t\geq 0$, by the following argument. The solution component $x_2(t)$ can be obtained (uniquely) for every $t\in[0,\tau]$ from the equation $x_2(t)=u_2(t)-cx_2(t-\tau)$. Having determined $x_2(t)$ for every $t\in[-r,\tau]$, we can compute the solution component $x_1(t)$ (uniquely) for every $t\in[0,\tau]$ by the equation

$$
x_1(t)=u_1(t)-\int_{t-r}^{t}\exp(a(t-r-w))x_2(w)dw
$$

and next we can repeat the process for all intervals of the form $[i\tau,(i+1)\tau]$, where $i\geq 0$ is any integer.

Using the functional $W(x_2):=\max_{-r\leq s\leq 0}\left(\exp(\sigma s)\,|x_2(s)|\right)$ with any choice $\sigma\in\left(0,\tau^{-1}\ln\left(c^{-1}\right)\right)$, we can prove that

$$
W(x_{2,t+h})\leq\exp(-\sigma h)W(x_{2,t})+\sup_{t\leq s\leq t+h}\left(|u_2(s)|\right)
$$

for all $h\in[0,\tau]$. The previous inequality (applied repeatedly) shows that

$$
W(x_{2,t})\leq\exp(-\sigma t)W(x_{2,0})+\frac{2-\exp(-\sigma\tau)}{1-\exp(-\sigma\tau)}\sup_{0\leq s\leq t}\left(|u_2(s)|\right)
$$

and consequently,

$$
\begin{aligned}
&\left\|x_{2,t}\right\|\leq\exp(-\sigma(t-r))\left\|x_{2,0}\right\|+\tfrac{2-\exp(-\sigma\tau)}{1-\exp(-\sigma\tau)}\exp(\sigma r)\sup\nolimits_{0\leq s\leq t}\left(|u_2(s)|\right),\\
&\qquad\qquad for\ all\ t\geq 0
\end{aligned}
\tag{48}
$$

An estimate like (48) for $\|x_{2,t}\|$ could also have been derived by using the theory of linear discrete-time systems (since only one delay is present). Using the inequality $|x_1(t)| \leq |u_1(t)| + r\,\|x_{2,t}\|$, we get the following for all $t \geq 0$:

$$\begin{aligned} &\|x_{1,t}\| \leq \exp\left(-\sigma(t-r)\right)\|x_{1,0}\| + \sup_{0\leq s\leq t}\left(|u_2(s)|\right) \\ &+r\exp\left(-\sigma(t-r)\right)\|x_{2,0}\| + r\tfrac{2-\exp(-\sigma\tau)}{1-\exp(-\sigma\tau)}\sup_{0\leq s\leq t}\left(|u_2(s)|\right) \end{aligned} \tag{49}$$

Estimates (48) and (49) show that the operator $K : C^0([-r,0];\mathbb{R}^2) \to \mathbb{R}^2$ is strongly stable. Finally, we show (30). Definition (41) and the feedback law (42) imply that

$$\begin{aligned} &\lim_{h\to 0^+}\left(h^{-1}\left(V\left(E_h(x;\,f(x))\right) - V(x)\right)\right) = \\ &q\left(x_2(0) + cx_2(-\tau) + kq\right) - cqx_2(-\tau) \\ &-(k-a)q^2 + \gamma x_2^2(0) - \gamma\exp(-p\tau)x_2^2(-\tau) - p\gamma\int_{-\tau}^{0}\exp\left(ps\right)x_2^2(s)ds \\ &-ML\left(x_2(0) + cx_2(-\tau) + kq\right)^2 \end{aligned} \tag{50}$$

where $q := \exp(ar)x_1(0) + \int_{-r}^{0}\exp(-as))x_2(s)ds$ and

$$f(x) = (ax_1(0) + x_2(-r),\ -k(a+L)q - (k+L)x_2(0) - Lcx_2(-\tau))$$

Completing the squares in (50), we get:

$$\begin{aligned} &\lim_{h\to 0^+}\left(h^{-1}\left(V\left(E_h(x;\,f(x))\right) - V(x)\right)\right) \leq \\ &-\left(k - a - \tfrac{c+1}{2} - \gamma k^2 - \gamma k - \gamma ck\right)q^2 \\ &-\left(\gamma\exp(-p\tau) - \tfrac{c}{2} - \gamma c^2 - \gamma c - \gamma ck\right)x_2^2(-\tau) - p\gamma\int_{-\tau}^{0}\exp\left(ps\right)x_2^2(s)ds \\ &-\left(ML - \tfrac{1}{2} - \gamma - \gamma k - \gamma c\right)\left(x_2(0) + cx_2(-\tau) + kq\right)^2 \end{aligned} \tag{51}$$

Since $V(x) = \frac{1}{2}q^2 + \gamma\int_{-\tau}^{0}\exp\left(ps\right)x_2^2(s)ds + \frac{M}{2}\left(x_2(0) + cx_2(-\tau) + kq\right)^2$ and

$$k \in \left(2a + c + 1,\ \frac{\exp(-p\tau)}{2c} - 1 - c\right),$$

the selections $\gamma = 1/(2(k+1+c))$ and $M > 1$ imply the existence of a constant $\mu > 0$ such that

$$\lim_{h\to 0^+}\left(h^{-1}\left(V\left(E_h(x;\,f(x))\right) - V(x)\right)\right) \leq -\mu V(x)$$

Thus (30) holds with $\rho(s) := \mu\,s$.$\triangleleft$

References

1. Driver, R.: Existence and stability of solutions of a delay-differential system. Arch. Ration. Mech. Anal. **10**(1), 401–426 (1962)
2. Hale, J., Verduyn Lunel, S.: Introduction to Functional Differential Equations. Springer, New York (1993)
3. Karafyllis, I., Jiang, Z.-P.: Stability and Stabilization of Nonlinear Systems. Springer, London, UK (2011)
4. Karafyllis, I., Pepe, P., Jiang, Z.-P.: Global output stability for systems described by retarded functional differential equations: Lyapunov characterizations. Eur. J. Control **14**(6), 516–536 (2008)
5. Karafyllis, I., Pepe, P., Jiang, Z.-P.: Input-to-output stability for systems described by retarded functional differential equations. Eur. J. Control **14**(6), 539–555 (2008)
6. Kolmanovskii, V., Myshkis, A.: Applied Theory of Functional Differential Equations. Kluwer, Dordrecht, The Netherlands (1992)
7. Pepe, P.: On Liapunov-Krasovskii functionals under Caratheodory conditions. Automatica **43**(4), 701–706 (2007)
8. Pepe, P.: On the actuator disturbance attenuation for systems described by neutral equations. IMA J. Math. Control Inf. **28**(2), 163–181 (2011)
9. Pepe, P., Karafyllis, I.: Converse Lyapunov-Krasovskii theorems for systems described by neutral functional differential equations in Hale's form. Int. J. Control **86**(2), 232–243 (2013)
10. Sontag, E.: Mathematical Control Theory: Deterministic Finite Dimensional Systems, 2nd edn. Springer, New York (1998)

Hybrid Dynamical Systems with Finite Memory

Jun Liu and Andrew R. Teel

Abstract Hybrid systems with memory are dynamical systems that exhibit both hybrid and delay phenomena, as seen in many physical and engineered applications. A prominent example is the use of delayed hybrid feedback in control systems. This chapter outlines a framework that allows studying hybrid systems with delays through generalized solutions and summarizes some recent results on basic existence and well-posedness of solutions and stability analysis using Lyapunov-based methods.

1 Introduction

Hybrid systems with memory refer to dynamical systems exhibiting both hybrid and delay phenomena. Control systems with delayed hybrid feedback and interconnected hybrid systems with network delays are particular examples of such systems. In fact, delays are often inevitable in many control applications [17] and often cause instability and/or loss of robustness [2].

Motivated by robust stability issues in hybrid feedback control systems, generalized solutions of hybrid inclusions defined on hybrid time domains have been proposed for hybrid systems [4, 5, 15]. These generalized solutions have led to most of the stability analysis tools and results for classical nonlinear systems, including converse Lyapunov theorems, being successfully extended to hybrid systems (see [5, 16] and references therein).

This work is supported, in part, by Royal Society grant IE130106, EU FP7 grant PCIG13-GA-2013-617377, US AFOSR grant FA9550-12-1-0127, and US NSF grant ECCS-1232035.

J. Liu (✉)
Department of Automatic Control and Systems Engineering, University of Sheffield, Sheffield S1 3JD, UK
e-mail: j.liu@sheffield.ac.uk

A.R. Teel
Center for Control, Dynamical Systems and Computation and Department of Electrical and Computer Engineering, University of California, Santa Barbara, CA 93106, USA
e-mail: teel@ece.ucsb.edu

I. Karafyllis et al. (eds.), *Recent Results on Nonlinear Delay Control Systems*,
Advances in Delays and Dynamics 4, DOI 10.1007/978-3-319-18072-4_13

In this chapter, we summarize some recent developments [9–11] towards a framework that allows studying hybrid systems with memory through generalized solutions. The main results recalled here include basic existence and well-posedness of solutions and Lyapunov-based sufficient conditions for stability. While asymptotic stability for hybrid systems with delays have been addressed in the past in various settings (e.g., in [8, 13, 18, 19]), general results on robust asymptotic stability along the lines of [5] have not been available. This is partially owing to the fact that most current tools and results for such systems rely on standard concepts like uniform convergence, while this concept is not well-suited to handle discontinuities caused by jumps in hybrid systems, especially when structural properties of the solutions are concerned. It is from this perspective that we believe it is necessary to formulate hybrid systems with delays using generalized concepts of solutions. The main purpose of this chapter is to recapitulate this newly developed framework of hybrid systems with memory and demonstrate the feasibility of establishing some notion of robustness for asymptotic stability within this framework.

2 Preliminaries

Notation:
$\mathbb{R}^n$ denotes the n-dimensional Euclidean space with its norm denoted by $|\cdot|$; $\mathbb{Z}$ denotes the set of all integers; $\mathbb{R}_{\geq 0} = [0, \infty)$, $\mathbb{R}_{\leq 0} = (-\infty, 0]$, $\mathbb{Z}_{\geq 0} = \{0, 1, 2, \cdots\}$, and $\mathbb{Z}_{\leq 0} = \{0, -1, -2, \cdots\}$; $C([a, b], \mathbb{R}^n)$ denotes the set of all continuous functions from $[a, b]$ to $\mathbb{R}^n$.

2.1 Hybrid Systems with Memory

We start with the definitions of hybrid time domains and hybrid arcs [3, 5] for hybrid systems with memory. All of the definitions in this section are recalled from [9–11].

Definition 1 Consider a subset $E \subseteq \mathbb{R} \times \mathbb{Z}$ with $E = E_{\geq 0} \cup E_{\leq 0}$, where $E_{\geq 0} := E \cap (\mathbb{R}_{\geq 0} \times \mathbb{Z}_{\geq 0})$ and $E_{\leq 0} := E \cap (\mathbb{R}_{\leq 0} \times \mathbb{Z}_{\leq 0})$. It is called a *compact hybrid time domain with memory* if

$$E_{\geq 0} = \bigcup_{j=0}^{J-1} ([t_j, t_{j+1}], j)$$

and

$$E_{\leq 0} = \bigcup_{k=1}^{K} ([s_k, s_{k-1}], -k + 1)$$

for some finite sequence of times $s_K \leq \cdots \leq s_1 \leq s_0 = 0 = t_0 \leq t_1 \leq \cdots \leq t_J$. It is called a *hybrid time domain with memory* if, for all $(T, J) \in E_{\geq 0}$ and all $(S, K) \in \mathbb{R}_{\geq 0} \times \mathbb{Z}_{\geq 0}$, the set

$$(E_{\geq 0} \cap ([0, T] \times \{0, 1, \cdots, J\})) \cup (E_{\leq 0} \cap ([-S, 0] \times \{-K, -K+1, \cdots, 0\}))$$

is a compact hybrid time domain with memory. The set $E_{\leq 0}$ is called a *hybrid memory domain*.

Definition 2 A *hybrid arc with memory* is a pair consisting of a domain *dom* x, which is a hybrid time domain with memory, and a function $x : \operatorname{dom} x \to \mathbb{R}^n$ such that $x(\cdot, j)$ is locally absolutely continuous on $I_j = \{t : (t, j) \in \operatorname{dom} x\}$ for each $j \in \mathbb{Z}$ such that I_j has nonempty interior. In particular, a hybrid arc x with memory is called a *hybrid memory arc* if $\operatorname{dom} x \subseteq \mathbb{R}_{\leq 0} \times \mathbb{Z}_{\leq 0}$. We shall simply use the term *hybrid arc* if we do not have to distinguish between the above two. The collection of all hybrid memory arcs is denoted by $\mathscr{M}$.

Definition 3 Given a hybrid arc and any $(t, j) \in \operatorname{dom} x$, we define an operator $\mathscr{A}_{[t,j]}$ that maps x to $\mathscr{A}_{[t,j]}x \in \mathscr{M}$ given by

$$\mathscr{A}_{[t,j]}x(s, k) = x(t+s, j+k),$$

for all $(s, k) \in \operatorname{dom} \mathscr{A}_{[t,j]}x$, where $\operatorname{dom} \mathscr{A}_{[t,j]}x$ is defined by $(s, k) \in \operatorname{dom} \mathscr{A}_{[t,j]}x$ if and only if $(t+s, j+k) \in \operatorname{dom} x$.

Definition 4 Data of a *hybrid system with memory* in $\mathscr{M}$ consists of four elements:

- a set $\mathscr{C} \subseteq \mathscr{M}$, called the *flow set*;
- a set-valued functional $\mathscr{F} : \mathscr{M} \rightrightarrows \mathbb{R}^n$, called the *flow map*;
- a set $\mathscr{D} \subseteq \mathscr{M}$, called the *jump set*;
- a set-valued functional $\mathscr{G} : \mathscr{M} \rightrightarrows \mathbb{R}^n$, called the *jump map*.

The system is denoted by $\mathscr{H}_{\mathscr{M}} = (\mathscr{C}, \mathscr{F}, \mathscr{D}, \mathscr{G})$.

Definition 5 A hybrid arc x is a *solution to the hybrid system* $\mathscr{H}_{\mathscr{M}}$ if $\mathscr{A}_{[0,0]}x \in \mathscr{C} \cup \mathscr{D}$ and:

(S1) for all $j \in \mathbb{Z}_{\geq 0}$ and almost all $t \in I_j$,

$$\mathscr{A}_{[t,j]}x \in \mathscr{C}, \quad \dot{x}(t, j) \in \mathscr{F}(\mathscr{A}_{[t,j]}x), \tag{1}$$

(S2) for all $j \in \mathbb{Z}_{\geq 0}$ and $(t, j) \in \operatorname{dom} x$ such that $(t, j+1) \in \operatorname{dom} x$,

$$\mathscr{A}_{[t,j]}x \in \mathscr{D}, \quad x(t, j+1) \in \mathscr{G}(\mathscr{A}_{[t,j]}x). \tag{2}$$

The solution x is called *nontrivial* if its positive domain $\operatorname{dom}_{\geq 0}(x) = \operatorname{dom} x \cap (\mathbb{R}_{\geq 0} \times \mathbb{Z}_{\geq 0})$ has at least two points. It is called *complete* if $\operatorname{dom}_{\geq 0}(x)$ is unbounded. It is called *maximal* if there does not exist another solution y to $\mathscr{H}_{\mathscr{M}}$ such that dom x

is a proper subset of dom y and $x(t, j) = y(t, j)$ for all $(t, j) \in \text{dom}\, x$. The set of all maximal solutions to $\mathcal{H}_{\mathcal{M}}$ is denoted by $\mathcal{S}_{\mathcal{H}_{\mathcal{M}}}$.

2.2 Phase Space

The solutions of a hybrid system with memory will be considered in the space of hybrid memory arcs, namely $\mathcal{M}$. First, note that $\mathcal{M}$ is not a vector space, since different hybrid arcs may have different domains. We use the notion of *graphical distance* (or the set distance between graphs [14]) to quantify the distance between and convergence of hybrid memory arcs in $\mathcal{M}$.

Let cl-sets$_{\neq\emptyset}(\mathbb{R}^n)$ denote the collection of all nonempty, closed subsets of $\mathbb{R}^n$. Given $\rho \geq 0$, for each pair $A, B \in$ cl-sets$_{\neq\emptyset}(\mathbb{R}^n)$, define

$$\mathbf{d}_\rho(A, B) := \max_{|z|\leq\rho} \left|d(z, A) - d(z, B)\right|.$$

where $d(z, H)$ for $z \in \mathbb{R}^n$ and $H \subseteq \mathbb{R}^n$ is defined by $\inf_{w\in H} |w - z|$. Furthermore, define

$$\mathbf{d}(A, B) := \int_0^\infty \mathbf{d}_\rho(A, B)e^{-\rho} d\rho,$$

which is called the *(integrated) set distance* between A and B [14].

We adopt this distance on graphs of hybrid arcs as follows. Given a hybrid arc $\varphi : \text{dom}\,\varphi \to \mathbb{R}^n$, the graph of x is defined by $\text{gph}\,\varphi := \{(t, j, x) : x = \varphi(s, k)\}$. Given $\rho \geq 0$, for a pair of hybrid arcs φ and ψ, define $\mathbf{d}_\rho(\varphi, \psi) := \mathbf{d}_\rho(\text{gph}\,\varphi, \text{gph}\,\psi)$ and $\mathbf{d}(\varphi, \psi) := \mathbf{d}(\text{gph}\,\varphi, \text{gph}\,\psi)$, which is called the *graphical distance* between hybrid arcs. Note that the same notion of graphical distance applies to both hybrid arcs and hybrid memory arcs.

We now focus on hybrid memory arcs in $\mathcal{M}$. As a consequence of Theorem 4.42 in [14] and the fact that the graph of a hybrid memory arc is a nonempty, closed subset of $\mathbb{R}^{n+2}$, we know that the space $(\mathcal{M}, \mathbf{d})$ is a separable metric space. However, $(\mathcal{M}, \mathbf{d})$ is not complete, since the limit of a sequence of graphically convergent hybrid memory arcs may not be a hybrid memory arc. The following subspace of $(\mathcal{M}, \mathbf{d})$ is often used where such compactness are needed. Given $b, \lambda \in \mathbb{R}_{\geq 0}$, define

$$\mathcal{M}_b := \left\{\varphi \in \mathcal{M} : \sup_{(s,k)\in\text{dom}\,\varphi} |\varphi(s, k)| \leq b\right\},$$

$$\mathcal{M}_{b,\lambda} := \left\{\varphi \in \mathcal{M}_b : \varphi \text{ is } \lambda\text{-Lipschitz}\right\},$$

where $\varphi \in$ is said to be *λ-Lipschitz* if $\left|\varphi(s', k) - \psi(s'', k)\right| \leq \lambda \left|s - s'\right|$ holds for all $(s, k), (s', k) \in \text{dom}\,\varphi$.

Proposition 1 [10] *The space* $(\mathscr{M}_{b,\lambda}, \mathbf{d})$ *is a separable, locally compact, and complete metric space.*

In the above proposition, b and λ are fixed constants. Motivated by the need to consider constrained but possibly unbounded solutions of hybrid systems, we also consider the following set. Let $\lambda : \mathbb{R}_{\geq 0} \to \mathbb{R}_{\geq 0}$ be a nondecreasing function and

$$\widehat{\mathscr{M}_\lambda} := \bigcup_{b \in \mathbb{R}_{\geq 0}} \mathscr{M}_{b,\lambda(b)}.$$

While $\widehat{\mathscr{M}_\lambda}$ is not closed under the graphical convergence topology, the set

$$\mathscr{M}_b \cap \widehat{\mathscr{M}_\lambda} = \mathscr{M}_{b,\lambda(b)}$$

is compact under the graphical convergence topology in $(\mathscr{M}, d)$ for each $b \geq 0$, according to Proposition 1.

2.3 Hybrid Basic Conditions

The regularity conditions on the data of a hybrid system with memory are formulated next, which are recalled from [10, 11].

Definition 6 (*Outer semicontinuous*) A set-valued functional $\mathscr{F} : \mathscr{M} \rightrightarrows \mathbb{R}^n$ is said to be *outer semicontinuous* at $\varphi \in \mathscr{M}$, if for all sequences of hybrid memory arcs in $\mathscr{M}_{b,\lambda}$ for some $b, \lambda \geq 0$ such that

$$\varphi_i \xrightarrow{\text{gph}} \varphi$$

and $y_i \to y$ with $y_i \in \mathscr{F}(\varphi_i)$, we have $y \in \mathscr{F}(\varphi)$.

Definition 7 (*Local boundedness*) A set-valued functional $\mathscr{F} : \mathscr{M} \rightrightarrows \mathbb{R}^n$ is said to be *locally bounded* at $\varphi \in \mathscr{M}$ if for each $b \geq 0$ there exists a neighborhood $\mathscr{U}_\varphi$ of φ (in graphical distance) such that the set $\mathscr{F}(\mathscr{U}_\varphi \cap \mathscr{M}_b) := \bigcup_{\psi \in \mathscr{U}_\varphi \cap \mathscr{M}_b} \mathscr{F}(\psi) \subseteq \mathbb{R}^n$ is bounded.

A stronger (but often easier to verify) condition for checking the local boundedness of $\mathscr{F}$ is that it maps every set $\mathscr{M}_b$, where $b \geq 0$, into a bounded set in $\mathbb{R}^n$.

In the above definitions, $\mathscr{F}$ is said to be outer semicontinuous (respectively, locally bounded) *relative* to a set $\mathscr{M}' \subseteq \mathscr{M}$ provided $\mathscr{F}|_{\mathscr{M}'}$ (defined by $\mathscr{F}|_{\mathscr{M}'}(\varphi) = \mathscr{F}(\varphi)$ if $\varphi \in \mathscr{M}'$ and $\mathscr{F}|_{\mathscr{M}'}(\varphi) = \emptyset$ elsewhere) is outer semicontinuous (respectively, locally bounded) at each $\varphi \in \mathscr{M}'$. Finally, the mapping $\mathscr{F}$ is said to be outer semicontinuous (respectively, locally bounded) if it is so relative to its domain.

The following are the basic conditions on the data of $\mathscr{H}_{\mathscr{M}} = (\mathscr{C}, \mathscr{F}, \mathscr{D}, \mathscr{G})$:

(A1) $\mathscr{C}$ and $\mathscr{D}$ are closed subsets of $\mathscr{M}$;
(A2) $\mathscr{F}$ is outer semicontinuous and locally bounded relative to the set $\mathscr{C}$ and $\mathscr{F}(\varphi)$ is nonempty and convex for each $\varphi \in \mathscr{C}$;
(A3) $\mathscr{G}$ is outer semicontinuous and locally bounded relative to $\mathscr{D}$, and $\mathscr{G}(\varphi)$ is nonempty for each $\varphi \in \mathscr{D}$.

The above definitions have been introduced and used in [10, 11] to establish basic existence and well-posedness for hybrid systems with memory.

2.4 Basic Existence and Continuation

This section recalls a result on the basic existence and continuation of solutions for hybrid systems with memory from [10, 11]. The result uses the notion of tangent cone, as in hybrid systems without memory [5]. The definition below is recalled from [10].

Definition 8 For any $\varphi \in \mathscr{K} \subseteq \mathscr{M}$, we define ${}^{\mathrm{T}}_{\mathscr{K}}(\varphi) \subseteq \mathbb{R}^n$ by $v \in {}^{\mathrm{T}}_{\mathscr{K}}(\varphi)$ if and only if, for any $\varepsilon > 0$, there exist $h \in (0, \varepsilon]$ and $x_h \in C([0, h], \mathbb{R}^n)$ such that
(1) $x_h(0) = \varphi(0, 0)$ and

$$\frac{x_h(h) - x_h(0)}{h} \in v + \varepsilon\mathbb{B};$$

(2) the hybrid memory arc ψ_{x_h} defined by

$$\psi_{x_h}(s, k) = \begin{cases} x_h(h + s), & \forall s \in [-h, 0],\ k = 0, \\ \varphi(h + s, k), & \forall (h + s, k) \in \operatorname{dom} \varphi, \end{cases} \tag{3}$$

lies in $\mathscr{K}$.

Theorem 1 [10, 11] *Let* $\mathscr{H}_{\mathscr{M}} = (\mathscr{C}, \mathscr{F}, \mathscr{D}, \mathscr{G})$ *satisfy the conditions* (A1)–(A3). *If, for every* $\xi \in \mathscr{C} \backslash \mathscr{D}$,

$$\mathscr{F}(\xi) \cap {}^{T}_{\mathscr{C}}(\xi) \neq \emptyset, \tag{4}$$

then there exists a nontrivial solution to $\mathscr{H}_{\mathscr{M}}$ *from every initial condition* $\varphi \in \mathscr{C} \cup \mathscr{D}$ *such that* $\varphi \in \mathscr{M}_{b,\lambda}$ *for some* $b, \lambda \in \mathbb{R}_{\geq 0}$. *Moreover, every such maximal solution* x *satisfies exactly one of the following conditions:*

(a) x *is complete, i.e.,* $\operatorname{dom}_{\geq 0}(x)$ *is unbounded;*
(b) $\operatorname{dom}_{\geq 0}(x)$ *is bounded, the interval* I_J *has nonempty interior, and*

$$\limsup_{t \to T^-} |x(t, J)| = \infty,$$

where $T = \sup_t \mathit{dom}\, x := \sup\{t : (t, j) \in \mathit{dom}\, x\}$ *and*

$$J = \sup_j \mathit{dom}\, x := \sup\{j : (t, j) \in \mathit{dom}\, x\};$$

(c) $\varphi(T, J) \notin \mathscr{C} \cup \mathscr{D}$*, where* (T, J) *is as defined in (b).*

Furthermore, if $\mathscr{G}(\varphi) \subseteq \mathscr{C} \cup \mathscr{D}$ *for all* $\varphi \in \mathscr{D}$*, then only (a) or (b) above occurs.*

3 Hybrid Systems with Finite Memory

The formulation of hybrid systems with memory presented in Sect. 2 does not bear any information on the size of the memory. The focus of this chapter is on hybrid systems with *finite* memory. Given $\Delta \geq 0$, let $\mathscr{M}^{\Delta}$ denote the collection of hybrid memory arcs φ satisfying the following two conditions: (1) $s + k \geq -\Delta - 1$ for all $(s, k) \in \mathrm{dom}\, \varphi$; and (2) there exists $(s', k') \in \mathrm{dom}\, \varphi$ such that $s' + k' \leq -\Delta$. We can similarly define

$$\mathscr{M}_b^{\Delta} := \mathscr{M}^{\Delta} \cap \mathscr{M}_b,\ \ \mathscr{M}_{b,\lambda}^{\Delta} := \mathscr{M}^{\Delta} \cap \mathscr{M}_{b,\lambda}, \text{ and}$$
$$\widehat{\mathscr{M}_{\lambda}^{\Delta}} := \mathscr{M}^{\Delta} \cap \widehat{\mathscr{M}_{\lambda}}.$$

Definition 9 A hybrid system $\mathscr{H}_{\mathscr{M}} = (\mathscr{C}, \mathscr{F}, \mathscr{D}, \mathscr{G})$ is said to be with *finite memory* of size $\Delta \geq 0$ if

$$\mathscr{C} \cup \mathscr{D} \cup \mathrm{dom}\, \mathscr{F} \cup \mathrm{dom}\, \mathscr{G} \subseteq \mathscr{M}^{\Delta}.$$

The system is denoted by $\mathscr{H}_{\mathscr{M}}^{\Delta} = (\mathscr{C}, \mathscr{F}, \mathscr{D}, \mathscr{G})$.

The solution operator $\mathscr{A}_{[t,j]}$ also has to be adapted so that it maps hybrid arcs into hybrid memory arcs of finite memory size.

Definition 10 [11] Given a hybrid arc and any $(t, j) \in \mathrm{dom}\, x$, we define an operator $\mathscr{A}_{[t,j]}^{\Delta}$ that maps x to $\mathscr{A}_{[t,j]}^{\Delta} x \in \mathscr{M}^{\Delta}$ given by

$$\mathscr{A}_{[t,j]}^{\Delta} x(s, k) = x(t + s, j + k),$$

for all $(s, k) \in \mathrm{dom}\, (\mathscr{A}_{[t,j]}^{\Delta} x)$, where

$$\mathrm{dom}\, (\mathscr{A}_{[t,j]}^{\Delta} x) := \Big\{(s, k) \in \mathbb{R}_{\leq 0} \times \mathbb{Z}_{\leq 0} : (t + s, j + k) \in \mathrm{dom}\, x,\ \ s + k \geq -\Delta_{\mathrm{inf}}\Big\},$$

where

$$\Delta_{\mathrm{inf}} := \inf\Big\{\delta \geq \Delta : \exists (t + s, j + k) \in \mathrm{dom}\, x \text{ s.t. } s + k = -\delta\Big\}.$$

4 Lyapunov Conditions for $\mathscr{K}\mathscr{L}$ Pre-asymptotic Stability

In this section, we discuss Lyapunov sufficient conditions for the asymptotic stability analysis of hybrid systems with memory. These results are adopted from [10, 12].

Definition 11 [10] Let $\mathscr{H}_{\mathscr{M}}^{\Delta}$ be a hybrid system in $\mathscr{M}^{\Delta}$ and $\mathscr{W} \subseteq \mathbb{R}^n$ be a closed set. The set $\mathscr{W}$ is said to be *$\mathscr{K}\mathscr{L}$ pre-asymptotically stable* for $\mathscr{H}_{\mathscr{M}}$ if there exists a $\mathscr{K}\mathscr{L}$ function β such that any solution x to $\mathscr{H}_{\mathscr{M}}^{\Delta}$ satisfies

$$|x(t,j)|_{\mathscr{W}} \le \beta(\|\mathscr{A}_{[0,0]}x\|_{\mathscr{W}}^{\Delta}, t+j), \tag{5}$$

where $\|\varphi\|_{\mathscr{W}}^{\Delta} = \sup_{\substack{(t,j)\in\operatorname{dom}\varphi \\ -\Delta\le t+j\le 0}} \inf_{y\in\mathscr{W}} |y-\varphi(t,j)|$ for $\varphi \in \mathscr{M}^{\Delta}$ and

$$|x|_{\mathscr{W}} := \inf_{y\in\mathscr{W}} |y-x|$$

for $x \in \mathbb{R}^n$.

4.1 Lyapunov–Krasovskii Functionals

As in functional differential equations, Lyapunov functionals can be used to formulate sufficient conditions for analyzing stability of hybrid systems with memory. The following result provides a set of such conditions, which resemble that for hybrid systems without memory.

Given a functional $V : \mathscr{M}^{\Delta} \to \mathbb{R}_{\ge 0}$, the upper right-hand derivative of V at $\varphi \in \mathscr{M}^{\Delta}$ along the solutions of $\mathscr{H}_{\mathscr{M}}^{\Delta}$ can be defined as follows:

$$D^+V(\varphi) := \sup_{\substack{x\in\mathscr{S}_{\mathscr{H}_{\mathscr{M}}^{\mathscr{D}}} \\ \mathscr{A}_{[0,0]}^{\Delta}x=\varphi}} \limsup_{h\to 0} \frac{V(\mathscr{A}_{[h,0]}x)-V(\varphi)}{h}.$$

Theorem 2 [12] *Let $\mathscr{H}_{\mathscr{M}}^{\Delta}$ be a hybrid system in $\mathscr{M}^{\Delta}$ and let $\mathscr{W} \subseteq \mathbb{R}^n$ be a closed set. If there exists a functional $V : \mathscr{M}^{\Delta} \to \mathbb{R}_{\ge 0}$ and $\mathscr{K}_{\infty}$ functions α_i $(i = 1, 2, 3)$ such that the following hold:*

(i) $\alpha_1(|\varphi(0,0)|_{\mathscr{W}}) \le V(\varphi) \le \alpha_2(\|\varphi\|_{W}^{\Delta})$ *for all* $\varphi \in \mathscr{C} \cup \mathscr{D} \cup \mathscr{G}(\mathscr{D})$*;*
(ii) $D^+V(\varphi) \le -\alpha_3(|\varphi(0,0)|_{\mathscr{W}})$ *for all* $\varphi \in \mathscr{C}$*;*
(iii) $V(\varphi_g^+) - V(\varphi) \le -\alpha_3(|\varphi(0,0)|_{\mathscr{W}})$ *for all* $\varphi \in \mathscr{D}$ *and* $g \in \mathscr{G}(\varphi)$*,*

where φ_g^+ is defined by

$$gph\,\varphi_v^+ = (gph\,\varphi + (0,-1,0)) \cup \{(0,0,g)\}$$

and D^+V is the upper right-hand derivative of V (defined below) along the solutions of $\mathcal{H}^{\Delta}_{\mathcal{M}}$, then $\mathcal{W}$ is $\mathcal{KL}$ pre-asymptotically stable for $\mathcal{H}_{\mathcal{M}}$.

The proof of Theorem 2, given in [12], is similar to that of Theorem 3.18 in [5] and draws inspiration from that of [7, Theorem 2.1, Chap. 5]. The main difference lies in that the solutions now evolve in $\mathcal{M}^{\Delta}$. We also note that a special version of Lyapunov–Krasovskii theorem for hybrid systems with delays was proved in [1, Proposition1] in the context of reset control systems, where the reset map only acts on the controller state and the emphasis on the delay is in continuous time. Theorem 2 above covers the general case of hybrid systems with delays, where the delay can be in both continuous and discrete time.

4.2 Lyapunov–Razumikhin Functions

It is also possible to formulate conditions for $\mathcal{KL}$ pre-asymptotical stability using Lyapunov functions defined on $\mathbb{R}^n$, as shown in the following result:

Theorem 3 [12] *Let $\mathcal{H}^{\Delta}_{\mathcal{M}} = (\mathcal{C}, \mathcal{F}, \mathcal{D}, \mathcal{G})$ be a hybrid system with finite memory (i.e., $\Delta < \infty$) and let $\mathcal{W} \subseteq \mathbb{R}^n$ be a closed set. If there exists a continuously differentiable function $V : \mathbb{R}^n \to \mathbb{R}_{\geq 0}$, $\mathcal{K}_\infty$ functions α_i $(i = 1, 2, 3)$, and continuous functions $p : \mathbb{R}_{\geq 0} \to \mathbb{R}_{\geq 0}$ and $\rho : \mathbb{R}_{\geq 0} \to \mathbb{R}_{\geq 0}$ with $p(s) > s$ and $\rho(s) < s$ for all $s > 0$ such that the following hold:*

(i) $\alpha_1(|\varphi(0, 0)|_{\mathcal{W}}) \leq V(\varphi(0, 0)) \leq \alpha_2(|\varphi(0, 0)|_{\mathcal{W}})$ *for all* $\varphi \in \mathcal{C} \cup \mathcal{D} \cup \mathcal{G}(\mathcal{D})$*;*
(ii) $\nabla V(\varphi(0, 0)) \cdot f \leq -\alpha_3(V(\varphi(0, 0)))$ *for all* $\varphi \in \mathcal{C}$ *such that* $p(V(\varphi(0, 0))) \geq \overline{V}(\varphi)$ *and all* $f \in \mathcal{F}(\varphi)$*;*
(iii) $V(g) \leq \rho(\overline{V}(\varphi))$ *for all* $\varphi \in \mathcal{D}$ *and all* $g \in \mathcal{G}(\varphi)$*,*

where $\overline{V}(\varphi) = \max_{-\Delta-1 \leq s+k \leq 0} V(\varphi(s, k))$, then $\mathcal{W}$ is $\mathcal{KL}$ pre-asymptotically stable for $\mathcal{H}^{\Delta}_{\mathcal{M}}$.

The following result, which first appeared in [10], can be seen as a corollary of the above theorem:

Corollary 1 [10] *Let $\mathcal{H}_{\mathcal{M}} = (\mathcal{C}, \mathcal{F}, \mathcal{D}, \mathcal{G})$ be a hybrid system with memory and let $\mathcal{W} \subseteq \mathbb{R}^n$ be a closed set. If there exist a continuously differentiable function $V : \mathbb{R}^n \to \mathbb{R}_{\geq 0}$, $\mathcal{K}_\infty$ functions α_i $(i = 1, 2)$, and positive constants $\mu > q$ and $\rho < 1$ such that*

(i) $\alpha_1(|\varphi(0, 0)|_{\mathcal{W}}) \leq V(\varphi(0, 0)) \leq \alpha_2(|\varphi(0, 0)|_{W})$ *for all* $\varphi \in \mathcal{C} \cup \mathcal{D} \cup \mathcal{G}(\mathcal{D})$*;*
(ii) $\nabla V(\varphi(0, 0)) \cdot f \leq -\mu V(\varphi(0, 0)) + q\overline{V}_{[0,0]}(\varphi)$ *for all* $\varphi \in \mathcal{C}$ *and* $f \in \mathcal{F}(\varphi)$*;*
(iii) $V(g) \leq \rho \overline{V}_{[0,0]}(\varphi)$ *for all* $\varphi \in \mathcal{D}$ *and* $g \in \mathcal{G}(\varphi)$*,*

all hold, where

$$\overline{V}_{[0,0]}(\varphi) = \max_{-\Delta \leq s+k \leq 0} V(\varphi(s, k)),$$

then $\mathcal{W}$ is $\mathcal{KL}$ pre-asymptotically stable for $\mathcal{H}_{\mathcal{M}}$.

The conditions above in Corollary 1 are more in the spirit of Halanay-type inequalities [6, p. 378] than the general Razumikhin-type conditions used in Theorem 3.

5 Well-Posed Hybrid Systems with Memory and Robustness of $\mathscr{K}\mathscr{L}$ Stability

In order to discuss robustness stability, we define perturbations of a hybrid system with memory as follows. It closely follows the notion of outer-perturbation of a hybrid system without memory [5], but formulated in a more restricted sense by making the following assumption on $\mathscr{H}^{\Delta}_{\mathscr{M}}$: there exists a nondecreasing function $\lambda : \mathbb{R}_{\geq 0} \to \mathbb{R}_{\geq 0}$ such that

$$\mathscr{C} \cup \mathscr{D} \subseteq \widehat{\mathscr{M}^{\Delta}_{\lambda}} := \bigcup_{b \in \mathbb{R}_{\geq 0}} \mathscr{M}^{\Delta}_{b,\lambda(b)},$$

and only defining perturbations within this set $\widehat{\mathscr{M}^{\Delta}_{\lambda}}$. The following definition is recalled from [11].

Definition 12 [11] Given a hybrid system with memory $\mathscr{H}^{\Delta}_{\mathscr{M}} = (\mathscr{C}, \mathscr{F}, \mathscr{D}, \mathscr{G})$ and a functional $\rho : \mathscr{M}^{\Delta} \to \mathbb{R}_{\geq 0}$, the ρ-perturbation of $\mathscr{H}^{\Delta}_{\mathscr{M}}$, denoted by $(\mathscr{H}^{\Delta}_{\mathscr{M}})_{\rho}$, is the hybrid system with data:

- $\mathscr{C}_{\rho} = \left\{\varphi \in \widehat{\mathscr{M}^{\Delta}_{\lambda}} : \widehat{\mathbb{B}}(\varphi, \rho(\varphi)) \cap \mathscr{C} \neq \emptyset\right\}$;
- $\mathscr{F}_{\rho}(\varphi) = \overline{\mathrm{con}}\mathscr{F}(\widehat{\mathbb{B}}(\varphi, \rho(\varphi)) \cap \mathscr{C}) + \rho(\varphi)\mathbb{B}$;
- $\mathscr{D}_{\rho} = \left\{\varphi \in \widehat{\mathscr{M}^{\Delta}_{\lambda}} : \widehat{\mathbb{B}}(\varphi, \rho(\varphi)) \cap \mathscr{D} \neq \emptyset\right\}$;
- $\mathscr{G}_{\rho}(\varphi) = \left\{y \in \mathbb{R}^n : y \in v + \rho(\varphi^+_v)\mathbb{B},\ v \in \mathscr{G}(\widehat{\mathbb{B}}(\varphi, \rho(\varphi)) \cap \mathscr{D})\right\}$, where φ^+_v is defined by $\mathrm{gph}\, \varphi^+_v = (\mathrm{gph}\, \varphi + (0, -1, 0)) \cup \{(0, 0, v)\}$.

where $\overline{\mathrm{con}}(E)$ is the closed convex hull of a set $E \subseteq \mathbb{R}^n$ and

$$\widehat{\mathbb{B}}(\varphi, \rho(\varphi)) := \left\{\psi \in \widehat{\mathscr{M}^{\Delta}_{\lambda}} : \psi \text{ and } \varphi \text{ are } \rho(\varphi)\text{-close}\right\}$$

and $\mathbb{B}$ is the closed unit ball in $\mathbb{R}^n$.

We can verify that continuous perturbations of a hybrid system still satisfy the regularity assumptions.

Proposition 2 [11] *Let ρ be a continuous functional $\rho : \mathscr{M}^{\Delta} \to \mathbb{R}_{\geq 0}$ and $\mathscr{H}^{\Delta}_{\mathscr{M}} = (\mathscr{C}, \mathscr{F}, \mathscr{D}, \mathscr{G})$ be a hybrid system satisfying assumptions* (A1)–(A3). *Then $(\mathscr{H}^{\Delta}_{\mathscr{M}})_{\rho} = (\mathscr{C}_{\rho}, \mathscr{F}_{\rho}, \mathscr{D}_{\rho}, \mathscr{G}_{\rho})$ satisfies assumptions* (A1)–(A3).

Definition 13 (*Well-posedness of hybrid systems with finite memory* [11]) A hybrid system $\mathscr{H}^{\Delta}_{\mathscr{M}}$ is said to be *well-posed* if the following properties hold: for any given continuous function $\rho : \mathscr{M}^{\Delta} \to \mathbb{R}_{\geq 0}$, a decreasing sequence $\{\delta_i\}_{i=1}^{\infty}$ in $(0, 1)$

with $\delta_i \to 0$ as $i \to \infty$, and for every graphically convergent sequence $\{x_i\}_{i=1}^{\infty}$ of solutions to $(\mathcal{H}_{\mathcal{M}}^{\Delta})_{\delta_i \rho}$ with $\mathcal{A}_{[0,0]} x_i \in \mathcal{M}_b^{\Delta}$ for some $b \geq 0$ and

$$\mathcal{A}_{[0,0]} x_i \xrightarrow{\text{gph}} \varphi \in \mathcal{M}^{\Delta},$$

we have

(a) if the sequence $\{x_i\}_{i=1}^{\infty}$ is locally eventually bounded (that is, for any $m > 0$, there exists $N > 0$ and $k > 0$ such that, for all $i > N$ and all $(t, j) \in \operatorname{dom} x_i$ with $t + j < m$, $|x_i(t, j)| < k$), then its graphical limit x is a solution to $\mathcal{H}_{\mathcal{M}}^{\Delta}$ with $\mathcal{A}_{[0,0]} x = \varphi$ and $\operatorname{dom} x = \lim_{i \to \infty} \operatorname{dom} x_i$;
(b) if the sequence $\{x_i\}_{i=1}^{\infty}$ is not locally eventually bounded, then there exist some T and J in $(0, \infty)$ and a sequence $\{t_i\}_{i=1}^{\infty}$ with $(t_i, J) \in \operatorname{dom} x_i$ for sufficiently large i such that $\lim_{i \to \infty} t_i = T$, $\lim_{i \to \infty} |x_i(t_i, J)| = \infty$, and the limit $x = \lim \operatorname{gph}_{i \to \infty} x_i$ restricted to the domain $\{(t, j) \in \operatorname{dom} x : t + j < T + J\}$ is a maximal solution to $\mathcal{H}_{\mathcal{M}}^{\Delta}$ with $\mathcal{A}_{[0,0]} x = \varphi$ and $\lim_{t \to T} |x(t, J)| = \infty$.

Theorem 4 [11] *If a hybrid system with finite memory* $\mathcal{H}_{\mathcal{M}}^{\Delta} = (\mathcal{C}, \mathcal{F}, \mathcal{D}, \mathcal{G})$ *satisfies Assumptions* (A1)–(A3), *then it is well-posed.*

One of the main consequences of well-posedness is that $\mathcal{KL}$ pre-asymptotic stability of hybrid systems with memory is robust in the following sense:

Definition 14 (*Robust* $\mathcal{KL}$ *pre-asymptotic stability* [11]) Let $\mathcal{W} \subseteq \mathbb{R}^n$ be a compact set and $\mathcal{H}_{\mathcal{M}}^{\Delta}$ be a hybrid system with memory.

(a) The set $\mathcal{W}$ is *robustly* $\mathcal{KL}$ *pre-asymptotically stable* for $\mathcal{H}_{\mathcal{M}}^{\Delta}$ if there exists a continuous function $\rho : \mathcal{M}^{\Delta} \to \mathbb{R}_{\geq 0}$ that is positive on

$$\mathcal{M}^{\Delta} \backslash \mathcal{W} := \left\{ \varphi \in \mathcal{M}^{\Delta} : \exists (s, k) \in \operatorname{dom} \varphi \text{ s.t. } \varphi(s, k) \notin \mathcal{W} \right\}$$

such that

$$\omega_2(x(t, j)) \leq \beta(\omega_1(\mathcal{A}_{[0,0]} x), t + j), \quad \forall (t, j) \in \operatorname{dom} x,$$

where

$$w_2(z) = |z|_{\mathcal{W}} := \inf_{y \in \mathcal{W}} |y - z|$$

for $z \in \mathbb{R}^n$ and

$$w_1(\varphi) = \|\varphi\|_{\mathcal{W}} = \sup_{(s,k) \in \operatorname{dom} \varphi} \inf_{y \in \mathcal{W}} |y - \varphi(s, k)|$$

for all $\varphi \in \mathcal{M}^{\Delta}$, holds for all solutions x of $(\mathcal{H}_{\mathcal{M}}^{\Delta})_{\rho}$.
(b) The set $\mathcal{W}$ is *semi-globally practically robustly* $\mathcal{KL}$ *pre-asymptotically stable* for $\mathcal{H}_{\mathcal{M}}^{\Delta}$ if there exists $\beta \in \mathcal{KL}$ such that

$$\omega_2(x(t, j)) \leq \beta(\omega_1(\mathscr{A}_{[0,0]}x), t + j), \quad \forall (t, j) \in \operatorname{dom} x,$$

holds for all solutions x of $\mathscr{H}^{\Delta}_{\mathscr{M}}$, and such that for every continuous function $\rho : \mathscr{M}^{\Delta} \to \mathbb{R}_{\geq 0}$ that is positive on $\mathscr{M}^{\Delta} \backslash \mathscr{W}$, the following holds: for every $\mathscr{M}^{\Delta}_{b}$ with $b \geq 0$ and every $\varepsilon > 0$, there exists $\delta \in (0, 1)$ such that every solution x to $(\mathscr{H}^{\Delta}_{\mathscr{M}})_{\delta\rho}$ satisfies

$$\omega_2(x(t, j)) \leq \beta(\omega_1(\mathscr{A}_{[0,0]}x), t + j) + \varepsilon, \quad \forall (t, j) \in \operatorname{dom} x.$$

Theorem 5 [11](*Robustness of pre-asymptotic stability*) *Let $\mathscr{H}^{\Delta}_{\mathscr{M}}$ be a well-posed hybrid system with memory and $\mathscr{W} \subseteq \mathbb{R}^n$ be a compact set. If $\mathscr{W}$ is $\mathscr{K}\mathscr{L}$ pre-asymptotically stable, then it is also robustly $\mathscr{K}\mathscr{L}$ pre-asymptotically stable.*

6 Conclusions

We have summarized a framework for studying hybrid dynamical systems with memory using generalized concepts of solutions. Such solutions are defined in the phase space of hybrid memory arcs equipped with graphical distance topology. Regularity conditions on the hybrid data are formulated in terms of the graphical topology, which allows us to establish basic existence and well-posedness of hybrid systems with memory, using ideas from functional viability theory. We then discussed Lyapunov-based sufficient conditions for the stability of hybrid systems with delays via generalized solutions. The main motivation for using generalized solutions for studying hybrid systems with memory is to establish well-posedness and its consequence in robust stability. We recalled some recent results in this direction, which show that pre-asymptotic stability of well-posed hybrid systems with memory is robust and, therefore, provide justification for using this generalized framework to study hybrid systems with memory.

References

1. Banos, A., Rubio, F., Tarbouriech, S., Zaccarian, L.: Delay-independent stability via reset loops. In: Seuret, A., Ozbay, H., Bonnet, C., Mounier, H. (eds.) Low-Complexity Controllers for Time-Delay Systems, pp. 111–125. Springer, New York (2014)
2. Cloosterman, M., van de Wouw, N., Heemels, W., Nijmeijer, H.: Stability of networked control systems with uncertain time-varying delays. IEEE Trans. Autom. Control **54**(7), 1575–1580 (2009)
3. Goebel, R., Teel, A.: Solutions to hybrid inclusions via set and graphical convergence with stability theory applications. Automatica **42**(4), 573–587 (2006)
4. Goebel, R., Hespanha, J., Teel, A., Cai, C., Sanfelice, R.: Hybrid systems: generalized solutions and robust stability. In: Proceedings of the 6th IFAC Symposium on Nonlinear Control Systems, pp. 1–12 (2004)
5. Goebel, R., Sanfelice, R., Teel, A.: Hybrid Dynamical Systems: Modeling, Stability, and Robustness. Princeton University Press, Princeton (2012)

6. Halanay, A.: Differential Equations: Stability, Oscillations. Time Lags. Academic Press, New York (1966)
7. Hale, J., Verduyn Lunel, S.: Introduction to Functional Differential Equations. Springer, New York (1993)
8. Liu, J., Liu, X., Xie, W.: Input-to-state stability of impulsive and switching hybrid systems with time-delay. Automatica **47**(5), 899–908 (2011)
9. Liu, J., Teel, A.: Generalized solutions to hybrid systems with delays. In: Proceedings of the IEEE Conference on Decision and Control, pp. 6169–6174 (2012)
10. Liu, J., Teel, A.: Hybrid systems with memory: modelling and stability analysis via generalized solutions. In: Proceedings of the 19th IFAC World Congress, pp. 6019–6024 (2014)
11. Liu, J., Teel, A.: Hybrid systems with memory: existence and well-posedness of generalized solutions. SIAM J. Control Optim. (2014) (submitted)
12. Liu, J., Teel, A.: Lyapunov-based sufficient conditions for stability of hybrid systems with memory. IEEE Trans. Autom. Control (2014) (submitted)
13. Liu, X., Shen, J.: Stability theory of hybrid dynamical systems with time delay. IEEE Trans. Autom. Control **51**(4), 620–625 (2006)
14. Rockafellar, R., Wets, J.: Variational Analysis. Series Grundlehren der mathematischen Wissenschaften, vol. 317. Springer, New York (1998)
15. Sanfelice, R., Goebel, R., Teel, A.: Generalized solutions to hybrid dynamical systems. ESAIM Control, Optimisation, and Calculus of Variations **14**(4), 699–724 (2008)
16. Sanfelice, R., Goebel, R., Teel, A.: Invariance principles for hybrid systems with connections to detectability and asymptotic stability. IEEE Transactions on Automatic ConTrol **52**(12), 2282–2297 (2007)
17. Sipahi, R., Niculescu, S.-I., Abdallah, C., Michiels, W., Gu, K.: Stability and stabilization of systems with time delay. IEEE Control Systems Magazine **31**(1), 38–65 (2011)
18. Yan, P., Ozbay, H.: Stability analysis of switched time delay systems. SIAM Journal on Control and Optimization **47**(2), 936–949 (2008)
19. Yuan, R., Jing, Z., Chen, L.: Uniform asymptotic stability of hybrid dynamical systems with delay. IEEE Transactions on Automatic Control **48**(2), 344–348 (2003)

Control-Oriented Modeling of Fluid Networks: A Time-Delay Approach

David Fernando Novella Rodriguez, Emmanuel Witrant and Olivier Sename

Abstract Fluid networks are characterized by complex interconnected flows, involving high order nonlinear dynamics and transport phenomena. Classical lumped models typically capture the interconnections and nonlinear effects but ignore the transport phenomena, which may strongly affect the transient response. To control such flows with regulators of reduced complexity, we improve a classical lumped model (obtained by combining Kirchhoff's laws and graph theory) by introducing the effect of advection as a time delay. The model is based on the isothermal Euler equations to describe the dynamics of the fluid through the pipe. The resulting hyperbolic system of partial differential equations (PDEs) is diagonalized using Riemann invariants to find a solution in terms of delayed equations, obtained analytically using the method of the characteristics. Conservation principles are applied at the nodes of the network to describe the dynamics as a set of (possibly non linear) delay differential equations. Both linearized and nonlinear Euler equations are considered.

1 Introduction

Modeling and control of fluid flow networks has been a challenging topic during the last decades. This research is motivated by engineering applications such as mine ventilation systems [16, 27], gas pipelines [2, 13], water channels [7, 20], traffic flow dynamics [26], cryogenic distribution lines [3], etc.

For example, considering mining ventilation engineering, the first approaches used a steady-state description of the pipe network and models were built using the Hardy Cross method [8]. With this method, the airflow in mine ventilation circuits is determined algebraically by combining graph theory and classical Kirchhoff's

D.F. Novella Rodriguez (✉)
IPN, ESIME Culhuacan, Santa Ana 1000, C.P. 44300 Mexico City, Mexico
e-mail: dnovellar@gmail.com

E. Witrant · O. Sename
Université de Grenoble, GIPSA-lab, Grenoble, France
e-mail: emmanuel.witrant@ujf-grenoble.fr

O. Sename
e-mail: olivier.sename@gipsa-lab.grenoble-inp.fr

I. Karafyllis et al. (eds.), *Recent Results on Nonlinear Delay Control Systems*,
Advances in Delays and Dynamics 4, DOI 10.1007/978-3-319-18072-4_14

laws to model the interconnection nodes [15]. In more recent approaches, a lumped parameter model for mine ventilation networks has been represented in terms of nonlinear ordinary differential equations [16, 21], with an extension to periodically forced networks in [18]. The model proposed in these works is based on Kirchhoff's voltage and current laws, combined with the fluid dynamical equations of individual branches. The branches are modeled by considering the incompressible Navier-Stokes equations as an electric equivalent RL circuit model with a nonlinear resistance. More precisely, the pressure drop over a branch is approximated to be proportional to the square of the air flow rate and to the air flow acceleration.

On the other hand, partial differential equations are often used to model fluid flows as hyperbolic conservation laws. For example, the Saint-Vennant equations are used to approximate 2-D shallow water phenomena with a 1-D PDE model [10]. Such models have been extensively used to control open channel networks [5, 7, 9, 19, 20]. Another example is provided in mine ventilation networks [27], where Euler equations are used to describe the gas flow dynamics in pipelines [2, 11, 13, 14]. In a different field, a first order PDE has been proposed to describe the traffic flow density on an homogeneous road, obtaining the Lighthill Whitham Richards model [26]. However, such detailed dynamic models are typically complex and often incompatible with real-time control objectives. To reduce the computational load, a 0-D approximation of the 1-D transport (with advection and sink) as a time-delay system has been proposed in [28, 29] and shown to be efficient as a reference model for feedback control of the large advective flows appearing in the mining ventilation problem. A similar approach was used in [4] to model the temperature in an SI engine exhaust catalyst. Nevertheless, such approximations did not take into account the occurrence of interconnected flows or the simultaneous transport of multiple variables.

The aim of this work is to present a time-delay model for fluid flow networks, leading to a classical state-space representation with delays to take into account the transport phenomena in the pipes of the network. Describing the flow in the branches of the network as isothermal Euler equations, we consider the transport of both density and momentum. The hyperbolic characteristics of the system of PDEs is used to find a solution described by delayed equations. Applying conservation laws in the nodes of the network, we finally obtain a delay differential equation describing the fluid dynamics of the complete system.

This chapter is organized as follows. Section 2 presents the physical equations that model the flow inside the pipelines. A time-delay model, found by the method of characteristics, describes the isothermal Euler equations in terms of delayed equations in Sect. 3. Both linear and nonlinear approximations are taken into account. The conservation principles are introduced at the nodes in Sect. 4, providing a state-space description in terms of delay differential equations from which we obtain the dynamics of the network. Numerical simulations of the time delay system are presented in Sect. 5.

2 Isothermal Euler Equations

A classical model for gas flow in pipe networks is provided by isothermal Euler equations [13, 14]. For such equations the temperature is constant and the pressure is given by the following state equation (e.g. for perfect gases):

$$P = \frac{ZRT}{M_g}\rho, \tag{1}$$

where Z is the natural gas compressibility factor, R the universal gas constant, T the absolute gas temperature, M_g the gas molecular weight, and $P(x,t)$ and $\rho(x,t)$ are the pressure and density of the gas, respectively. We consider a mono-dimensional description of the flow in terms of the space variable x. To simplify the notation, we define the speed of sound constant $a^2 = ZRT/Mg$ and the isothermal Euler equations are:

$$\frac{\partial \rho}{\partial t} + \frac{\partial q}{\partial x} = 0, \tag{2}$$

$$\frac{\partial q}{\partial t} + \frac{\partial}{\partial x}\left(\frac{q^2}{\rho} + a^2\rho\right) = -f_g \frac{q \mid q \mid}{2D\rho}, \tag{3}$$

where $q(x,t)$ is the momentum, f_g is the friction factor, and D is the diameter of the pipe. The first equation states the conservation of mass and the second equation is the momentum equation.

The flux of the Euler equations is thus defined as [25]:

$$F = \begin{pmatrix} q \\ \frac{q^2}{\rho} + a^2\rho \end{pmatrix},$$

and its Jacobian is

$$A(\rho, q) = \begin{pmatrix} 0 & 1 \\ a^2 - \frac{q^2}{\rho^2} & 2\frac{q}{\rho}\cdot \end{pmatrix}. \tag{4}$$

The eigenvalues of the Jacobian matrix $A(\rho, q)$, namely the characteristic velocities, are

$$\lambda_{1,2} = \frac{q}{\rho} \pm a. \tag{5}$$

The system of isothermal Euler equations (2) and (3) can be diagonalized using the Riemann invariants, some quantities that have the interesting property of remaining constant along special trajectories called the characteristic curves. This invariance property is crucial in the control design. The Riemann invariants are defined by the change of coordinates [14]:

$$\xi_{1,2}(\rho, q) = -\frac{q}{\rho} \mp a\ln(\rho), \tag{6}$$

and we assume that the characteristic velocities (5) satisfy $\lambda_2 < 0 < \lambda_1$ (subsonic case), which characterizes the system as a strictly hyperbolic PDE. We can also express the physical variables ρ and q in terms of the Riemann coordinates as follows:

$$\rho = \exp\left(\frac{\xi_2 - \xi_1}{2a}\right), \tag{7}$$

$$q = -\frac{\xi_1 + \xi_2}{2}\exp\left(\frac{\xi_2 - \xi_1}{2a}\right). \tag{8}$$

With the new coordinates $(\xi_1,\ \xi_2)$, the system (2)–(3) can be written in the following diagonal form:

$$\frac{\partial \Xi(x,t)}{\partial t} + \Lambda(\Xi)\frac{\partial \Xi(x,t)}{\partial x} = S(\Xi), \tag{9}$$

where $\Xi(x,t) \doteq [\xi_1\ \xi_2]^T$, and

$$\Lambda(\Xi) = \begin{bmatrix} -\frac{\xi_1+\xi_2}{2} + a & 0 \\ 0 & -\frac{\xi_1+\xi_2}{2} - a \end{bmatrix}. \tag{10}$$

The source term is then defined by:

$$S(\Xi) = -\frac{f_g}{8D}(\xi_1 + \xi_2)|\xi_1 + \xi_2|\begin{pmatrix}1\\1\end{pmatrix}, \tag{11}$$

The initial and boundary conditions are given by

$$\xi_1(x,0) = \varphi_1(x), \tag{12}$$

$$\xi_2(x,0) = \varphi_2(x), \tag{13}$$

$$\xi_1(0,t) = u_1(t), \tag{14}$$

$$\xi_2(L,t) = u_2(t). \tag{15}$$

3 Time Delay Model

There is a connection between the boundary problems for one-dimensional hyperbolic partial differential equations, namely with a single space coordinate accounting for wave propagation, and functional equations. These functional equations may be defined as difference, delay-differential or even integral or integro-differential

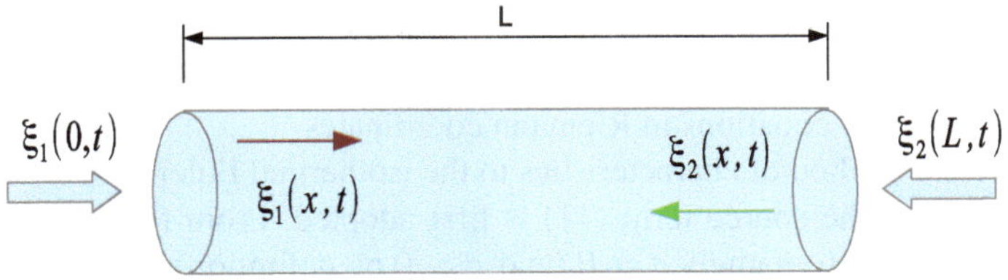

Fig. 1 Wave propagation

equations [22, 23]. Conservation laws are typically described by nonlinear hyperbolic PDE belonging to the "lossless" (or conservative) class.

This connection between two different mathematical objects, the hyperbolic PDEs and the functional equations, has been considered in many studies. For instance, in [6] delayed differential equations are derived to describe the propagation phenomena in power lines. On the other hand, in [24] neutral differential models are developed for different hyperbolic conservation laws (namely, circulating fuel in nuclear reactors, control of an overhead crane with flexible cable, etc.). From a different perspective, [29] proposes a distributed time-delay system to describe large convective flows modeled by linear time-varying partial differential equations (e.g. Euler or Navier-Stokes equations). In the recent work [17] the equivalence between systems described by a single first-order hyperbolic partial differential equation and systems described by integral delay equations is stated.

Consider the flow propagation in a pipe described in Fig. 1, with boundary conditions $\xi_1(0, t)$ and $\xi_2(L, t)$. In this section, we develop a model based on the nonlinear hyperbolic system described by isothermal Euler equations expressed in terms of the Riemann invariants:

$$\frac{\partial}{\partial t}\begin{bmatrix}\xi_1\\ \xi_2\end{bmatrix}+\begin{bmatrix}\lambda_1 & 0\\ 0 & \lambda_2\end{bmatrix}\frac{\partial}{\partial x}\begin{bmatrix}\xi_1\\ \xi_2\end{bmatrix}=\begin{bmatrix}-\alpha(\xi_1+\bar{\xi}_2)^2\\ -\alpha(\bar{\xi}_1+\xi_2)^2\end{bmatrix}, \tag{16}$$

where λ_1 and λ_2 are the characteristic velocities of the hyperbolic system. We averaged the quantities ξ_1 and ξ_2 as $\bar{\xi}_1$ and $\bar{\xi}_2$ (i.e., average equilibrium values in the pipe volume), respectively, to avoid the internal coupling between the two waves and allow finding a solution in terms of a time-delay equation. Note that this approximation is not strictly necessary, as an equivalent functional equation with a delayed kernel can be derived for the coupled case, but we adopt it here to simplify the derivations and focus on the major transport effects. We first consider the linearized case and then investigate the impact of the squared term.

3.1 Linear Approximation

Our approach extends the results in [29] to the multiple variable case and is based on the method of characteristics. This method makes it possible to reduce a partial differential equation to a system of ordinary differential equations; see for instance [12] for more details. Here, this method provides the functional equations related to the isothermal Euler equations in Riemann coordinates.

To apply the method of characteristics to the isothermal Euler equations, a linear approximation of the source term (11) is first adopted. From (6) and assuming a mono-directional flow, namely $q > 0$ (and $\rho > 0$ by definition), we have:

$$|\xi_1 + \xi_2| = -(\xi_1 + \xi_2). \tag{17}$$

Taking into account the averaged terms in (16) and from the assumption (17), it is possible to obtain a linearization around the equilibrium point $(\bar{\xi_1}, \bar{\xi_2})$ as:

$$S(\Xi) \simeq 2\alpha \begin{bmatrix} (\bar{\xi_1} + \bar{\xi_2}) & 0 \\ 0 & (\bar{\xi_1} + \bar{\xi_2}) \end{bmatrix} \begin{bmatrix} \xi_1 \\ \xi_2 \end{bmatrix}. \tag{18}$$

where $\alpha = \frac{f_g}{8D}$. We thus obtain the following decoupled PDE system:

$$\frac{\partial}{\partial t} \begin{bmatrix} \xi_1 \\ \xi_2 \end{bmatrix} + \begin{bmatrix} \lambda_1 & 0 \\ 0 & \lambda_2 \end{bmatrix} \frac{\partial}{\partial x} \begin{bmatrix} \xi_1 \\ \xi_2 \end{bmatrix} = 2\alpha \begin{bmatrix} (\bar{\xi_1} + \bar{\xi_2}) & 0 \\ 0 & (\bar{\xi_1} + \bar{\xi_2}) \end{bmatrix} \begin{bmatrix} \xi_1 \\ \xi_2 \end{bmatrix}. \tag{19}$$

With the previous assumptions and taking into account the PDE system (19) the solution for each wave can be found by the *method of characteristics* [12, 29]. First, the propagation wave ξ_1 satisfies:

$$\frac{\partial \xi_1}{\partial_t} + \lambda_1 \frac{\partial \xi_1}{\partial_x} = 2\alpha(\bar{\xi_1} + \bar{\xi_2})\xi_1, \tag{20}$$

with the initial and boundary conditions (12) and (14), respectively. We can construct a characteristic curve s_1 emanating from $(0,\ t_0,\ u_1(t_0))$ and look for a solution to the following characteristic equations:

$$\frac{dt}{ds_1} = 1, \tag{21}$$

$$\frac{dx}{ds_1} = \lambda_1, \tag{22}$$

$$\frac{dz_1}{ds_1} = 2\alpha(\bar{\xi_1} + \bar{\xi_2})z_1(s_1), \tag{23}$$

which satisfies the initial conditions $t(0) = t_0$, $x(0) = 0$, and $z_1(0) = u_1(t_0)$. A simple integration of the ODE system leads to

$$t = s_1 + t_0, \tag{24}$$

$$x = \lambda_1 s_1, \tag{25}$$

$$z_1 = u_1(t_0)e^{2\alpha(\bar{\xi_1}+\bar{\xi_2})s_1}. \tag{26}$$

After an appropriate change of variables, we obtain the following solution:

$$\xi_1(x,t) = \xi_1\left(0, t - \frac{x}{\lambda_1}\right) e^{2\alpha(\bar{\xi_1}+\bar{\xi_2})\frac{x}{\lambda_1}}. \tag{27a}$$

Similarly, applying the method of characteristics to the second propagation wave leads to:

$$\xi_2(x,t) = \xi_2\left(L, t - \frac{(L-x)}{\lambda_2}\right) e^{2\alpha(\bar{\xi_1}+\bar{\xi_2})\frac{(L-x)}{\lambda_2}}. \tag{27b}$$

Note that λ_1 and λ_2 are of opposite sign and ξ_1 and ξ_2 are propagated in opposite directions. Note also that these results still holds for time-varying characteristic velocities, which would result in time-varying time-delays.

From (27a) to (27b) at their boundaries, it is thus possible to describe the flow transport by the following difference equations:

$$\xi_1(L,t) = \xi_1(0, t - h_1)\, e^{2\alpha(\bar{\xi_1}+\bar{\xi_2})h_1}, \tag{28a}$$

$$\xi_2(0,t) = \xi_2(L, t - h_2)\, e^{2\alpha(\bar{\xi_1}+\bar{\xi_2})h_2}, \tag{28b}$$

where $h_i = L/\lambda_i$.

3.2 Nonlinear Approximation

Calculating the characteristics of (16) without linearizing, we obtain a solution for the PDE hyperbolic system (16) using the following ODE system:

$$\frac{dt}{ds_1} = 1, \tag{29}$$

$$\frac{dx}{ds_1} = \lambda_1, \tag{30}$$

$$\frac{dz_1}{ds_1} = -\alpha(z_1 + \bar{\xi_2})^2, \tag{31}$$

which satisfies the initial conditions

$$t(0) = t_0, \quad x(0) = 0, \quad z_1(0) = u_1(t_0). \tag{32}$$

Integrating this ODE system implies that:

$$t = s_1 + t_0, \tag{33}$$

$$x = \lambda_1 s_1, \tag{34}$$

$$z_1(s_1) = -\bar{\xi_2} + \frac{z_1(0) + \bar{\xi_2}}{\alpha s_1\left(z_1(0) + \bar{\xi_2}\right) + 1}. \tag{35}$$

Finally substituting the initial conditions (32) in (35) leads to:

$$\xi_1(x,t) = -\bar{\xi_2} + \frac{\xi_1(0, t - \frac{x}{\lambda_1}) + \bar{\xi_2}}{\alpha \frac{x}{\lambda_1}\left(\xi_1(0, t - \frac{x}{\lambda_1}) + \bar{\xi_2}\right) + 1} \tag{36a}$$

Similarly, for the nonlinear approximation of the second wave, we obtain the following result:

$$\xi_2(x,t) = -\bar{\xi_1} + \frac{\xi_2(L, t - \frac{(L-x)}{\lambda_2}) + \bar{\xi_1}}{\alpha \frac{(L-x)}{\lambda_2}\left(\xi_2(L, t - \frac{(L-x)}{\lambda_2}) + \bar{\xi_1}\right) + 1}. \tag{36b}$$

Taking into account the flow propagation of the waves described in Fig. 1 and Eqs. (36a) and (36b) computed at the boundaries, we finally obtain the following delayed nonlinear equations:

$$\xi_1(L,t) = -\bar{\xi_2} + \frac{\xi_1(0, t - h_1) + \bar{\xi_2}}{\alpha h_1\left(\xi_1(0, t - h_1) + \bar{\xi_2}\right) + 1} \tag{37a}$$

$$\xi_2(0,t) = -\bar{\xi_1} + \frac{\xi_2(L, t - h_2) + \bar{\xi_1}}{\alpha h_2\left(\xi_2(L, t - h_2) + \bar{\xi_1}\right) + 1}. \tag{37b}$$

4 Network Model

The previous approximation of Euler's equations describes the flow transport in each pipe. The network model is obtained by considering a node as a finite control volume to which the pipes are connected. It is then possible to obtain a dynamical model for the fluid network, in terms of the transported variables. Figure 2 shows a possible geometry of the networks studied in this work. Analyzing the dynamics of the transported variables in a specific node N of the network, the flow enters from the pipes connected to the nodes X_k and goes out to the pipes connected to the nodes Y_k. The propagation delay

$$h_1^{(X_k,\,N)} \left(\text{respectively, } h_2^{(N,\,Y_k)}\right)$$

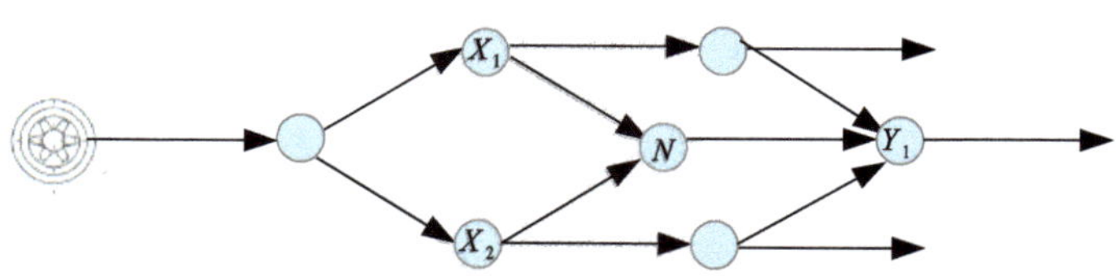

Fig. 2 Network topology

is linked to the pipe length L and the characteristic velocity

$$\lambda_1^{(X_k,\, N)} \left(\text{respectively, } \lambda_2^{(N,\, Y_k)}\right).$$

4.1 Conservation at the Nodes and Fluid Capacitance

Taking into account the conservation for the physical variables ρ and q at a node N with n_{in} ingoing pipes and n_{out} outgoing pipe, the mass conservation at the intersection is stated as follows:

$$\sum_{i=1}^{n_{in}} q^{(X_i,N)}(L,t) = \sum_{i=1}^{n_{out}} q^{(N,Y_i)}(0,t), \quad \forall t > 0. \tag{38}$$

This condition is similar to Kirchhoff's law and is usually referred to as a Rankine-Hugoniot condition at the node [1, 2]. An additional coupling condition for the intersections is that the pressure inside each node is uniform (and thus the same at each extremity of the connected pipes), namely:

$$a^2\rho^{(X_i,N)}(L,t) = a^2\rho^{(N,Y_k)}(0,t) = constant \quad \forall i = 1,\ldots,n_{in},\ \forall k = 1,\ldots,n_{out} \tag{39}$$

With the transformations (7) and (8), we can use (39) to state that $\forall i = 1,\ldots,n_{in}$ and $\forall k = 1,\ldots,n_{out}$, we have:

$$\xi_1^{(X_i,N)}(L,t) - \xi_2^{(X_i,N)}(L,t) = \xi_1^{(N,Y_k)}(0,t) - \xi_2^{(N,Y_k)}(0,t) = constant, \tag{40}$$

The constraint (38) implies that

$$\sum_{i=1}^{n_{in}} \xi_1^{(X_i,N)}(L,t) + \xi_2^{(X_i,N)}(L,t) = \sum_{k=1}^{n_{out}} \xi_1^{(N,Y_k)}(0,t) + \xi_2^{(N,Y_k)}(0,t). \tag{41}$$

A dynamics is introduced to model the fluid capacitance as a time-varying state associated with the node. Denoting the fluid capacitance as C_f, it follows that a node N:

$$\dot{P}(t)^N = q^N/C_f \tag{42}$$

$$\Leftrightarrow \dot{\rho}(t)^N = -\frac{M_g}{ZRTC_f}q^N = -\frac{M_g}{ZRTC_f}\left(\sum_{i=1}^{n_{in}} \xi_1^{(X_i,N)}(L,t) + \xi_2^{(X_i,N)}(L,t) - \sum_{k=1}^{n_{out}} \xi_1^{(N,Y_k)}(0,t) + \xi_2^{(N,Y_k)}(0,t)\right)\rho^N \tag{43}$$

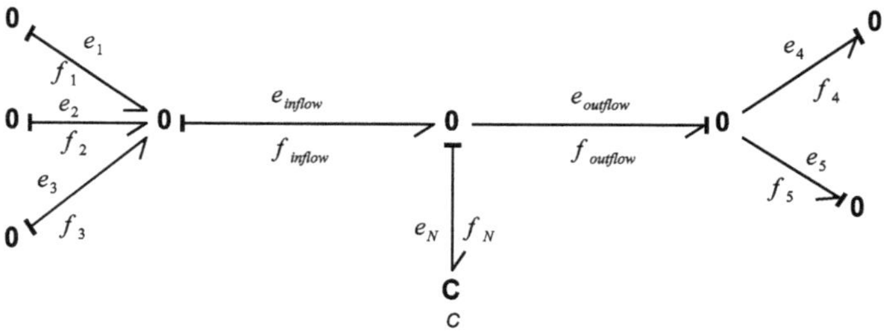

Fig. 3 Expanded bond graph representation of the node

where we used (1) with the isothermal hypothesis, (7), (8) and (41). Note that C_f could also be node-dependent to capture large changes in the network configuration. It is then possible to represent the node with an equivalent Bond graph description composed of a 0-junction (using constant effort and the fact that the sum of the inflows is equal to the sum of the outflows) and a C-element. Figure 3 shows the equivalent Bond graph representation for a node. The constraints (40) and (43) are then satisfied if we define the effort as $e = \xi_1 - \xi_2$ and the flow as $f = \xi_1 + \xi_2$ for each junction.

Choosing the internal state of the node as e^N and expressing (43) in terms of e^N using (7)–(8), we have that:

$$\dot{e}^N = -\alpha_c \left(\sum_{i=1}^{n_{in}} \xi_1^{(X_i,N)}(L,t) + \xi_2^{(X_i,N)}(L,t) - \sum_{k=1}^{n_{out}} \xi_1^{(N,Y_k)}(0,t) + \xi_2^{(N,Y_k)}(0,t) \right) \tag{44}$$

$$\xi_1^{(X_i,N)}(L,t) - \xi_2^{(X_i,N)}(L,t) = \xi_1^{(N,Y_k)}(0,t) - \xi_2^{(N,Y_k)}(0,t) = e^N \tag{45}$$

where $\alpha_c = aM_g/ZRTC_f$. The node is thus well defined by an algebro-differential system of equations and the corresponding Riemann invariants are computed in each branch by adding or subtracting the effort and flow. Note that the flow inertia and friction are already taken into account within the Euler equations with friction.

4.2 State-Space Representation

The previous bond-graph description can be expressed as a classical state-space representation, as follows. Considering that ξ_1 enters the node from the left and that ξ_2 enters the node from the right, we wish to establish a mapping between the inputs

$$\{\xi_1^{(X,N)}(L,t),\ \xi_2^{(N,Y)}(0,t)\}$$

and the outputs

$$\{\xi_1^{(N,Y)}(0,t),\ \xi_2^{(X,N)}(L,t)\},$$

where we used the notation X and Y (without subscripts) to denote the column vectors containing all the inflow or outflow components, respectively. The effort relationships (40) and (45) imply that:

$$\begin{aligned}&\begin{bmatrix}\xi_1^{(X,N)}(L,t)\\ \xi_1^{(N,Y)}(0,t)\end{bmatrix}-\begin{bmatrix}\xi_2^{(X,N)}(L,t)\\ \xi_2^{(N,Y)}(0,t)\end{bmatrix}=\mathbf{1}^{(n_{in}+n_{out})\times 1}e^N\\ \Leftrightarrow&\begin{bmatrix}\xi_2^{(X,N)}(L,t)\\ -\xi_1^{(N,Y)}(0,t)\end{bmatrix}=-\mathbf{1}^{(n_{in}+n_{out})\times 1}e^N+\begin{bmatrix}\xi_1^{(X,N)}(L,t)\\ -\xi_2^{(N,Y)}(0,t)\end{bmatrix}\end{aligned}\tag{46}$$

where $\mathbf{1}^{i\times j}$ is a vector of ones of size $i\times j$. Considering (44), the node dynamics is:

$$\begin{aligned}&\dot{e}(t)^N\\ &=-\alpha_c\left(\left[\mathbf{1}^{1\times n_{in}}\ \mathbf{1}^{1\times n_{out}}\right]\begin{bmatrix}\xi_1^{(X,N)}(L,t)\\ -\xi_2^{(N,Y)}(0,t)\end{bmatrix}+\left[\mathbf{1}^{1\times n_{in}}\ \mathbf{1}^{1\times n_{out}}\right]\begin{bmatrix}\xi_2^{(X,N)}(L,t)\\ -\xi_1^{(N,Y)}(0,t)\end{bmatrix}\right)\\ &=-\alpha_c e^N-2\alpha_c\left[\mathbf{1}^{1\times n_{in}}\ \mathbf{1}^{1\times n_{out}}\right]\begin{bmatrix}\xi_1^{(X,N)}(L,t)\\ -\xi_2^{(N,Y)}(0,t)\end{bmatrix}\end{aligned}$$

We thus obtained a state-space description where (47) describes the state dynamics and (46) determines the output

$$\begin{bmatrix}\xi_2^{(X,N)}(L,t)\\ -\xi_1^{(N,Y)}(0,t)\end{bmatrix}$$

from the input (direct feedthrough)

$$\begin{bmatrix}\xi_1^{(X,N)}(L,t)\\ -\xi_2^{(N,Y)}(0,t)\end{bmatrix}$$

and the state e^N.

4.3 Interconnections and Delays

We can now consider the interconnections on node N with its predecessors X and successors Y. The linear approximation of the pipe propagation waves and the delayed equations (28a) and (28b) along with the previous node description give:

$$\dot{e}(t)^N = -\alpha_c e^N - 2\alpha_c \left[\mathbf{1}^{1\times n_{in}}\ \mathbf{1}^{1\times n_{out}}\right] u(t) \tag{47}$$

$$\begin{bmatrix} \xi_2^{(X,N)}(L,t) \\ -\xi_1^{(N,Y)}(0,t) \end{bmatrix} = -\mathbf{1}^{(n_{in}+n_{out})\times 1} e^N + u(t) \tag{48}$$

with

$$u(t) = \begin{bmatrix} \xi_1^{(X,N)}\left(0, t - h_1^{(X,N)}\right) e^{2\alpha(\bar{\xi}_1^{(X,N)} + \bar{\xi}_2^{(X,N)}) h_1^{(X,N)}} \\ -\xi_2^{(N,Y)}\left(L, t - h_2^{(N,Y)}\right) e^{2\alpha(\bar{\xi}_1^{(N,Y)} + \bar{\xi}_2^{(N,Y)}) h_2^{(N,Y)}} \end{bmatrix} \tag{49}$$

Each node is defined by a delay differential equation with direct feedthrough of the interconnection variable $u(t)$ on the output. This interconnection variable is found from the outputs of predecessors (terms in $\xi_1^{(X,N)}$) and successors (terms in $\xi_2^{(N,Y)}$).

The nonlinear approximation is obtained in a similar way by taking into account the functional equations (37a) and (37b). The delayed differential equation is then given by (47)–(48) with the interconnection:

$$u(t) = \begin{bmatrix} -\bar{\xi}_2^{(X,N)} + \dfrac{\xi_1^{(X,N)}(0,t-h_1^{(X,N)}) + \bar{\xi}_2^{(X,N)}}{\alpha h_1^{(X,N)}\left(\xi_1^{(X,N)}(0,t-h_1^{(X,N)}) + \bar{\xi}_2^{(X,N)}\right)+1} \\ \bar{\xi}_1^{(N,Y)} - \dfrac{\xi_2^{(N,Y)}(L,t-h_2^{(N,Y)}) + \bar{\xi}_1^{(N,Y)}}{\alpha h_2^{(N,Y)}\left(\xi_2^{(N,Y)}(L,t-h_2^{(N,Y)}) + \bar{\xi}_1^{(N,Y)}\right)+1}. \end{bmatrix} \tag{50}$$

5 Simulation Results

Consider the network configuration shown in Fig. 4, corresponding to an air ventilation network, where the speed of sound constant is $a = 347$ m/s. The network parameters are given in Table 1.

The inflow (i.e. boundary conditions) is considered as constant and the initial conditions are $\rho(t,0) = 1.16$ kg/m^3 and $q(t,0) = 23.2$ kg/m^2 s. We perform

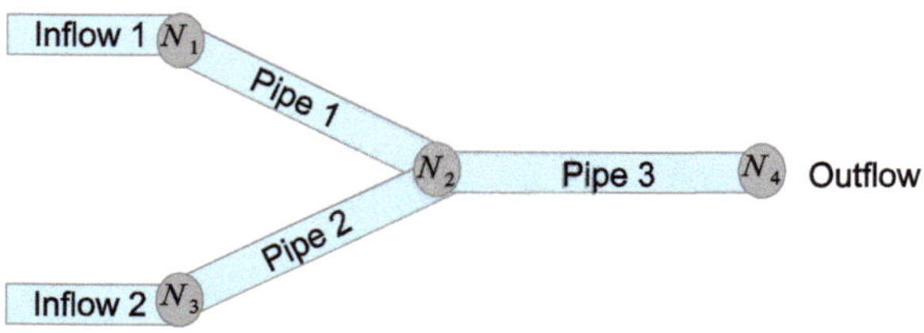

Fig. 4 Network example

Table 1 Pipe parameters of the network

Parameter	Pipe 1	Pipe 2	Pipe 3
Length (L)	100 m	200 m	300 m
Diameter (D)	0.4 m	0.2 m	0.4 m
Friction factor (f_g)	0.008	0.0012	0.003

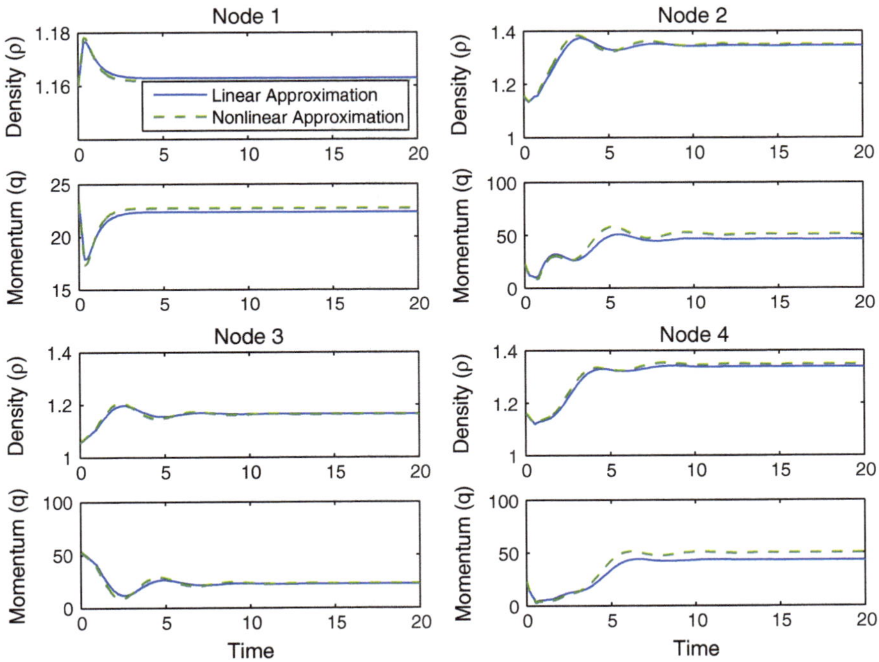

Fig. 5 Dynamics of the physical variables in the network nodes

numerical simulations of the network model derived in the previous sections and compare the linear and nonlinear delay differential equations in Fig. 5.

The results are presented for each node of the network in terms of the physical variables, density and momentum, which can be found from the Riemann invariants by using the transformation (7) and (8), respectively. According to the network parameters, the time delays and friction terms are in Table 2.

We observe the propagation delays in the simulations, and oscillations due to wave reflections and coupling at the nodes. It is also interesting to note the closeness between the linear and the nonlinear models, which suggests that linear approaches may provide satisfactory results despite the nonlinear characteristic of friction.

Table 2 Simulation network parameters

Parameter	Value	Parameter	Value	Parameter	Value	Parameter	Value
$h_1^{(1)}$	0.2725	$h_2^{(1)}$	0.3058	$\beta^{(1)}$	0.9470	$\beta^{(5)}$	0.9407
$h_1^{(2)}$	0.5450	$h_2^{(2)}$	0.6116	$\beta^{(2)}$	0.9678	$\beta^{(6)}$	0.9640
$h_1^{(3)}$	0.8174	$h_2^{(3)}$	0.9174	$\beta^{(3)}$	0.9521	$\beta^{(7)}$	0.9464
$h_1^{(4)}$	0.1362	$h_2^{(4)}$	0.1529	$\beta^{(4)}$	0.9967	$\beta^{(8)}$	0.9963

6 Conclusion

A time-delay approach is used in this work to find a model for the flow in fluid networks. The model of the flow is taken from the isothermal Euler equations. The hyperbolic characteristics of the PDE system are taken into account in order to diagonalize the system. The method of characteristics is considered to find functional equations related to the isothermal Euler equations. By decoupling the incoming and outgoing waves and averaging the source term on the appropriate components, two approximations are adopted to compute a delay model of the flow, namely a linear approximation and a nonlinear one. Taking into account the conservative characteristics at the nodes, a delay differential system of equations for the flow network is finally given.

References

1. Banda, M., Herty, M., Klar, A.: Coupling conditions for gas networks governed by the isothermal euler equations. Netw. Heterogenous Media **1**(2), 295–314 (2006)
2. Banda, M., Herty, M., Klar, A.: Gas flow in pipeline networks. Netw. Heterogenous Media **1**(1), 41–56 (2006)
3. Bradu, B., Gayet, P., Niculescu, S.-I., Witrant, E.: Modeling of the very low pressure helium flow in the lhc cryogenic distribution line after a quench. Cryogenics **50**(2), 71–77 (2010)
4. Bresch-Pietri, D., Leroy, T., Petit, N.: Control-oriented time-varying input-delayed temperature model for si engine exhaust catalyst. In: Proceedings of the American Control Conference, pp. 2189–2195 (2013)
5. Colombo, R., Guerra, G., Herty, M., Schleper, V.: Optimal control in networks of pipes and canals. SIAM J. Control Optim. **48**(3), 2032–2050 (2009)
6. Cooke, K., Krumme, D.: Differential-difference equations and non-linear partial-boundary value problems for linear hyperbolic partial differential equations. J. Math. Anal. Appl. **24**(2), 362–387 (1968)
7. Corriga, G., Sanna, S., Usai, G.: Sub-optimal constant-volume control for open channel networks. Appl. Math. Model. **7**(4), 262–267 (1983)
8. Cross, H.: Analysis of flow in networks of conduits or conductors. University of Illinois Engineering Experiment Station Bulletin 286 (1936)
9. de Halleux, J., Prieur, C., Coron, J.-M., d'Andrea Novel, B., Bastin, G.: Boundary feedback control in networks of open channels. Automatica **39**(8), 1365–1376 (2003)

10. de Saint-Venant, B.: Theorie du mouvement non-permanent des eaux avec applications aux crues des rivieres et a l'introduction des marees dans leur lit. Comptes-rendus de l'Academie des Sciences **73**, 148–154 (1871)
11. Dick, M., Gugat, M., Herty, M., Steffensen, S.: On the relaxation approximation of boundary control of the isothermal Euler equations. Int. J. Control **8**(11), 1766–1778 (2012)
12. Evans, L.: Partial Differential Equations. Series Graduate Studies in Mathematics, vol. 19. American Mathematical Society, Providence (1998)
13. Gugat, M., Herty, M., Schleper, V.: Flow control in gas networks: exact controllability to a given demand. Math. Methods Appl. Sci. **34**(7), 745–757 (2011)
14. Gugat, M., Dick, M.: Time-delayed boundary feedback stabilization of the isothermal Euler equations with friction. Math. Control Relat. Fields **1**(4), 469–491 (2011)
15. Hartman, H., Mutmansky, J., Ramani, R., Wang, Y.: Mine Ventilation and Air Conditioning, 3rd edn. Wiley, New York (1997)
16. Hu, Y., Koroleva, O., Krstic, M.: Nonlinear control of mine ventilation networks. Syst. Control Lett. **49**(4), 239–254 (2003)
17. Karafyllis, I., Krstic, M.: On the relation of delay equations to first-order hyperbolic partial differential equations. ESAIM Control Optim. Calc. Var. **20**(3), 894–923 (2014)
18. Koroleva, O., Krstic, M.: Averaging analysis of periodically forced fluid networks. Automatica **41**(1), 129–135 (2005)
19. Litrico, X., Georges, D.: Robust continuous-time and discrete-time flow control of a dam-river system. Appl. Math. Model. **23**(11), 809–827 (1999)
20. Litrico, X., Fromion, V.: Analytical approximation of open-channel flow for controller design. Appl. Math. Model. **28**(7), 677–695 (2004)
21. Petrov, N., Shishkin, M., Dmitriev, V., Shadrin, V.: Modeling mine aerology problems. J. Min. Sci. **28**(2), 185–191 (1992)
22. Rasvan, V.: Functional differential equations and one-dimensional distortionless propagation. Tatra Mountains Math. Publ. **43**(1), 215–228 (2009)
23. Rasvan, V.: Delays. Propagation. Conservation Laws. In: Sipahi, R., Vyhlidal, T., Niculescu, S-I., Pepe, P. (eds.) Time delay systems: Methods, Applications and New Trends. Series Lecture Notes in Control and Information Sciences, vol. 423, pp. 147–159. Springer, New York (2012)
24. Rasvan, V.: Three lectures on neutral functional differential equations. J. Control Eng. Appl. Inf. **11**(9), 49–55 (2009)
25. Toro, E.: Riemann Solvers and Numerical Methods for Fluid Dynamics, 3rd edn. Springer, New York (2009)
26. Treiber, M., Kesting, A.: Traffic Flow Dynamics: Data. Models and Simulation. Springer, Berlin Heidelberg (2013)
27. Witrant, E., Johansson, K.: Air flow modeling in deepwells: Application to mining ventilation. In: Proceedings of the IEEE International Conference on Automation Science and Engineering, pp. 845–850 (2008)
28. Witrant, E., Marchand, N.: Modeling and feedback control for air flow regulation in deep pits. In: Sivasundaram, S. (ed.) Mathematical Problems in Engineering. Aerospace and Sciences. Cambridge Scientific Publishers, Cambridge (2011)
29. Witrant, E., Niculescu, S.-I.: Modeling and control of large convective flows with time-delays. Math. Eng. Sci. Aerosp. **1**(2), 191–205 (2010)

Finite Spectrum Assignment for Nonlinear Time-Delay Systems

Toshiki Oguchi

Abstract We consider the stabilization problem for nonlinear time-delay systems. First, we review the finite spectrum assignment method for a class of linear retarded systems and its extension for nonlinear systems. Next, relaxing required conditions for the method, we propose a new stabilizing technique for nonlinear delay systems. The technique includes the use of a state predictor which is based on the idea of anticipating synchronization. Furthermore, we discuss the relationship among other control design methods for a wider class of retarded systems and this technique.

1 Introduction

Recently the study of time-delay systems has attracted a growing interest in various fields, such as mathematics, information technology, applied physics, biology and social science. Of course, control of time-delay systems has continued to be a hot topic in control engineering and science. In particular, a large number of useful and important results have been obtained for linear time-delay systems [1, 4, 6, 7, 9, 21, 22]. While in practice most systems contain nonlinear elements and are described as nonlinear time-delay systems, the study of control theory for such systems is in progress [2, 5, 8, 14, 18, 19]. For nonlinear retarded systems, the author and others have already proposed an extension of the finite spectrum assignment for linear retarded systems which are controllable over the polynomial ring. However the applicable systems are relatively confined a narrow class of retarded systems. In particular, if systems have time-delay at the input or they can be transformed into input-delay systems, the condition is unsatisfied.

On the other hand, in the interdisciplinary fields including applied physics, mathematical biology and control science, synchronization of coupled chaotic systems has been studied from both theory and experiments [10–12, 20]. As an extension of synchronization of coupled systems, Voss [23] discovered the occurrence of antici-

T. Oguchi (✉)
Department of Mechanical Engineering, Graduate School of Science and Engineering, Tokyo Metropolitan University, 1-1 Minami-osawa, Hachioji-shi, Tokyo 192-0397, Japan
e-mail: t.oguchi@tmu.ac.jp

I. Karafyllis et al. (eds.), *Recent Results on Nonlinear Delay Control Systems*, Advances in Delays and Dynamics 4, DOI 10.1007/978-3-319-18072-4_15

pated synchronization of chaotic systems with time-delay in a unidirectional coupling configuration. The anticipated synchronization can be recognized as a state prediction method which does not require numerical integration. Therefore this prediction scheme may be applicable for wilder classes of nonlinear systems.

In this chapter, we consider the stabilization problem of nonlinear retarded systems. In particular, we attempt to extend applicability of the finite spectrum assignment approach by using a state prediction based on synchronization. The state prediction scheme based on synchronization [15] has been developed by the author and others in order to predict the behavior of chaotic systems with time-delay and it is designed by using delay-dependent stability or stabilizability conditions based on the Lyapunov-Razumikhin theorem and the Lyapunov-Krasovskii theorem. This method can be considered as an extension of anticipating synchronization of chaotic systems proposed in [23]. Combining this predictor with the state linearization technique, we propose a new stabilizing controller for nonlinear retarded systems.

The rest of this chapter is organized as follows. Section 2 discusses the finite spectrum assignment for a class of linear retarded systems and its extension for nonlinear systems. In Sect. 3, after explaining the motivation for this study and introducing a state predictor based on anticipating synchronization, we attempt to extend the finite spectrum assignment for nonlinear systems with time-delays. Section 4 shows two numerical examples that illustrate the effectiveness of the proposed methods. Finally some concluding remarks are drawn in Sect. 5.

2 Preliminaries

2.1 Mathematical Definitions

Firstly, we review the definitions of the delayed state derivative and the delayed state bracket which are extensions of the Lie derivative and the Lie bracket into nonlinear difference-differential equations. Though the following definitions and the approach based on them can be extended for systems with non-commensurate delays, it will be considered that all the delays are multiples of an elementary delay τ for the sake of simplicity.

Let $\tau > 0$ be a time delay, then σ denotes the delay operator that shifts the time from t to $t - \tau$ and is defined as

$$\sigma\lambda(t) := \lambda(t - \tau)$$

where $\lambda(t)$ is a function defined on the interval $[t - \tau, t]$. If $\varphi(\cdot)$ is a function of $\lambda(t)$ defined on the interval $[t - \tau, t]$, σ operates as $\sigma\varphi(\lambda(t)) := \varphi(\lambda(t - \tau))$.

In addition, let $\mathscr{K}$ be a field of meromorphic functions of a finite number of variables in $\{x(t), x(t - k\tau) | k \in \mathbb{N}, \tau \in \mathbb{R}_+\}$, and $\mathscr{K}[\sigma]$ be the ring of polynomials of operator σ over $\mathscr{K}$.

Definition 1 (*Delayed State Derivative*) [14] Let $\bar{f}(x,\sigma)$ be an n-vector whose components belong to $\mathscr{K}$ (or $\mathscr{K}[\sigma]$), and let $\bar{\varphi}(x,\sigma) := \varphi(x(t), x(t-\tau), \ldots, x(t-m\tau))$ be a real-valued function of $x(t)$ and $x(t-j\tau)$, $1 \leq j \leq m$. Then, the derivative of $\bar{\varphi}(x,\sigma)$ along $\bar{f}(x,\sigma)$ is defined as

$$L_{\bar{f}}\bar{\varphi}(x,\sigma) := \frac{\partial\bar{\varphi}}{\partial x(t)}\bar{f}(x,\sigma) + \sum_{i=1}^{m}\frac{\partial\bar{\varphi}}{\partial\sigma^i x}\sigma^i\bar{f}(x,\sigma).$$

For $\gamma \geq 2$, the general form of higher-order derivatives can be denoted by $L_{\bar{f}}^{\gamma+1}\bar{\varphi}(x,\sigma) = L_{\bar{f}}L_{\bar{f}}^{\gamma}\bar{\varphi}(x,\sigma)$.

Definition 2 (*Delayed State Bracket*) [14] Let $\bar{f}(x,\sigma)$ and $\bar{g}(x,\sigma)$ be vectors whose entries may belong to $\mathscr{K}$ (or $\mathscr{K}[\sigma]$). Then the delayed-state bracket $[\bar{f},\bar{g}](x,\sigma) \in \mathscr{K}$ (correspondingly $\mathscr{K}[\sigma]$) of $\bar{f}(x,\sigma)$ and $\bar{g}(x,\sigma)$ is defined as

$$[\bar{f},\bar{g}](x,\sigma) := \sum_{i=0}^{q}\frac{\partial\bar{g}}{\partial\sigma^i x}(\sigma^i\bar{f}) - \sum_{j=0}^{p}\frac{\partial\bar{f}}{\partial\sigma^j x}(\sigma^j\bar{g}).$$

where $\sigma^0 : x(t) \mapsto x(t)$, and p and q correspond to the maximum delays of delayed states $\sigma^p x$ and $\sigma^q x$ included in $\bar{f}$ and $\bar{g}$, respectively.

Repeated bracketing of $\bar{g}(x,\sigma)$ with $\bar{f}(x,\sigma)$ is recursively defined by

$$\begin{aligned}
&ad_{\bar{f}}^{0}\bar{g}(x,\sigma) := \bar{g}(x,\sigma),\\
&ad_{\bar{f}}^{i}\bar{g}(x,\sigma) := [\bar{f}, ad_{\bar{f}}^{i-1}\bar{g}(x,\sigma)] \ \text{ for } \ i \geq 1.
\end{aligned}$$

If $\tau = 0$, the delayed state derivative and the delayed state bracket coincide with the conventional Lie derivative and the conventional Lie bracket, respectively. The delayed state derivative and the delayed state bracket satisfy the following relation.

$$L_{ad_{\bar{f}}\bar{g}}\bar{\varphi}(x,\sigma) = L_{\bar{f}}L_{\bar{g}}\bar{\varphi}(x,\sigma) - L_{\bar{g}}L_{\bar{f}}\bar{\varphi}(x,\sigma)$$

2.2 *Retarded Linear Systems*

The finite spectrum assignment for retarded nonlinear systems considered herein is closely related to the finite spectrum assignment for $\mathbb{R}[\sigma]$-controllable linear systems and the exact linearization for finite dimensional nonlinear systems. Therefore, in this subsection, we briefly review the finite spectrum assignment for $\mathbb{R}[\sigma]$-controllable linear systems [9, 22]. For simplicity, we consider the following linear system with a commensurate time delay:

$$\dot{x}(t) = \sum_{i=0}^{k} A_i x(t - i\tau) + bu(t) \tag{1}$$

where $x \in \mathbb{R}^n$ and $u \in \mathbb{R}$. Using the pure delay operator $\sigma : \sigma x(t) \to x(t - \tau)$, the system (1) can be rewritten as

$$\dot{x}(t) = A(\sigma)x + bu(t),$$

where $A(\sigma) = \sum_{i=0}^{k} A_i \sigma^i$ and the characteristic polynomial is given by

$$|sI - A(\sigma)| = s^n + \alpha_1(\sigma)s^{n-1} + \cdots + \alpha_n(\sigma).$$

The coefficients $\alpha_i(\sigma)$ for $i = 1, \ldots, n$ of this characteristic polynomial are given by polynomials of σ. If the system (1) is controllable over the polynomial ring $\mathbb{R}[\sigma]$, that is

$$\operatorname{rank} \mathscr{C}(\sigma) := \operatorname{rank} \left(b\, A(\sigma)b\, \cdots\, A(\sigma)^{n-1}b\right) = n$$

for any $\sigma \in \mathbb{C}$, there exists a feedback $u(t) = F(\sigma)x(t)$ such that the closed loop system has a characteristic polynomial with real coefficients:

$$|sI - A(\sigma) - bF(\sigma)| = s^n + \beta_1 s^{n-1} + \cdots + \beta_n.$$

Hence this system can have n poles at arbitrarily preassigned position. This method is called a *"finite spectrum assignment"* procedure. Here, $F(\sigma)$ is given by

$$F(\sigma) = \left(\alpha_n(\sigma) - \beta_n\, \cdots\, \alpha_1(\sigma) - \beta_1\right) M^{-1}(\sigma),$$

in which

$$M(\sigma) = \mathscr{C}(\sigma) \begin{pmatrix} \alpha_{n-1}(\sigma) & \cdots & \alpha_1(\sigma) & 1 \\ \vdots & \iddots & \iddots & 0 \\ \alpha_1(\sigma) & \iddots & \iddots & \vdots \\ 1 & 0 & \cdots & 0 \end{pmatrix}.$$

Note that if the system (1) is controllable over the polynomial ring $\mathbb{R}[\sigma]$, the matrix $M(\sigma)$ is unimodular, i.e. the determinant of $M(\sigma)$ is a non-zero real number, and $M^{-1}(\sigma)$ is also a polynomial matrix of σ.

This procedure can be interpreted as follows. By applying a variable transformation

$$z(t) = M(\sigma)^{-1}x(t)$$

which contains the past values of the state variables, the system (1) is transformed into the form

$$\begin{cases} \dot{z}_i(t) = z_{i+1}(t), i = 1, \dots, n-1 \\ \dot{z}_n(t) = -\alpha_n(\sigma)z_1(t) - \alpha_{n-1}(\sigma)z_2(t) - \cdots - \alpha_1(\sigma)z_n(t) + u(t) \end{cases}$$

and then the remained delay elements in the nth row are canceled by the feedback $u(t) = f(\sigma)x(t)$. As a result, the original system (1) can be changed into a finite dimensional system:

$$\begin{cases} \dot{z}_i(t) = z_{i+1}(t), i = 1, \dots, n-1 \\ \dot{z}_n(t) = -\beta_n z_1(t) - \beta_{n-1}z_2(t) - \cdots - \beta_1 z_n(t) \end{cases}$$

This procedure can be considered to be a simple extension of the conventional pole placement for finite dimensional controllable systems achieved by replacing constant matrices with polynomial matrices.

2.3 Retarded Nonlinear Systems

Next, we review the finite spectrum assignment problem for retarded nonlinear systems considered in [14]. For simplicity of description, we consider the following single input nonlinear system with commensurate delays throughout this paper:

$$\begin{cases} \dot{x}(t) = f(x(t), x(t-\tau), \dots, x(t-p\tau)) + g(x(t), \dots, x(t-q\tau))u(t) \\ x(t) = \varphi(t), \ t \in [-\max\{p, q\}\tau, 0], \quad p, q \in \mathbb{N}, \end{cases} \tag{2}$$

in which $x(t) \in \mathbb{R}^n$, $u \in \mathbb{R}$, τ is a fixed time delay, and $\varphi : \mathbb{R} \to \mathbb{R}^n$ is a continuous vector valued function specifying the initial state of the system. It is assumed that $f(\cdot)$ and $g(\cdot)$ are real smooth vector fields with $f(0) = 0$ and $g(0) \neq 0$.

Introducing the pure delay operator $\sigma : \lambda(t) \mapsto \lambda(t-\tau)$, we rewrite system (2) as

$$\begin{aligned} \dot{x}(t) = & \ f(x, \sigma x, \dots, \sigma^p x) + g(x, \dots, \sigma^q x)\sigma^0 u(t) \\ := & \ \bar{f}(x, \sigma) + \bar{g}(x, \sigma)u(t) \end{aligned} \tag{3}$$

where $\bar{f}(x, \sigma) := f(x, \sigma x, \dots, \sigma^p x)$ and $\bar{g}(x, \sigma) := g(x, \dots, \sigma^q x)\sigma^0$.

Remark 1 Note that from the above definition the entries of $\bar{f}(x, \sigma)$ belong to $\mathcal{K}$ and those of $\bar{g}(x, \sigma)$ to $\mathcal{K}[\sigma]$. When the pure delay operator σ is applied into $\bar{g}(x, \sigma)u(t)$, this operation means that

$$\begin{aligned} \sigma\bar{g}(x, \sigma)u(t) = & \sigma\{g(x, \sigma x \dots, \sigma^q x)\sigma^0\}u(t) = g(\sigma x, \dots, \sigma\sigma^q x)\sigma u(t) \\ = & g(x(t-\tau), \dots, x(t-q\tau-\tau))u(t-\tau). \end{aligned}$$

Note that $\sigma^k \bar{g}(x, \sigma) = \{\sigma^k g(x, \dots, \sigma^q x)\}\sigma^k$. In addition, since σ^0 is identified with an identity mapping, σ^0 may be omitted in the description.

For the systems mentioned above, a finite spectrum assignment problem is formally stated as follows:

Definition 3 *A finite spectrum assignment problem for retarded nonlinear systems:* Let a plant of a nonlinear system with time delays in state variables (3) be given. Then, if possible, find a feedback with delayed-state variables, $u(t) = \alpha(x, \sigma)$, and a variable transformation with delayed-state variables, $z(t) = \Phi(x, \sigma)$, such that the resulting closed loop system

$$\dot{x}(t) = \bar{f}(x, \sigma) + \bar{g}(x, \sigma)\alpha(x, \sigma)$$

is linear in the variable $z(t) = \Phi(x, \sigma)$ and has spectra at an arbitrarily preassigned self-conjugate set of n points $\lambda_i, i = 1, \dots, n$, in the complex plane, i.e. it is denoted by $\dot{z}(t) = Az(t)$ for some suitable matrix $A \in \mathbb{R}^{n\times n}$ which has totally n eigenvalues at $\lambda_i, i = 1, \dots, n$.

For this problem, we obtain the following result:

Theorem 1 [14] *The finite spectrum assignment problem for retarded nonlinear systems described by* (3) *is accomplished by a static feedback such as* $u(t) = \alpha(x, \sigma) \in \mathscr{K}$ *if, and only if there exists a* C^∞*-class real-valued function* $\bar{\varphi}(x, \sigma) \in \mathscr{K}$ *of the state* $x(t)$ *and the past value of the state* $\sigma^k x$ *for* $k \in \mathbb{N}$ *satisfying the following conditions for all* $x(t)$ *and* $\sigma^k x$ *in a neighborhood* $U \subseteq \mathbb{R}^n$ *of the origin:*

(a) $L_{\bar{g}} L_{\bar{f}}^i \bar{\varphi}(x, \sigma) = 0 \quad \text{for} \ \ i = 0, \dots, n-2$

(b) $\dfrac{\partial L_{\bar{f}}^{n-1} \bar{\varphi}(x, \sigma)}{\partial x(t)} \bar{g}(x, \sigma) \neq 0$

(c) $\dfrac{\partial L_{\bar{f}}^{n-1} \bar{\varphi}(x, \sigma)}{\partial \sigma^k x} \sigma^k \Big(\bar{g}(x, \sigma)\Big) = 0$ for any $\sigma^k x$ included in $L_{\bar{f}}^{n-1} \bar{\varphi}(x, \sigma)$.

When there exists $\bar{\varphi}(x, \sigma)$ *which satisfies these conditions, the required feedback law defined for any* x *and* $\sigma^k x \in U$ *is given by*

$$u(t) = \frac{-L_{\bar{f}}^n \bar{\varphi}(x, \sigma) + v(t)}{L_{\bar{g}} L_{\bar{f}}^{n-1} \bar{\varphi}(x, \sigma)} \tag{4}$$

and the required variable transformation $\Phi : U^{L+1} \to W \subseteq \mathbb{R}^n$ *is*

$$z(t) = \Phi(x, \sigma) := \begin{pmatrix} \bar{\varphi}(x, \sigma) \\ \vdots \\ L_{\bar{f}}^{n-1} \bar{\varphi}(x, \sigma) \end{pmatrix} \in \mathbb{R}^n$$

where L denotes that the maximal length of time-delay included in $\Phi(x,\sigma)$ *is* $\sigma^L x = x(t-L\tau)$, *and* $v(t) = \sum_{j=1}^{n} a_j z_j(t)$. *Here,* a_j *are chosen so that the closed loop system has n poles in the preassigned position.*

Remark 2 This design method is based on the continuity of the solution of system. However, since the applied variable transformation and feedback contain the delayed state variables, the initial condition of the system must be directly involved in them for a while after beginning of control. Therefore, note that the desired response appears after eliminating the effect of the initial condition.

To show the existence of the scalar function satisfying the conditions (a)–(c) in Theorem 1, we introduce an extension of the involutive condition.

Definition 4 (σ*-involutive*) [14]
Let $\mathcal{D}_{\mathcal{K}}$ denote a subspace spanned by $\bar{\xi}_1(x,\sigma), \ldots, \bar{\xi}_d(x,\sigma)$ over $\mathbb{R}$ as follows:

$$\mathcal{D}_{\mathcal{K}} := span\{\bar{\xi}_1(x,\sigma), \cdots, \bar{\xi}_d(x,\sigma)\} = \left\{\sum_{i=1}^{d} a_i\bar{\xi}_i(x,\sigma)\big|a_i \in \mathbb{R}\right\}$$

Then $\mathcal{D}_{\mathcal{K}}$ is σ-involutive, or involutive over the field $\mathcal{K}[\sigma]$ if $\mathcal{D}_{\mathcal{K}}$ is embedded in a distribution spanned by $\gamma_1(x), \ldots, \gamma_d(x)$ over the field $\mathcal{K}[\sigma]$, i.e. if there exist d vector fields $\gamma_1(x), \ldots, \gamma_d(x)$ such that

$$\mathcal{D}_{\mathcal{K}} \subseteq span_{\mathcal{K}[\sigma]}\{\gamma_1(x), \ldots, \gamma_d(x)\} = \left\{\sum_{i=1}^{d} \gamma_i(x)\alpha_i(x,\sigma)\big|\alpha_i \in \mathcal{K}[\sigma]\right\}$$

and the delayed state brackets $[\bar{\xi}_i, \bar{\xi}_j]$ of any pair of vector fields $\bar{\xi}_i(x,\sigma)$ and $\bar{\xi}_j(x,\sigma)$ belonging to $\mathcal{D}_{\mathcal{K}}$ are vector fields which belong to $\mathcal{D}_{\mathcal{K}}$, i.e.

$$\bar{\xi}_i(x,\sigma) \in \mathcal{D}_{\mathcal{K}},\ \bar{\xi}_j(x,\sigma) \in \mathcal{D}_{\mathcal{K}} \Longrightarrow [\bar{\xi}_i, \bar{\xi}_j] \in \mathcal{D}_{\mathcal{K}},$$

where $[\cdot,\cdot]$ denotes the delayed state bracket.

Then, concerning to the solvability of the finite spectrum assignment problem, we obtain the following theorem:

Theorem 2 [14] *Suppose system* (3) *is given. There exists a scalar function* $\varphi(x)$ *which consists of* $x(t)$ *and which accomplishes the finite spectrum assignment for the system if and only if the following conditions are satisfied:*

(i) The distribution

$$\text{span}\{\bar{g}(x,\sigma), \ldots, ad_{\bar{f}}^{n-1}\bar{g}(x,\sigma)\}$$

evaluated by any $\sigma^k x(t)$ *and* σ^k *for any* $k \in \mathbb{N}$ *has rank n for any x in the neighborhood U of the origin.*

(ii) The distribution

$$\text{span}\{\bar{g}(x,\sigma), ad_{\bar{f}}\bar{g}(x,\sigma), \ldots, ad_{\bar{f}}^{n-2}\bar{g}(x,\sigma)\}$$

is σ-involutive.

The first condition corresponds to the controllability over the polynomial ring $\mathbb{R}[\sigma]$ for linear retarded systems and requires that the approximately linearized system around the origin is controllable over the polynomial ring. However, this condition is relatively restrictive and the finite spectrum assignment for linear retarded systems is extended for a wider class of systems. Therefore in the remains of this chapter we attempt to mainly relax the first condition of Theorem 2.

3 Main Results

3.1 Motivation

Consider a linear retarded systems as follows:

$$\begin{cases} \dot{x}_1(t) = u(t) \\ \dot{x}_2(t) = x_1(t-\tau) \end{cases}, \quad y(t) = x_2(t)$$

This system is mentioned as an example of systems which does not satisfy the solvability condition of the finite spectrum assignment in [14]. Applying the variable transformation

$$\begin{bmatrix} z_1(t) \\ z_2(t) \end{bmatrix} = \begin{bmatrix} x_2(t) \\ x_1(t-\tau) \end{bmatrix},$$

transforms the system into the following system with time-delay at the input for $t \geq \tau$:

$$\dot{z}(t) = \begin{bmatrix} 0 & 1 \\ 0 & 0 \end{bmatrix} z(t) + \begin{bmatrix} 0 \\ 1 \end{bmatrix} u(t-\tau), \quad y(t) = \begin{bmatrix} 1 & 0 \end{bmatrix} z(t)$$

This means that the retarded system is transformed into an input-delay system. If the input $u(t)$ could somehow realize such that $u(t-\tau) = -\alpha_1 z_1(t) - \alpha_2 z_2(t)$, the closed-loop system would be changed into a finite dimensional system

$$\dot{z} = \begin{bmatrix} 0 & 1 \\ -\alpha_1 & -\alpha_2 \end{bmatrix} z(t),$$

but this input is not applicable because the input violates the causality principle.

This example suggests that applying the variable transformation containing time-delay, retarded systems may be transformed into systems with input delay and even if the finite spectrum assignment is not applicable for the original system, the transformed system may be accomplished the finite spectrum assignment by combining with some sort of state predictor.

3.2 Synchronization-Based Predictor

This subsection introduces a state predictor based on synchronization of coupled systems. In the first contribution concerning anticipating synchronization [23], which is a kind of master-slave synchronization, Voss considers anticipating synchronization of the Ikeda equation, but the fundamental idea of anticipating synchronization can be extended for more general nonlinear systems [15].

The scheme of anticipating synchronization is briefly summarized as follows. We consider the following possible nonlinear system:

$$\Sigma_M : \begin{cases} \dot{x}(t) = A_0 x(t) + A_1 x(t-\tau) + f(x(t)) \\ y(t) = Cx(t), \quad x(t) = \varphi(t), \ t \in [-\tau, 0] \end{cases} \tag{5}$$

where $x \in \mathbb{R}^n$, $y \in \mathbb{R}^m$, A_0 and A_1 are $n \times n$ matrices, C is a $m \times n$ matrix and $f : \mathbb{R}^n \to \mathbb{R}^n$ is a smooth vector field.

For the system (5), we construct the predictor

$$\Sigma_s : \begin{cases} \dot{z}(t) = A_0 z(t) + A_1 z(t-\tau) + f(z(t)) + K\{Cz(t-\tau) - y(t)\} \\ z(t) = z_0, \ t \in [-2\tau, 0] \end{cases} \tag{6}$$

where $z \in \mathbb{R}^n$ and $K \in \mathbb{R}^{n\times m}$ is a constant matrix. For simplicity of notation, we will often suppress the explicit dependence of time t when no confusion can arise. The configuration of the prediction scheme is shown in Fig. 1. In this figure, $e^{-\tau s}$ denotes a time-delay element. Then the dynamics of prediction error $e(t) := z(t-\tau) - x(t)$ is given by

$$\begin{aligned} \dot{e}(t) &= A_0 e(t) + (A_1 + KC)e(t-\tau) + f(e+x) - f(x) \\ &:= A_0 e(t) + A_d e(t-\tau) + \varphi(x, e) \end{aligned} \tag{7}$$

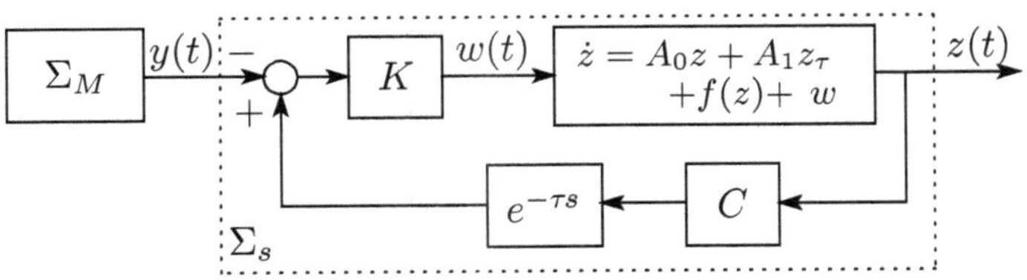

Fig. 1 Configuration of the prediction scheme

with $\varphi(x, e) = f(e+x) - f(x)$. Clearly, $e \equiv 0$ is a solution of Eq. (7), and the system (6) acts as a predictor for (5) if the error dynamics (7) has $e = 0$ as asymptotically stable equilibrium.

3.3 Predictor-Based Control

In this subsection, we consider a predictor based control method for nonlinear systems with time-delay at the input. Now we assume that the system is transformed into systems with input delay by applying a variable transform $z(t) = \Phi(x, \sigma)$. Then as the transformed system, we consider the following system

$$\begin{cases} \dot{z}(t) = f(z(t), u(t-\tau)) \\ y(t) = Cz(t) \end{cases} \tag{8}$$

where f is smooth with respect to z and u. We assume that when $\tau = 0$ holds, i.e. the system is delay-free, a feedback $u = \gamma(z(t))$ is so designed that the closed-loop system achieves satisfactory stability. However if $\tau \neq 0$, the closed-loop system becomes $\dot{z}(t) = f(z(t), \gamma(z(t-\tau)))$ and may be destabilized by the existence of the delay. While if we could obtain the future value $z(t+\tau)$ of z, by applying a feedback $u(t) = \gamma(z(t+\tau))$, the closed-loop system would be given by $\dot{z}(t) = f(z(t), \gamma(z(t)))$. From such a viewpoint, we propose a control scheme in which a predictor having the structure described in Sect. 2 is used to estimate the future value of a given system and the output of the predictor is used in place of the actual state of the system in a feedback. The synchronization-based predictor is described by

$$\begin{aligned} \dot{\tilde{z}}(t) &= f(\tilde{z}(t), u(t)) + K\{C\tilde{z}(t-\tau) - y(t)\} \\ &= f(\tilde{z}(t), u(t)) + KC\{\tilde{z}(t-\tau) - \Phi(x, \sigma)\} \end{aligned}$$

and the controller is given by $u(t) = \gamma(\tilde{z}(t))$.

The configuration of the proposed control scheme is shown in Fig. 2. In this figure, the system to be controlled is given by $\dot{x} = F(x_t, u)$ and $y(t) = Cx(t)$ and this system is assumed to be transformed into the Eq. (8).

Note that to realize this control scheme, the solution of system (8) must remain bounded at least until the control input comes into effect at $t = \tau$.

3.4 Extended Separation Principle

We consider the local stability of the total system. For the system (8), a predictor and a control law are given by

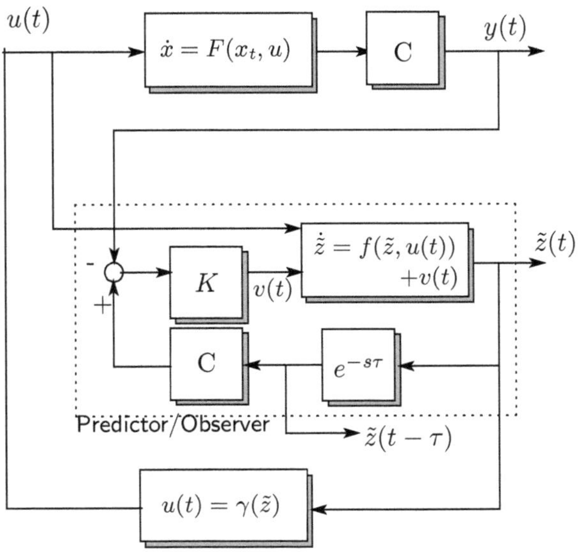

Fig. 2 Configuration of the proposed control scheme

$$\dot{\tilde{z}}(t) = f(\tilde{z}(t), u(t)) + KC\{\tilde{z}(t-\tau) - \Phi(x,\sigma)\} \tag{9}$$

$$u(t) = \gamma(\tilde{z}(t)). \tag{10}$$

Substituting (10) into (8) yields

$$\dot{z}(t) = f(z(t), \gamma(\tilde{z}_\tau)) := f(z, \gamma(z+e)) \tag{11}$$

where

$$\tilde{z}_\tau := \tilde{z}(t-\tau) \text{ and } e(t) := \tilde{z}(t-\tau) - z(t).$$

The dynamics of the prediction error $e(t)$ is given by

$$\begin{aligned} \dot{e} &= f(\tilde{z}_\tau, \gamma(\tilde{z}_\tau)) - f(z, \gamma(\tilde{z}_\tau)) + KCe(t-\tau) \\ &= f(e+z, \gamma(\tilde{z}_\tau)) - f(z, \gamma(\tilde{z}_\tau)) + KCe_\tau \end{aligned} \tag{12}$$

If f and γ are sufficiently smooth to permit Taylor series expansion, we obtain

$$f(z, \gamma(z+e)) = f(z, \gamma(z)) + \left(\frac{\partial f}{\partial \gamma}\right)\left(\frac{\partial \gamma}{\partial e}\right)e + \mathcal{O}(e^2) \tag{13}$$

$$f(z+e, \gamma(\tilde{z}_\tau)) = f(z, \gamma(\tilde{z}_\tau)) + \left(\frac{\partial f}{\partial z}\right)e + \mathcal{O}(e^2) \tag{14}$$

where $\mathcal{O}(e^2)$ represents terms that are smaller in absolute value than some constant times $|e^2|$ when e is close enough to zero. Substituting (13) and (14) into (11) and (12) respectively, we obtain

$$\dot{z} = f(z, \gamma(z)) + \left(\frac{\partial f}{\partial \gamma}\right)\left(\frac{\partial \gamma}{\partial e}\right)e + \mathcal{O}(e^2)$$
$$\dot{e} = \left(\frac{\partial f}{\partial z}\right)e + KCe_\tau + \mathcal{O}(e^2). \tag{15}$$

As a result, when the prediction error converges to zero, the effect of the prediction error in Eq. (15) vanishes and then the behavior of z coincides with $\dot{z}(t) = f(z, \gamma(z))$.

If the system is linear, this result corresponds to a counterpart of so-called the "separation principle" of controller and observer.

Remark 3 If the variable transformation is not invertible, the transformation is described by a difference equation and it causes the internal dynamics. The related issue is discussed in [3].

We consider the proposed control scheme for some special cases in the following subsections.

3.5 FSA for Linearizable Systems with Input Delay

At the beginning, we consider a case in which delay-free systems are linearizable by change of coordinates. Consider the following input delay system:

$$\dot{x}(t) = f(x) + g(x)u(t - \tau) \tag{16}$$

where $x \in \mathbb{R}^n$, $u \in \mathbb{R}$ and f and g are smooth vector fields with $f(0) = 0$ and $g(0) \neq 0$. Since this system has a time-delay at the input, the first condition of Theorem 2 is obviously unsatisfied for this system. Then as is well known, if f and g satisfy the following conditions

1. $\mathrm{span}\{g, ad_f g, \ldots, ad_f^{n-1} g\} = \mathbb{R}^n$
2. $[ad_f^i g, ad_f^j g] = 0$ for $0 \leq i, j \leq n - 1$

in a neighborhood of the origin, the system (16) is locally state linearizable, i.e. the system (16) is transformed into a linear controllable system

$$\dot{z}(t) = Az(t) + bu(t - \tau)$$

in a neighborhood of the origin by applying a local diffeomorphism $z = \Phi(x)$ with $\Phi(0) = 0$. For such a system, the proposed synchronization-based predictor is given by

$$\dot{\tilde{z}}(t) = A\tilde{z}(t) + bu(t) + KC(\tilde{z}(t-\tau) - \Phi(x(t)))$$

where $C = I_{n\times n}$ and K is chosen such that the origin of the prediction error dynamics

$$\dot{e}(t) = Ae(t) + KCe(t-\tau)$$

is asymptotically stable, where the prediction error e is defined by

$$e(t) = \tilde{z}(t-\tau) - z(t) = \tilde{z}(t-\tau) - \Phi(x(t)).$$

For linear retarded systems, a large number of delay-dependent stability criteria has been proposed in the form of LMI conditions derived from the Lyapunov-Razumikhin theorem and the Lyapunov-Krasovskii theorem. For instance, the coupling gain K is designed as follows. If there exist a symmetric and positive-definite matrix $P > 0$ and a matrix Y such that the following linear matrix inequality holds:

$$\begin{bmatrix} \Xi_{11} + \Xi_{11}^T & A^T P & C^T Y & YC \\ PA & -\frac{1}{\tau}P & 0 & 0 \\ YC & 0 & -\frac{1}{\tau}P & 0 \\ C^T Y^T & 0 & 0 & -\frac{1}{2\tau}P \end{bmatrix} < 0 \tag{17}$$

where $\Xi_{11} = PA + YC$, then choosing $K = P^{-1}Y$ the zero solution of system

$$\dot{e}(t) = Ae(t) + KCe(t-\tau)$$

is asymptotically stable. As a result, applying the stabilizing controller summarized by

$$\begin{cases} u(t) = -\sum_{i=1}^{n} \alpha_i \tilde{z}_i(t) \\ \dot{\tilde{z}}(t) = A\tilde{z}(t) + bu(t) + KC(\tilde{z}(t-\tau) - \Phi(x(t))) \end{cases},$$

the finite spectrum assignment can be accomplished.

Thus the design problem of the coupling gain matrix K can be reduced to LMI problems by applying such stability conditions.

3.6 FSA for Nonlinear Retarded Systems

Next, we extend the above control scheme for nonlinear retarded systems.

Consider the following single input nonlinear system with commensurate delays:

$$\begin{cases} \dot{x}(t) = f(x(t), x(t-\tau)) + gu(t) \\ x(t) = \varphi(t), \ t \in [-\tau, 0] \end{cases} \tag{18}$$

in which $x(t) \in \mathbb{R}^n$, $u \in \mathbb{R}$, τ is a fixed time delay, and $\varphi : \mathbb{R} \to \mathbb{R}^n$ is a continuous vector valued function specifying the initial state of the system. It is assumed that $f(\cdot)$ is a real smooth vector field with $f(0) = 0$ and $g \neq 0$ is a constant vector.

Using the delay operator $\sigma : \lambda(t) \mapsto \lambda(t - \tau)$, system (18) can be rewritten as

$$\dot{x}(t) = f(x, \sigma x) + g\sigma^0 u(t) := \bar{f}(x, \sigma) + \bar{g}(x, \sigma)u(t) \tag{19}$$

where $\bar{g}(x, \sigma) = g\sigma^0$. Assume that the above system satisfies the following conditions:

1. the distribution

$$\mathscr{C}(x, \sigma) = \left[\bar{g}(x, \sigma) \cdots ad_{\bar{f}}^{n-1}\bar{g}(x, \sigma)\right]$$

   evaluated by any $\sigma^k x(t)$ and non-zero σ^k for any $k \in \mathbb{N}$ has rank n for any x in the neighborhood U of the origin.
2. the delayed state brackets satisfy $[ad_{\bar{f}}^i \bar{g}, ad_{\bar{f}}^j \bar{g}] = 0$ for $0 \leq i, j \leq n - 1$.

Under these conditions, the system (19) can be transformed into a linear systems with input delay by applying a variable transformation $z(t) = \Phi(x, \sigma)$:

$$\dot{z}(t) = Az(t) + bu(t - L)$$

where L is integral multiple length of τ and the length depends on the variable transformation. As a result, since the nonlinear retarded system to be stabilized is transformed into a linear systems with input delay, in the same way as the forgoing subsection, this system can be stabilized using the state predictor. The corresponding predictor is given by

$$\dot{\tilde{z}}(t) = A\tilde{z}(t) + bu(t) + K(\tilde{z}(t - L) - \Phi(x, \sigma))$$

and the coupling gain K is chosen such that the prediction error dynamics

$$\dot{e}(t) = Ae(t) + Ke(t - L)$$

is asymptotically stable.

4 Examples

To illustrate the effectiveness of the proposed control strategies, we show two typical examples corresponding to the foregoing section. For simplicity, all states are assumed to be measurable for both examples.

4.1 Linearizable System

As the first example, we consider the following nonlinear system with input delay.

$$\begin{cases} \dot{x}_1(t) = x_2(t) + x_2^2(t)u(t-\tau) \\ \dot{x}_2(t) = u(t-\tau) \end{cases}$$

where $x \in \mathbb{R}^2$, $u \in \mathbb{R}$, $f(x) = \left(x_2, 0\right)^T$ and $g(x) = \left(x_2^2, 1\right)^T$. Since

$$ad_f g = \begin{pmatrix} -1 \\ 0 \end{pmatrix}, \; span\{g, ad_f g\} = \left\{\begin{pmatrix} x_2^2 \\ 1 \end{pmatrix}, \begin{pmatrix} -1 \\ 0 \end{pmatrix}\right\}$$

and $[g, ad_f g] = \left(0, 0\right)^T$, this system is linearized by applying a nonlinear coordinate transformation. In fact, applying the coordinate transformation defined as

$$z(t) = \begin{pmatrix} x_1 - \frac{1}{3}x_2^3 \\ x_2 \end{pmatrix} := \Phi(x(t)),$$

the system is transformed into the following linear system with input-delay.

$$\begin{aligned} \dot{z}(t) &= \begin{bmatrix} 0 & 1 \\ 0 & 0 \end{bmatrix} z(t) + \begin{bmatrix} 0 \\ 1 \end{bmatrix} u(t-\tau) \\ &:= A_0 z(t) + bu(t-\tau) \end{aligned}$$

Therefore a state predictor based on anticipating synchronization is designed as

$$\dot{\tilde{z}}(t) = A_0\tilde{z}(t) + bu(t) + K\left(\tilde{z}(t-\tau) - \Phi(x(t))\right)$$

and a stabilizing feedback is designed as $u(t) = -\alpha_1\tilde{z}_1(t) - \alpha_2\tilde{z}_2(t)$ where K is chosen such that the prediction error dynamics

$$\dot{e}(t) = A_0 e(t) + Ke(t-\tau)$$

is asymptotically stable, where $e(t) = \tilde{z}(t-\tau) - \Phi(x(t))$. Figures 3 and 4 show a simulation result.

The coupling gain is obtained by solving the LMI condition (17). From these figures, we know that the proposed synchronization-based predictor plays full part to estimate the future value of $x(t)$ and the total system is stabilized.

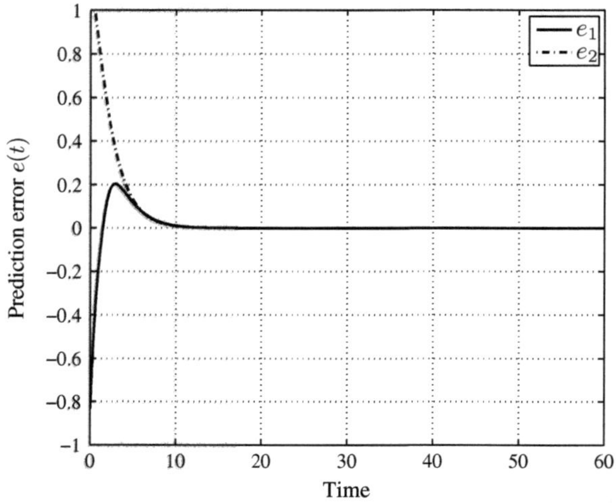

Fig. 3 Error $e(t)$

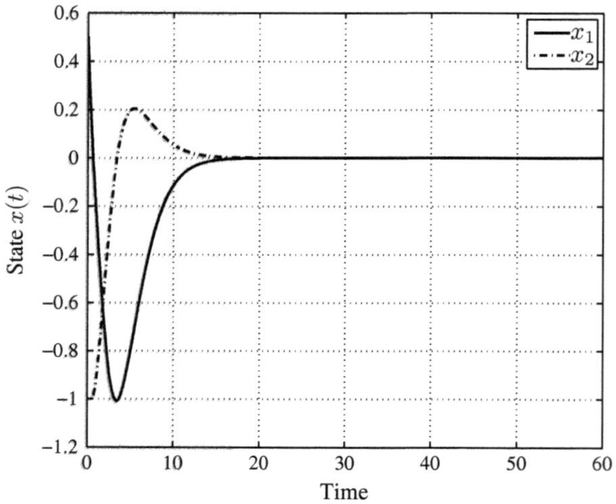

Fig. 4 Behavior of $x(t)$

4.2 FSA for Nonlinear Retarded Systems

Consider the following nonlinear retarded system:

$$\dot{x}(t) = \begin{pmatrix} x_2(t) \\ x_3(t-\tau) + \sin x_1(t-\tau) \\ -x_2(t)\cos x_1(t) \end{pmatrix} + \begin{pmatrix} 0 \\ 0 \\ 1 \end{pmatrix} u(t)$$

where $\tau = 0.3$. Using the delay operator, this system can be rewritten as

$$\dot{x}(t) = \begin{pmatrix} x_2 \\ \sigma x_3 + \sin(\sigma x_1) \\ -x_2 \cos x_1 \end{pmatrix} + \begin{pmatrix} 0 \\ 0 \\ 1 \end{pmatrix} u(t) = \bar{f}(x, \sigma) + gu(t).$$

Appropriate calculations show that

$$ad_{\bar{f}} g(x, \sigma) = (0, -\sigma, 0)^T \text{ and } ad_{\bar{f}}^2 g(x, \sigma) = (\sigma, 0, -(\cos x_1)\sigma)^T.$$

The matrix $\mathscr{C}(x, \sigma)$ is given by

$$\mathscr{C}(x, \sigma) = \begin{bmatrix} 0 & 0 & \sigma \\ 0 & -\sigma & 0 \\ 1 & 0 & -(\cos x_1)\sigma \end{bmatrix}$$

and it has rank 3 for any $x(t)$, σx and non-zero σ. Therefore condition 1 shown in Sect. 3.6 is satisfied. In addition, $\bar{g}$ and $ad_{\bar{f}} g(x, \sigma)$ satisfy Condition 2. Therefore, this system can be transformed into a linear system with input delay. In fact, applying a variable transformation $z(t) = \Phi(x, \sigma)$ given by

$$z(t) = (x_1, x_2, \sigma x_3 + \sin(\sigma x_1))^T,$$

the system is transformed into

$$\dot{z}(t) = \begin{bmatrix} 0 & 1 & 0 \\ 0 & 0 & 1 \\ 0 & 0 & 0 \end{bmatrix} z(t) + \begin{bmatrix} 0 \\ 0 \\ \sigma \end{bmatrix} u(t) := Az(t) + bu(t - \tau).$$

Therefore, by applying

- **Predictor**: $\dot{\tilde{z}}(t) = A\tilde{z}(t) + bu(t) + K(\tilde{z}(t - \tau) - \Phi(x, \sigma))$
- **Feedback**: $u(t) = -\sum_{i=1}^{3} \alpha_i \tilde{z}_i(t)$

to the system, the finite spectrum assignment is accomplished and the system can be stabilized.

The left panel of Fig. 5 shows a numerical simulation of the trajectory of x, and the right panel of Fig. 5 shows the prediction error. In this simulation, the initial condition of x is $x(0) = \begin{pmatrix} 1 & 0.5 & -0.3 \end{pmatrix}^T$ and $x(\theta) = 0$ for $-\tau \le \theta < 0$ and the initial condition of the predictor is $\tilde{z}(\theta) = 0$ for $\theta \le 0$. The synchronization based predictor estimates the future value of x, and the proposed control scheme works well. As a result, the origin of the original system is stabilized by the proposed controller.

Of course, there are limitations to the length of time delay that is allowable for this prediction method but this predictor can be used in the same way as the conventional observer. In addition, it is worth mentioning that the inverse mapping of the applied variable transformation is

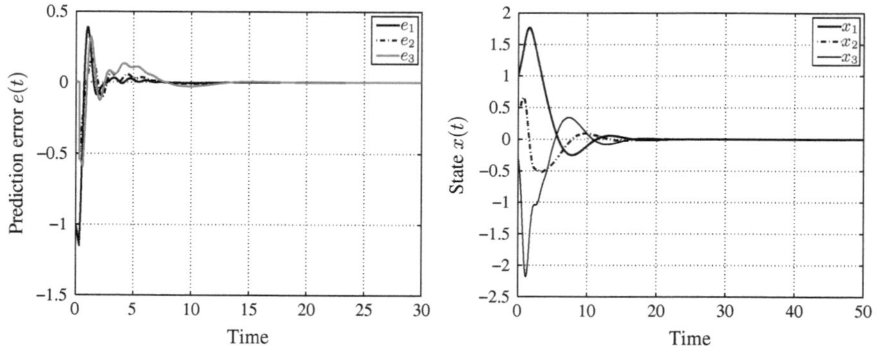

Fig. 5 Prediction error $e(t)$ (*left*) and behavior of $x(t)$ (*right*)

$$\begin{pmatrix} x_1(t) \\ x_2(t) \\ x_3(t-\tau) \end{pmatrix} = \begin{pmatrix} z_1(t) \\ z_2(t) \\ z_3(t) - \sin z_1(t-\tau) \end{pmatrix},$$

which acts counter to the causal law. However, since this method does not use the inverse transformation, this control strategy itself is feasible.

5 Conclusions

We considered an extension of the finite spectrum assignment for nonlinear systems and discussed the finite spectrum assignment by combining it with the state predictor based on anticipating synchronization. The proposed approach transforms retarded systems into input delay systems by applying variable transformations with delays and combines linearization with the synchronization-based predictor. For simplicity of discussion, we focused on linearizable systems by variable transformations with or without delay, but the restriction on the applicable systems comes from the stability of the prediction error dynamics. Even if the transformed system is not linear, the predictor is available if the nonlinear terms have good properties such as the sector condition. For instance, when the transformed systems is nonlinear, the predictor may be found using the transformed system with correction term $KC\,(\tilde{z}(t-L) - \Phi(x, \sigma))$ if it is guaranteed that the trivial solution of the prediction error dynamics is asymptotically stable. As another option, even if the transformed system is nonlinear, a linear predictor may be given by neglecting the nonlinear terms. In such a case, the stability of the prediction error dynamics has been discussed in [16, 17].

References

1. Boukas, E.-K., Liu, Z.-K.: Deterministic and Stochastic Time Delay Systems. Birkhauser, Boston (2001)
2. Germani, A., Manes, C., Pepe, P.: Linearization of input-output mapping for nonlinear delay systems. In: Proceedings of the IMACS Multiconference on Computational Engineering in Systems Applications, pp. 599–602 (1996)
3. Germani, A., Manes, C., Pepe, P.: Input-output linearization with delay cancellation for nonlinear delay systems: the problem of the internal stability. Int. J. Robust Nonlinear Control **13**(9), 909–937 (2003)
4. Gu, K., Kharitonov, K., Chen, J.: Stability of Time-Delay Systems. Birkhauser, Boston (2003)
5. Krstic, M.: Delay Compensation for Nonlinear, Adaptive, and PDE Systems. Birkhauser, Boston (2009)
6. Mahmoud, M.: Robust Control and Filtering for Time-delay Systems. Marcel Dekker, New York (2000)
7. Manitius, A., Olbrot, A.: Finite spectrum assignment problem for systems with delays. IEEE Trans. Autom. Control **24**(4), 541–552 (1979)
8. Moog, C., Castro-Linares, R., Velasco-Villa, M., Marquez-Martinez, L.: The disturbance decoupling problem for time-delay nonlinear systems. IEEE Trans. Autom. Control **45**(2), 305–309 (2000)
9. Morse, A.: Ring models for delay-differential systems. Automatica **12**(5), 529–531 (1976)
10. Nijmeijer, H.: A dynamical control view on synchronization. Physica D **154**(3–4), 219–228 (2001)
11. Nijmeijer, H., Mareels, I.: An observer looks at synchronization. IEEE Trans. Circuits Syst. I: Fundam. Theory Appl. **40**(10), 882–890 (1997)
12. Ogorzalek, M.: Taming chaos. Part I: Synchronization. IEEE Trans. Circuits Syst. I: Fundam. Theory Appl. **40**(10), 693–699 (1993)
13. Oguchi, T.: Finite spectrum assignment for nonlinear time-delay systems using synchronization-based predictor. In: Proceedings of the 51st IEEE Conference on Decision and Control, pp. 3874–3879 (2006)
14. Oguchi, T.: A finite spectrum assignment for retarded non-linear systems and its solvability condition. Int. J. Control **80**(6), 898–907 (2007)
15. Oguchi, T., Nijmeijer, H.: Prediction of chaotic behavior. IEEE Trans. Circuit Syst. I: Regul. Pap. **52**(11), 2464–2472 (2005)
16. Oguchi, T., Nijmeijer, H.: Control of nonlinear systems with time-delay using state prediction based on synchronization. In: Proceedings of 5th EUROMECH Nonlinear Dynamics Conference, pp. 1150–1156 (2005)
17. Oguchi, T., Nijmeijer, H.: Anticipating synchronization of nonlinear systems with uncertainties. In: Proceedings of 6th IFAC Workshop on Time Delay Systems, pp. 290–295 (2006)
18. Oguchi, T., Watanabe, A.: Input-output linearization of nonlinear systems with time-delays in state variables. Int. J. Syst. Sci. **29**(6), 573–578 (1998)
19. Oguchi, T., Watanabe, A., Nakamizo, T.: Input-output linearization of retarded non-linear systems by using an extension of Lie derivative. Int. J. Control **75**(8), 582–590 (2002)
20. Pyragas, L.: Synchronization of coupled time-delay systems: analytical estimation. Phys. Rev. E **58**(3), 3067–3071 (1998)
21. Smith, O.: A controller to overcome dead time. ISA J. **6**(2), 28–33 (1959)
22. Sontag, E.: Linear systems over commutative rings: a survey. Ricerche di Automatica **7**(1), 1–34 (1976)
23. Voss, H.: Anticipating chaotic synchronization. Physical Review E **61**(5), 5115–5119 (2000)

Improved Stability Criteria for Sampled-Data Systems with Input Saturation

Alexandre Seuret, Frédéric Gouaisbaut, Sophie Tarbouriech and Joao M Gomes da Silva Jr.

Abstract In this of either the controller or the network is addressed for sampled-data systems with input saturation. Using modified sector conditions, an adequate looped functional, and the Wirtinger-based integral inequality, quasi-LMI conditions, with a scalar parameter to tune, are proposed in the regional (or local) context for both design problems. The associated convex optimizations are briefly described. Finally, some examples show the efficiency of the methods with respect to existing results.

1 Introduction

In the past decade, much attention has been devoted to the stability analysis of sampled-data systems [8]. This area addresses the problem of stability and stabilization of systems evolving in continuous time whereas the control input evolves in discrete time. Although the theory for linear sampled-data control systems is

This work was supported by ANR project LimICoS contract number 12 BS03 005 01 and by STIC-AmSud project ADNEC.

A. Seuret (✉) · S. Tarbouriech · F. Gouaisbaut
CNRS, LAAS, 7 avenue du Colonel Roche, 31400 Toulouse, France
e-mail: seuret@laas.fr

A. Seuret · S. Tarbouriech · F. Gouaisbaut
Univ de Toulouse, LAAS, 31400 Toulouse, France

S. Tarbouriech
e-mail: tarbour@laas.fr

F. Gouaisbaut
UPS, LAAS, 7 avenue du Colonel Roche, 31400 Toulouse, France
e-mail: gouaisb@laas.fr

F. Gouaisbaut
Univ de Toulouse, UPS, 31400 Toulouse, France

J.M. Gomes da Silva Jr.
UFRGS Departamento de Engenharia Elétrica, Avenida Osvaldo Aranha, Porto Alegre-rs 90035-051, Brazil
e-mail: jmgomes@ece.ufrgs.br

I. Karafyllis et al. (eds.), *Recent Results on Nonlinear Delay Control Systems*, Advances in Delays and Dynamics 4, DOI 10.1007/978-3-319-18072-4_16

well established for constant sampling period [1] through the discretization of the continuous-time systems, there is still much ongoing research activity that studies the case of asynchronous sampling. Indeed, in practice, a heavy temporary load in the hardware which computes the control inputs can lead to significant variations in the sampling time. These variations can dramatically affect the stability properties of a control system. That is then clearly the motivation for the design of robust stability and stabilization conditions which take into account the variations on the sampling period.

Among the possible solutions, the sampling in a control loop can be modeled as a time-varying delay on the plant control input while considering a continuous-time dynamics. In this case, the delay variation rate between two sampling instants is equal to one. Thus, a fundamental problem is the determination of bounds on the delay variation for which stability of the closed-loop system can be kept. Considering this perspective, we can cite for example the approach proposed in [6], and the improvements made in [5, 13].

On the other hand, many techniques have been proposed in the literature to deal with the stabilization of time-delay systems subject to actuator saturation. In this context, we can cite, for instance: [14, 15], where globally stabilizing control laws are proposed; [2, 6, 20], where the regional stabilization problem is considered; [11, 12], where the authors consider the problem of globally stabilizing systems with bounded feedbacks in the case where an arbitrarily long delay is present in the input. These works are mainly concerned with state delayed systems. The stabilization conditions consider delay independent approaches, and delay dependent conditions with fixed delays. Recently, in [4, 22], results on systems with time-varying delays on the states have been proposed. On the other hand, we can note a lack of results considering input delays, and, in particular, time-varying ones. In [3, 21], anti-windup techniques are proposed for systems with fixed input delays.

In this chapter, we are interested in the problem of stabilizing a sampled-data system taking into account the possible variations in the sample intervals which may be due to packet losses in a communication network and also the fact that the signals provided by the actuators are bounded (i.e., saturating plant inputs). The method is based on the use of a particular functional that (unlike the Lyapunov-Krasovskii based approaches adopted for instance in [5, 6, 13]) does not need to be positive definite. It is shown that if the time-derivative of this functional along all trajectories of the continuous-time model is strictly negative, then a quadratic Lyapunov function is strictly decreasing for the discrete-time asynchronous system. Such an analysis has already been addressed in [17]. In the present chapter, we aim to present an improved version of the stability conditions in [17] by using of a less conservative integral inequality than the usual Jensen inequality. This new integral inequality is inspired by [18], and is adapted to our purpose in order to fit with the proposed analysis. On the other hand, the control saturation is taken into account from the use of a generalized sector condition.

The chapter is organized as follows. Section 2 describes the problem formulation. Section 3 presents some preliminary lemmas on a LMI manipulation and on the generalized sector condition. Section 4 presents the main stabilization theorem. In Sect. 5, some optimization problems are proposed. Some numerical examples illustrating the potentialities of the method are provided in Sect. 6.

Notation. Throughout the chapter, the sets $\mathbb{N}$, $\mathbb{R}^+$, $\mathbb{R}^n$, $\mathbb{R}^{n\times n}$ and $\mathbb{S}^n$ denote the sets of all positive integers, positive scalars, n-dimensional vectors, $n \times n$ matrices and symmetric matrices in $\mathbb{R}^{n\times n}$, respectively. For a given positive scalar $T_{\max}$, define $\mathbb{K}$ to be the set of all continuous functions from an interval $[0,\ T]$ to $\mathbb{R}^n$, where T is a positive scalar less than $T_{\max}$. The notation $|\cdot|$ and $\|\cdot\|$ stand for the absolute value of a scalar and for the Euclidean norm of a vector, respectively. The superscript 'T' stands for matrix or vector transposition. The notation $P > 0$ for $P \in \mathbb{S}^n$ means that P is symmetric positive definite. For any positive integer $j \leq n$, any vector $x \in \mathbb{R}^n$, and any matrix $A \in \mathbb{R}^{n\times n}$. Then the notation A_j and x_j refer to the jth line of matrix A and the jth component of vector x, respectively. The symbols I and 0 represent the identity and zero matrices of appropriate dimensions, and $Co\{\cdot\}$ denotes a convex hull. For any matrix A in $\mathbb{R}^{n\times n}$, we define $\text{He}\{A\}$ as $A + A^T$.

2 Problem Formulation

Let $\{t_k\}_{k\in\mathbb{N}}$ be an increasing sequence of positive scalars such that $\bigcup_{k\in\mathbb{N}}[t_k,\ t_{k+1}[=[0,\ +\infty[$. Assume that there exist two positive scalars $T_{\min} \leq T_{\max}$ such that the difference between two successive sampling instants $T_k = t_{k+1} - t_k$ satisfies

$$\forall k \in \mathbb{N}, \quad 0 \leq T_{\min} \leq T_k \leq T_{\max}. \tag{1}$$

Consider the linear system

$$\forall t \in [t_k,\ t_{k+1}[, \quad \dot{x}(t) = Ax(t) + Bu(t_k) \tag{2}$$

with a sampled-data input, where $x \in \mathbb{R}^n$ and $u \in \mathbb{R}^m$ represent the state variable and the input vector. The matrices A and B are assumed to be constant, known, and of appropriate dimensions. We suppose that the input vector u is subject to amplitude limitations defined by

$$|u_i| \leq u_{0i}, \quad u_{0i} > 0,\ i = 1, \ldots, m. \tag{3}$$

Consider a linear state feedback control law $u(t) = Kx(t)$, where $K \in \mathbb{R}^{m\times n}$. Due to the control bounds defined in (3), the effective control signal to be applied to the system is given by

$$u(t) = sat(Kx(t)), \tag{4}$$

where $u_i(t) = sat(K_i x(t)) = \text{sign}(K_i x(t)) \min\{u_{0i}, |K_i x(t)|\}$, $i = 1, \ldots, m$, with u_{0i}, the ith level of saturation. Hence, the closed-loop system reads

$$\dot{x}(t) = Ax(t) + Bsat(Kx(t_k)). \tag{5}$$

Although the system dynamics are considered to be linear, the closed-loop system is nonlinear, due to the control saturation. Hence, the determination of a global stabilizing controller is only possible when some stability assumptions are verified by the open-loop system [10, 19]. When these hypotheses are not verified, it is only possible to achieve semi-global, or local or regional stabilization. In this case, given a stabilizing matrix K, we associate a region of attraction to the equilibrium point $x_e \equiv 0$ of the system (5). The region of attraction corresponds to all initial conditions $x_0 \in \mathbb{R}^n$ such that the corresponding trajectories of the system (5) converge asymptotically to the origin [9]. Since the determination of the exact region of attraction is practically impossible, a problem of interest is to ensure asymptotic stability for a set of admissible initial conditions x_0. Hence, from the considerations above, this chapter is concerned with studying the stabilization problems stated as follows.

P1 Maximize the bound on the maximal allowable sampling period T_{max}, for which asymptotic stability of the closed-loop system (5) can be ensured for some set of admissible initial conditions.

P2 Given T_{min} and T_{max}, find K and a set of admissible initial conditions, which are as large as possible, for which asymptotic stability of the closed-loop system (5) is ensured.

Regarding a networked control system, problem **P2** can be seen as a "controller" design problem, which is defined from the network schedule constraints. On the other hand, problem **P1** can be seen as a "network" design problem. In this case, the network designer will consider the maximum allowable jitter (for which stability of the system can be guaranteed) as a constraint in the network scheduling. Of course, when possible, the objective is the global stabilization of the closed-loop system. Otherwise, a set of admissible initial conditions, included in the region of attraction of the closed-loop system, has to be defined. This set can be seen as an estimate of the actual region of attraction and defines a region of "safe initialization" for the system.

In order to develop conditions to solve problems **P1** and **P2**, we adopt a particular notation in the sequel. For all integers $k \in \mathbb{N}$, a function $\chi_k \in \mathbb{K}$ can be defined such that Eq. (5) can be equivalently represented by:

$$\forall \tau \in [0,\ T_k], \quad \begin{cases} \chi_k(0) = x(t_k), \\ \chi_k(\tau) = x(t_k + \tau), \\ \dot{\chi}_k(\tau) = \frac{d}{d\tau}\chi_k(\tau) = A\chi_k(\tau) + Bsat(K\chi_k(0)). \end{cases} \tag{6}$$

In the sequel, we aim to design constructive LMI conditions guaranteeing the stabilization of such systems.

3 Preliminary Lemmas and Theorems

3.1 Modified Sector Condition

Using the notation χ_k given in (6), the following dead-zone function is defined:

$$\psi(K\chi_k(0)) = K\chi_k(0) - sat(K\chi_k(0)). \tag{7}$$

Note that $\psi(K\chi_k(0))$ corresponds to a decentralized dead-zone nonlinearity. Considering the function $\psi(K\chi_k(0))$, the closed-loop system can be re-written as

$$\dot{\chi}_k(\tau) = A\chi_k(\tau) + BK\chi_k(0) - B\psi(K\chi_k(0)), \quad \forall\tau \in [0,\ T_k].$$

Now consider a matrix $G \in \mathbb{R}^{m\times n}$, and define the polyhedral set

$$\mathscr{S} = \{x \in \mathbb{R}^n;\ |(K_i - G_i)x| \leq u_{0i},\ i = 1, \ldots, m\}.$$

The next lemma is from [3], and concerns the nonlinearity $\psi(K\chi_k(0))$.

Lemma 1 *Consider the function $\psi(K\chi_k(0))$ defined in (7). If $\chi_k(0) \in \mathscr{S}$, then the relation*

$$\psi^T(K\chi_k(0))U[\psi(K\chi_k(0)) - G\chi_k(0)] \leq 0, \tag{8}$$

is verified for any matrix $U \in \mathbb{R}^{m\times m}$ that is diagonal and positive definite.

The result in Lemma 1 can be seen as a generalized sector condition. In contrast with the classical sector condition (used for instance in [21]), this condition allows us to obtain stability conditions directly in an LMI form, as will be seen in the sequel. For notational simplicity, we denote $\psi(K\chi_k(0))$ by ψ_k in the sequel.

3.2 Saturated and Sampled-Data Systems

Let us first recall the following theorem on asymptotic stability of saturated sampled-data systems from [17]:

Theorem 1 *Consider given matrices K and G in $\mathbb{R}^{m\times n}$ and any positive definite diagonal matrix U in $\mathbb{R}^{m\times m}$. Let T_{min} and T_{max} be two positive scalars such that $T_{min} > T_{max}$, and $V : \mathbb{R}^n \to \mathbb{R}^+$ be a function for which there exist real numbers satisfying $0 < \mu_1 < \mu_2$ such that*

$$\forall x \in \mathbb{R}^n, \quad \mu_1|x|^2 \leq V(x) \leq \mu_2|x|^2. \tag{9}$$

and such that for all $i = 1, \ldots, m$ *and all* $x \in \mathbb{R}^n$, *we have*

$$x^T(K_i - G_i)^T(K_i - G_i)x \le u_{0i}^2. \tag{10}$$

Then, the following two statements are equivalent:

(i) *For all* $k \in \mathbb{N}$, *the increment* $T_k \in [T_{min},\ T_{max}]$ *of the Lyapunov function satisfies*

$$\Delta V(k) - 2T_k\psi_k^T U[\psi_k - G\chi_k(0)] < 0,$$

where $\Delta V(k) = V(\chi_k(T_k)) - V(\chi_k(0))$ *and where* $\psi_k = \psi(K\chi_k(0))$, *defined in* (7).

(ii) *There exists a continuous functional* $\mathscr{V}_0 : [0,\ T_{max}] \times \mathbb{K} \to \mathbb{R}$ *which satisfies*

$$\forall T_k \in [T_{min},\ T_{max}], \quad \mathscr{V}_0(T_k, z) = \mathscr{V}_0(0, z) \tag{11}$$

for all $z \in \mathbb{K}$, *and such that for all* $k \in \mathbb{N}$, $T_k \in [T_{min},\ T_{max}]$, *and* $\tau \in [0\ T_k[$, *we have*

$$\dot{\mathscr{W}}(\tau, \chi_k) = \frac{d}{d\tau}[V(\chi_k(\tau)) + \mathscr{V}_0(\tau, \chi_k)] - 2\psi_k^T U[\psi_k - G\chi_k(0)] < 0. \tag{12}$$

Moreover, if one of these two statements is satisfied, then for all initial conditions $x(0) = \chi_0(0)$ *in the set*

$$\mathscr{E} = \left\{x \in \mathbb{R}^n;\quad V(x) \le 1\right\}, \tag{13}$$

the solutions to system (2) *with the saturated and sampled control law* (4) *converge asymptotically to the origin.*

Under the conditions of Theorem 1, it follows that the set $\mathscr{E}$ is included in the region of attraction of the closed-loop system (5). Note that this set is not necessarily positively invariant for the continuous-time system, but it is indeed for the discrete-time system, i.e., if $\chi_0(0) \in \mathscr{E}$, then it follows that $\chi_k(0) \in \mathscr{E}$ for all $k > 0$. Nonetheless, the convergence of the continuous-time system trajectories to the origin is guaranteed for any initial condition belonging to $\mathscr{E}$, i.e., regional asymptotic stability is ensured.

3.3 Integral Inequality

We next recall a result proved in [18], which provides a lower bound on a certain integral quadratic term.

Lemma 2 *Let $\omega : [a, b] \to \mathbb{R}^n$ be a differentiable function over (a, b) having a square integrable first order derivative. Then, for all $R \in \mathbb{S}_+^n$ and all matrices $N \in \mathbb{R}^{3n\times 2n}$, the inequality*

$$\int_a^b \dot{\omega}^T(u)R\dot{\omega}(u)du \geq \Omega^T \left[He\{NM\} - (b-a)N \begin{bmatrix} R & 0 \\ 0 & 3R \end{bmatrix}^{-1} N^T \right] \Omega, \qquad (14)$$

holds, where

$$\Omega = \left[\omega^T(b)\ \omega^T(a)\ \tfrac{1}{b-a}\int_a^b \omega^T(u)du \right]^T \text{ and } M = \begin{bmatrix} I & -I & 0 \\ I & I & -2I \end{bmatrix}.$$

Proof This lemma refers to the Wirtinger-based integral inequality stated in [18]. Define the function z and the associated quantity $\mathscr{I}(z)$ as follows:

$$z(u) := \dot{\omega}(u) - R^{-1}\left[I \ \left(\frac{2u-b-a}{b-a}\right) I \right] N^T \Omega \text{ and } \mathscr{I}(z) := \int_a^b z^T(u)Rz(u)du.$$

Some elementary integral calculus and an integration by parts show that

$$\mathscr{I}(z) = \int_a^b \dot{\omega}^T(u)R\dot{\omega}(u)du - \Omega^T \left[\text{He}\{NM\} - (b-a)N \begin{bmatrix} R & 0 \\ 0 & 3R \end{bmatrix}^{-1} N^T \right] \Omega.$$

Finally, the positive definiteness of matrix R ensures that $\mathscr{I}(z)$ is positive definite, which yields the result of Lemma 2. □

Remark 1 The previous lemma presents another formulation of the Wirtinger-based integral inequality. Indeed, choosing the particular matrix $N = M^T R/(b-a)$ allows us to recover the original Wirtinger-based integral inequality from [18].

The objective of this chapter is to design new stability and stabilization conditions expressed in terms of LMI, based on Theorem 1, as proposed in [17], together with the use of the Wirtinger-based integral inequality proposed in Lemma 2.

4 Stabilization of Sampled-Data Systems with Input Saturation

In this section, we present asymptotic stabilization conditions for the sampled-data system (2) in a regional or local context, with the saturating control law given by (4). Our conditions are based on the results of Theorem 1. These conditions allow the computation of a gain K that ensures the asymptotic stability of the closed-loop system.

Theorem 2 *For given positive scalars T_{max} and $T_{min} > T_{max}$ and u_{0j} for $j = 1, \ldots, m$, assume that there exist positive definite matrices P and r in $\mathbb{S}^n$, a positive definite diagonal matrix $U \in \mathbb{S}^m$, two matrices S_1 and Q_1 in $\mathbb{S}^n$, a symmetric matrix $X \in \mathbb{S}^{n+m}$, matrices Y, S_2, and Q_2 in $\mathbb{R}^{n\times n}$ and $N \in \mathbb{R}^{(4n+m)\times 2n}$, two matrices $\tilde{K}$ and $\tilde{G} \in \mathbb{R}^{m\times n}$, and a positive scalar ε that satisfy the following for $\theta = T_{min}$ and $\theta = T_{max}$:*

$$\begin{aligned}\Psi_1(\theta) &= \Pi_1(\theta) + \theta(\Pi_2 + \Pi_4) < 0,\\ \Psi_2(\theta) &= \begin{bmatrix} \Pi_1(\theta) + \theta(\Pi_3 - \Pi_4) & \theta N \\ * & -\theta \begin{bmatrix} R & 0 \\ 0 & 3R \end{bmatrix} \end{bmatrix} < 0,\end{aligned} \tag{15}$$

$$\Psi_{3j}(u_{0j}) = \begin{bmatrix} P & (\tilde{K}_j - \tilde{G}_j)^T \\ * & u_{0j}^2 \end{bmatrix} \geq 0, \quad \forall j = 1, \ldots, m, \tag{16}$$

where

$$\begin{aligned}\Pi_1(\theta) &= \text{He}\left\{M_1^T P M_4 - M_6^T S_2 M_2 - N M_8 + M_5^T \tilde{G} M_2\right\} - M_6^T S_1 M_6\\ &\quad - \theta M_3 Q_1 M_3 - 2M_5^T U M_5\\ &\quad + \text{He}\left\{(\varepsilon M_1^T + M_4^T)(AYM_1 + B\tilde{K}M_2 - YM_4 - BUM_5)\right\}\\ \Pi_2 &= M_4^T R M_4 + \text{He}\left\{M_4^T(S_1 M_6 + S_2 M_2) + M_1^T(Q_1 M_3 + Q_2 M_2)\right\}\\ \Pi_3 &= He\{M_3 Q_2 M_2\}, \quad \Pi_4 = M_7^T X M_7,\end{aligned} \tag{17}$$

where

$$\begin{aligned}M_1 &= \begin{bmatrix} I\,0\,0\,0\,0 \end{bmatrix}, \; M_2 = \begin{bmatrix} 0\,I\,0\,0\,0 \end{bmatrix}, \; M_3 = \begin{bmatrix} 0\,0\,I\,0\,0 \end{bmatrix},\\ M_4 &= \begin{bmatrix} 0\,0\,0\,I\,0 \end{bmatrix}, \; M_5 = \begin{bmatrix} 0\,0\,0\,0\,I \end{bmatrix}, \; M_6 = \begin{bmatrix} I\,-I\,0\,0\,0 \end{bmatrix},\\ M_7 &= \begin{bmatrix} 0\,I\,0\,0\,0 \\ 0\,0\,0\,0\,I \end{bmatrix}, \; \text{and} \; M_8 = \begin{bmatrix} I & -I & 0 & 0\,0 \\ I & I & -2I & 0\,0 \end{bmatrix}\end{aligned}$$

(which are matrices that are not of the same dimensions). Then, for all initial conditions $x_0 = x(0)$ belonging to the ellipsoidal set

$$\mathscr{E}_P = \left\{x \in \mathbb{R}^n;\; x^T \left(Y^{-T} P Y^{-1}\right) x \leq 1\right\},$$

the corresponding trajectories of the system (2) *under the saturated control law defined in* (4) *with* $K = \tilde{K}Y^{-1}$ *converge asymptotically to the origin for any asynchronous sampling satisfying* (1).

Proof Introduce a quadratic Lyapunov function candidate that is defined for any x in $\mathbb{R}^n$ by $V(x) = x^T\tilde{P}x$, where $\tilde{P}$ is a symmetric positive definite matrix from $\mathbb{S}^n$. Thus, the function V satisfies (9) since it has a quadratic form.

Considering the result of Theorem 1, the idea is to prove that $\Delta V_k = V(\chi_k(T_k)) - V(\chi_k(0)) - 2T_k\psi_k^T\tilde{U}[\psi_k - G\chi_k(0)] < 0$, for all $k \in \mathbb{N}$. With this aim, an appropriate functional $\mathscr{V}_0$ satisfying (11) and (12) must be chosen. A candidate of such a class of functionals is defined for all $\tau \in [0,\ T_k]$, as follows:

$$\begin{aligned}\mathscr{V}_0(\tau,\chi_k) &= (T_k-\tau)(\chi_k(\tau)-\chi_k(0))^T[\tilde{S}_1(\chi_k(\tau)-\chi_k(0))+2\tilde{S}_2\chi_k(0)]\\ &\quad +(T_k-\tau)\tau v_k^T(\tau)[\tilde{Q}_1 v_k(\tau)+2\tilde{Q}_2\chi_k(0)]\\ &\quad +(T_k-\tau)\tau\begin{bmatrix}\chi_k(0)\\ \psi_k\end{bmatrix}^T\tilde{X}\begin{bmatrix}\chi_k(0)\\ \psi_k\end{bmatrix} + (T_k-\tau)\int_0^\tau \dot{\chi}_k^T(\theta)\tilde{R}\dot{\chi}_k(\theta)d\theta,\end{aligned}$$

where $\tilde{S}_1 > 0$, $\tilde{Q}_1 > 0$, and $\tilde{R} > 0$ are in $\mathbb{S}^n$; $\tilde{S}_2$ and $\tilde{Q}_2$ are in $\mathbb{R}^{n\times n}$; $\tilde{X} \in \mathbb{S}^{n+m}$; and

$$v_k(\tau) := \frac{1}{\tau}\int_0^\tau \chi_k(s)ds.$$

The only difference from the functional employed in [17] relies on the introduction of the function v_k. Introducing this function to the looped functional is necessary to fully take advantage of the Wirtinger-based integral inequality provided in Lemma 2. It is clear that v_k is continuous and differentiable over the interval $(0,\ T_k]$ and can be extended to the closed interval $[0,\ T_k]$, by noting that its limit as τ goes to 0 is $\chi_k(0)$. Moreover simple calculations show that

$$\frac{d}{d\tau}v_k(\tau) = \frac{1}{\tau}\left(\chi_k(\tau) - v_k(\tau)\right).$$

Obviously, this expression is only valid over the interval $(0,\ T_k]$ but can be extended by continuity to the interval $[0,\ T_k]$.

Since the functional $\mathscr{V}_0(\tau,\chi_k)$ is equal to zero at $\tau = 0$ and $\tau = T_k$, the functional $\mathscr{V}_0$ satisfies the looped condition (11). The end of the proof consists of showing that inequality (12) holds. Introduce the extended vector

$$\xi_k(\tau) = [\chi_k^T(\tau)\ \chi_k^T(0)\ v_k^T(\tau)\ \dot{\chi}_k^T(\tau)\ \psi_k^T]^T,$$

for all $\tau \in [0,\ T_k[$. Using this augmented vector, an upper bound on the derivative of $\mathcal{W}$ is given by

$$\dot{\mathcal{W}}(\tau, \chi_k) \leq \tilde{\xi}_k^T(\tau)[\tilde{\Pi}_1(T_k) + (T_k - \tau)\tilde{\Pi}_2 + \tau\tilde{\Pi}_3 + (T_k - 2\tau)\tilde{\Pi}_4]\tilde{\xi}_k(\tau), \quad (18)$$

where

$$\begin{aligned}
\tilde{\Pi}_1(T_k) &= \mathrm{He}\{M_1^T\tilde{P}M_4 - M_6^T\tilde{S}_2M_2 - \tilde{N}M_8\} - M_6^T\tilde{S}_1M_6 - T_kM_3\tilde{Q}_1M_3 \\
&\quad + \mathrm{He}\{(M_1^T\tilde{Y}_1^T + M_4^T\tilde{Y}_2^T)M_0 + M_5^T\tilde{U}GM_2\} - 2M_5^T\tilde{U}M_5 \\
\tilde{\Pi}_2 &= M_4^T\tilde{R}M_4 + \mathrm{He}\{M_4^T(\tilde{S}_1M_6 + \tilde{S}_2M_2)\} + \mathrm{He}\{M_1^T(\tilde{Q}_1M_3 + \tilde{Q}_2M_2)\} \quad (19) \\
\tilde{\Pi}_3 &= \mathrm{He}\{M_3\tilde{Q}_2M_2\} + \tilde{N}\begin{bmatrix}\tilde{R} & 0 \\ 0 & 3\tilde{R}\end{bmatrix}^{-1}\tilde{N}^T, \quad \tilde{\Pi}_4 = M_7^T\tilde{X}M_7,
\end{aligned}$$

and $M_0 = \begin{bmatrix}A & BK & 0 & -I & -B\end{bmatrix}$, $\tilde{Y}_1$ and Y_2 are in $\mathbb{R}^{n\times n}$, $\tilde{N} \in \mathbb{R}^{4n+m\times 2n}$, and the matrices M_i are defined in Theorem 2. This inequality has been obtained by (a) an application of Lemma 2, which ensures that for all matrices $\tilde{N}$ in $\mathbb{R}^{4n+m\times 2n}$, we have

$$-\int_0^\tau \dot{\chi}_k^T(\theta)\tilde{R}\dot{\chi}_k(\theta)d\theta \leq \tilde{\xi}_k^T(\tau)\left[-\mathrm{He}\{\tilde{N}M_6\} + \tau\tilde{N}^T\begin{bmatrix}\tilde{R} & 0 \\ 0 & 3\tilde{R}\end{bmatrix}^{-1}\tilde{N}\right]\tilde{\xi}_k(\tau), \quad (20)$$

and (b) by noting that

$$2(\chi_k^T(\tau)\tilde{Y}_1^T + \dot{\chi}_k^T(\tau)\tilde{Y}_2^T)(A\chi_k(\tau) - \dot{\chi}_k(\tau) + BK\chi_k(0) - B\psi_k) = 0,$$

for any square matrices $\tilde{Y}_1$ and $\tilde{Y}_2$ in $\mathbb{R}^{n\times n}$. This manipulation can be interpreted as the use of the descriptor approach introduced in [7] or the Finsler Lemma.

Since the matrices K and G are decision variables, the matrix inequality issued from (18) is not an LMI. Indeed, the matrix Π_1 contains products of matrices variables (namely, $\tilde{Y}_1$ with K and $\tilde{Y}_2$ with K). There are several transformations available to derive an LMI condition that allows us to compute a gain K, while guaranteeing that $\dot{\mathcal{W}}(\tau, \chi_k) < 0$. In this chapter, a method inspired from [7] is applied. It relies on the fact that an underlying necessary condition for the previous condition to hold is the nonsingularity of the matrix Y_2. Additionally, U is assumed to be a diagonal positive definite matrix. Thus, it is possible to define the matrices $Y = \tilde{Y}_2^{-1}$, $U = \tilde{U}^{-1}$, and $\Xi = diag\{Y, Y, Y, Y, U\}$. Consider the vector

$$\xi_k(\tau) = \Xi^{-1}\tilde{\xi}_k(\tau).$$

Rewriting (18) using the new variable ξ leads to

$$\dot{\mathcal{W}}(\tau, \chi_k) \leq \xi_k^T(\tau)\Xi^T[\Pi_1 + (T_k - \tau)\Pi_2 + \tau\Pi_3 + (T_k - 2\tau)\Pi_3]\Xi\xi_k(\tau).$$

From the definition of the matrices M_i for $i = 1, \dots, 8$, one has

$$M_i \Xi = YM_i \text{ for } i = 1, 2, 3, 4, 6,\ M_5\Xi = UM_5,\ M_7\Xi = \begin{bmatrix} Y & 0 \\ 0 & U \end{bmatrix} M_7,$$
$$M_8\Xi = \begin{bmatrix} Y & 0 \\ 0 & Y \end{bmatrix} M_8, \text{ and } M_0\Xi = AYM_1 + BKYM_2 - YM_4 - BUM_5.$$

Now setting $Y_1 = \varepsilon Y_2$ and considering the change of variables given by $P = Y^T\tilde{P}Y$ and $S_i = Y^T\tilde{S}_iY$ and $Q_i = Y^T\tilde{Q}_iY$, for $i = 1, 2$, and

$$X = \begin{bmatrix} Y & 0 \\ 0 & U \end{bmatrix}^T \tilde{X} \begin{bmatrix} Y & 0 \\ 0 & U \end{bmatrix},\quad N = \Xi^T\tilde{N}\begin{bmatrix} Y & 0 \\ 0 & Y \end{bmatrix},$$

and $R = Y^T\tilde{R}Y$, $K = \tilde{K}Y$, and $G = \tilde{G}Y$, the following inequality is obtained:

$$\dot{\mathscr{W}}(\tau, \chi_k) \le \xi_k^T(\tau)[\Pi_1 + (T_k - \tau)\Pi_2 + \tau\bar{\Pi}_3 + (T_k - 2\tau)\Pi_4]\xi_k(\tau), \tag{21}$$

where

$$\bar{\Pi}_3 = \Pi_3 + N^T \begin{bmatrix} R & 0 \\ 0 & 3R \end{bmatrix}^{-1} N$$

and Π_1, Π_2, Π_3 and Π_4 are defined in (17). To prove that $\dot{\mathscr{W}}$ is negative definite for all τ, note that the right hand side of Eq. (21) is affine with respect to the variable τ in $[0,\ T_k]$. Then, by convexity, it suffices to ensure that the right hand side of (21) is negative for $\tau = 0$ and $\tau = T_k$; see [13] for more details. This fact leads to the inequalities

$$\Pi_1 + T_k(\Pi_2 + \Pi_3) < 0, \text{ and } \Pi_1 - T_k\Pi_3 + T_kNR^{-1}N^T < 0.$$

Applying the same argument on T_k in the interval $[T_{\min},\ T_{\max}]$ and using the Schur's complement, conditions $\Psi_1(\theta) < 0$ and $\Psi_2(\theta) < 0$ given in (15) are obtained for $\theta = T_{\min}, T_{\max}$.

Right and left-multiplying (16) by $diag\{Y_2, I\}$ and next applying Schur's complement, we conclude that (16) is equivalent to (10). Hence, it follows that $\mathscr{E}_P \subseteq \mathscr{S}$ and the condition (8) is verified for all $\chi_k(0) \in \mathscr{E}_P$. Therefore, it follows from Theorem 1 that conditions (15)–(16) ensure the asymptotic convergence of the trajectories to the origin, provided that $x(0) \in \mathscr{E}_P$. □

Remark 2 The conditions from Theorem 2 are linear with respect to the matrices A and B, which characterizes the dynamics of the system. Then a direct extension of this result can be provided in the case of systems with polytopic uncertainties defined by some positive scalars λ_i's such that $\sum_{i=1}^{M} \lambda_i = 1$ and $[A\ B] = \sum_{i=1}^{M} \lambda_i [A_i\ B_i]$. Note that the λ_i's could either be constant parameters or time-varying but in both cases they are considered unknown.

5 Optimization Problems

P1. Optimization of $T_{\max}$ for a Given Set of Initial Conditions: Using Theorem 2, it is possible to include an optimization scheme in order to maximize the upper bound of the sampling period for a given set of initial conditions. Define a region of admissible initial states by

$$\mathscr{E}_{P_0} = \left\{ x \in \mathbb{R}^n \quad x^T P_0 x \leq 1 \right\}.$$

Then, it is clear that the condition $\tilde{P} = Y^{-T} P Y^{-1} < P_0$ ensures that $\mathscr{E}_{P_0} \subset \mathscr{E}_{\tilde{P}}$. The following lemma is employed in the sequel:

Lemma 3 *Let P be a positive definite matrix and Y be a nonsingular matrix. Then the inequality $P_0 > Y^{-T} P Y^{-1}$ holds if*

$$\begin{bmatrix} P_0 & I \\ I & Y + Y^T - P \end{bmatrix} > 0. \tag{22}$$

Proof The proof can be found in [17], for example. □

Hence, given $T_{\min}$ and u_0, the idea is to find the maximal $T_{\max}$ for which it is possible to compute K such that asymptotic stability is ensured for all initial conditions belonging to $\mathscr{E}_{P_0}$. An upper-bound on the maximal $T_{\max}$ can therefore be obtained from the following optimization problem:

$$\begin{array}{c} \max \; T_{\max} \\ \text{subject to} \\ (15), (16), \text{ and } (22). \end{array} \tag{23}$$

Note that from ensures that $P < P_0$, which ensures that $\mathscr{E}_{P_0} \subseteq \mathscr{E}_P$.

P2. Maximization of the Estimate of the Region of Attraction: For a given u_0 and an asynchronous sampling satisfying (1) with given $T_{\min}$ and $T_{\max}$, the objective is to determine K that leads to the largest set of initial conditions for which the conditions of Theorem 2 are satisfied. In other words, we should find K such that $\mathscr{E}_P$ is maximized in the sense of some size criterion. For instance, we can maximize the minimal axis of $\mathscr{E}_P$, which corresponds to minimizing the maximal eigenvalue of P. This can be accomplished from the following optimization problem:

$$\begin{array}{c} \min \; \delta \\ \text{subject to} \\ (15), (16), \text{ and} \\ \begin{bmatrix} \delta I & I \\ I & \tilde{Y} + \tilde{Y}^T - P \end{bmatrix} > 0. \end{array} \tag{24}$$

Note that from Lemma 3, the last inequality above ensures that $\tilde{P} < \delta I$, which ensures that $\lambda_{\max}(\tilde{P}) < \delta$.

Remark 3 For a fixed ε, the constraints in (23) are LMIs. Then the optimal solution of the previous problem can be found by solving LMI-based problems on a grid in ε.

Remark 4 Theorem 2 concerns the synthesis of a stabilizing gain K. Nonetheless, conditions (15), and (16) can also be used for analytic purposes, when the gain K is given. It suffices to replace $\tilde{K}$ by KY. In this case, the problem (23) can be solved, to determine an estimate of the region of attraction of the closed-loop system or to determine a bound on the admissible sampling period jitter, respectively. Note that the conditions are still LMIs for a fixed ε.

6 Illustrative Examples

6.1 Example 1: Static Output Feedback Control

We consider (2) with the following matrices, taken from [17]:

$$\dot{x}(t) = \begin{bmatrix} 0 & 1 \\ -2 & 0.1 \end{bmatrix} x(t) + \begin{bmatrix} 0 \\ 1 \end{bmatrix} sat(u(t_k))$$

where $u_0 = 5$. The control gain K is chosen as $[1\ 0]$. This system is unstable with a continuous-time state feedback control $u(t) = Kx(t)$. However, it was proven in [16] that the closed-loop system with a sampled version of the same control law becomes stable, if the sampling period is sufficiently large. We will now show that the same behavior appears even if the system is subject to input saturation.

In order to assess the stability of the closed-loop system with the given gain K, we can consider the two optimization problems (23) and (24) with the variable $\tilde{K}$ replaced by KY; see Remark 4.

Consider the first optimization problem **P1**. With $P_0 = I$, the results of the optimization problem (23) are summarized in Table 1. One can clearly see the considerable improvements with respect to [17] based on Jensen's inequality.

Table 1 Maximal allowable sampling period $T_{\max}$ for Example 1

$T_{\min}$	0.25	0.4	0.8	1.2	1.5
[6]	–	–	–	–	–
[17]	0.37	0.71	0.97	–	–
Theorem 2	0.44	1.01	1.33	1.48	1.54

Next, consider the second optimization problem **P2**. We take $T_{\min} = 0.8, T_{\max} = 1$ and $u_0 = 5$. We aim to optimize the size of the set of allowable initial conditions. Solving problem (23), we obtained $\delta_{\min} = 1.2$ while in [17], the same minimization problem leads to $\delta_{\min} = 6.37$. Again, we note the improvements by the method proposed in the present chapter (based on the Wirtinger-based integral) compared with the stability conditions from [17] that were based on the Jensen's inequality.

6.2 Example 2: Static Feedback Control

We consider (2) with the following matrices (which are taken from [2], where $h = 0$):

$$\dot{x}(t) = \begin{bmatrix} 1.1 & -0.6 \\ 0.5 & -1.0 \end{bmatrix} x(t) + \begin{bmatrix} 1 \\ 1 \end{bmatrix} sat(u(t_k)),$$

where $u_0 = 5$ and $T_{\max} = 1$. Here we aim to optimize the controller gain, in order to maximize the set of allowable initial conditions. Applying the conditions from [17] with $\varepsilon = 1$, it follows that the asymptotic stability of the system is ensured with the controller gain $K = [-1.7483\ 0.5391]$ for any asynchronous sampling period characterized by $T_{\min} = 0$ and $T_{\max} = 1$. The set of admissible initial conditions obtained in this case is given by

$$\mathscr{E}_1 = \left\{ x \in \mathbb{R}^2;\ x^T \begin{bmatrix} 0.4320 & -0.1332 \\ -0.1332 & 0.0411 \end{bmatrix} x \leq 1 \right\}, \quad \delta_{\min 1} = 0.497.$$

Now using the conditions in the present chapter with $\varepsilon = 1$, the optimal controller gain is $K = [-1.5355\ 0.4735]$ and the set of acceptable initial conditions is

$$\mathscr{E}_2 = \left\{ x \in \mathbb{R}^2;\ x^T \begin{bmatrix} 0.4319 & -0.1332 \\ -0.1332 & 0.0411 \end{bmatrix} x \leq 1 \right\}, \quad \delta_{\min 2} = 0.497.$$

Since the results of the two lemmas are very close to each other, we can conclude that the use of the Wirtinger-based integral inequality does not help in improving this optimization problem for the stabilization.

7 Conclusion

This chapter proposes a novel constructive stabilization criterion for sampled and saturated controlled systems based on a discrete-time Lyapunov theorem. The difference from previous research is in the use of the Wirtinger-based integral inequality in order to reduce the conservatism of the conditions. Moreover, as illustrated by an example, the method provides less conservative results than existing approaches that were reported in the literature, at least for the analysis.

References

1. Astrom, K., Wittenmark, B.: Computer Controlled Systems. Prentice-Hall, Englewood Cliffs (1984)
2. Cao, Y., Lin, Z., Hu, T.: Stability analysis of linear time-delay systems subject to input saturation. IEEE Trans. Circuit Syst. I Fundam. Theory Appl. **49**(2), 233–240 (2002)
3. da Silva J.M.G. Jr., Tarbouriech, S.: Anti-windup design with guaranteed regions of stability: an LMI-based approach. IEEE Trans. Autom. Control **50**(1), 106–111 (2005)
4. da Silva J.M.G. Jr., Seuret, A., Fridman, E., Richard, J.-P.: Stabilisation of neutral systems with saturating control inputs. Int. J. Syst. Sci. **42**(7), 1093–1103 (2011)
5. Fridman, E.: A refined input delay approach to sampled-data control. Automatica **46**(2), 421–427 (2010)
6. Fridman, E., Seuret, A., Richard, J.-P.: Robust sampled-data stabilization of linear systems: an input delay approach. Automatica **40**(8), 1141–1446 (2004)
7. Fridman, E., Shaked, U.: A descriptor system approach to H^{∞} control of linear time-delay systems. IEEE Trans. Autom. Control **47**(2), 253–270 (2002)
8. Hespanha, J., Naghshtabrizi, P., Xu, Y.: A survey of recent results in networked control systems. Proc. IEEE **95**(1), 138–162 (2007)
9. Khalil, H.: Nonlinear Systems, 2nd edn. Prentice Hall, Upper Saddle River (1996)
10. Lin, Z., Saberi, A.: Semi-global exponential stabilization of linear systems subject to input saturation via linear feedback. Syst. Control Lett. **21**(3), 225–239 (1993)
11. Mazenc, F., Mondie, S., Niculescu, S.-I.: Global asymptotic stabilization for chains of integrators with a delay in the input. IEEE Trans. Autom. Control **48**(1), 57–63 (2003)
12. Mazenc, F., Mondie, S., Niculescu, S.-I.: Global stabilization of oscillators with bounded delayed input. Syst. Control Lett. **53**(5), 415–422 (2004)
13. Naghshtabrizi, P., Hespanha, J., Teel, A.: Exponential stability of impulsive systems with application to uncertain sampled-data systems. Syst. Control Lett. **57**(5), 378–385 (2008)
14. Niculescu, S.-I., Dion, J.-M., Dugard, L.: Robust stabilization for uncertain time-delay systems containing saturating actuators. IEEE Trans. Autom. Control **41**(5), 742–747 (1996)
15. Oucheriah, S.: Global stabilization of a class of linear continuous time delay systems with saturating controls. IEEE Trans. Circuit Syst. I Fundam. Theory Appl. **43**(12), 1012–1015 (1996)
16. Seuret, A.: A novel stability analysis of linear systems under asynchronous samplings. Automatica **48**(1), 177–182 (2012)
17. Seuret, A., da Silva Jr, J.M.G.: Taking into account period variations and actuator saturation in sampled-data systems. Syst. Control Lett. **61**(12), 1286–1293 (2012)
18. Seuret, A., Gouaisbaut, F.: Wirtinger-based integral inequality: application to time-delay systems. Automatica **49**(9), 2860–2866 (2013)
19. Sussmann, H., Sontag, E., Yang, Y.: A general result on the stabilization of linear systems using bounded controls. IEEE Trans. Autom. Control **39**(12), 2411–2425 (1994)
20. Tarbouriech, S., da Silva Jr, J.M.G.: Synthesis of controllers for continuous-time delay systems with saturating controls via LMIs. IEEE Trans. Autom. Control **45**(1), 105–111 (2000)
21. Tarbouriech, S., da Silva Jr, J.M.G., Garcia, G.: Delay-dependent anti-windup loops for enlarging the stability region of time-delay systems with saturating inputs. ASME. J. Dyn. Syst. Meas. Control **125**(1), 265–267 (2003)
22. Zhang, L., Boukas, E., Haidar, A.: Delay-range-dependent control synthesis for time-delay systems with actuator saturation. Automatica **44**(10), 2691–2695 (2008)

Inversion of State-Dependent Delay

Erik I. Verriest

Abstract The exact determination of the state of a finite dimensional linear system where a variable of interest, say $y(t)$, and its delayed versions appear in an implicit relation with the delays themselves is addressed. It is assumed that the observed signal is not this variable of interest, but rather the delay (or multiple delay vector) itself. This implies that this delay must be state dependent (through y). An implicit relation

$$\overline{\tau}(t) \stackrel{\text{def}}{=} [\tau_1(t), \ldots, \tau_N(t)]^\top = F(y(t), y(t-\tau_1(t)), \ldots, y(t-\tau_N(t))),$$

with known $F : \mathbb{R}^{N+1} \to \mathbb{R}^N$ is assumed. If $x(t) \in \mathbb{R}^n$ is the state of the linear system, the observability problem is to determine this state $x(t)$ from the knowledge of the system input $u(t)$ and the delays $\overline{\tau}(t)$. This differs form the well known observability problem where $x(t)$ is to be determined from $u(t)$ and $y(t)$. In the problem at hand, an *inversion* is involved, to obtain $y(t)$ from $\overline{\tau}(t)$, rendering the problem nonlinear. Such problems are relevant when dealing with sonar, pertinent in robotics, where mobile systems must avoid hitting walls, and in underwater vehicles, for instance the soft "landing" problem on an ocean floor. The requisite observability/invertibility conditions are derived. The relevance of the restrictions $\dot{\tau}_i < 1$ for the problem to be well posed is illuminated from the physical context in the problem. In addition, the inversion of a special 'autoregressive' relation (in iterated function sense) obeyed by a delay, is solved. This is of interest in singular perturbation approaches to systems with state dependent delay.

1 Observation Models

Problems with state dependent delays are notoriously hard. Yet, these problems are not simply advanced academic problems, but are encountered in many physical situations. Signal transmission incurs a delay, which depends on the distance between

E.I. Verriest (✉)
Georgia Tech, Atlanta, GA 30332-0250, USA
e-mail: erik.verriest@ece.gatech.edu

I. Karafyllis et al. (eds.), *Recent Results on Nonlinear Delay Control Systems*,
Advances in Delays and Dynamics 4, DOI 10.1007/978-3-319-18072-4_17

transmitter and receiver. If these are mobile, the delay is state dependent [11]. In addition, the medium that facilitates the transmission may not be uniform, suggesting a path-dependent signalling speed. State dependence of the delays occurs in extrusion processes and in supply networks, due to transport constraints. In control and communication over a network, the delay is affected by the congestion state in the network. Delays also appear in milling processes, where the delay depends on the speed due to the deformation of the cutting tool. Other instances are known in chemical process control, irrigation problems and population dynamics. Problems with implicitly defined delays are relevant in the physics of gravitation, electrodynamics and acoustics. Much of the theoretical aspects of systems with state-dependent delays are discussed in [4]. The recent monograph [2] elaborates on predictor feedback laws for nonlinear systems with state-dependent delays.

In [1] and [9], preliminary results for the problem of echo location for a mobile unit with simple second order Newtonian dynamics, possibly with viscous friction

$$\dot{y} = v, \tag{1}$$
$$\dot{v} = -\alpha v + u. \tag{2}$$

were reported. This paper extends these and focuses on a linear system model

$$\dot{x} = Ax + bu \tag{3}$$
$$y = cx. \tag{4}$$

Although the problem can easily be generalized to accommodate a more general nonlinear and time variant system, we feel that not much is gained from this additional complexification. We shall assume that the observation is not the signal $y(t)$, but rather a set of delays, $\overline{\tau}(t) = [\tau_1(t), \dots, \tau_N(t)]^\top$, that are implicitly related to this $y(t)$:

$$G(\overline{\tau}) = F(y(t), y(t-\tau_1), \dots, y(t-\tau_N)), \tag{5}$$

for some $F \in C(\mathbb{R}^{N+1}, \mathbb{R}^m)$ and $G \in C(\mathbb{R}^N, \mathbb{R}^m)$. The scalar model where $F \in C(\mathbb{R}^{N+1}, \mathbb{R})$ is the linear function $\tau(t) = \sum_{k=0}^{N} a_k y(t - k\tau(t))$, with commensurate delays, is appropriate in echo-location via sonar, and was treated in [9]. Multiple beam (over different paths, bouncing off different scatterers) echo-location leads to the more general model (5). Such systems can be simulated with an echo chamber consisting of a tape in an infinite loop, on which a signal $x(t)$ can be recorded. Let a number of pick-off taps be located at multiples of a distance Δ from the write-head. Assume that the tape moves at a speed c, and that the weight on the tap at distance $i\Delta$ is a_i. A linear combination of the signals picked off at these taps is computed to yield $z(t) = \sum_{i=0}^{N-1} a_i y(t - i\tau)$, where $\tau = \Delta/c$. Now let the speed of the tape be governed by a function of this $z(t)$. If this function is inversely proportional to $z(\cdot)$, then one precisely has the situation modeled in the linear implicit delay equation.

It is a fundamental property that the state $x(t)$ of the linear system (3)–(4) can be retrieved from knowledge of the input $u(t)$ and the output $y(t)$ if the system is *observable*, that is, the observability matrix

$$\mathbf{O}(A, c) = [c^\top, A^\top c^\top, \ldots (A^{n-1})^\top c^\top]^\top,$$

where n is the order of the system (dimension of x), has full rank. So this condition remains necessary. However this may no longer be sufficient, as the implicit relation between the output y and the observed delays will need to be made explicit. We label this problem the *inversion* of the delay.

2 Exact Inversion

Consider a mobile unit (MU) of mass m, moving in a viscous fluid with friction coefficient α. The MU emits a continuous time-stamped signal $s(t)$. By the latter it is meant that if the signal $s(t)$ is transmitted at time t_y, and observed after propagating with a speed γ for a time $t - t_y$, the transmission time t_y can be detected. This can be accomplished by transmitting a chirp signal of slowly changing frequency. The *passive* ranging problem limited to one spatial dimension is then this: Suppose that the signal $s(t_y)$ is emitted by the MU, when it is at position $y(t_y)$ and detected by a stationary observer located at the origin at time t. Since the signal has traveled for a distance $\gamma(t - t_y) = y(t_y)$, it reveals an earlier position of the MU to this stationary observer. In this example we assume that the receiver sits at an impenetrable wall so that we may assume that $y(\cdot) \geq 0$. This could model the (one-dimensional) vertical motion of a submersible, with the detector at the bottom of the ocean. Letting $t - t_y = \tau(t)$, this gives the implicit relation

$$y(t - \tau(t)) = \gamma\tau(t), \tag{6}$$

which corresponds to

$$F(y(t), y(t - \tau)) = y(t - \tau) \text{ and } (\tau) = \gamma\tau \tag{7}$$

in the general model (5).

In the *active* observation model, both transmitter and receiver are located on the MU, and ranging is accomplished by observing the echo after the signal bounces off the wall (located at $y = 0$). This leads to the special case

$$y(t) + y(t - \tau(t)) = \gamma\tau(t), \tag{8}$$

accounting for forward and reflected signal wave propagation. Note that more realism is added by modeling the dependency of the speed of sound in water on salinity, temperature and pressure, which are all location dependent (in a stationary

Fig. 1 Multi-path problem

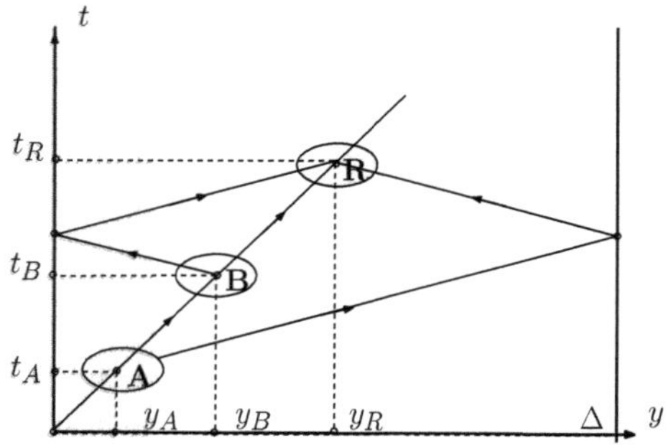

ocean). Roughly speaking, increasing pressure (hence depth), salinity and temperature increase the speed of underwater sound [12]. Thus a more realistic version, at the cost of higher complexity, replaces (8) with

$$\tau = \int_0^{y(t-\tau)} \frac{\mathrm{d}y}{\gamma(y)} + \int_0^{y(t)} \frac{\mathrm{d}y}{\gamma(y)}. \tag{9}$$

Here a functional takes the role of the function F in (5). The multidimensional case occurs when multi-path is involved. In Fig. 1 a space-time diagram is sketched for the active ranging of an MU moving with uniform speed to the right. When in position y_R, at time t_R, i.e., point R in the space-time diagram, two earlier emitted signals are received. The first was emitted in retrograde when the MU was in position y_B at time t_B, and bounced off an obstacle at the origin, the other when it was in position y_A at time t_A in prograde direction. This signal bounces off a wall at position Δ. The smaller the slope, the faster the motion. Here the MU moves at about half the speed of sound. Labeling the delay of the retrograde signal by τ_r and the prograde signal by τ_p, the relevant vector relation for

$$\tau_r = t_R - t_B \text{ and } \tau_p = t_R - t_A$$

is

$$\gamma \begin{bmatrix} \tau_r \\ \tau_p \end{bmatrix} = \begin{bmatrix} y(t) + y(t-\tau_r) \\ -y(t) - y(t-\tau_p) \end{bmatrix} + \begin{bmatrix} 0 \\ 2\Delta \end{bmatrix}. \tag{10}$$

First we shall discuss what can be inferred from such implicit observation model (5) without any knowledge of the dynamics involved. There are two subproblems: determining the delay(s) from position and its reverse: determining the position from the delay(s). The latter is the real observation problem.

2.1 Analytic Determination of Delay from Position

It is fairly simple to derive $\overline{\tau}(\cdot)$ from knowledge of $y(\cdot)$ by invoking the implicit function theorem. Assume that $y(t)$ is known for all time. For simplicity of notation let us focus on the scalar version of the implicit relation (5):

$$G(\tau) = F(y(t), y(t-\tau)). \tag{11}$$

Define $R_y(t,\tau) \stackrel{\text{def}}{=} G(\tau) - F(y(t), y(t-\tau))$. Let $\tau(t_0) = \tau_0$ for known (t_0, τ_0). If $F \in C^{k_1}$, $G \in C^{k_2}$ and $y \in C^{k_3}$, with $k = \min\{k_1, k_2, k_3\} \geq 1$, then $R_y \in C^k$, and if the Jacobian $\partial R_y/\partial\tau$ does not vanish at (t_0, τ_0), by the implicit function theorem there exists a C^k-function, $T : \mathbb{R} \to \mathbb{R}$, such that $T(t) = \tau$ in some neighborhood about (t_0, τ_0). Note that

$$\left.\frac{\partial R_y}{\partial \tau}\right|_{(t_0,\tau_0)} = \left.\frac{\mathrm{d}G(\tau)}{\mathrm{d}\tau}\right|_{\tau_0} + F_2(t_0,\tau_0)\left.\frac{\mathrm{d}y}{\mathrm{d}t}\right|_{(t_0-\tau_0)},$$

where $F_2(t_0, \tau_0)$ denotes the partial derivative of F with respect to its second argument, evaluated at (t_0, τ_0).

If F, G and y are *analytic*, much more can be invoked than just the above existence theorem.

Theorem 1 *Given* (11) *with F, G and y analytic. Let* (t_0, τ_0) *satisfying* $\tau(t_0) = \tau_0$ *be known. Then if $G'(\tau_0) + F_2(y(t), y(t-\tau_0))$ is nonzero at t, the inversion at t is explicitly given by the following series for $\tau(t)$:*

$$\tau_0 + \sum_{k\geq 1} \frac{(-G(\tau_0) + F(y(t), y(t-\tau_0)))^k}{k!} \left\{ \left(\frac{\mathrm{d}}{\mathrm{d}s}\right)^{k-1} \left(\frac{s}{f_t(\tau_0+s,\tau_0)}\right)^k \right\}_{s=0}, \tag{12}$$

where

$$f_t(\tau;\tau_0) = G(\tau) - F(y(t), y(t-\tau)) - G(\tau_0) + F(y(t), y(t-\tau_0)). \tag{13}$$

Proof The auxiliary function (13) of τ, parameterized by t, satisfies $f_t(\tau;\tau_0) = R_y(t,\tau) - R_y(t,\tau_0)$, and thus for all t, it holds that $f_t(\tau_0;\tau_0) = 0$. Also, for all t,

$$\frac{\partial}{\partial\tau} f_t(\tau_0;\tau_0) = \left.\frac{\mathrm{d}G(\tau)}{\mathrm{d}\tau}\right|_{\tau_0} + F_2(y(t), y(t-\tau_0))$$

is nonzero by assumption. The implicit relation between y and τ implies that $R_y(t_0,\tau_0) = 0$ and that $R_y(t,\tau) = 0$. Hence, $f_t(\tau;\tau_0) = -G(\tau_0) + F(y(t), y(t-\tau_0)) = F(y(t), y(t-\tau_0)) - F(y(t_0), y(t_0-\tau_0)) =: z$. The Lagrange-Bürmann inversion theorem may be invoked to give an explicit solution for $f_t(\tau_0+\tilde{\tau};\,\tau_0)=z$ as a power series

$$\tilde{\tau}=\sum_{k\geq 1}\frac{z^k}{k!}\left\{\left(\frac{\mathrm{d}}{\mathrm{d}s}\right)^{k-1}\left(\frac{s}{f_t(\tau_0+s;\,\tau_0)}\right)^k\right\}_{s=0},$$

from which the statement follows. □

Unfortunately, methods based on the implicit function theorem and the Lagrange-Bürmann inversion are limited by the need of the explicit functional form of $y(t)$.

3 Examples

Consider the simple cases $F(\xi,\eta)=\eta$ with $G(\tau)=\tau$, and $F(\xi,\eta)=\eta$ with $G(\tau)=\tau-\frac{\mu}{4}\tau^2$ respectively. The first corresponds to the simple model for passive echo localization (ranging), with the signal speed, γ, normalized to 1. The second adds a nonlinear perturbation to the active echo-localization problem.

Passive ranging for uniform motion

Applying Theorem 1, the auxiliary function (13) is $f_t(\tau;0)=\tau+y(t)-y(t-\tau)$ and hence, the first three terms give

$$\tau(t)=\frac{y(t)}{1+y'(t)}+\frac{1}{2}\frac{y''(t)y^2(t)}{(1+y'(t))^3}+\frac{1}{6}\frac{(-y'''(t)-y'''(t)y'(t)+3y''(t)^2)}{(1+y'(t))^5}+\cdots$$

For a uniform motion of the MU: $y(t)=y_0+v_0t$. This was treated in [9]. Using the Kronecker delta,

$$\left(\frac{\mathrm{d}}{\mathrm{d}s}\right)^{k-1}\left(\frac{s}{f(s)}\right)^k=\frac{1}{1+v_0}\delta_{1k}.$$

Since $z=-G(0)+F(y_0+v_0t,\,y_0+v_0t)=2(y_0+v_0t)$, the inversion yields

$$\tau(t)=\frac{y_0+v_0t}{1+v_0},$$

which is proportional to the present position $y(t)$.

Echo localization for uniformly accelerated vertical underwater motion.

Here we let the vertical position be given by $y(t)=y_0+v_0t+\frac{1}{2}u_0t^2$, measured from the bottom up: $y=0$ at the ocean floor. The water pressure in excess of atmospheric pressure is proportional to the depth, and since the speed of sound is approximately a linear function of pressure (see [12]), we obtain for the speed of sound as function of y, the height above the ocean floor, a linear relation $\gamma(y) = \gamma_0 \frac{1-\mu y}{1-\mu d}$, where γ_0 is the speed of sound at the ocean surface, and d is the ocean depth. Thus we get the form $\gamma(y) = c(1-\mu y)$ where $c = \gamma_0/(1-\mu d)$. The integrals in (9) reduce to

$$\int_0^Y \frac{\mathrm{d}y}{c(1-\mu y)} = -\frac{1}{\mu c}\int_1^{1-\mu Y} \mathrm{d}\ln x = -\frac{1}{\mu c}\ln(1-\mu Y)$$

so that the nonlinear relation is given by

$$\tau(t) = -\frac{1}{\mu c}\ln[(1-\mu y(t))(1-\mu y(t-\tau))].$$

Since $\mu \approx 10^{-5}\ m^{-1}$, the right hand side can be approximated by

$$-\frac{1}{\mu c}\ln(1-\mu y(t))(1-\mu y(t-\tau)) \approx \frac{1}{c}\left[(y(t)+y(t-\tau)) + \frac{\mu}{2}(y^2(t)+y^2(t-\tau))\right],$$

up to second order in μ. If the speed of the MU is significantly less than the speed of sound, $\dot{y} \ll c$, then since $y(t-\tau) = y(t) - \tau\dot{y}(t-\theta\tau)$ for some $\theta \in [0, 1]$, it follows that

$$y(t) + y(t-\tau) = 2y(t)\left(1 - \frac{\tau\dot{y}(t-\theta\tau)}{2y(t)}\right),$$

and

$$\frac{\tau\dot{y}(t-\theta\tau)}{2y(t)} \ll \frac{\tau c}{2y(t)} \approx 1.$$

Consequently, the nonlinear (τ, y) relation may be approximated by

$$c\tau(t) \approx [y(t)+y(t-\tau)]\left(1 + \frac{\mu}{4}(y(t)+y(t-\tau))\right),$$

leading to the approximation $c\left(\tau(t) - \frac{\mu}{4}\tau^2(t)\right) = y(t) + y(t-\tau)$. This motivates the perturbed form for $G(\tau)$ as suggested above, normalizing c to 1.

Armed with this new model, let us compute the inversion for $G(\tau) = \tau - \frac{\mu}{4}\tau^2$. The auxiliary function (13) is

$$f_t(\tau, \tau_0) = \tau - \frac{\mu}{4}\tau^2 + y(t-\tau_0) - y(t-\tau),$$

and $z(t) = 2y(t)$. For $t_0 = \tau_0 = 0$, thus $y(0) = 0$, one gets

$$\tau(t) = \frac{2y(t)}{1+y'(t)} + \frac{1}{2}\frac{(y''(t)+\frac{\mu}{2})(2y(t))^2}{(1+y'(t))^3} + \\ +\frac{1}{6}\frac{(-y'''(t) - y'''(t)y'(t) + 3y''(t)^2 + 3\mu y''(t) + \frac{3}{4}\mu^2)}{(1+y'(t))^5} + \cdots$$

With $y(t) = y_0 + v_0 t + \frac{u_0}{2} t^2$ and

$$\frac{s}{f_t(s;0)} = \left(1 + v_0 + u_0 t - \frac{2u_0 + \mu}{4} s\right)^{-1},$$

this yields

$$\tau(t) = \frac{2v_0 t + u_0 t^2}{1 + v_0 + u_0 t} + \frac{(\mu_0 + 2u_0)(2v_0 t + u_0 t^2)^2}{4(1 + v_0 + u_0 t)^3} + \cdots$$

Note that $\tau(t)$ is exactly solvable by a quadratic. In many situations only empirical or graphical information may be available. Such methods are developed below.

3.1 Graphical Determination of Delay from Position

Consider the passive ranging problem (6). Let $y(t)$ be given in $[t_0, t_1] = [0, 1]$, with $y(t_0) = y(t_1) = 0$; see Fig. 2. Consider the point A with coordinates $(t, y(t)/\gamma)$ on the graph of y/γ. Construct the line with slope -1 through A, which intersects the time axis in B, with coordinates $(t + y(t)/\gamma, 0)$. The horizontal line through A and the vertical line through B intersect in C, which has coordinates $(t + xy(t)/\gamma, y(t)/\gamma)$ and lies on the graph of τ. In fact, if $t' = t + y(t)/\gamma$, then $\tau(t') = y(t)/\gamma$, which gives a parameterized form of the graph of τ. Moreover, if y is differentiable, then two neighboring points on the graph of y/γ, say $(t, y(t)/\gamma)$ and $(t + \mathrm{d}t, (y(t) + \dot{y}(t)\mathrm{d}t)/\gamma)$, map to $(t + y(t)/\gamma, y(t)/\gamma) = (t', \tau(t'))$ and $(t + \mathrm{d}t + (y(t) + \dot{x}(t)\,\mathrm{d}t)/\gamma) = (t' + \mathrm{d}t', \tau(t') + \dot{\tau}(t')\mathrm{d}t')$, which gives

$$\dot{\tau}(t') = \frac{\dot{y}(t)}{\gamma + \dot{y}(t)}. \tag{14}$$

The *causality* constraint $\dot{\tau} < 1$, see [6], imposes then a constraint on the feasible functions y, namely $\dot{y} > -\gamma$. When this constraint is violated, a unique τ cannot be constructed, as illustrated in Fig. 3. For $t \geq 1$, two compatible delay values occur.

3.2 Graphical Determination of Position from Delay

The graphical construction discussed in Sect. 3.1 can be inverted. Let the delay $\tau(\cdot)$ be specified and strictly positive in the interval (t_0, t_1), where it satisfies the causality constraint. Then for $t_0 \leq t \leq t_1$, the parameterized point $(t - \tau(t), \gamma\tau(t))$ lies on the graph of y; see Fig. 4. Point A has coordinates $(t, \tau(t))$. The line AB has slope 1, so that B has coordinates $(t - \tau(t), 0)$. The vertical through B intersects

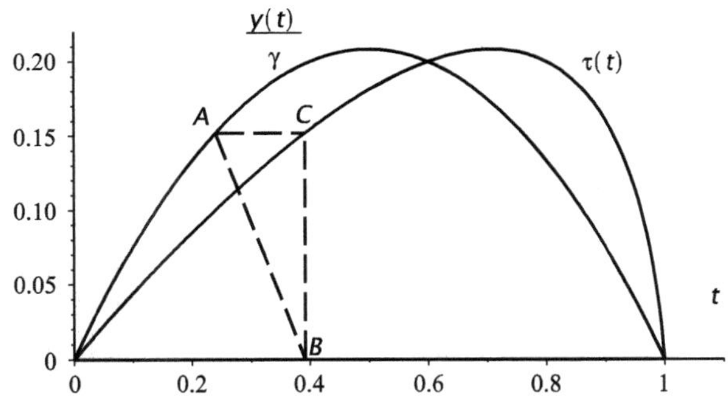

Fig. 2 Constructing $\tau(\cdot)$ from $y(\cdot)$ in the passive ranging problem

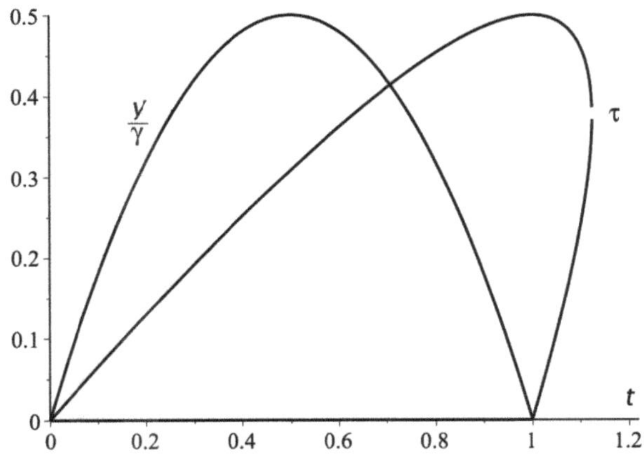

Fig. 3 Non-unique $\tau(\cdot)$ if $\dot{y} < -\gamma$

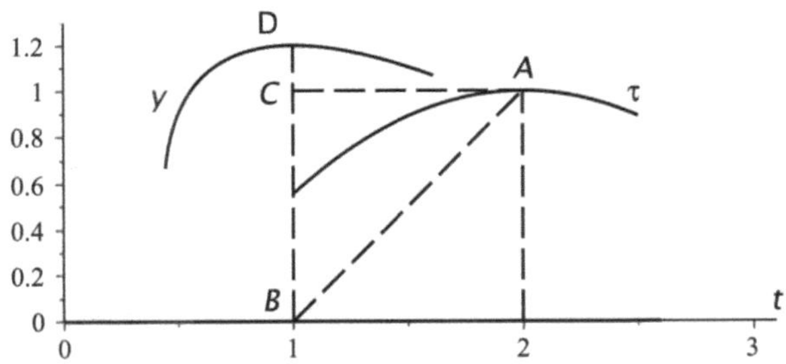

Fig. 4 Construction of y from τ (for $\gamma = 1.2$)

the horizontal through A to give C with coordinates $(t - \tau(t), \tau(t))$. The length of BD is γ times the length of BC, thus D has coordinates $(t - \tau(t), \gamma\tau(t))$ and therefore lies on the graph of y. Finally, note that $y(\cdot)$ can only be determined in the interval $(t_0 - \tau(t_0), t_1 - \tau(t_1))$. Graphical inversion for the active ranging system (8) is discussed in [9].

3.3 Analytic Determination of Position from Delay

It is desired to reconstruct the state x from observations of the delay τ, given by (5) for $N = 1$. Consider the subproblem of determining the signal y from the delay, in order to proceed with the usual observation problem. If $F_2(y(t), y(t-\tau)) \neq 0$ in some domain of $\mathbb{R}^2$, then by the implicit function theorem a function, f, exists such that in some neighborhood $y(t-\tau) = f(y(t), G(\tau))$. This is a difference equation modeling a discrete system, where $G(\tau)$ can be thought of as an input. More precisely, consider the second order (backward) recursion

$$\begin{aligned} t_{k+1} &= t_k - \tau(t_k) \\ y_{k+1} &= f(y_k, G(\tau(t_k))) \end{aligned}$$

initialized at $(t_0, y_0) = (t, y(t))$. Note that $y(t)$ is actually unknown. This iteration produces the sequence $t_{k+1} \leq t_k \leq \cdots \leq t_0 = t$. Each term in this sequence is a function of t, and this can be computed for each t' in $(t_1(t), t)$. That way, the function segment of y in $(t_1(t), t)$ acts as an initial condition (in function space) for the backwards evolution. As described in [7] and [8], certain restrictions must be in order to make the problem well-posed. See [6]. For instance, for the model (8) one finds

$$t_k(t) = t - [\tau(t) + \tau(t - \tau(t)) + \tau(t - \tau(t - \tau(t))) + \cdots].$$

The two step recursion $y_{k+2} = -y_{k+1} + \gamma\tau(t_{k+1}) = y_k - \gamma\tau(t_k) + \gamma\tau(t_{k+1})$ leads then to a solution in iterated form

$$y_{2k}(t) = y(t) - \gamma[\tau(t) - \tau(t - \tau(t)) + \tau(t - \tau(t - \tau(t))) + \cdots].$$

3.3.1 Causality

The aforementioned condition for causality states that at all times $\dot{\tau} < 1$. Hence if the commensurate condition

$$G(\tau) = F(y(t), y(t-\tau), \cdots, y(t - \tau_N)))$$

holds, one finds by differentiation that

$$G'(\tau)\dot{\tau} = \sum_{i=1}^{N} F_i(t, \tau) y'(t - i\tau)(1 - i\dot{\tau}).$$

from which the causality condition follows as

$$\dot{\tau} = \frac{\sum_{i=1}^{N} F_i y'(t - i\tau)}{G'(\tau) + \sum_{i=1}^{N} i F_i y'(t - i\tau)} < 1. \tag{15}$$

This may not be easy to check in general, but the simple case, $N = 1$, gives

$$\dot{\tau} = \frac{F_1 y'(t) + F_2 y'(t - i\tau)}{G'(\tau) + F_2 y'(t - i\tau)} = 1 + \frac{F_1 y'(t) - G'(\tau)}{G'(\tau) + F_2 y'(t - i\tau)},$$

yielding the equivalent quadratic condition

$$[F_1 y'(t) - G'(\tau)][F_2 y'(t - \tau) + G'(\tau)] < 0.$$

Hence, causality holds if $G'(\tau)$ lies outside the interval $[\mu_1, \mu_2]$, where the values μ_1 and μ_2 are the minimum and maximum elements of $\{-F_2 y'(t - \tau), F_1 y'(t)\}$, respectively.

In the echo localization example treated in Sect. 2 the causality condition is

$$(G'(\tau) + v(t - \tau))G'(\tau) > 0.$$

With $G'(\tau) = 1 - \mu\tau$, and $\tau, \mu > 0$ causality means $1 - \mu\tau > \max\{-v(t - \tau), 0\}$, which is implied if at any time the MU speed is below the effective speed of sound.

3.3.2 Analyticity

Theorem 2 *Let $y(\cdot)$ and $\tau(\cdot)$ be two scalar functions of t, satisfying the relation*

$$F(y(t), y(t - \tau)) = G(\tau), \tag{16}$$

where F and G are analytic functions of their arguments. Then if

$$G'(\tau_0) + F_2(y(t_0), y(t_0 - \tau_0)) \neq 0$$

at (t_0, τ_0) satisfying the relation (16)*, it holds that the analyticity of y implies the analyticity of τ in a neighborhood of t_0.*

Proof It follows from the premises that $F(y(t), y(t - \tau))$ is analytic in t and τ. Since also G is analytic in τ, it follows that the function

$$R(t, \tau) = F(y(t), y(t - \tau)) - G(\tau)$$

is analytic in t and τ. By the inverse function theorem this implies that if $\frac{\partial R}{\partial \tau}$ is nonzero at (t_0, τ_0) satisfying the relation (16), then in a neighborhood of (t_0, τ_0), τ can be expressed as an explicit function of t; moreover, this function is analytic. □

We point out that the converse may not hold. A counterexample is given by $F(\xi, \eta) = \eta^2$ and $x(t) = |t|$. Here $F(y(t), y(t - \tau(t))) = |t - \tau(t)|^2$, which is an analytic function of t if τ is, but clearly y fails to be analytic at the origin. However, in the special linear case $F(\xi, \eta) = \eta$, the converse also holds.

Theorem 3 *Consider the equation* $y(t - \tau(t)) = \gamma\tau(t)$. *Then* y *is analytic if and only if* τ *is analytic.*

Proof 1. If $\tau(\cdot)$ is analytic, then $t - \tau(t)$ is analytic. Suppose now that y were not analytic, then $y(t - \tau(t))$ is also not analytic, which contradicts the analyticity of $\gamma\tau(t)$.
2. If $y(\cdot)$ is analytic, then let $y(t') = \tau(t)$ and $t' = t - \tau(t)$. Thus

$$t = t' + \frac{1}{\gamma} y(t'),$$

so that t is an analytic function of t'. Since

$$\tau(t(t')) = \frac{1}{\gamma} y(t')$$

is an analytic function of t', and $t(t')$ is analytic it *must* follow that $\tau(\cdot)$ is analytic. The latter follows by contradiction: Suppose that $\tau(\cdot)$ were not analytic, then $\tau(t(t')) = \frac{1}{\gamma} y(t')$ is not analytic, which is a contradiction. □

Note that x is generated by a finite dimensional linear time invariant ODE. Hence if the driving force u is an analytic function of time, so will be x, and by the theorem therefore also the delay τ.

4 Iterated Functional Equations

In this section a different problem, also leading to an implicit delay relation is solved. Consider the scalar state-dependent differential delay equation with a small parameter $\varepsilon > 0$:

$$\varepsilon \dot{x}(t) = ax(t) + bx(t - x(t)) + ct. \tag{17}$$

Here the delay τ is directly related to the state, $\tau(t) = x(t)$. Of course we should guarantee that the physical constraints, $x(t) > 0$, $\dot{x}(t) < 1$ hold [6]. In the limit for $\varepsilon = 0$, this system behaves as a *state-dependent difference equation*, for which the solutions are typically easier to derive. Under some stability conditions the solution for small ε may be close to the difference equation solution. Here we focus on the $\varepsilon = 0$ case. Equation (17) may be rewritten as

$$a[x(t) - t] + b[x(t - x(t)) - (t - x(t))] + b(t - x(t)) + (a + c)t = 0. \quad (18)$$

Defining $\theta(t) = t - x(t)$, equation (18) transforms to

$$(b - a)\theta(t) - b\theta(\theta(t)) + (a + c)t = 0. \quad (19)$$

For $b = 0$, this reduces to the trivial $a\theta(t) = (a + c)t$ giving $ax(t) = -ct$. For $b \neq 0$ one gets an iterated function equation

$$\theta(\theta(t)) + \frac{a - b}{b}\theta(t) - \frac{a + c}{b}t = 0,$$

the solution of which seems less obvious. Thus motivated, let us investigate the general higher order *iterated function functional* (IFF) of the form discussed in [7],

$$\theta^{\circ n}(t) + a_1\theta^{\circ(n-1)}(t) + \cdots + a_{n-1}\theta(t) + a_n t = 0. \quad (20)$$

where $\theta^{\circ n}(t) = \theta(\theta^{\circ(n-1)}(t))$ are the iterates to be initialized by $\theta^{\circ 0}(t) = t$. Equation (20) can be streamlined in a form, valid for all $\lambda \in \mathbb{C}$,

$$\begin{aligned}
&[\theta(t_{n-1}) - \lambda t_{n-1}] + [\theta(t_{n-2}) - \lambda t_{n-2}]\ell_{n-1}(\lambda) + \\
&\quad \cdots + [\theta(t_0) - \lambda t_0]\ell_1(\lambda) + \ell_0(\lambda) = 0.
\end{aligned} \quad (21)$$

where we set $t_k = \theta^{\circ k}(t)$, and the $\ell_i(\cdot)$ are the Leverrier polynomials (see [3])

$$\begin{aligned}
\ell_0(\lambda) &= \lambda^n + a_1\lambda^{n-1} + \cdots + a_n \\
\ell_1(\lambda) &= \lambda^{n-1} + a_1\lambda^{n-2} + \cdots + a_{n-1} \\
&\vdots \\
\ell_{n-1}(\lambda) &= \lambda + a_1 \\
\ell_n &= 1.
\end{aligned}$$

In analogy to the LTI systems, $\ell_0(\cdot)$ may be called the 'characteristic polynomial', $a(\lambda)$, of the IFF. It follows from (21) that if λ is a root of the characteristic equation, then $\theta(t) = \lambda t$ is a solution to the IFF, and it is C^∞, in fact analytic. Hence there are n such solutions, one associated with each root, if these roots are disjoint. We referred to these "eigen-solutions" as *ray solutions*, as they emanate from the origin in the space-time, or (t, θ)-diagram. We also note that, again using (21), the affine function $\theta(t) = \lambda t + \mu$ is a solution of the IFF it it holds that

$$t\ell_0(\lambda) + \mu\sum_{i=1}^{n}\ell_i(\lambda) = 0. \quad (22)$$

Since (22) must hold for all t, this means $\ell_0(\lambda) = 0$ and $\sum_{i=1}^{n} \ell_i(\lambda) = 0$. The following lemma gives a refinement of this:

Lemma 1 *Given a polynomial $\ell_0(s)$, of degree $n \geq 2$, let $\ell_i(s)$; $i = 1, \ldots, n$ be its sequence of Leverrier polynomials. If $\ell_0(1) = 0$, then $\ell_0(\lambda) = 0$ for $\lambda \neq 1$ implies $\sum_{i=1}^{n} \ell_i(\lambda) = 0$. If $\ell_0'(1) = 0$, then also $\sum_{i=1}^{n} \ell_i(1) = 0$.*

Proof Without loss of generality, we may assume that ℓ_0 is monic. It follows from $\ell_0(1) = 0$ that $\ell_0(\lambda) = (\lambda - 1)\overline{\ell}(\lambda)$ for some polynomial $\overline{\ell}_0(\lambda) = \lambda^{n-1} + \alpha_1 \lambda^{n-2} + \cdots + \alpha_{n-1}$. The condition $\ell_0'(1) = 0$ implies that 1 is a root of ℓ_0 of higher multiplicity, so that 1 is then a root of $\overline{\ell}_0$ as well. But then

$$\ell_0(\lambda) = \lambda^n + (\alpha_1 - 1)\lambda^{n-1} + (\alpha_2 - \alpha_1)\lambda^{n-2} \cdots + (\alpha_{n-1} - \alpha_{n-2})\lambda - \alpha_{n-1}.$$

It is now readily verified that the Leverrier polynomials satisfy

$$\sum_{i=1}^{n} \ell_i(\lambda) = \lambda^n + \alpha_1 \lambda^{n-1} + \cdots + \alpha_{n-1} = \overline{\ell}_0(\lambda).$$

Hence, if $\lambda = 1$ is a root of higher multiplicity or if λ is any other root of ℓ_0, it follows that $\ell_0(\lambda) = 0$, thus proving the lemma. □

Thus if $\ell_0(\lambda) = 0$ has a root $\lambda = 1$, it follows from (22) and Lemma 1 that

$$\theta(t) = \lambda t + \mu$$

are solutions of the IFF for all $\mu \in \mathbb{R}$ and all roots λ of $\overline{\ell}_0$.

It remains to investigate the degenerate case when the characteristic polynomial of the IFF has roots of higher multiplicity. What additional solutions does this entail? Let, for $k \geq 0$, $t_k = k\lambda^k$ and $x_k = \lambda^{k+1}(1+k)$, then the pointwise solution sequence through the event (x_0, t_0) is given by

$$\begin{aligned} x_k &= \lambda^k [x_0 - k(x_0 - \lambda t_0)] \\ t_k &= \lambda^k \left[t_0 + k \frac{x_0 - \lambda t_0}{\lambda} \right]. \end{aligned}$$

The analytic solution, $x(t)$ is obtained by eliminating k from the above equations. For instance the second order case with double root λ has its solution through the initial event (x_0, t_0) expressible in terms of the LambertW function:

$$x(t) = \lambda t + x_0 \exp\left\{ \left(\frac{\lambda t}{x_0} \log \lambda \right) \exp\left[-\text{LambertW}\left(\frac{\lambda t}{x_0} \log \lambda \right) \right] \right\}. \tag{23}$$

Note that

$$\lambda^{-k}(x_k + \lambda t_k) = x_0 + \lambda t_0$$

is invariant, and thus characterizes the analytic solution through (x_0, t_0). In the special case $\lambda = 1$, these solutions simplify to the affine form,

$$x(t) = t + x_0 - t_0.$$

For characteristic roots of multiplicity m, the eigen modes are of the form $x_k = \lambda^k P_{m-1}(k)$, where $P_m(\cdot)$ is a polynomial of degree m.

5 Inversion of the Delay for Known Input

Analogous to the derivation of the observability condition for a linear system (e.g., see [5]), apply successive differentiation to τ (see also the appendix). These equations can be streamlined in matrix form. Let $\mathcal{T}$ denote the vector of successive derivatives of τ, and define for each $k = 0, 1, \ldots, N-1$, the vector $\mathcal{U}(t - k\tau)$ by

$$\mathcal{T}(t) = \begin{bmatrix} \tau \\ \dot{\tau} \\ \ddot{\tau} \\ \vdots \\ \tau^{(n-1)} \end{bmatrix}, \quad \mathcal{U}(t - k\tau) = \begin{bmatrix} u(t - k\tau) \\ \dot{u}(t - k\tau) \\ \ddot{u}(t - k\tau) \\ \vdots \\ u^{(n-1)}(t - k\tau) \end{bmatrix} \tag{24}$$

Define the matrix of powers and derivatives of $(1 - k\dot{\tau})$ by $\mathbb{T}_k(\tau)$,

$$\mathbb{T}_k(\tau) = \begin{bmatrix} 1 & & & \\ 0 & 1 - k\dot{\tau} & & \\ 0 & -k\ddot{\tau} & (1 - k\dot{\tau})^2) & \\ \vdots & \vdots & \ddots & \\ 0 & -k\tau^{(n-1)} & \ldots & (1 - k\dot{\tau})^{n-1} \end{bmatrix}. \tag{25}$$

Then in compact format:

$$\mathcal{T}(t) = \sum_{k=0}^{N-1} a_k \left[\mathbb{T}_k(\tau)\mathbf{O}(\Sigma)x(t - k\tau) + \mathbf{T}(\Sigma)\mathcal{U}(t - k\tau)\right].$$

where $\mathbf{O}(\Sigma)$ and $\mathbf{T}(\Sigma)$ are respectively the observability and Toeplitz matrix associated with the system $\Sigma = (A, b, c)$:

$$\mathbf{O}(\Sigma) = \begin{bmatrix} c \\ cA \\ cA^2 \\ \vdots \\ cA^{n-1} \end{bmatrix}, \quad \mathbf{T}(\Sigma) = \begin{bmatrix} 0 & & & \\ cb & 0 & & \\ cAb & cb & 0 & \\ \vdots & & \ddots & \ddots \\ cA^{n-2}b & \dots & cb & 0 \end{bmatrix}. \tag{26}$$

The explicit solution of the system equation yields the convolution expression

$$x(t-k\tau) = \mathrm{e}^{-Ak\tau}x(t) - \int_{t-k\tau}^{t} \mathrm{e}^{A(t-k\tau-\theta)}bu(\theta)\,\mathrm{d}\theta \stackrel{\text{def}}{=} \mathrm{e}^{-Ak\tau}x(t) - J_k(\{u\}_{t-k\tau}^{t})$$

Hence we deduce that

$$\begin{aligned} \mathcal{T}(\tau) + \sum_{k=0}^{N-1} a_k \left[\mathbb{T}_k \mathbf{O}(\Sigma) J_k\left(\{u\}_{t-k\tau}^{t}\right) - \mathbf{T}(\Sigma)\mathcal{U}(t-k\tau)\right] \\ = \left[\sum_{k=0}^{N-1} a_k \mathbb{T}_k \mathbf{O}(\Sigma)\mathrm{e}^{-Ak\tau}\right] x(t). \end{aligned}$$

It follows that $x(t)$ can be retrieved from the input and delay history if the matrix

$$\mathbf{O}_1(\Sigma, a, \tau(t)) \stackrel{\text{def}}{=} \left[\sum_{k=0}^{N-1} a_k \mathbb{T}_k \mathbf{O}(\Sigma)\mathrm{e}^{-Ak\tau(t)}\right] \tag{27}$$

is nonsingular for t. Denoting the sum which depends on u simply by $\mathcal{A}(\{u\}, \tau))$, we get thus the dynamic inversion of the delay:

$$y(t) = c\mathbf{O}_1^{-1}(\Sigma, a, \tau)\left[\mathcal{T}(\tau) + \mathcal{A}(\{u\}, \tau)\right]. \tag{28}$$

Theorem 4 *Observability of the system Σ with output y is necessary for state-observability from τ.*

Proof By contradiction. If $\mathbf{O}(\Sigma)$ does not have full rank, then by the PBH-test, there exists an eigenvector v of A which is orthogonal to the rows of c. But then, letting λ be the corresponding eigenvalue, $Av = \lambda v$, it follows that

$$\begin{aligned} \mathbf{O}(\Sigma)\mathrm{e}^{-Ak\tau}v = \mathbf{O}(\Sigma)\mathrm{e}^{-\lambda k\tau}v = \mathrm{e}^{-\lambda k\tau}\mathbf{O}(\Sigma)v \\ = \mathrm{e}^{-\lambda k\tau}\begin{bmatrix} cv \\ cAv \\ \vdots \\ cA^{n-1}v \end{bmatrix} = 0. \end{aligned}$$

Hence for all t,

$$\mathbf{O}_1(\Sigma, a, \tau(t))v = 0,$$

which contradicts observability from τ. □

The above then proves:

Theorem 5 *The system Σ is state observable from the delay τ if the matrix in* (27) *is nonsingular for all t.*

The next section shows that under a certain condition observation of x may be possible, even if the input is unknown.

6 Simultaneous Observability and Input Recovery

Given the scalar LTI system $\dot{x} = Ax + bu$ with output $y = cx + du$, when can the exact initial state x_0 *and* the exact input $u(\cdot)$ be recovered for the observation of the output, $y(\cdot)$? Applying successive differentiation of the output, y, gives

$$y^{(n)} = cA^n x + cA^{n-1}bu + \cdots du^{(n-1)}$$

Invoking the Cayley-Hamilton theorem, the n-th derivative can be expressed as

$$y^{(n)} = -\sum_{i=1}^{n} a_i cA^{n-i}x + cA^{n-1}bu + \cdots du^{(n-1)}$$

leading to a specific polynomial input-output relation in the differential operator $\mathbf{D}$ $a(\mathbf{D})y = b(\mathbf{D})u$. It is clear that if $y(\cdot)$ is known, then $a(\mathbf{D})y = \omega$ is known, and hence $u(\cdot)$ must satisfy $b(\mathbf{D})u = \omega$. This has a unique solution only if and only if the degree of the polynomial $b(\cdot)$ is zero. Equivalently, $c\text{Adj}\,(sI - A)^{-1}b + d\,\det(sI - A)$ must be independent of s. By the resolvent identity,

$$\text{Adj}\,(sI - A)^{-1} = s^{n-1}I + s^{n-2}(A + a_1 I) + \cdots + (A^{n-1} + a_1 A^{n-2} + \cdots + a_{n-1}I)$$

this leads to the system of equations

$$d = cb = cAb = \cdots = cA^{n-2}b = 0, \quad \text{and} \quad cA^{n-1}b \neq 0.$$

Thus we have proved the following theorem:

Theorem 6 *The input and the state of the system are uniquely recoverable if $h_0 = \cdots = h_{n-1}$ and $h_n \neq 0$, where $h_0 = d$ and $h_k = cA^{k-1}b$ for $k \geq 1$ are the systems Markov parameters, and n is the order of the realization.*

Note that it is not necessary to specify the observability of the system, as the latter is already implied by the Markov parameter conditions. Indeed, the Hankel matrix is

$$H = \mathbf{O}(A, c)\mathbf{R}(A, b) = \begin{bmatrix} cb & \cdots & cA^{n-1}b \\ \vdots & \ddots & \vdots \\ cA^{n-1}b & \cdots & cA^{2n-2}b \end{bmatrix}.$$

This is an invertible matrix if $h_1 = h_2 = h_{n-1} = 0$ and $h_n \neq 0$, hence $O(A, c)$ must be invertible. It follows from all this that if the Markov parameter conditions of the system hold, state and input recovery may be combined with delay inversion to yield full observability from the delay. In practice, differentiation should be avoided as it enhances the ever-present noise. For this reason, it is more practical to develop asymptotic state observers instead of the exact observers we treated here. In [1] an asymptotic observer, based on *delay injection* was introduced. This observer is itself a delay system, hence infinite dimensional. A simpler solution, using a finite dimensional observer was presented in [9, 10].

Appendix: Successive Differentiation

Here we establish some useful identities for differentiation of concatenated functions. Let σ_t be the *evaluation functional* on $C(\mathbb{R}, \mathbb{R})$ defined by $\sigma_t \varphi = \varphi(t)$. This implies that for all φ and y in $C(\mathbb{R}, \mathbb{R})$, $\sigma_{\varphi(t)} y = y(\varphi(t)) = \sigma_t(y \circ \varphi)$. Consequently, by the chain rule, letting $\mathbf{D}$ be the differentiation operator,

$$\sigma_t \mathbf{D}(y \circ \varphi) = \frac{\mathrm{d}}{\mathrm{d}t} y(\varphi(t)) = \sigma_{\varphi(t)} \mathbf{D} y \cdot \sigma_t \mathbf{D}\varphi. = \sigma_t[\mathbf{D}\varphi \cdot (\mathbf{D} y \circ \varphi)]$$

This can be lifted to the domain of functions $\mathbf{D}(y \circ \varphi) = \mathbf{D}\varphi \cdot (\mathbf{D} y \circ \varphi)$, giving a more abstract representation of the chain rule. Iteration, using the product rule yields

$$\mathbf{D}^2(y \circ \varphi) = \mathbf{D}(\mathbf{D}\varphi \cdot (\mathbf{D} y \circ \varphi)) = \mathbf{D}^2\varphi \cdot (\mathbf{D} y \circ \varphi) + (\mathbf{D}\varphi)^2 \cdot (\mathbf{D}^2 y \circ \varphi).$$

Likewise, one finds

$$\mathbf{D}^3(y \circ \varphi) = (\mathbf{D}\varphi)^3 \cdot (\mathbf{D}^3 y \circ \varphi) + 3(\mathbf{D}\varphi)(\mathbf{D}^2\varphi) \cdot (\mathbf{D}^2 y \circ \varphi) + (\mathbf{D}^3\varphi) \cdot (\mathbf{D} y \circ \varphi).$$

Let's hypothesize that for some functions $\Phi_{n,i}$, $i = 0, \ldots, n-1$,

$$\mathbf{D}^n(y \circ \varphi) = \sum_{i=0}^{n-1} \Phi_{n,i} \cdot (\mathbf{D}^{n-i} y \circ \varphi),$$

then the following 2-D recursions can be shown by induction

$$\begin{aligned}\Phi_{n+1,0} &= \mathbf{D}\varphi \cdot \Phi_{n,0}\\ \Phi_{n+1,i} &= \mathbf{D}\varphi \cdot \Phi_{n,i} + \mathbf{D}\Phi_{n,i-1}, \quad i = 1, \ldots, n-1\\ \Phi_{n+1,n} &= \mathbf{D}\Phi_{n,n-1}.\end{aligned}$$

The initialization is $\Phi_{1,0} = \mathbf{D}\varphi$ and for all $n > 0$, $\Phi_{n,n} = 0$. The first few terms in the doubly indexed sequence are

$$\begin{aligned}\{\Phi_{2,0}, \Phi_{2,1}\} &= \{(\mathbf{D}\varphi)^2, \mathbf{D}^2\varphi\}\\ \{\Phi_{3,0}, \Phi_{3,1}, \Phi_{3,2}\} &= \{(\mathbf{D}\varphi)^3, 3(\mathbf{D}\varphi)(\mathbf{D}^2\varphi), \mathbf{D}^3\varphi\}\\ \{\Phi_{4,0}, \Phi_{4,1}, \Phi_{4,2}, \Phi_{4,3}\} &= \{(\mathbf{D}\varphi)^4, 6(\mathbf{D}\varphi)^2(\mathbf{D}^2\varphi)^2, 4(\mathbf{D}\varphi)(\mathbf{D}^3\varphi)+3(\mathbf{D}\varphi)^2, \mathbf{D}^4\varphi\}.\end{aligned}$$

In fact this is simply the known Faà di Bruno's formula

$$\mathbf{D}^n(y \circ \varphi) = \sum_{i=0}^{n} \frac{n!}{m_1!1!^{m_1} m_2!2!^{m_2} \cdots m_n!n!^{m_n}} \cdot \left[\prod_{j=1}^{n} \mathbf{D}^j\varphi\right]^{m_j} \cdot (\mathbf{D}^{\sum_{i=1}^{n} m_i}) y \circ \varphi,$$

where the sum ranges over all n-tuples of nonnegative integers $(m_1, \ldots, m_n)$ satisfying $1 \cdot m_1 + 2 \cdot m_2 + 3 \cdot m_3 + \cdots + n \cdot m_n = n$. It follows that the functions $\Phi_{n,i} = B_{n,n-i}(\mathbf{D}\varphi, \mathbf{D}^2\varphi, \ldots, \mathbf{D}^{i+1}\varphi)$ are the Bell-polynomials.

Acknowledgments This work was made possible through the RIP programme of the Mathematisches Forschunginstitüt Oberwolfach, Germany, March 2013. The author is also indebted to Prof. Bernhard Lampe for suggesting the problem of echo location, and to Prof Anatoli Ivanov for discussions leading to the preliminary results.

References

1. Ahmed, A., Verriest, E.: Estimator design for a subsonic rocket car (soft landing) based on state-dependent delay measurement. In: Proceedings of the 52nd IEEE Conference on Decision and Control, pp. 5698–5703 (2013)
2. Bekiaris-Liberis, N., Kristic, M.: Nonlinear Control under Nonconstant Delays. SIAM, Philadelphia (2013)
3. Gantmacher, F.: The Theory of Matrices. Chelsea, New York (1960)
4. Hartung, F., Krisztin, T., Walther, H., Wu, J.: Functional differential equations with state-dependent selays: theory and applications. In: Handbook of Differential Equations, vol. 3, pp. 435–545. Elsevier, New York (2006)
5. Kailath, T.: Linear Systems. Prentice-Hall, Englewood-Cliffs, NJ (1980)
6. Verriest, E.: Inconsistencies in systems with time-varying delays and their resolution. IMA J. Math. Control Inf. **28**(2), 147–162 (2011)
7. Verriest, E.: State space realization for a continuous difference equation with varying delay. In: Proceedings of the 10th IFAC Workshop on Time-Delay Systems, pp. 242–247 (2012)

8. Verriest, E.: State space realization for systems with state-dependent delay. In: Proceedings of the 11th IFAC Workshop on Time-Delay Systems, pp. 446–451 (2013)
9. Verriest E., Ivanov, A.: Observation and observers for systems from delay convoluted observation. In: Proceedings of the 19th IFAC World Congress, pp. 3720–3725 (2014)
10. Verriest E., Ivanov, A.: Observability of systems with delay convoluted observation. Oberwolfach Preprints OWP 2014–10. Mathematisches Forschungsinstitut Oberwolfach, Germany (2014)
11. Walther, H.: On a model for soft landing with state-dependent delay. J. Dyn. Differ. Equ. **19**(3), 593–622 (2007)
12. Wilson, W.: Speed of sound in water as a function of temperature, pressure, and salinity. J. Acoust. Soc. Am. **32**(6), 641–644 (1960)

Sliding Mode Observer Based-Controller Design for Nonlinear Systems with Time Varying Delay

Xing-Gang Yan, Sarah K. Spurgeon and Yury Orlov

Abstract A class of nonlinear time varying delay systems in the presence of time delay uncertainties is considered in this chapter. The entries of the system input distribution matrix may be nonlinear functions of the outputs and time. Under mild limitations on the uncertainty, an observer is synthesised using sliding mode techniques such that the error dynamics are ultimately uniformly bounded in the presence of uncertainties and time delay. Then, a nonlinear control scheme is developed based on the estimated states, and a set of sufficient conditions is presented such that the corresponding closed-loop systems are uniformly ultimately bounded, using the well-known Lyapunov-Razumikhin approach. It is not required that the structure of the uncertainty be known. Finally, a numerical example is presented to demonstrate the approach and simulation results show the effectiveness of the developed paradigm.

1 Introduction

It is often assumed in theoretical studies that all system states are available for the control design. This assumption is not valid for many real systems. In order to implement such control schemes, a pertinent way forward is to construct an appropriate dynamic system which is called an observer, to estimate the state variables. Unfortunately, the traditional separation principle for linear control systems usually does not hold for the nonlinear counterpart, which implies that for nonlinear systems, the closed-loop properties achieved by the state feedback control law may not be achieved when the

X.-G. Yan (✉) · S.K. Spurgeon
Instrumentation, Control and Embedded Systems Research Group,
School of Engineering and Digital Arts, University of Kent,
Canterbury, Kent CT2 7NT, UK
e-mail: x.yan@kent.ac.uk

S.K. Spurgeon
e-mail: S.K.Spurgeon@kent.ac.uk

Y. Orlov
CICESE Research Center, P.O. Box 434944, San Diego, CA 92143-4944, USA
e-mail: yorlov@cicese.mx

I. Karafyllis et al. (eds.), *Recent Results on Nonlinear Delay Control Systems*,
Advances in Delays and Dynamics 4, DOI 10.1007/978-3-319-18072-4_18

control law is implemented with the estimated states [28]. Therefore, it is necessary to develop and formally analyze control strategies based on observer state estimates in this case.

During recent decades, several approaches have been developed for observer design, such as the geometric approach, high-gain techniques and error linearisation [16, 27]. The earliest work can be traced to the well-known Luenberger observer for linear systems. In Luenberger's approach, the observer dynamics are driven by the system input and the difference between the output of the system considered and the output of the observer being designed. This output error should be zero in the ideal case. The need to achieve zero output error naturally suggests generating a sliding motion on the subspace for which the output is zero, which has motivated the development of sliding mode observers. Although sliding mode control has been widely studied due to its high robustness, observer design using sliding mode techniques is much less mature especially for nonlinear time delay systems; see the survey paper [22].

Time delay systems widely exist in the practical world. Such systems have been studied extensively (e.g., in [21] and the references therein) since Krasovskii extended Lyapunov theory to time-delay systems and Razumikhin proposed a method to avoid the functional in Lyapunov stability analysis. Although the problem of observer design for time delay systems has been studied for a relatively long period [3, 4, 25], results concerning sliding mode observer design for time delay systems are very few and only a very limited literature is available [2, 13, 18]. Two integral sliding mode control compensators were designed to suppress disturbances for stochastic systems with input and observation delays in [2]. Later, a sliding mode observer was proposed for a class of systems with parametric uncertainty in [18]. However, in both [2, 18], the systems are linear. Higher order sliding mode techniques are employed in [5, 7], where time delay is not considered. Jaforov proposed a sliding mode observer for both delayed and non-delayed systems in [13] but only matched uncertainty and matched nonlinearities are considered. A sliding mode observer has been designed for nonlinear systems in [24], but time delay is not considered. More recently, [30] proposed a sliding mode observer for nonlinear time delay systems where the focus was on state and parameter estimation. Adaptive techniques were utilised to estimate unknown parameters but the control problem was not considered in [30].

Observer-based control for time delay systems has received much attention; see, e.g., [11, 12, 15, 17, 32]. The backstepping approach is employed in [11], where it is required that the nominal system have a triangular structure. By choosing an appropriate Lyapunov-Krasovskii functional, a high gain linear controller is presented in [12]. In both [11, 12], it is required that the systems have a particular structure. An observer-based sliding mode control is proposed in [17], where it is required that the nonlinear term be matched. Also, [15] studied a class of time-delay systems using static and dynamic output feedback but it is required that the uncertainty be matched. Recently, a sliding mode control scheme has been proposed for a class of nonlinear systems in [32], where it is required that the input distribution matrix be constant and the designed observer is actually not a sliding mode observer. A finite-time stabilization scheme is proposed using observer based output feedback control in [1, 20], where the systems are linear time-invariant with matched disturbances and delay is not considered.

It should be pointed out that in most of the results mentioned above, in order to obtain asymptotic properties, strong conditions are imposed on the systems. It is required that the structure of the mismatched uncertainty be known and that it satisfies strong structural conditions [17, 32], and/or the uncertainties vanish at the origin [31, 32]. However, in practical cases, the structure of the uncertainties may not be known and even if it is known, it may not satisfy the structural conditions. Specifically, due to the disturbances or uncertainties, the system may have no equilibrium point [14]. In connection with this, the study of uniform ultimate boundedness is meaningful where relaxed limitations on the uncertainties are possible, which may overcome these disadvantages.

In this chapter, an observer-based feedback controller is synthesized for a class of nonlinear time varying delay systems which experience uncertainties. The known nonlinear term and unknown nonlinear term are dealt with separately to reduce conservatism. It is required that the bound on the uncertainty is known but it does not require the uncertainty to vanish at the origin and it does not require that the structure of the uncertainty be known. A robust sliding mode observer is designed for the system by employing the system structure to ensure that the error dynamics are uniformly ultimately bounded. Unlike the work of [30] where an adaptive law is proposed to estimate uncertain parameters, the control synthesis problem (instead of state estimation) is the focus of the chapter. Then, observer based controllers are proposed such that the corresponding closed-loop systems are uniformly ultimately bounded in the presence of the uncertainties and time delay. The well known Lyapunov-Razumikhin approach is used to deal with the time delay in the analysis of the closed-loop system formed by the system, the observer, and the proposed control law. It is not required that either the nonlinear term or the uncertainty act on the input channel, and thus they are mismatched. The input distribution matrix is a nonlinear function matrix. The only limitation on the time varying delay is that it is continuous and bounded. It should be noted that in [8], it is required that the time derivative of the time delay is less than 1. It is shown in [23] that problems with causality, minimality, and inconsistency can arise in delay systems when the derivative of the time delay exceeds one. This provides the motivation for this chapter to develop results which are applicable to the case when the derivative of the time delay is larger than 1. In connection with this, this chapter will develop results which have no limitation on the rate of change (i.e., time derivative) of the delay. A numerical example is presented to illustrate the developed control scheme and simulation results reflect the effectiveness of the approach we propose. It should be pointed out that the main difference between this chapter and the work in [33] is that uniform ultimate boundedness is considered in this chapter, while asymptotic stability was the concern in [33]. The limitation on the uncertainty in this chapter is largely reduced when compared with the work in [33].

2 System Description and Preliminaries

We use the following notation. The set of all $n \times m$ matrices with elements defined in $\mathbb{R}$ will be denoted by $\mathbb{R}^{n\times m}$. For $A \in \mathbb{R}^{n\times n}$, we let $A > 0$ denote a symmetric positive definite matrix, and $\lambda_{\min}(A)$ (resp., $\lambda_{\max}(A)$) denotes the minimum (resp., maximum) eigenvalue of A. The symbol I_n represents the nth order unit matrix and $\mathbb{R}^+$ represents the set of non-negative real numbers. A function $f(x_1, \ldots, x_{n_1}, y_1, \ldots, y_{n_2})$ is also written as $f(x, y)$ where

$$x = \begin{bmatrix} x_1 & \ldots & x_{n_1} \end{bmatrix}^T \in \mathbb{R}^{n_1} \quad \text{and} \quad y = \begin{bmatrix} y_1 & \ldots & y_{n_2} \end{bmatrix}^T \in \mathbb{R}^{n_2}.$$

The Lipschitz constant or the generalized Lipschitz bound of a function f will be written as $\mathscr{L}_f$. Finally, $\|\cdot\|$ denotes the Euclidean norm or its induced norm.

Definition 1 A continuous function $\alpha : [0, a) \mapsto [0, \infty)$ is called a class K function if it is strictly increasing and $\alpha(0) = 0$ [14].

Definition 2 A function vector or matrix $f(x_1, x_2)$ (with $x_i \in \Omega_i \subseteq \mathbb{R}^{n_i}$ for $i = 1, 2$) is said to satisfy the generalized Lipschitz condition with respect to (w.r.t.) x_2 in Ω_2 for $x_1 \in \Omega_1$ if there exists a function $\mathscr{L}_f(\cdot)$ defined for all $x_1 \in \Omega_1$ such that for any $x_2, \hat{x}_2 \in \Omega_2$, we have

$$\|f(x_1, x_2) - f(x_1, \hat{x}_2)\| \leq \mathscr{L}_f(x_1)\|x_2 - \hat{x}_2\|, \qquad x_1 \in \Omega_1,$$

where the function $\mathscr{L}_f(\cdot)$ is called the generalized Lipschitz bound.

Remark 1 It should be noted that the generalized Lipschitz condition defined in Definition 2 is for partial variables. It can be considered as an extension of the normal Lipschitz condition. The generalized Lipschitz bound $\mathscr{L}_f(\cdot)$ is usually a function instead of a constant. However for simplicity, the symbol $\mathscr{L}_f$ is used instead of $\mathscr{L}_f(\cdot)$ throughout the chapter unless it is necessary.

Consider nonlinear systems described by

$$\dot{x} = Ax + G(t, y)u + \Phi(t, x, x_d) + \Psi(t, x, x_d) \tag{1}$$

$$y = Cx, \tag{2}$$

where $x \in \Omega \subset \mathbb{R}^n$, u and y are valued in $\mathbb{R}^m$ (where $m < n$), which are the system states, inputs and outputs, respectively, and where the domain of interest Ω is a neighbourhood of the origin in $\mathbb{R}^n$; A and C are constant matrices of appropriate dimensions; the nonlinear function matrix $G(\cdot) \in \mathbb{R}^{n\times m}$ is assumed to be known and full rank; the nonlinear term $\Phi(\cdot)$ is known and satisfies a generalized Lipschitz condition w.r.t. the variables x and x_d for $t \in \mathbb{R}^+$; and the term $\Psi(\cdot)$ includes all the uncertainties and which satisfies a global uniform bound

$$\|\Psi(t, x, x_d)\| \leq \beta \tag{3}$$

where $\beta > 0$ is a constant. The symbol $x_d := x(t-d)$ represents the delayed state where $d := d(t)$ is the time varying delay which is assumed to be known, continuous, nonnegative and bounded in $\mathbb{R}^+ := \{t \mid t \geq 0\}$, that is

$$d^* := \sup_{t \in \mathbb{R}^+} \{d(t)\} < \infty.$$

The initial condition related to the delay is given by

$$x(t) = \varphi(t), \ \text{ for all } t \in [-d^*, 0] \tag{4}$$

where $\varphi(\cdot)$ is continuous in $[-d^*, 0]$. It is assumed that all of the nonlinear functions are smooth enough for the subsequent analysis, which guarantees that the unforced system has a unique continuous solution in $t \in \mathbb{R}^+$ for each initial condition.

Firstly, the following assumptions are imposed on the system (1)–(2).

Assumption 1 The matrix pair (A, C) is observable with C being of full rank.

Assumption 2 There exist a continuous function $u^a(\cdot) : \mathbb{R}^+ \times \mathbb{R}^n \mapsto \mathbb{R}^m$ which is generalized Lipshitz w.r.t x in Ω for $t \in \mathbb{R}^+$, and a C^1 function $V_0(t, x) : \mathbb{R}^+ \times \mathbb{R}^n \mapsto \mathbb{R}^+$ such that

(i) $\alpha_1 \|x\|^2 \leq V_0(t, x) \leq \alpha_2 \|x\|^2$

(ii) $\frac{\partial V_0}{\partial t} + \left(\frac{\partial V_0}{\partial x}\right)^T \left(Ax + G(\cdot)u^a(t, x)\right) \leq -\alpha_3 \|x\|^2$

(iii) $\left\|\frac{\partial V_0}{\partial x}\right\| \leq \alpha_4 \|x\|$

hold for positive constants α_i for $i = 1, \ldots, 4$.

As before, our notation in the preceding assumption is $x := \text{col}\,(x_1, \cdots, x_n)$ and

$$\frac{\partial V_0}{\partial x} := \left[\frac{\partial V_0}{\partial x_1} \cdots \frac{\partial V_0}{\partial x_n}\right]^T.$$

Remark 2 Assumption 2 has been used in converse Lyapunov theorems; see pp. 162–163 in [14]. Due to the complexity of the nonlinear input channel $G(t, y)$, Assumption 2 is introduced to guarantee that the system $\dot{x} = Ax + G(t, y)u$ is stabilizable using state feedback $u = u^a(t, x)$.

Assumption 3 There exist continuous function matrices $N(\cdot)$ and $M(\cdot)$ where $M(\cdot)$ is nonsingular such that

$$G^T(t, y)\frac{\partial V_0}{\partial x} = M(t, y)y \tag{5}$$

$$\Phi^T(t, x, x_d)\frac{\partial V_0}{\partial x} = N(t, x, x_d)y \tag{6}$$

and where $V_0(\cdot)$ is given in Assumption 2 and $N(\cdot) \in \mathbb{R}^{n \times m}$ is generalized Lipschitz w.r.t. x and x_d for $t \in \mathbb{R}^+$.

Remark 3 Assumptions 2 and 3 together can be considered as an extension of the constrained Lyapunov problem (CLP) to the nonlinear case. For the linear case, a similar condition has been imposed by many authors; see, e.g., [13, 29]. Necessary and sufficient conditions for solving the CLP can be found in [6, 9]. It is straightforward to see that the Eq. (6) will be satisfied if (5) holds and $\Phi(\cdot)$ is matched (i.e., $\Phi(\cdot) = G(\cdot)\bar{\Phi}(\cdot)$ for some continuous function $\bar{\Phi}(\cdot)$). However, condition (6) does not imply that the nonlinear term $\Phi(\cdot)$ is matched; see, e.g., the simulation example in Sect. 5.

The following result will be presented which will be used in the subsequent analysis. Consider a nonlinear time-delay system

$$\dot{x}(t) = \tilde{f}(t, x(t - d(t)) \tag{7}$$

with an initial condition

$$x(t) = \varphi(t), \text{ for all } t \in [-d^*, 0],$$

where $\tilde{f} : \mathbb{R}^+ \times \mathscr{C}_{[-d^*,0]} \mapsto \mathbb{R}^n$ takes $\mathbb{R} \times$ (bounded sets of $\mathscr{C}_{([-d^*,0]}$) into bounded sets in $\mathbb{R}^n$. The symbol $\mathscr{C}_{[-d^*,0]}$ denotes the set of all continuous functions defined on interval $[-d^*, 0]$); the continuous function $d(t)$ is the time-varying delay and $d^* := \sup_{t\in\mathbb{R}^+}\{d(t)\} < \infty$.

Lemma 1 *Suppose $\gamma_i : \mathbb{R}^+ \mapsto \mathbb{R}^+$ for $i = 1, 2, 3$ are continuous nondecreasing functions, where $\gamma_1(\tau) \to \infty$ as $\tau \to \infty$. If there exist a continuous nondecreasing function $\gamma_4 : \mathbb{R}^+ \mapsto \mathbb{R}^+$ satisfying $\gamma_4(\tau) > \tau$ for $\tau > 0$, a constant γ_0, and a continuous function $V_1(\cdot) : \mathbb{R}^+ \times \mathbb{R}^n \mapsto \mathbb{R}^+$ satisfying*

$$\gamma_1(\|x\|) \le V_1(t, x) \le \gamma_2(\|x\|), \qquad t \in \mathbb{R}^+, \quad x \in \mathbb{R}^n$$

such that the time derivative of V_1 along the solution of system (7) satisfies

$$\dot{V}_1(t, x) \le -\gamma_3(\|x\|),$$

whenever $\|x\| \ge \gamma_0$ and $V_1(t - \vartheta, x(t - \vartheta)) \le \gamma_4(V_1(t, x(t)))$ for any $\vartheta \in [0, d^*]$, then the system (7) is uniformly ultimately bounded.

For the proof of the preceding lemma, see Theorem 4.3 on p. 159 in [10].

3 Sliding Mode Observer Design

In this section, a sliding mode observer will be proposed to estimate the system state. Assume that Assumption 1 holds and so (A, C) is observable. Then, from the analysis in Section II in [30], there exists a linear coordinate transformation $z = Tx$ such that in the new coordinate system z, the system (1)–(2) can be described by

$$\dot{z}_1 = A_1 z_1 + A_2 z_2 + G_1(t, y)u + \Phi_1(t, z, z_d) + \Psi_1(t, z, z_d) \quad (8)$$

$$\dot{z}_2 = A_3 z_1 + A_4 z_2 + G_2(t, y)u + \Phi_2(t, z, z_d) + \Psi_2(t, z, z_d) \quad (9)$$

$$y = z_2, \quad (10)$$

where $z = \text{col}\,(z_1, z_2)$ with $z_1 \in \mathbb{R}^{n-m}$, A_i are constant matrices of appropriate dimensions, and A_1 is Hurwitz stable, which implies that the inequality $A_1^T P_1 + P_1 A_1 < 0$ is solvable for $P_1 > 0$. The nonlinear functions $\Phi_i(\cdot)$ for $i = 1, 2$ are known and Lipschitz in $z \in T(\Omega)$; the nonlinear term $\Psi_i(\cdot)$ for $i = 1, 2$ represents uncertainties. From (3), there exist constants β_1 and β_2 which are dependent on the transformation matrix T and the bound β in (3) such that the following uniform global bounds hold:

$$\|\Psi_1(t, z, z_d)\| \leq \beta_1 \text{ and } \|\Psi_2(t, z, z_d)\| \leq \beta_2. \quad (11)$$

From (4), the initial condition related to the delay is given by

$$z(t) = T\varphi(t) := \rho_1(t), \quad t \in [-\bar{d}, 0] \quad (12)$$

Remark 4 The coordinate transformation $z = Tx$ can be obtained using basic matrix theory [30]. Thus, the system (8)–(10) is well defined and can be obtained directly from system (1)–(2) and the transformation matrix T.

Consider system (8)–(10). Construct dynamical systems

$$\dot{\hat{z}}_1 = A_1 \hat{z}_1 + A_2 y + G_1(t, y)u + \Phi_1(t, \hat{z}_y, \hat{z}_{yd}) \quad (13)$$

$$\dot{\hat{z}}_2 = A_3 \hat{z}_1 + A_4 \hat{z}_2 + D(y - \hat{z}_2) + G_2(t, y)u + \Phi_2(t, \hat{z}_y, \hat{z}_{yd}) + v(\cdot) \quad (14)$$

where

$$\hat{z}_y := \begin{bmatrix} \hat{z}_1 \\ y \end{bmatrix} \text{ and } \hat{z}_{yd} := \begin{bmatrix} \hat{z}_{1d} \\ y_d \end{bmatrix}, \quad (15)$$

the matrix D is chosen such that $A_4 - D$ is Hurwitz stable, and the term $v(\cdot)$ is defined by

$$v(\cdot) = (A_4 - D)\left(y - \hat{z}_2\right) + k(\cdot)\,\text{sgn}(y - \hat{z}_2) \quad (16)$$

where $\text{sgn}(\cdot)$ denotes the usual sign vector function and $k(\cdot)$ is to be determined later. The initial condition related to the delay is given by

$$\hat{z}(t) = \rho_2(t), \quad t \in [-\bar{d}, 0] \quad (17)$$

where $\rho_2(\cdot)$ can be chosen as any continuous function such that

$$\|\rho_1(t) - \rho_2(t)\| \leq b_0 \tag{18}$$

holds for all t and some constant b_0, where $\rho_1(\cdot)$ is given in (12).

Let $e_{z_1} = z_1 - \hat{z}_1$ and $e_{z_2} = z_2 - \hat{z}_2$. Then by comparing (8)–(10) with (13)–(14), the error dynamical equation is described by

$$\dot{e}_{z_1} = A_1 e_{z_1} + \Phi_1(t, z, z_d) - \Phi_1(t, \hat{z}_y, \hat{z}_{yd}) + \Psi_1(\cdot) \tag{19}$$

$$\dot{e}_{z_2} = A_3 e_{z_1} + (A_4 - D)\, e_{z_2} + \Phi_2(t, z, z_d) - \Phi_2(t, \hat{z}_y, \hat{z}_{yd}) + \Psi_2(\cdot) - \nu(\cdot) \tag{20}$$

where $\nu(\cdot)$ is defined by (16). For the error dynamics (19)–(20), consider a sliding surface

$$S := \left\{(e_{z_1}, e_{z_2}) \mid e_{z_2} = 0\right\}. \tag{21}$$

The following result is ready to be presented:

Theorem 1 *It is assumed that* Assumption 1 *holds and the uncertainty* $\Psi(\cdot)$ *satisfies* (3). *Then, the sliding motion of system* (19)–(20) *associated with the sliding surface* (21) *is uniformly ultimately bounded if there exists a constant* $q_0 > 1$ *such that the matrix inequality*

$$Q := -\left(A_1^T P_1 + P_1 A_1 + 2\mathcal{L}_{\Phi_1}\left(1 + \sqrt{q_0 \tfrac{\lambda_{\max}(P_1)}{\lambda_{\min}(P_1)}}\right) I_{n-m}\right) > 0 \tag{22}$$

is solvable for matrix $P > 0$.

Proof From the definition of the sliding surface in (21), it is clear that system (19) is the sliding mode dynamics which governs the sliding motion, and so it is only necessary to prove that (19) is uniformly ultimately bounded. For system (19), consider the Lyapunov function candidate

$$V_e = e_{z_1}^T P_1 e_{z_1}, \tag{23}$$

where P_1 satisfies (22). Then, the time derivative of V_e along all trajectories of system (19) is given by

$$\begin{aligned}
\dot{V}_e &= e_{z_1}^T \left(A_1^T P_1 + P_1 A_1\right) e_{z_1} + 2e_{z_1}^T \left(\Phi_1(t, z, z_d) - \Phi_1(t, \hat{z}_y, \hat{z}_{yd})\right) \\
&\quad + 2e_{z_1}^T \Psi_1(t, z, z_d) \\
&\leq e_{z_1}^T \left(A_1^T P_1 + P_1 A_1\right) e_{z_1} + 2\|e_{z_1}\|\, \mathcal{L}_{\Phi_1} \left\| \begin{bmatrix} z - \hat{z}_y \\ z_d - \hat{z}_{yd} \end{bmatrix} \right\| \\
&\quad + 2\|e_{z_1}\|\, \|\Psi_1(t, z, z_d)\| \\
&\leq e_{z_1}^T \left(A_1^T P_1 + P_1 A_1\right) e_{z_1} + 2\mathcal{L}_{\Phi_1} \left(\|e_{z_1}\|^2 + \|e_{z_1}\|\, \|e_{z_1 d}\|\right) + 2\beta_1 \|e_{z_1}\|
\end{aligned} \tag{24}$$

where $\beta_1 > 0$ is the constant satisfying (11).

If there is a constant $q_0 > 1$ such that $V_e(e_{z_1 d}) \leq q_0 V_e(e_{z_1})$, then,

$$\lambda_{\min}(P_1)\|e_{z_1 d}\|^2 \leq e_{z_1 d} P_1 e_{z_1 d} \leq q_0 e_{z_1}^T P_1 e_{z_1} \leq q_0 \lambda_{\max}(P_1)\|e_{z_1}\|^2$$

and thus

$$\|e_{z_1 d}\| \leq \sqrt{q_0 \frac{\lambda_{\max}(P_1)}{\lambda_{\min}(P_1)}}\, \|e_{z_1}\|. \tag{25}$$

Then from (24) and (25), it follows that when $V_e(e_{z_1 d}) \leq q_0 V_e(e_{z_1})$ with $q_0 > 1$, we have

$$\begin{aligned}
\dot{V}_e &\leq e_{z_1}^T \Big(A_1^T P_1 + P_1 A_1 + 2\mathscr{L}_{\Phi_1}\big(1 + \sqrt{q_0 \tfrac{\lambda_{\max}(P_1)}{\lambda_{\min}(P_1)}}\,\big) I_{n-m}\Big) e_{z_1} + 2\beta_1 \|e_{z_1}\| \\
&\leq -\lambda_{\min}(Q)\|e_{z_1}\|^2 + 2\beta_1 \|e_{z_1}\| \\
&\leq -\lambda_{\min}(Q)\Big(\|e_{z_1}\| - \frac{2\beta_1}{\lambda_{\min}(Q)}\Big)\|e_{z_1}\|.
\end{aligned} \tag{26}$$

It is clear that $\dot{V}_e < 0$ if $\|e_{z_1}\| > 2\beta_1/\lambda_{\min}(Q)$. Hence the conclusion follows from Lemma 1. ∇

Remark 5 Theorem 1 has shown that $e_{z_1}(t)$ is bounded, and its bound can be estimated based on (26).

Assume that

$$\|e_{z_1}(t)\| \leq b_1(t), \text{ for all } t \geq 0, \tag{27}$$

where b_1 can be calculated using the approach given in Sect. 4.8 in [14]. From (18) and (27), we have

$$\|e_{z_1 d}(t)\| \leq \max\{b_1(t), b_0\} =: b_2(t) \tag{28}$$

where b_0 is given in (18).

Theorem 2 *Under* Assumption 1, *the error dynamical system* (19)–(20) *is driven to the sliding surface* (21) *in finite time and remains on it thereafter if* $k(\cdot)$ *in* (16) *is chosen as*

$$k = \|A_3\| b_1 + \mathscr{L}_{\Phi_2}(b_1 + b_2) + \beta_2 + \eta, \tag{29}$$

where b_1 *and* b_2 *satisfy* (27) *and* (28) *respectively, and* $\eta > 0$ *is any constant.*

Proof From Eq. (20), we have

$$e_{z2}^T \dot{e}_{z2} = e_{z2}^T \Big(A_3 e_{z_1} + (A_4 - D)\, e_{z_2} + \big(\Phi_2(t, z, z_d) - \Phi_2(t, \hat{z}_y, \hat{z}_{yd})\big) \\ + \Psi_2(t, z, z_d) \Big) - e_{z_2}^T v(\cdot). \tag{30}$$

It is clear that for any vector e_{z_2}, we have

$$e_{z_2}^T \operatorname{sgn}(y - \hat{z}_2) = e_{z_2}^T \operatorname{sgn}(e_{z_2}) \geq \|e_{z_2}\|. \tag{31}$$

Then, by applying (11), (31) and (16)–(30), we get

$$\begin{aligned} e_{z_2}^T(t)\dot{e}_{z_2}(t) &\leq \|e_{z_2}\| \left(\|A_3\| \, \|e_{z_1}\| + \mathscr{L}_{\Phi_2} \left(\|e_{z_1}\| + \|e_{z_1 d}\| \right) + \|\Psi_2(\cdot)\right) \\ &\quad - k(\cdot) e_{z_2}^T \operatorname{sgn}(y - \hat{z}_2) \\ &\leq \|e_{z_2}\| \left(\|A_3\| \, \|e_{z_1}\| + \mathscr{L}_{\Phi_2} \left(\|e_{z_1}\| + \|e_{z_1 d}\| \right) + \beta_2 \right) - k(\cdot)\|e_{z_2}\| \\ &\leq \left(\|A_3\| \, b_1(t) + \mathscr{L}_{\Phi_2} \left(b_1(t) + b_2(t) \right) + \beta_2 \right) \|e_{z_2}\| - k(\cdot)\|e_{z_2}\| \\ &= -\eta \|e_{z_2}\| \end{aligned} \tag{32}$$

which shows that the reachability condition is satisfied. Hence the conclusion follows. ∇

By combining Theorems 1 and 2, it follows from sliding mode theory that the error dynamical system (19)–(20) is uniformly ultimately bounded. Therefore, (13)–(14) is an approximate sliding mode observer for the system (8)–(10). Clearly, the formula

$$\hat{x} = T^{-1} \hat{z}_y \tag{33}$$

provides an estimate for the states x of the dynamical system (1), where $\hat{z}_y$ is defined in (15) and $\hat{z}_1$ is given by (13)–(14). In fact, from $z = Tx$, we have

$$\|x - \hat{x}\| = \left\| T^{-1} z - T^{-1} \hat{z}_y \right\| \tag{34}$$

$$= \left\| T^{-1} \begin{bmatrix} e_{z_1} \\ 0 \end{bmatrix} \right\| \leq \|T^{-1}\| \, \|e_{z_1}\| \tag{35}$$

and thus $\hat{x}$ defined in (33) gives an estimate for the state x.

4 Stabilizing Controller Synthesis

In this section, it is assumed that the observer (13)–(14) has been designed. A control law based on the associated state estimates will be proposed for system (1)–(2) such that the corresponding closed-loop system is uniformly ultimately bounded.

For system (1)–(2), consider the control law

$$u := u^a(t, \hat{x}) + u^b(t, y, \hat{x}, \hat{x}_d), \quad (36)$$

where $u^a(\cdot)$ satisfies Assumption 2 and $u^b(\cdot)$ is defined by

$$u^b(\cdot) := \begin{cases} -M^{-T}(t, y)y\left(\frac{\varepsilon}{2} + \frac{\|y^T N(t, \hat{x}, \hat{x}_d)\|}{\|y\|^2}\right), & y \neq 0 \\ 0 & y = 0 \end{cases} \quad (37)$$

where $\hat{x}$ is given by (33), ε is a positive constant, and $M(\cdot)$ satisfies (5).

Remark 6 Consider the control (36). The value of the control component $u^b(t, y, \hat{x}, \hat{x}_d) = 0$ at $y = 0$ has been pre-specified in (37) according to the equivalent control method. The extension of this method to time delay systems has been justified in [19].

Theorem 3 *Under Assumptions* 1–3, *the closed-loop system formed by applying the control* (36) *to system* (1)–(2) *is uniformly ultimately bounded if the matrix*

$$W := \begin{bmatrix} w_{11} & 0 & w_{13} & 0 \\ 0 & w_{22} & 0 & 0 \\ w_{31} & 0 & w_{33} & w_{34} \\ 0 & 0 & w_{43} & w_{44} \end{bmatrix}$$

is positive definite and $\gamma_0 := \inf\{\lambda_{\min}(W(\cdot))\} > 0$, *where*

$$\begin{aligned} w_{11} &:= \alpha_3 - \alpha_2\gamma, \quad w_{22} := \alpha_1, \\ w_{33} &:= \lambda_{\min}(A_1^T P_1 + P_1 A) - \frac{\mathcal{L}_N^2}{2\varepsilon}\|T^{-1}\|^2 - 2\mathcal{L}_\Phi - \gamma\lambda_{\max}(P_1), \\ w_{44} &:= \lambda_{\min}(P_1) - \frac{\mathcal{L}_N^2}{2\varepsilon}\|T^{-1}\|^2, \\ w_{13} &= w_{31} := -\tfrac{1}{2}\alpha_4\|T^{-1}\|\|G(t, y)\|\mathcal{L}_{u^a}, \text{ and} \\ w_{34} &= w_{43} := -\mathcal{L}_{\Phi_1} \end{aligned}$$

for some $\varepsilon > 0$, $\gamma > 1$ *and* $P_1 > 0$.

Proof It is clear that the closed loop system obtained by applying the control law in (36) to system (1)–(2) is described by system (13)–(14) and the system

$$\dot{x} = Ax + G(t, y)(u^a(t, \hat{x}) + u^b(t, y, \hat{x}, \hat{x}_d)) + \Phi(t, x, x_d) + \Psi(t, x, x_d) \quad (38)$$

where $\hat{x}$ is determined by (33). Let $e_{z_1} = z_1 - \hat{z}_1$ and $e_{z_2} = z_2 - \hat{z}_2$. Based on the analysis in Sect. 3, the closed-loop system in $\mathrm{col}(x, e_{z_1}, e_{z_2})$ coordinates can be described by (38), (19) and (20). For the closed-loop system, consider the sliding surface $S := \{(x, e_{z_1}, e_{z_2}) \mid e_{z_2} = 0\}$. It follows that the sliding mode dynamics are described by (19) and (38). Theorem 2 has provided a reachability condition. It remains to prove that the sliding mode dynamics (19) and (38), which govern the sliding motion, are uniformly ultimately bounded.

Consider the Lyapunov candidate function

$$V(t, x, e_{z_1}, e_{z_2}) = V_0(t, x) + e_{z_1}^T P_1 e_{z_1} = V_0(t, x) + V_e(e_{z_1}, e_{z_2}), \tag{39}$$

where $V_0(\cdot)$ satisfies Assumption 2, $V_e(\cdot)$ is defined in (23), and $P_1 > 0$. Then, the time derivative of $V(\cdot)$ along all trajectories of the closed-loop system is given by

$$\begin{aligned}\dot{V} = \tfrac{\partial V_0}{\partial t} &+ \left(\tfrac{\partial V_0}{\partial x}\right)^T \big(Ax + G(t, y)u^a(t, \hat{x})\big) + \left(\tfrac{\partial V_0}{\partial x}\right)^T \Psi(t, x, x_d) \\ &+ \left(\tfrac{\partial V_0}{\partial x}\right)^T G(t, y)u^b(\cdot) + \left(\tfrac{\partial V_0}{\partial x}\right)^T \Phi(t, x, x_d) + \dot{V}_e\end{aligned} \tag{40}$$

From Assumption 2 and (5), we get

$$\begin{aligned}&\tfrac{\partial V_0}{\partial t} + \left(\tfrac{\partial V_0}{\partial x}\right)^T \big(Ax + G(t, y)u^a(t, \hat{x})\big) \\ &= \tfrac{\partial V_0}{\partial t} + \left(\tfrac{\partial V_0}{\partial x}\right)^T \big(Ax + G(t, y)u^a(t, x)\big) \qquad (41)\\ &\quad + \left(\tfrac{\partial V_0}{\partial x}\right)^T G(t, y)\Big(u^a(t, \hat{x}) - u^a(t, x)\Big) \\ &\le -\alpha_3\|x\|^2 + \alpha_4 \|x\| \mathscr{L}_{u^a} \|G(t, y)\| \|x - \hat{x}\| \\ &\le -\alpha_3\|x\|^2 + \alpha_4 \mathscr{L}_{u^a} \|G(t, y)\| \|T^{-1}\| \|e_{z_1}\| \|x\|, \qquad (42)\end{aligned}$$

where (34) is employed above. From Assumption 2 and the inequality (3), we get

$$\left(\frac{\partial V_0}{\partial x}\right)^T \Psi(t, x, x_d) \le \alpha_4 \beta \|x\| \tag{43}$$

Let $\delta(N) := N(t, x, x_d) - N(t, \hat{x}, \hat{x}_d)$. From $z = Tx$ and the Eq. (33), we get

$$\begin{aligned}\|\delta(N)\| &= \left\|N(t, T^{-1}z, T^{-1}z_d) - N(t, T^{-1}\hat{z}_y, T^{-1}\hat{z}_{yd})\right\| \\ &\le \mathscr{L}_N \left\| \operatorname{diag}\{T^{-1}, T^{-1}\} \begin{bmatrix} z - \hat{z}_y \\ z_d - \hat{z}_{yd} \end{bmatrix} \right\| \\ &\le \mathscr{L}_N \|T^{-1}\| \left\| \begin{bmatrix} z_1 - \hat{z}_1 \\ 0 \\ z_{1d} - \hat{z}_{1d} \\ 0 \end{bmatrix} \right\| \\ &\le \mathscr{L}_N \|T^{-1}\| \big(\|z_1 - \hat{z}_1\| + \|z_{1d} - \hat{z}_{1d}\|\big) \\ &= \mathscr{L}_N \|T^{-1}\| \big(\|e_{z_1}\| + \|e_{z_{1d}}\|\big)\end{aligned} \tag{44}$$

From (6), (34), and Young's inequality $ab \leq \frac{1}{2\varepsilon}a^2 + \frac{\varepsilon}{2}b^2$ for any $\varepsilon > 0$, it follows that for any $\varepsilon > 0$, we have

$$\begin{aligned}\left(\frac{\partial V_0}{\partial x}\right)^T \Phi(t,x,x_d) &= y^T\left(N(t,x,x_d) - N(t,\hat{x},\hat{x}_d)\right) + y^T N(t,\hat{x},\hat{x}_d)\\ &\leq \|y\|\,\|\delta(N)\| + \|y^T N(t,\hat{x},\hat{x}_d)\|\\ &\leq \frac{1}{2\varepsilon}\|\delta(N)\|^2 + \frac{\varepsilon}{2}\|y\|^2 + \|y^T N(t,\hat{x},\hat{x}_d)\|\\ &\leq \frac{1}{2\varepsilon}\|T^{-1}\|^2 \mathscr{L}_N^2\left(\|e_{z_1}\|^2 + \|e_{z_1 d}\|^2\right) + \frac{\varepsilon}{2}\|y\|^2\\ &\quad + \|y^T N(t,\hat{x},\hat{x}_d)\| \end{aligned} \tag{45}$$

where the inequality (44) is used above. From (5), we conclude as follows:
(i) if $y = 0$, then

$$\left(\frac{\partial V_0}{\partial x}\right)^T G(t,y)u^b(\cdot) + \left(\frac{\partial V_0}{\partial x}\right)^T \Phi(t,x,x_d) = y^T M^T(\cdot)u^b(\cdot) + y^T N(t,x,x_d) = 0.$$

(ii) if $y \neq 0$, then from (45), the definition of $u^b(\cdot)$ in (37), and by similar reasoning as for (43), it follows that

$$\left(\frac{\partial V_0}{\partial x}\right)^T G(t,y)u^b(\cdot) + \left(\frac{\partial V_0}{\partial x}\right)^T \Phi(t,x,x_d) \leq \frac{1}{2\varepsilon}\|T^{-1}\|^2 \mathscr{L}_N^2\left(\|e_{z_1}\|^2 + \|e_{z_1 d}\|^2\right).$$

The analysis in (i) and (ii) implies that the inequality

$$\begin{aligned}&\left(\frac{\partial V_0}{\partial x}\right)^T G(t,y)u^b(\cdot) + \left(\frac{\partial V_0}{\partial x}\right)^T \Phi(t,x,x_d)\\ &\leq \frac{1}{2\varepsilon}\|T^{-1}\|^2 \mathscr{L}_N^2\left(\|e_{z_1}\|^2 + \|e_{z_1 d}\|^2\right)\end{aligned} \tag{46}$$

holds. From the analysis in (24), we get

$$\begin{aligned}\dot{V}_e(e_{z_1}, e_{z_2}) \mid &\leq e_{z_1}^T\left(A_1^T P_1 + P_1 A_1\right)e_{z_1} + 2\mathscr{L}_{\Phi_1}\left(\|e_{z_1}\|^2 + \|e_{z_1}\|\,\|e_{z_1 d}\|\right)\\ &\quad + 2\beta_1\|e_{z_1}\|.\end{aligned} \tag{47}$$

Substituting (41), (43), (46) and (47) into (40) yields

$$\begin{aligned}\dot{V} \leq& -\alpha_3\|x\|^2 + \alpha_4\,\mathscr{L}_{u^a}\|G(t,y)\|\,\|T^{-1}\|\,\|e_{z_1}\|\,\|x\| + \alpha_4\beta\|x\|\\ &+ \frac{1}{2\varepsilon}\|T^{-1}\|^2\mathscr{L}_N^2\left(\|e_{z_1}\|^2 + \|e_{z_1 d}\|^2\right) + e_{z_1}^T\left(A_1^T P_1 + P_1 A_1\right)e_{z_1}\\ &+ 2\mathscr{L}_{\Phi_1}\left(\|e_{z_1}\|^2 + \|e_{z_1}\|\,\|e_{z_1 d}\|\right) + 2\beta_1\|e_{z_1}\|\end{aligned} \tag{48}$$

In order to apply the Lyapunov-Razumikhin approach, it is assumed that for any $d \in [0,\ \bar{d}]$, we have $V(t-d, x_d, e_{z_1d}) \leq \gamma V(t, x, e_{z_1})$, where $\gamma > 1$. Then, from Assumption 2 and the definition of $V(\cdot)$ in (39), we have

$$\begin{aligned} 0 &\leq \gamma V(t, x, e_{z_1}) - V(t-d, x_d, e_{z_1d}) \\ &= \gamma V_0(t, x) + \gamma e_{z_1}^T P_1 e_{z_1} - V_0(t-d, x_d) - e_{z_1d}^T P_1 e_{z_1d} \\ &\leq \alpha_2 \gamma \|x\|^2 - \alpha_1 \|x_d\|^2 + \gamma \lambda_{\max}(P_1) \|e_{z_1}\|^2 - \lambda_{\min}(P_1) \|e_{z_1d}\|^2. \end{aligned} \tag{49}$$

From (49) and (48), it follows that

$$\begin{aligned} \dot{V} &\leq -\alpha_3 \|x\|^2 + \alpha_4 \mathscr{L}_{u^a} \|G(t, y)\| \, \|T^{-1}\| \, \|e_{z_1}\| \, \|x\| + \alpha_4 \beta \|x\| \\ &\quad + \frac{1}{2\varepsilon} \|T^{-1}\|^2 \mathscr{L}_N^2 \left(\|e_{z_1}\|^2 + \|e_{z_1d}\|^2 \right) + e_{z_1}^T \left(A_1^T P_1 + P_1 A_1 \right) e_{z_1} \\ &\quad + 2\mathscr{L}_{\Phi_1} \left(\|e_{z_1}\|^2 + \|e_{z_1}\| \, \|e_{z_1d}\| \right) + 2\beta_1 \|e_{z_1}\| + \alpha_2 \gamma \|x\|^2 \\ &\quad - \alpha_1 \|x_d\|^2 + \gamma \lambda_{\max}(P_1) \|e_{z_1}\|^2 - \lambda_{\min}(P_1) \|e_{z_1d}\|^2 \\ &\leq (\alpha_2 \gamma - \alpha_3) \|x\|^2 - \alpha_1 \|x_d\|^2 + \Big(- \lambda_{\min} \left(A_1^T P_1 + P_1 A_1 \right) \\ &\quad + \frac{1}{2\varepsilon} \|T^{-1}\|^2 \mathscr{L}_N^2 + 2\mathscr{L}_{\Phi_1} + \gamma \lambda_{\max}(P_1) \| \Big) \|e_{z_1}\|^2 \\ &\quad + \left(\frac{1}{2\varepsilon} \|T^{-1}\|^2 \mathscr{L}_N^2 - \lambda_{\min}(P_1) \right) \|e_{z_1d}\|^2 \\ &\quad + \alpha_4 \mathscr{L}_{u^a} \|G(t, y)\| \, \|T^{-1}\| \, \|e_{z_1}\| \, \|x\| + 2\mathscr{L}_{\Phi_1} \|e_{z_1}\| \, \|e_{z_1d}\| \\ &\quad + \alpha_4 \beta \|x\| + 2\beta_1 \|e_{z_1}\| \\ &\leq - X^T W(\cdot) X + \alpha_4 \beta \|x\| + 2\beta_1 \|e_{z_1}\| \\ &\leq - \gamma_0 \|X\|^2 + \max\{\alpha_4 \beta, 2\beta_1\} \|x\|^2 \\ &= - \gamma_0 \left(\|x\| - \frac{\max\{\alpha_4 \beta, 2\beta_1\}}{\gamma_0} \right) \|x\|, \end{aligned}$$

where $X := \begin{bmatrix} \|x\| & \|x_d\| & \|e_{z_1}\| & \|e_{z_1d}\| \end{bmatrix}^T$. It is straightforward to see that $\dot{V} < 0$ when $\|x\| > \max\{\alpha_4 \beta, 2\beta_1\}/\gamma_0$. Hence, the conclusion follows from Lemma 1. ∇

Remark 7 If $\Phi(t, x, x_d) = \Phi(t, x_1, x_2, x_{1d}, x_{2d})$ where $y = x_2$, then the condition that $\Phi(\cdot)$ is generalized Lipschitz w.r.t. x and x_d can be relaxed to the condition that $\Phi(t, x_1, x_2, x_{1d}, x_{2d})$ is generalized Lipschitz w.r.t. the variables x_1 and x_{1d} for the variables t, x_2 and x_{2d}. This is applicable to all nonlinear functions which are required to satisfy the generalized Lipschitz condition throughout the chapter.

5 Illustrative Example

Consider a nonlinear time varying delay system

$$\dot{x} = \underbrace{\begin{bmatrix} -5 & 0 \\ 0 & 1 \end{bmatrix}}_{A} x + \underbrace{\begin{bmatrix} 0 \\ \frac{1}{1+\sin^2(t+x_2)} \end{bmatrix}}_{G(\cdot)} u(t) + \underbrace{\begin{bmatrix} 0.2x_2 \\ 0.2x_{1d}x_2\exp\{-t\} \end{bmatrix}}_{\Phi(\cdot)} + \underbrace{\begin{bmatrix} \Psi_1 \\ \Psi_2 \end{bmatrix}}_{\Psi(\cdot)} \tag{50}$$

$$y = \underbrace{\begin{bmatrix} 0 \; 1 \end{bmatrix}}_{C} x \tag{51}$$

where $x = \text{col}(x_1, x_2) \in \mathbb{R}^2$, $u \in \mathbb{R}$, and $y \in \mathbb{R}$ are respectively the states, input and output of the system. The term $\Psi(\cdot) \in \mathbb{R}^2$ includes all uncertainties and satisfies $\|\Psi_1(\cdot)\| \leq 0.5$ and $\|\Psi_2(\cdot)\| \leq 1$. The domain considered here is $\Omega = \{(x_1, x_2) \mid x_1 \in \mathbb{R}, |x_2| < 18.15\}$. It is straightforward to check that (A, C) is observable, and that $\Phi(\cdot)$ is generalized Lipschitz w.r.t. x_{1d} for t and x_2 with $\mathcal{L}_\Phi = 0.2|y|\exp\{-t\}$. It is clear that the system (50)–(51) is already in the form of (13)–(14) with

$$\left[\begin{array}{c|c} A_1 & A_2 \\ \hline A_3 & A_4 \end{array}\right] := \left[\begin{array}{c|c} -5 & 0 \\ \hline 0 & 1 \end{array}\right],$$

$$\left[\begin{array}{c} G_1 \\ \hline G_2 \end{array}\right] := \left[\begin{array}{c} 0 \\ \hline \frac{1}{1+\sin^2(t+x_2)} \end{array}\right], \quad \text{and} \quad \left[\begin{array}{c} \Phi_1 \\ \hline \Phi_2 \end{array}\right] := \left[\begin{array}{c} 0.2x_2 \\ \hline 0.2x_{1d}x_2\exp\{-t\} \end{array}\right].$$

One can also check that $\mathcal{L}_{\Phi_1} = 0$ and $\mathcal{L}_{\Phi_2} = 0.2|y|\exp\{-t\}$. Let

$$u^a = -6(1 + \sin^2(t + y))y$$

and $V_0 = 0.1(x_1^2 + x_2^2)$. It follows that Assumption 2 holds with $\alpha_1 = \alpha_2 = 0.1$, $\alpha_3 = 1$, and $\alpha_4 = 0.2$. Let

$$M(\cdot) = \frac{0.2}{1 + \sin^2(t + y)} \quad \text{and} \quad N(\cdot) = 0.04\,(x_1 + x_{1d}x_2\exp\{-t\}).$$

It is straightforward to check that Assumptions 3 holds. Choose $q_0 = 1.01$ and $P = 1$. By direct computation, we have

$$\mathcal{L}_N = 0.04\sqrt{1 + x_2^2\exp\{-2t\}}, \quad \mathcal{L}_{u^a} = 1 + \sin^2(t + y), \quad \mathcal{L}_\Phi = 0.2|y|\exp\{-t\},$$
$$\beta_1 = 0.5, \quad \beta_2 = 1, \quad \text{and} \quad \beta = 1.12$$

and the entries of the matrix W are given by

$$\begin{aligned}
w_{11} &= 1 - 0.1\gamma, \quad w_{22} = 0.1,\\
w_{33} &= 10 - \tfrac{0.04^2}{2\varepsilon}\left(1 + x_2^2 \exp\{-2t\}\right) - 0.4|x_2|\exp\{-t\} - \gamma,\\
w_{44} &= 1 - \tfrac{0.04^2}{2\varepsilon}\left(1 + x_2^2 \exp\{-2t\}\right),\\
w_{13} &= w_{31} = -0.1, \quad \text{and} \quad w_{34} = w_{43} = 0.
\end{aligned}$$

By direct computation, all assumptions in Theorems 1–3 are satisfied in the domain Ω with $\gamma = 1.01$. Both the observer (13)–(14) and the controller (36) are well defined. According to (36) and (37), the designed control is given by

$$u = -6(1 + \sin^2(t + y))y + u^b(t, y, \hat{x}, \hat{x}_d),$$

where u^b is defined by

$$u^b = \begin{cases} -\dfrac{1 + \sin^2(t + y)}{0.2}\left(\dfrac{\varepsilon}{2} + \dfrac{0.04|\hat{x}_1 y_d + \hat{x}_{1d} y \exp\{-t\}|}{|y|}\right)y, & y \neq 0 \\ 0, & y = 0 \end{cases}.$$

For implementation purposes, choose $\eta = 5$, $b_0 = 5$ and the uncertainty

$$\Psi = \begin{bmatrix} 0.5 \sin t \\ -\sin t \end{bmatrix}.$$

The initial states are given by $x_0 = \text{col}(10, -6)$ and $z_0 = \text{col}(0, -2)$. The time-varying delay $d(t)$ is chosen as

$$d(t) = 5 + 2 \sin t. \tag{52}$$

The delay related initial condition is chosen as $\varphi(t) = \text{col}(\cos(t), 1-\sin(t))$. Figure 1 shows the time response of the states of the closed-loop system and the estimated states, and Fig. 2 shows the estimation error and the control signal. The simulation results confirm that the proposed approach is effective.

Remark 8 For linear mechanical systems where position and velocity are viewed as states, the system input distribution matrix is usually given by $B = [0\ 1]^T$. In this case, Assumption 3 requires that the velocity is measurable using an appropriate sensor. Alternatively, Assumption 3 may hold if a combination of position and velocity is

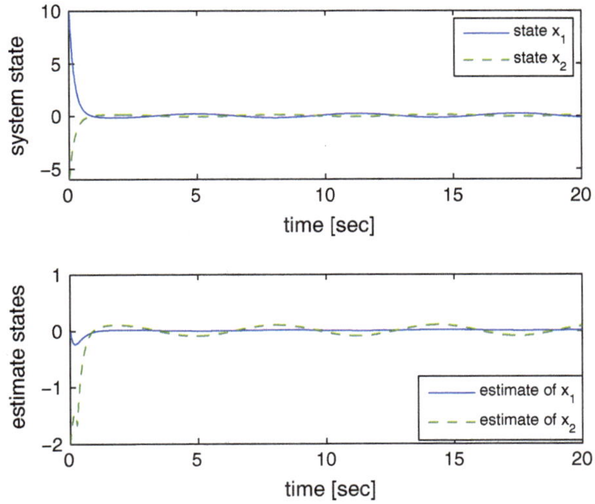

Fig. 1 Time responses of system states (*upper*) and estimated states (*lower*)

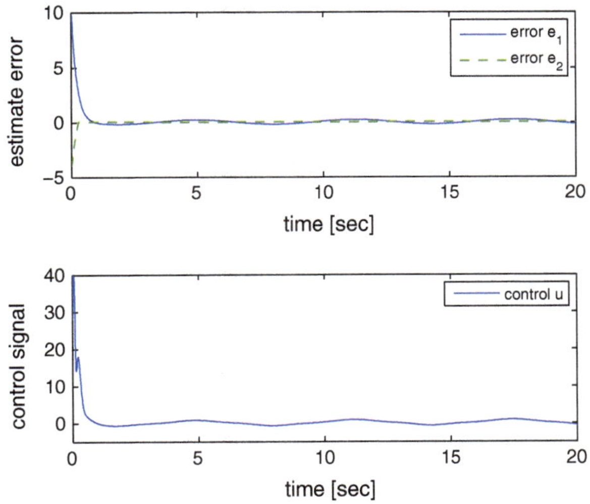

Fig. 2 Time responses of estimation errors (*upper*) and control signal (*lower*)

measurable. This may occur in some real systems such as certain remote control applications where the number of transmission and receive lines and frequencies are limited [26].

Remark 9 In the simulation, the discontinuous input (16), regulating (13)–(14), has been replaced by the saturation function to avoid chattering.

6 Conclusion

A sliding mode based control design approach has been proposed for a class of nonlinear time delay systems. A controller-observer pair is designed. Sufficient conditions have been derived using the Lyapunov-Razumikin approach under which the corresponding closed-loop system is uniformly ultimately bounded. The input matrix of the system is nonlinear. The known and unknown nonlinear terms are dealt with separately, to reduce conservatism. There is no limitation on the rate of change of the time delay. The approach is applicable where uncertainties have unknown structure and do not vanish at the origin.

References

1. Angulo, M., Fridman, L., Levant, A.: Output-feedback finite-time stabilization of disturbed LTI systems. Automatica **48**(4), 606–611 (2012)
2. Basin, M., Rodriguez-Gonzalez, J., Fridman, L., Acosta, P.: Integral sliding mode design for robust filtering and control of linear stochastic time-delay systems. Int. J. Robust Nonlinear Control **15**(9), 407–421 (2005)
3. Bhat, K., Koivo, H.: Observer theory for time-delay systems. IEEE Trans. Autom. Control **21**(2), 266–269 (1976)
4. Cacace, F., Germani, A., Manes, C.: An observer for a class of nonlinear systems with time varying observation delay. Syst. Control Lett. **59**(5), 2305–2312 (2010)
5. Dinuzzo, F., Ferrara, A.: Higher order sliding mode controllers with optimal reaching. IEEE Trans. Autom. Control **54**(9), 2126–2136 (2009)
6. Edwards, C., Yan, X.-G., Spurgeon, S.: On the solvability of the constrained Lyapunov problem. IEEE Trans. Autom. Control **52**(10), 1982–1987 (2007)
7. Efimov, D., Fridman, L.: Global sliding-mode observer with adjusted gains for locally Lipschitz systems. Automatica **47**(3), 565–570 (2011)
8. Fridman, E., Dambrine, M.: Control under quantization, saturation and delay: a LMI approach. Automatica **45**(10), 2258–2264 (2009)
9. Galimidi, A., Barmish, B.: The constrained Lyapunov problem and its application to robust output feedback stabilization. IEEE Trans. Autom. Control **31**(5), 410–419 (1986)
10. Hale, J., Lunel, S.: Introduction to Functional Differential Equations. Springer, New York (1993)
11. Hua, C., Guan, X., Shi, P.: Robust backstepping control for a class of time-delayed systems. IEEE Trans. Autom. Control **50**(6), 894–899 (2005)
12. Ibrir, S.: Observer-based control of a class of time-delay nonlinear systems having triangular structure. Automatica **47**(2), 388–394 (2011)
13. Jafarov, E.M.: Design modification of sliding mode observers for uncertain MIMO systems without and with time-delay. Asian J. Control **7**(4), 380–392 (2005)
14. Khalil, H.: Nonlinear Systems, 3rd edn. Prentice-Hall, Englewood Cliffs (2002)
15. Luo, N., De La Sen, M., Rodellar, J.: Robust stabilization of a class of uncertain time delay systems in sliding mode. Int. J. Robust Nonlinear Control **7**(1), 59–74 (1997)
16. Marino, R., Tomei, P.: Adaptive observers with arbitrary exponential rate of convergence for nonlinear systems. IEEE Trans. Autom. Control **40**(7), 1300–1304 (1995)
17. Niu, Y., Lam, J., Wang, X., Ho, D.: Observer-based sliding mode control for nonlinear state-delayed systems. Int. J. Syst. Sci. **35**(2), 139–150 (2004)
18. Nounou, M., Nounou, H., Mahmoud, M.: Robust adaptive sliding-mode control for continuous time-delay systems. IMA J. Math. Control Inf. **24**(3), 299–313 (2007)

19. Orlov, Y.: Discontinuous Systems: Lyapunov Analysis and Robust Synthesis Under Uncertainty Conditions. Springer, London (2009)
20. Orlov, Y., Aoustin, Y., Chevallereau, C.: Finite time stabilization of a perturbed double integrator part I: continuous sliding mode-based output feedback synthesis. IEEE Trans. Autom. Control **56**(3), 614–618 (2011)
21. Richard, J.-P.: Time-delay systems: an overview of some recent advances and open problems. Automatica **39**(10), 1667–1694 (2003)
22. Spurgeon, S.: Sliding mode observers: a survey. Int. J. Syst. Sci. **39**(8), 751–764 (2008)
23. Verriest, E.: Inconsistencies in systems with time-varying delays and their resolution. IMA J. Math. Control Inf. **28**(2), 147–162 (2011)
24. Veluvolu, K., Soh, Y.: Fault reconstruction and state estimation with sliding mode observers for Lipschitz non-linear systems. IET Control Theory Appl. **5**(11), 1255–1263 (2011)
25. Wang, Z., Huang, B., Unbehauen, H.: Robust H_∞ observer design of linear state delayed systems with parametric uncertainty: the discrete-time case. Automatica **35**(6), 1161–1167 (1999)
26. Walcott, B., Zak, S.: Combined observer- controller synthesis for uncertain dynamical systems with application. IEEE Trans. Syst. Man Cybern. **18**(1), 88–104 (1988)
27. Xia, X.-H., Gao, W.-B.: Nonlinear observer design by observer error linearization. SIAM J. Control Optim. **27**(1), 199–216 (1989)
28. Yan, X.-G., Lam, J., Li, H.-S., Chen, I.-M.: Decentralized control of nonlinear large-scale systems using dynamic output feedback. J. Optim. Theory Appl. **104**(2), 459–475 (2000)
29. Yan, X.-G., Lam, J., Xie, L.: Robust observer design for non-linear interconnected systems using structural characteristics. Int. J. Control **76**(7), 741–746 (2003)
30. Yan, X.-G., Spurgeon, S., Edwards, C.: State and parameter estimation for nonlinear delay systems using sliding mode techniques. IEEE Trans. Autoam. Control **58**(4), 1023–1029 (2013)
31. Yan, X.-G., Spurgeon, S., Edwards, C.: Static output feedback sliding mode control for time-varying delay systems with time-delayed nonlinear disturbances. Int. J. Robust Nonlinear Control **20**(7), 777–788 (2010)
32. Yan, X.-G., Spurgeon, S., Edwards, C.: Sliding mode control for time-varying delayed systems based on a reduced-order observer. Automatica **46**(8), 1354–1362 (2010)
33. Yan, X-G., Spurgeon, S., Orlov, Y.: Sliding mode observer based control for a class of nonlinear time delay systems with delayed uncertainties. In: Proceedings of the 50th IEEE Conference on Decision and Control, pp. 7611–7616 (2011)

An LMI Approach to Control of Exponentially Unstable Systems Subject to Saturation and Time-Varying Delay in the Input

Se Young Yoon, Parinya Anantachaisilp and Zongli Lin

Abstract This chapter studies the stabilization of exponentially unstable linear systems subject to saturation and time-varying delay in the input. The proposed stabilizing controller is developed in a two step process. First, the stabilizing controller for the system without saturation is obtained from the optimal solution to an iterative LMI problem, such that the upper bound of the delay function is maximized. Second, the input saturation is included in the analysis, and the previous iterative LMI problem is updated in order to maximize the domain of attraction of the closed-loop system. The effectiveness of the proposed methods is demonstrated through numerical examples, and by experimental validation on a test rig that captures the main operating characteristics of active magnetic bearings.

1 Introduction

The presence of time delays in the control of modern industrial applications is nearly unavoidable. For example, the gas/fluid flow in piping of chemical plants and transmission lines of pneumatic systems are common sources of time delay in control applications. Moreover, with the introduction of digital controllers and powerful processors, controller implementations have gradually become more complex. Industrial controllers has moved to include advance capabilities such as self system monitoring/diagnosis, remote control, and coordinated control in network systems. All these advancements in control applications have benefited the adaptability and reliability of the final controlled system, but at the cost of increased computation

S.Y. Yoon (✉) · P. Anantachaisilp · Z. Lin
Department of Electrical and Computer Engineering,
University of New Hampshire, Kingsbury Hall W201,
33 Academic Way, Durham, NH 03824-2619, USA
e-mail: SeYoung.Yoon@unh.edu

P. Anantachaisilp
e-mail: pa7d@virginia.edu

Z. Lin
e-mail: zl5y@virginia.edu

I. Karafyllis et al. (eds.), *Recent Results on Nonlinear Delay Control Systems*,
Advances in Delays and Dynamics 4, DOI 10.1007/978-3-319-18072-4_19

and communication time delays. As a result, the need of a reliable solution to the problem of controlling systems with large time delays has become eminent.

Indeed, the control of linear and nonlinear systems with time-delays has been an active field of research for many years, and the literature reviewed in this chapter only represents a very small sample of what is available. Introductions to the problem of stabilizing time-delay systems were provided in great detail in [9, 12]. References [7, 8] studied the stability of linear systems with time-delay in the states and saturation in the actuator. Employing the linear matrix inequality (LMI) technique, an optimal controller was developed to maximize the domain of attraction of the closed-loop system. Results on the stabilization of a generic type of linear oscillators with bounded delayed input were presented in [10, 21], where global stability of the system was guaranteed for an arbitrarily large delay. The stabilization of delayed integrators was studied in [3], and the chain of integrators with delayed input was considered in [20].

A popular approach in the literature on the stabilization of systems with arbitrarily large time delays in the input is the predictor feedback control (see, for example, [5, 15, 24]). For systems with time-varying delays, the predictor feedback method was originally introduced by Artstein in [1], and it has become a very active topic of research ever since. For the cases when the delay function is known before hand, a sample of the literature concerning the stabilization of linear and nonlinear systems can be found in [4, 16, 18, 26]. On the other hand, for systems where the delay function is unknown, adaptive predictor feedback methods were developed for linear systems in [2] and nonlinear systems in [6].

The use of low gain feedback for the stabilization of linear systems with time delays in the input was initiated in [18]. For open loop systems that are not exponentially unstable, the authors of [18] show that a stabilizable and detectable linear system with an arbitrarily large time delay in the input can be asymptotically stabilized either by linear state feedback or by linear output feedback using an eigenstructure assignment based low gain feedback design [17]. A salient feature of this low gain design is that it takes the structure of a predictor feedback control law but with the distributed portion of the predictor feedback control dropped, resulting in a finite dimensional feedback law. A simple example was also constructed in [18] to show that such a result would not be true if the open loop system is exponentially unstable. Another byproduct of this low gain feedback design is that, with no additional condition, the resulting linear feedback laws would also semi-globally asymptotically stabilize such systems when they are subject to input saturation.

The authors of [26] further developed the low gain feedback design approach introduced in [18]. In [26] a parametric Lyapunov equation based low gain feedback design is developed and the design method is termed "truncated predictor feedback (TPF)." In addition, time-varying delays are allowed. The results of [18, 26] were recently extended in [22] to exponentially unstable linear systems, that is, systems with poles in the open right-half plane.

In this chapter we extend the results presented in [22] by further increasing the bound on the time delay that the closed-loop system can tolerate, and including the input saturation in the analysis and design. First, we consider the time-delay system without saturation. The Riccati equation used for the synthesis of the stabilizing controller in [22] is generalized to include a larger set of feasible solutions. The resulting set of stabilizing controllers for the time-delay system is defined by a set of nonlinear matrix inequalities, which are employed to search for the optimal control law that maximizes the bound on the time-varying delay function. The path-following method [13] is employed to derive an LMI approximation from the original nonlinear constraints, and the optimal solution that maximizes the upper bound of the input delay is found iteratively around an initial feasible solution. Next, the input saturation is included in the time-delay system, and the LMI problem is updated to maximize the domain of attraction of the closed-loop system. A preliminary version of this chapter was presented in [23], which only included a sketch of the necessary proofs and the input saturation was not considered.

The remainder of this chapter is organized in the following sections. Section 2 introduces the control problem and the TPF method considered in this chapter. Some preliminary results are presented in Sect. 3. Section 4 develops stability conditions for the input delayed system under the TPF control law, and Sect. 5 implements the path-following method for solving the nonlinear matrix optimization problem. In Sect. 6 the input saturation is included in the stability analysis, and the results are employed in Sect. 7 to enlarge the domain of attraction of the closed-loop system. We verify the theoretical results through numerical examples included in Sect. 8, and by the experimental testing in Sect. 9. Finally, we draw our conclusions in Sect. 10.

Notation: We use fairly standard notation throughout this chapter. Let $\mathscr{C}_{n,\tau} = \mathscr{C}([-\tau, 0], \mathbf{R}^n)$ stand for the Banach space of continuous vector functions mapping the interval $[-\tau, 0]$ into R^n, and define $x_t \in \mathscr{C}_{n,\tau}$ to be the restriction of $x(t)$ to the interval $[t-\tau, t]$ translated to $[-\tau, 0]$, or in other words $x_t(\theta) = x(t+\theta)$ for $\theta \in [-\tau, 0]$. The ∞-norm of a vector u is denoted as $\|u\|_\infty$, and we define the standard saturation function as $\sigma : \mathbf{R} \to \mathbf{R}$, such that $\sigma(s) = \operatorname{sign}(s)\min\{1, |s|\}$. With a slight abuse of notation, we also denote the vector valued saturation function as

$$\sigma(u) = [\sigma(u_1)\ \sigma(u_2)\ \cdots\ \sigma(u_m)]^{\mathrm{T}},$$

where u_i is the ith element of a vector $u \in \mathbf{R}^m$. Without loss of generality, we assume a unity saturation level.

2 Problem Definition—Truncated Predictor Feedback Control

We first consider a linear time-invariant system with input delay but without input saturation, represented by the state space equation,

$$\dot{x}(t) = Ax(t) + Bu\left(\varphi(t)\right), \tag{1}$$

for a time-dependent function $\varphi(t)$. The input saturation will be treated later in Sect. 6. The state vector in the above system is $x(t) \in \mathbf{R}^n$, and the input vector is $u(t) \in \mathbf{R}^m$.

The matrix pair (A, B) is assumed to be controllable. Without loss of generality, we assume that (1) is exponentially unstable and all the eigenvalues of A are in the closed right-half plane.

The function $\varphi(t)$ represents the delay present in the system and is assumed to be known, continuously differentiable and invertible. Additionally, as discussed in [4], the assumption that $\frac{\mathrm{d}}{\mathrm{d}t}\varphi(t) > 0$ for all $t > 0$ is needed to guarantee that the control signal stays effective after a finite time. In this chapter we will define the time-varying delay to be in the standard form $\varphi(t) = t - D(t)$, where the delay function $D(t) : \mathbf{R}^+ \to \mathbf{R}^+$ is bounded as $0 \leq D(t) \leq \bar{D}$.

One approach for the stabilization of time-delay systems such as (1) is the truncated predictor feedback (TPF) control method [18, 26], which is based on the prediction of the system future states. In the TPF method, however, the prediction of the states is truncated to the finite dimensional component. This simplifies the controller equation by eliminating the input dependent term in the prediction. The stabilizing control law resulting from the TPF method is given as

$$u(t) = K\mathrm{e}^{A\left(\varphi^{-1}(t)-t\right)}x(t). \tag{2}$$

We base the stabilizing controller design on the TPF feedback method.

The equation for the delayed input $u(\varphi(t))$ can be determined analytically by solving (1) with the feedback control in (2) and initial condition $x(\varphi(t))$. The resulting expression for the input is $u(\varphi(t)) = K(x(t) - \lambda(t))$, where

$$\lambda(t) = \int_{\varphi(t)}^{t} \mathrm{e}^{A(t-s)}BK\mathrm{e}^{A(s-\varphi(s))}x(\varphi(s))\,\mathrm{d}s. \tag{3}$$

Then, the state space equation (1) with the truncated predictor feedback controller is $\dot{x}(t) = (A + BK)x(t) - BK\lambda(t)$. The design objective of the TPF controller is to synthesize the feedback gain K such that the above system is asymptotically stable.

3 Preliminary Results

In this section we introduce some preliminary results, which are used to derive the stability conditions of the closed-loop system under the TPF control law. The following lemma is adapted from [25].

Lemma 1 *Assume that $P > 0$ is the solution to the Riccati equation*

$$PA + A^TP - PBB^TP = -Q, \tag{4}$$

for some $Q > 0$, where the controllable pair (A, B) are the matrices of the state space equation (1). *Then for any scalar $\omega \geq 0$, it holds that*

$$e^{A^T t} P e^{At} - e^{\omega t} P = -e^{\omega t} \int_0^t e^{-\omega s} e^{A^T s} R e^{As} ds, \tag{5}$$

where the matrix R is defined as

$$R = Q + \omega P - PBB^T P. \tag{6}$$

Remark 1 Lemma 1 was slightly modified from the original lemma in [25]. The proof, however, remains mostly unchanged. A direct result from both lemmas is that, if matrix $R > 0$, then

$$\mathrm{e}^{A^T t} P \mathrm{e}^{At} \leq \mathrm{e}^{\omega t} P. \tag{7}$$

We will also need the following lemma from [11], which is a special application of the Jensen's inequality.

Lemma 2 *For any positive definite matrix $P > 0$, scalars γ_1 and γ_2 such that $\gamma_2 \geq \gamma_1$, and a vector valued function $x : [\gamma_1, \gamma_2] \to \boldsymbol{R}^n$ such that the integrals in the following are well defined, then*

$$\left(\int_{\gamma_1}^{\gamma_2} x^T(s) ds\right) P \left(\int_{\gamma_1}^{\gamma_2} x(s) ds\right) \leq (\gamma_2 - \gamma_1) \int_{\gamma_1}^{\gamma_2} x^T(s) P x(s) ds. \tag{8}$$

Finally, we recall the Razumikhin stability theorem. This theorem is used in developing the stability conditions for the time-delay systems we consider in this chapter.

Theorem 1 (Razumikhin Stability Theorem [7]) *Consider the functional differential equation*

$$\dot{x}(t) = f(t, x_t), \quad x(t) \in \boldsymbol{R}^n, \ t \geq 0 \tag{9a}$$

$$x(t) = \psi(t), \quad t \in [-\tau, 0] \tag{9b}$$

Let $u(s)$, $v(s)$, $w(s)$ and $p(s) \in \boldsymbol{R}^+ \to \boldsymbol{R}^+$ be positive, continuous and nondecreasing functions, such that $u(0) = v(0)$ for $s = 0$, v is strictly increasing, and $p(s) > s$ for $s > 0$. If there exist a continuous function $V : \boldsymbol{R}^n \to \boldsymbol{R}$ and a scalar $\rho > 0$ such that

1. $u(\|x\|) \leq V(s) \leq v(\|x\|)$,
2. $\dot{V}(x(t)) \leq -\omega(\|x(t)\|)$, *if* $V(x(t+\theta)) < p(V(x(t))$, $\forall \theta \in [-\tau, 0]$,

hold for all $x_t \in M_V(\rho) = \{\psi \in \mathscr{C}_{n,\tau} : V(\psi(\theta)) \leq \rho, \forall \theta \in [-\tau, 0]\}$, then the trivial solution $x(t) \equiv 0$ of (9) is asymptotically stable. Furthermore, the set $M_V(\rho)$ is an invariant set inside the domain of attraction of (9).

4 Stabilization of Input Delayed Systems

In this section we identify the stability conditions for the closed-loop system of (1) under the TPF control law. The stability conditions derived in this section will be used in the search of the optimal control law in the sections to follow.

Theorem 2 *Consider the input delayed system in* (1), *where the pair* (A, B) *is controllable. Furthermore, for an arbitrary matrix* $Q > 0$, *let* $P > 0$ *be the unique solution to* (4). *Then, the truncated predictor feedback control in* (2) *with* $K = -B^T P$ *asymptotically stabilizes the time-delay system* (1) *if the conditions*

$$\alpha W \geq BB^T, \tag{10a}$$

$$\left(A - \frac{1}{2}\omega I\right) W + W^T \left(A - \frac{1}{2}\omega I\right)^T < 0, \tag{10b}$$

$$W \left(A + \frac{\alpha^3 \bar{D}}{2\omega} e^{\omega \bar{D}} \left(e^{\omega \bar{D}} - 1\right) I\right)^T + \left(A + \frac{\alpha^3 \bar{D}}{2\omega} e^{\omega \bar{D}} \left(e^{\omega \bar{D}} - 1\right) I\right) W < BB^T, \tag{10c}$$

are satisfied for $W = P^{-1}$ *and some real scalars* $\omega, \alpha > 0$.

Proof We consider the following Lyapunov function,

$$V(x) = x^{\mathrm{T}} P x, \tag{11}$$

where $P > 0$ is the unique solution to the Riccati equation (4). The time derivative of this Lyapunov function along the trajectory of system (1) under the TPF control (2) and $K = -B^{\mathrm{T}} P$ is found to be

$$\begin{aligned} \dot{V}(x) &= \dot{x}^{\mathrm{T}} P x + x P \dot{x} \\ &= x^{\mathrm{T}} (PA + A^{\mathrm{T}} P - 2PBB^{\mathrm{T}} P) x + x^{\mathrm{T}} PBB^{\mathrm{T}} P \lambda \\ &\leq -x^{\mathrm{T}} Q x + \lambda^{\mathrm{T}} PBB^{\mathrm{T}} P \lambda, \end{aligned} \tag{12}$$

where λ was defined in (3). Let $\alpha > 0$ be a scalar such that

$$\alpha I \geq P^{\frac{1}{2}} BB^{\mathrm{T}} P^{\frac{1}{2}},$$

which is equivalent to (10a). The inequality (12) then simplifies to

$$\dot{V}(x) \leq -x^{\mathrm{T}} Q x + \alpha \lambda^{\mathrm{T}} P \lambda. \tag{13}$$

The first term on the right-hand side of the above inequality is negative definite. The second term, however, is positive definite. Using Lemma 2, we can determine an upper bound of the term $\lambda^{\mathrm{T}}P\lambda$ to be,

$$\lambda^{\mathrm{T}}P\lambda \leq (t-\varphi(t)) \int_{\varphi(t)}^{t} x^{\mathrm{T}}(\varphi(s))\, \mathrm{e}^{A^{\mathrm{T}}(s-\varphi(s))} PBB^{\mathrm{T}} \mathrm{e}^{A^{\mathrm{T}}(t-s)} \times P\mathrm{e}^{A(t-s)} BB^{\mathrm{T}} P\mathrm{e}^{A(s-\varphi(s))} x(\varphi(s))\, \mathrm{d}s. \tag{14}$$

By Remark 1, $\mathrm{e}^{A^{\mathrm{T}}t}P\mathrm{e}^{At} \leq \mathrm{e}^{\omega t}P$ for some $\omega > 0$ if R in (6) is positive definite. Substituting Q on the right-hand side of (6) with the left-hand side of (4), we obtain that R is positive definite if,

$$R = -A^{\mathrm{T}}P - PA + \omega P > 0.$$

The above condition can be rewritten by multiplying W from both sides, and the resulting expression is given as (10b).

Assuming that (10b) is satisfied for some $\omega > 0$, we obtain from Lemma 1 and the condition on α that (14) can be simplified to

$$\lambda^{\mathrm{T}}P\lambda \leq \alpha^2 \bar{D} \mathrm{e}^{\omega\bar{D}} \int_{t-\bar{D}}^{t} \mathrm{e}^{\omega(t-s)} V(x(\varphi(s)))\, \mathrm{d}s. \tag{15}$$

At this point, we want to eliminate the Lyapunov function inside the integral of the above expression. To this end, we observe that, by the definition of $\varphi(t)$, it must hold that $\varphi(t-\bar{D}) = t - \bar{D} - D(t) \geq t - 2\bar{D}$. If we assume that the Lyapunov function admits a constant $\eta > 1$ such that $V(x(t+\theta)) < \eta V(x(t))$ holds for all $\theta \in \left[-2\bar{D}, 0\right]$, then $V(x(\varphi(s))) < \eta V(x(t))$ for $s \in \left[t-\bar{D}, t\right]$, and (15) becomes

$$\lambda^{\mathrm{T}}P\lambda \leq \alpha^2 \bar{D} \mathrm{e}^{\omega\bar{D}} \eta \int_{t-\bar{D}}^{t} \mathrm{e}^{\omega(t-s)} \mathrm{d}s V(x(t)).$$

Next, we substitute the above inequality into the derivative of the Lyapunov equation in (13), and obtain,

$$\dot{V}(x) \leq -x^{\mathrm{T}} \left(Q - \eta \frac{\alpha^3 \bar{D}}{\omega} \mathrm{e}^{\omega\bar{D}} \left(\mathrm{e}^{\omega\bar{D}} - 1 \right) P \right) x. \tag{16}$$

Because η can be chosen arbitrarily close to one, the derivative of the Lyapunov function is then negative definite if

$$Q - \frac{\alpha^3 \bar{D}}{\omega} \mathrm{e}^{\omega\bar{D}} \left(\mathrm{e}^{\omega\bar{D}} - 1 \right) P > 0.$$

The above expression can be further expanded by substituting Q with the left-hand side of the Riccati equation in (4), and multiplying the inequality by W both to the left and to the right. The resulting inequality can be written as (10c).

Finally, under the assumption that (10) is satisfied, the right-hand side of (16) is a negative definite function, and by Theorem 1, the linear time-delay system in (1) under the truncated predictor feedback control (2) is asymptotically stable. □

5 Maximization of the Delay Bound by Solving LMI Problems

The stability conditions given in Theorem 2 involve nonlinear matrix inequalities. These nonlinear matrix inequalities cannot be solved using traditional LMI solvers. Instead, an iterative method may be used to convert the stability conditions into an LMI problem, and an LMI solver may be used to find a local optimal solution around an initial feasible solution $(W_0, \alpha_0, \omega_0, \bar{D}_0)$. In this section, we develop an algorithm to find a stabilizing control law in the form of (2) such that the delay bound $\bar{D}$ the controlled system can tolerate is maximized.

The path-following method [13] incrementally searches for a local optimal solution to a bilinear matrix inequality (BMI) by iteratively solving an LMI approximation of the original problem. The search region of the optimal solution in the linearized optimization problem is limited to a small neighborhood around an initial feasible solution. The initial feasible solution is updated at the beginning of each iteration, and a slightly improved controller is designed at each successful iteration. While the method was originally developed for solving BMI problems, the approach can easily be applied to solve more general nonlinear matrix inequality problems. In particular, as we will see later in this section, (10) is an ideal candidate for the path-following method since all decision variables except W are scalars.

We consider the linearized decision variables as

$$W = W_0 + \delta W, \ \alpha = \alpha_0 + \delta\alpha, \ \omega = \omega_0 + \delta\omega, \ \text{and } \bar{D} = \bar{D}_0 + \delta\bar{D}, \tag{17}$$

W_0, α_0, ω_0 and $\bar{D}_0$ satisfy the matrix inequalities (10). The perturbations $\delta W, \delta\alpha, \delta\omega$ and $\delta\bar{D}$ are the incremental changes at each iteration of the path-following method. The linearized decision variables are substituted into the matrix inequalities in (10). After expanding these matrix inequalities, the second and higher order terms of the perturbations are neglected under the assumption that the magnitudes of the perturbations are small.

After eliminating the higher order perturbation terms, we obtain the following linearized matrix inequalities as the constraints to the optimization problem:

$$\alpha_0 \delta W + W_0 \delta\alpha + \alpha_0 W_0 - BB^{\mathrm{T}} > 0, \tag{18a}$$

$$A\delta W + \delta W A^{\mathrm{T}} - \delta\omega W_0 - w_0 \delta W + AW_0 + W_0 A^{\mathrm{T}} - w_0 W_0 < 0, \tag{18b}$$

$$\begin{aligned}&\left(AW_0 + W_0A^{\mathrm{T}} - BB^{\mathrm{T}} + \frac{h_1}{\omega_0}\alpha_0 \bar{D}_0 W_0\right)\\ &+ \left(A\delta W + \delta W A^{\mathrm{T}} + \frac{h_1}{\omega_0}\bar{D}_0 \alpha_0^3 \delta W\right) + \left(\frac{h_1}{\omega_0}\alpha_0^3 + h_2 \alpha_0^3 \bar{D}_0\right) W_0 \delta\bar{D}\\ &+ \left(3\frac{h_1}{\omega_0}\bar{D}_0 \alpha_0^2 W_0\right)\delta\alpha + \left(\frac{h_2}{\omega_0}\alpha_0^3 \bar{D}_0^2 W_0 - \frac{h_1}{\omega_0^2}\alpha_0^3 \bar{D}_0 W_0\right)\delta\omega < 0,\end{aligned} \tag{18c}$$

$$\begin{bmatrix} h_{\text{step}} W_0 & \delta W \\ \delta W^{\mathrm{T}} & h_{\text{step}} W_0 \end{bmatrix} > 0, \tag{18d}$$

$$\begin{bmatrix} h_{\text{step}} \bar{D}_0 & \delta\bar{D} \\ \delta\bar{D} & h_{\text{step}} \bar{D}_0 \end{bmatrix} > 0, \tag{18e}$$

$$\begin{bmatrix} h_{\text{step}} \alpha_0 & \delta\alpha \\ \delta\alpha & h_{\text{step}} \alpha_0 \end{bmatrix} > 0, \tag{18f}$$

$$\begin{bmatrix} h_{\text{step}} \omega_0 & \delta\omega \\ \delta\omega & h_{\text{step}} \omega_0 \end{bmatrix} > 0. \tag{18g}$$

The values of h_1 and h_2 are constants within each iteration, and they are defined as $h_1 = e^{2\omega_0 \bar{D}_0} - e^{\omega_0 \bar{D}_0}$ and $h_2 = 2e^{2\omega_0 \bar{D}_0} - e^{\omega_0 \bar{D}_0}$. The first three conditions in (18) come from the linearization of the matrix conditions in (10a), (10b) and (10c), respectively. The last four conditions, for an $h_{\text{step}} > 0$, are added to the LMI problem in order to limit the maximum step size of the perturbations variables.

The overall objective is to maximize the upper bound $\bar{D}$ of the input time delay. A straightforward approach is to maximize $\delta\bar{D}$ at every iteration. However, it was found during the numerical testing that this objective function drives the solution of the path-following method rapidly toward a local minimum, limiting the improvement from the initial solution. A remedy for this problem is to perturb the objective function by including conditions that indirectly contribute towards the end goal. We define the objective function J_i to be,

$$J_i = \begin{cases} -\delta\bar{D} \text{ for } i = 0, \\ \delta\omega \text{ for } i = 1, \\ \delta\alpha \text{ for } i = 2, \end{cases} \tag{19}$$

for a variable integer i. At each step of the iteration, we switch the objective function to perturb the LMI solution out of the region of attraction of a local minimum, and toward an improved solution.

The algorithm for finding a stabilizing controller to the linear input delayed system in (1) that maximizes upper bound of the delay can be summarized as follows.

1. Find a feasible solution $(W_0, \alpha_0, \omega_0, \bar{D}_0)$ for the stabilization problem using the method described in [22]. Since the pair (A, B) is controllable, the Riccati equation (4) has a unique solution for any $Q > 0$, and a feasible solution to (10) can always be found for a small enough $\bar{D}$. Select an initial value for h_{step}, and set $i = 0$.
2. With the initial feasible solution, we minimize J_i over $(\delta W, \delta\alpha, \delta\omega, \delta\bar{D})$, subject to the LMI conditions in (18).
3. For the newly found solution $(W, \alpha, \omega, \bar{D})$, check if the original stability conditions in (10) are satisfied. If (10) is not satisfied, then reduce h_{step} and return to Step 2. On the other hand, if (10) is satisfied, then continue to the next step.
4. Update the initial feasible solution $(W_0, \alpha_0, \omega_0, \bar{D}_0)$ to the newly found solution $(W, \alpha, \omega, \bar{D})$. Set $i \equiv i + 1 \pmod 3$ and go to Step 2.

A trigger should be included in the above algorithm to stop the iteration in case h_{step} falls below a set minimum step size, and the initial feasible solution cannot be further improved. Once again, the optimal control law is given as (2) with the feedback gain $K = -B^{\mathrm{T}} W^{-1}$.

6 Stabilization of Input Delayed Systems with Input Saturation

We now consider the linear system

$$\begin{aligned} \dot{x}(t) &= Ax(t) + B\sigma\left(u\left(\varphi(t)\right)\right), \\ x(t) &= \psi(t), \ t \in [-\bar{D}, 0], \end{aligned} \tag{20}$$

with both time-varying input delay and input saturation and with an initial condition $\psi \in \mathscr{C}_{n,\bar{D}}$. We let $\sigma : \mathbf{R}^m \to \mathbf{R}^m$ denote the standard vector-valued saturation function, where the ith element of $\sigma(u)$ is defined as

$$\sigma(u_i) = \mathrm{sign}(u_i) \min\{1, |u_i|\}.$$

In order to develop the stability conditions for this time-delay system with input saturation, we use some of the notation and results presented in [7, 14]. For a positive definite matrix $P \in \mathbf{R}^{n\times n}$ and a scalar $\rho > 0$, we define the ellipsoid $\Omega(P, \rho) = \{x \in \mathbf{R}^n : x^{\mathrm{T}} P x \le \rho\}$. For an $m \times n$ matrix F with the ith row denoted by f_i, we define the symmetric polyhedron $\mathscr{L}(F)$ as

$$\mathscr{L}(F) = \{x \in \mathbf{R}^n : |f_i x| \le 1, i = 1, 2, \cdots, m\}.$$

Then, if $x(t) \in \mathscr{L}(B^{\mathrm{T}} P \mathrm{e}^{A(\varphi^{-1}(t)-t)})$ for $t \in [-\bar{D}, \infty)$, the truncated predictor feedback control (2) with $K = -B^{\mathrm{T}} P$ does not saturate, and the nonlinear system (20) is equivalent to its linear counterpart in (1). In other words, the stability conditions in

Theorem 3 applies for the saturated system (20) as long as the state trajectories stay within $\mathscr{L}(B^{\mathrm{T}}P\mathrm{e}^{A(\varphi^{-1}(t)-t)})$. This is formally stated in the following theorem.

Theorem 3 *Consider the time-delay system with input saturation as in* (20)*, where the pair* (A, B) *is controllable. Furthermore, for an arbitrary matrix* $Q > 0$*, let* $P > 0$ *be the unique solution to* (4)*. Then, the truncated predictor feedback control in* (2) *with* $K = -B^T P$ *asymptotically stabilizes the time-delay system with input saturation* (20) *and with invariant set*

$$M_V(\rho) = \{\psi \in \mathscr{C}_{n,\tau} : \psi(\theta) \in \Omega(P, \rho), \forall \theta \in [-\bar{D}, 0]\}$$

inside the domain of attraction of the closed-loop system if (10) and

$$\rho^{-1} - \alpha \mathrm{e}^{\omega \bar{D}} \geq 0 \tag{21}$$

are satisfied.

Proof Consider the TPF control law (2), with $K = -B^{\mathrm{T}}P$ satisfying the stability conditions (10) for the saturation-free system (1). Then, Theorem 2 yields that the same control law asymptotically stabilizes the system with input saturation (20) if

$$\|u(\varphi(t))\|_\infty = \|B^{\mathrm{T}}P\mathrm{e}^{A(t-\varphi(t))}x(\varphi(t))\|_\infty \leq 1, \tag{22}$$

for all $t \geq 0$, or simply $x(\varphi(t)) \in \mathscr{L}(B^{\mathrm{T}}P\mathrm{e}^{AD(t)})$. It follows from Lemma 1 that,

$$\mathrm{e}^{A^{\mathrm{T}}D(t)}PBB^{\mathrm{T}}P\mathrm{e}^{AD(t)} \leq \alpha \mathrm{e}^{\omega \bar{D}}P, \tag{23}$$

and therefore $\|u(\varphi(t))\|_\infty \leq 1$ if

$$x(\varphi(t)) \in \Omega(\alpha \mathrm{e}^{\omega \bar{D}}P, 1) \subset \mathscr{L}(B^{\mathrm{T}}P\mathrm{e}^{AD(t)}). \tag{24}$$

Finally, Theorem 1 with the Lyapunov function (11) in Theorem 2 yields that $M_V(\rho)$ is an invariant set within the domain of attraction of the closed-loop system (20) under the TPF control if $\Omega(P, \rho) \subset \Omega(\alpha \mathrm{e}^{\omega \bar{D}}P, 1)$, which is equivalent to the condition (21). This concludes the proof. □

7 Maximization of the Delay Bound and the Domain of Attraction by Solving LMI Problems

The "size" of the invariant set $M_V(\rho)$ inside the domain of attraction of the closed-loop system is determined by the size of the ellipsoid $\Omega(P, \rho)$. In this section we enlarge our approximation of the domain of attraction by enlarging $M_V(\rho)$. Let $S > 0$ be a prescribed $n \times n$ matrix and γ be a positive scalar. Then, the size of the invariant set $M_V(\rho)$ is increased by increasing the magnitude of γ and enforcing the condition $\Omega(S, \gamma^2) \subset \Omega(P, \rho)$. With the help of the Schur complement, this set inclusion condition can be translated to the matrix form

$$\begin{bmatrix} \rho S & \gamma I \\ \gamma I & W \end{bmatrix} \geq 0. \tag{25}$$

Condition (21) in Theorem 3 for the stability of the system with input saturation (20) can be rewritten in the form

$$\rho - \mathrm{e}^{-\omega \bar{D}} \alpha^{-1} \leq 0. \tag{26}$$

Note that inequalities (25) and (26) are linear with respect to the decision variables γ and ρ. As in Sect. 4, the path-following method can be employed to synthesize the control law that maximizes the size of invariant set $M_V(\rho)$ and the delay bound $\bar{D}$ by iteratively solving the linearized LMI problem of (10), (25) and (26). The decision variables of the linearized LMI problem are once again δW, $\delta\alpha$, $\delta\omega$, $\delta\bar{D}$, ρ and γ. The objective function of the path-following algorithm is modified as

$$J_i = \begin{cases} -\delta\bar{D} - \mu\gamma \text{ for } i = 0, \\ \delta\omega - \mu\gamma \text{ for } i = 1, \\ \delta\alpha - \mu\gamma \text{ for } i = 2, \end{cases} \tag{27}$$

for some small scalar $\mu > 0$. This forces the path-following method to work towards a larger invariant set $M_V(\rho)$ at each iteration.

8 A Numerical Example

Consider the delayed double oscillator system with an additional unstable positive real pole, originally considered in [22]. The system state space equation are given as (1) with matrices,

$$A = \begin{bmatrix} 0.1 & 1 & 0 & 0 & 0 \\ 0 & 0 & 1 & 0 & 0 \\ 0 & -1 & 0 & 1 & 0 \\ 0 & 0 & 0 & 0 & 1 \\ 0 & 0 & 0 & -1 & 0 \end{bmatrix}, \quad B = \begin{bmatrix} 0 \\ 0 \\ 0 \\ 0 \\ 1 \end{bmatrix}. \tag{28}$$

The initial condition of the system is defined as

$$x(\theta) = [-1 \ 2 \ 2 \ -1 \ 2]^{\mathrm{T}}$$

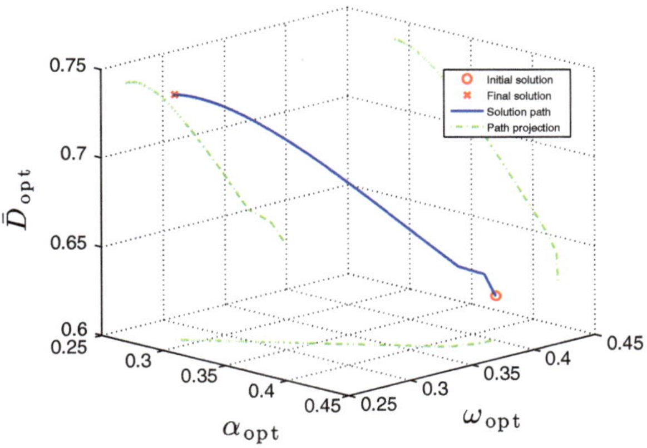

Fig. 1 Path of the optimal solution found through the path following method. The initial feasible solution is marked by '∘' and the final optimal solution is marked by '×'

for all $\theta \in [-\bar{D}, 0]$. The time-varying delay is defined by the inverse function

$$\varphi^{-1}(t) = t + \frac{2}{3}\bar{D}\left(1 + \frac{1}{2}\cos(t)\right). \tag{29}$$

This corresponds to an oscillatory delay function with a maximum delay of $\bar{D}$.

8.1 Maximizing the Delay Bound

Using the control law derived in the numerical example of [22] as the initial feasible solution, we synthesized a TPF controller by following the algorithm presented in Sect. 5. The trajectory of the optimal solution at each iteration is shown in Fig. 1. We observe that the optimal value of $\bar{D}$ increases rapidly at first, but then slowly settles at a local optimum. The upper bound of the delay function with the optimal control is $\bar{D}_{\mathrm{opt}} = 0.74$ s. The feedback gain K corresponding to the optimal solution is $K = [-0.224 \ \ -0.0209 \ \ -0.222 \ \ -0.039 \ \ -0.289]$. Shown in the left and right panels of Fig. 2 are the simulated responses of the closed-loop system under the optimal TPF control with $\bar{D} = 0.5$ and 0.74 s, respectively. It is observed that the TPF controller is able to stabilize the trajectory of the states. The amplitudes of the oscillations in Fig. 2 increase initially because of the delay in the activation of the control input, but they eventually approach zero as $t \to \infty$.

8.2 Enlarging the Domain of Attraction

Assume that (28) is also subject to input saturation as in (20). We incorporate ideas from Sects. 6, 7 to increase the size of the invariant set corresponding to the

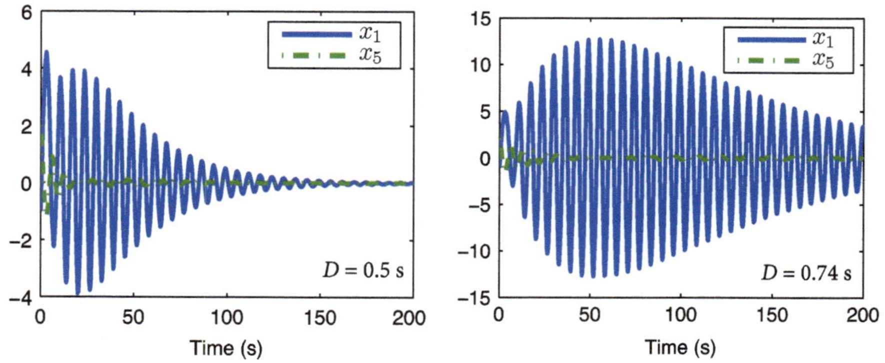

Fig. 2 Simulated time responses of states x_1 and x_5 for the delay system without input saturation with $\bar{D} = 0.5$ (*Left*) and $\bar{D} = 0.74$ (*Right*)

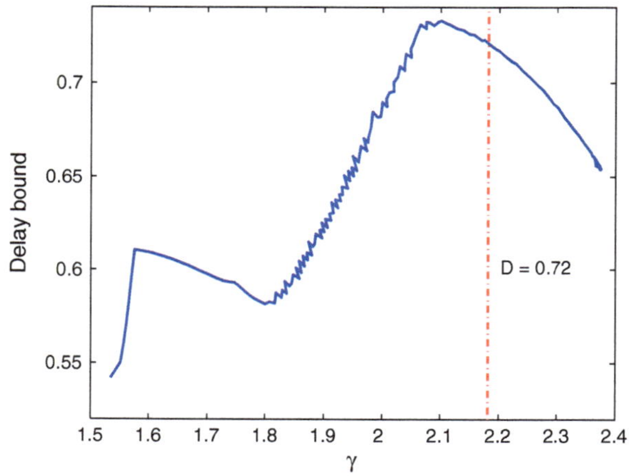

Fig. 3 $\bar{D}$ versus γ from the path of the optimal solution for the delay system with input saturation

closed-loop system under the TPF control. For this numerical example, we let $S = I$ and $\mu = 0.5$ in the objective function of the path-following method.

Figure 3 shows the relationship between γ and $\bar{D}$ after each successful iteration of the path-following method. Starting from $\gamma = 1.53$, the path-following method can initially increase both $\bar{D}$ and γ. Once the maximum delay bound of $\bar{D} = 7.33$ s is reached, we see a trade-off between the delay bound and the size of the invariant set. Figure 4 shows the simulated state response of the system with saturation and under the control law corresponding to $\bar{D} = 0.72$ s in Fig. 3. Compared to the case without saturation in Fig. 2, the convergence rates of the states in Fig. 4 are slower.

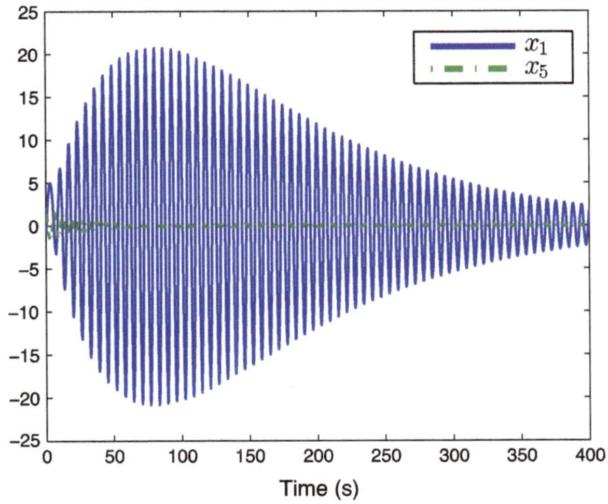

Fig. 4 Simulated time responses of states x_1 and x_5 for the delay system with input saturation

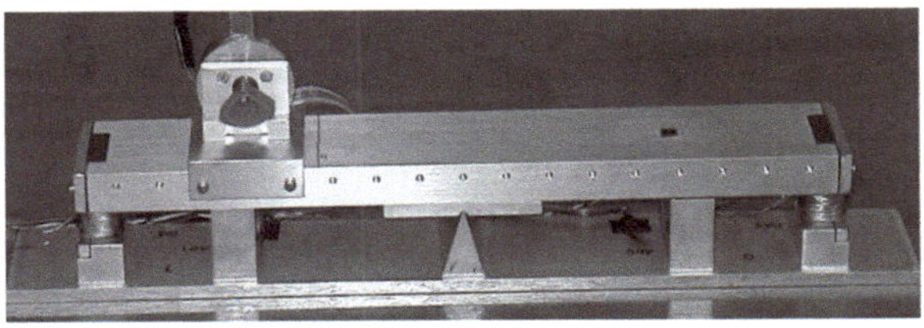

Fig. 5 The balancing beam test rig [19]

9 Experimental Testing

In this section we present experimental results from the testing of the TPF controller on the balancing beam test rig shown in Fig. 5. The test rig consists of two magnetic coils acting at each end of a pivoted beam. The magnetic coils generate magnetic forces to balance the beam at the center pivot point. The test rig has mechanical stops, which limit the range of angular motion of the beam to ± 0.013 rad, to protect the coils from damage. Proximity sensors measure the angle of the beam with respect to the fixed base. A detailed description of the test rig can be found in [19].

The states of the balancing beam are the beam angle θ and the angular velocity $\dot{\theta}$. The input to the test rig is the control current to the coil, which is limited to 3 A by the maximum output of the current amplifier. A constant delay is added at the output of the controller implementation to simulate an input delay. The dynamics of the balancing beam with delayed input are given by the state space equation,

$$\begin{bmatrix} \dot{\theta} \\ \ddot{\theta} \end{bmatrix} = \begin{bmatrix} 0 & 1 \\ 9248 & -1.635 \end{bmatrix} \begin{bmatrix} \theta \\ \dot{\theta} \end{bmatrix} + \begin{bmatrix} 0 \\ 281.9 \end{bmatrix} \sigma(u(t-D)). \tag{30}$$

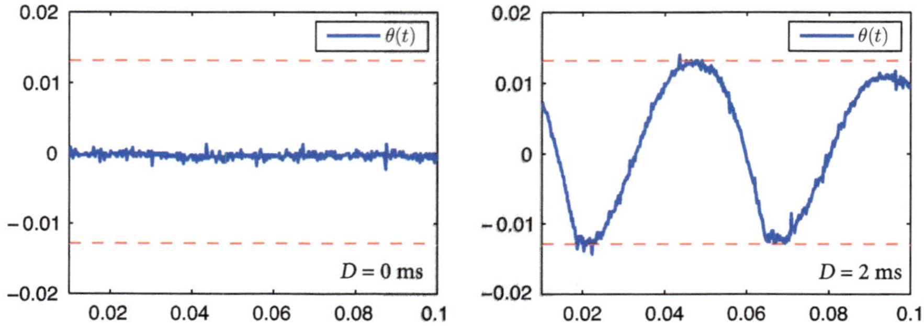

Fig. 6 Beam angle response with a nominal control law $u(t) = Kx(t)$ for the cases $D = 0$ ms (*Left*) and $D = 2$ ms (*Right*)

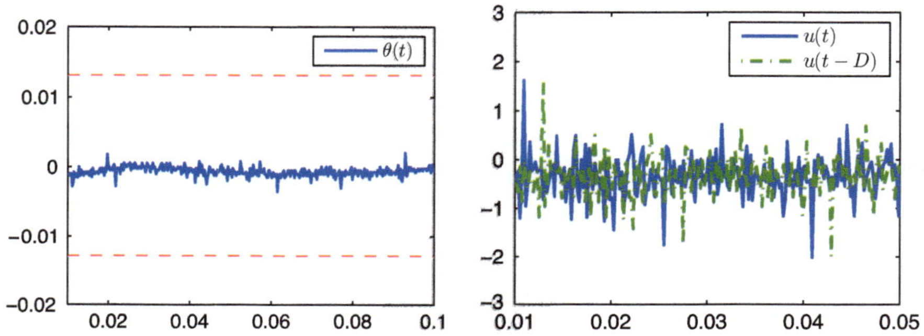

Fig. 7 Beam angle and control input responses with truncated predictor control law (2) and input delay of $D = 2$ ms

The path-following method was employed to synthesize the TPF controller. Following the procedure described in Sect. 7, we obtain the optimal feedback gain as $K = [85.12 \;\; 0.8776]$, corresponding to a maximum input delay of $\bar{D}_{\text{opt}} = 1.54$ ms.

Figure 6 shows the experimental response of the balancing beam test rig with the nominal control law

$$u(t) = Kx(t). \tag{31}$$

The response in the left panel in Fig. 6 corresponds to the system without input delay, and it shows the stable regulation of the beam angle near the horizontal position. The response of the balancing beam with a constant input delay of $D = 2$ ms is presented in right panel of Fig. 6. The figure shows that the delayed system is unstable, and the beam bounces between the mechanical stops limiting the motion of the beam.

The responses of the balancing beam test rig with the controller (2) and input time delay of $D = 2$ ms are presented in Fig. 7. The response of the beam angle is presented in the left panel of Fig. 7, and it shows the stable regulation of the beam angle. The control current input to the magnetic actuators is shown in the right panel of Fig. 7, which shows both the control $u(t)$ and the delayed signal $u(t - D)$.

10 Conclusions

The stabilization of exponentially unstable linear systems subject to time-varying delay and saturation in the input was examined in this chapter. For the stabilization of exponentially unstable systems with input delay, we derived conditions in the form of matrix inequalities that guarantee the stability of the closed-loop system under the truncated predictor feedback control. These nonlinear matrix inequalities were linearized, and the path-following method was employed to construct a linearized LMI problem that iteratively solves for the optimal control law that maximizes the bound of the delay function that the closed-loop system can tolerate. It was demonstrated in Sects. 8 and 9 that this approach can significantly increase the maximum allowable delay when compared to the results in [22]. This linearized LMI problem was updated for systems with input saturation, such that the optimal control law also enlarges the domain of attraction of the closed-loop system.

Acknowledgments This work was supported in part by the U.S. National Science Foundation under grant CMMI-1129752.

References

1. Artstein, Z.: Linear systems with delayed controls: a reduction. IEEE Trans. Autom. Control **27**(4), 869–876 (1982)
2. Bekiaris-Liberis, N., Krstic, M.: Delay-adaptive feedback for linear feedforward systems. Syst. Control Lett. **59**(5), 277–283 (2010)
3. Bekiaris-Liberis, N., Krstic, M.: Stabilization of linear strict-feedback systems with delayed integrators. Automatica **46**(11), 1902–1910 (2010)
4. Bekiaris-Liberis, N., Krstic, M.: Compensation of time-varying input and state delays for nonlinear systems. J. Dyn. Syst. Measur. Control **134**(1), 011009 (2011)
5. Bekiaris-Liberis, N., Krstic, M.: Lyapunov stability of linear predictor feedback for distributed input delays. IEEE Trans. Autom. Control **56**(3), 655–660 (2011)
6. Bresch-Pietri, D., Krstic, M.: Delay-adaptive predictor feedback for systems with unknown long actuator delay. IEEE Trans. Autom. Control **55**(9), 2106–2112 (2010)
7. Cao, Y.-Y., Lin, Z., Hu, T.: Stability analysis of linear time-delay systems subject to input saturation. IEEE Trans. Circuits Syst. I: Fundam. Theory Appl. **49**(2), 233–240 (2002)
8. Chen, B.-S., Wang, S.-S.: Lu. H-C.: Stabilization of time-delay systems containing saturating actuators. Int. J. Control **47**(3), 867–881 (1988)
9. Dugard, L., Verriest, E. (eds.): Stability and control of time-delay systems. Series Lecture Notes in Control and Information Sciences, vol. 228. Springer, London (1988)
10. Fang, H., Lin, Z.: A further result on global stabilization of oscillators with bounded delayed input. IEEE Trans. Autom. Control **51**(1), 121–128 (2006)
11. Gu, K.: An integral inequality in the stability problem of time-delay systems. In: Proceedings of the 39th IEEE Conference on Decisision and Control, pp. 2805–2810 (2000)
12. Gu, K., Kharitonov, V., Chen, J.: Stability of Time-Delay Systems. Birkhauser, Boston (2003)
13. Hassibi, A., How, J., Boyd, S.: A path-following method for solving BMI problems in control. In: Proceedings of the American Control Conference, pp. 1385–1389 (1999)
14. Hu, T., Lin, Z.: Control Systems with Actuator Saturation: Analysis and Design. Birkhauser, Boston (2001)

15. Krstic, M.: Compensation of infinite-dimensional actuator and sensor dynamics. IEEE Control Syst. Mag. **30**(1), 22–41 (2010)
16. Krstic, M.: Lyapunov stability of linear predictor feedback for time-varying input delay. IEEE Trans. Autom. Control **55**(2), 554–559 (2010)
17. Lin, Z.: Low Gain Feedback. Springer, London (1988)
18. Lin, Z., Fang, H.: On asymptotic stabilizability of linear systems with delayed input. IEEE Trans. Autom. Control **52**(6), 998–1013 (2007)
19. Lin, Z., Glauser, M., Hu, T., Allaire, P.: Magnetically suspended balance beam with disturbances: A test rig for nonlinear output regulation. In: Proceedings of the 43rd IEEE Conference on Decision and Control, pp. 4577–4582 (2004)
20. Mazenc, F., Mondie, S., Niculescu, S.-I.: Global asymptotic stabilization for chain of integrators with a delay in the input. IEEE Trans. Autom. Control **48**(1), 57–63 (2003)
21. Mazenc, F., Mondie, S., Niculescu, S.-I.: Global stabilization of oscillators with bounded delayed input. Syst. Control Lett. **53**(5), 415–422 (2004)
22. Yoon, S.-Y., Lin, Z.: Truncated predictor feedback control for exponentially unstable linear systems with time-varying input delay. Syst. Control Lett. **62**(10), 837–844 (2013)
23. Yoon, S-Y., Anantachaisilp. P., Lin, Z.: An LMI approach to the control of exponentially unstable systems with input time delay. In: Proceedings of the 52nd IEEE Conference on Decision and Control, pp. 312–317 (2013)
24. Yoon, S-Y., Lin, Z.: Predictor based control for linear systems with both state and input delays. In: Proceedings of the American Control Conference, pp. 4169–4174 (2014)
25. Zhou, B., Lin, Z., Duan, G.: Stabilization of linear systems with input delay and saturation—a parametric Lyapunov equation approach. Int. J. Robust Nonlinear Control **20**(13), 1502–1519 (2010)
26. Zhou, B., Lin, Z., Duan, G.: Truncated predictor feedback for linear systems with long time-varying input delay. Automatica **48**(10), 2387–2399 (2012)

Identifiability and Observability of Nonlinear Time-Delay Systems with Unknown Inputs

Gang Zheng and Jean-Pierre Richard

Abstract Using the theory of non-commutative rings, the delay identification problem of nonlinear time-delay systems with unknown inputs is studied. Necessary and sufficient conditions are proposed to judge the identifiability of the delay, where two different cases are discussed for the dependent and independent outputs, respectively. After that, necessary and sufficient conditions are given to analyze the causal and non-causal observability for nonlinear time-delay systems with unknown inputs.

1 Introduction

Time-delay systems are widely used to model concrete systems in engineering and sciences such as biology, chemistry, mechanics and so on [16, 26, 31]. Many results have been reported for the purpose of stability and observability analysis, by assuming that the delay of the studied systems is known. This makes the delay identification one of the most important topics in the field of time-delay systems. Up to now, various techniques have been proposed for the delay identification problem, such as identification by using variable structure observers [15, 34, 35], modified least squares techniques [36], neural network algorithms [45], the convolution approach [4], the algebraic fast identification technique [6, 8] as initiated in [18], and so on; see [5, 15] for additional references. Most of the papers on identification under delays concern linear models. Another source of complexity comes from the presence of feedback loops involving delays. Indeed, when the delay only appears in the inputs or outputs, the system has finite dimension. When the delays are involved in a

G. Zheng (✉) · J.-P. Richard
Non-A Team, Inria Lille and LAGIS CNRS 8219, 40 Avenue Halley,
59650 Villeneuve d'Ascq, France
e-mail: gang.zheng@inria.fr

J.-P. Richard
e-mail: jean-pierre.richard@ec-lille.fr

I. Karafyllis et al. (eds.), *Recent Results on Nonlinear Delay Control Systems*,
Advances in Delays and Dynamics 4, DOI 10.1007/978-3-319-18072-4_20

closed-loop manner, the resulting model has delayed states and becomes a functional differential equation, which has infinite dimension [7, 37].

Besides identifiability, the observability property has been exhaustively studied for nonlinear systems without delays. It has been characterized in [22, 27, 42, 43] from a differential point of view, and in [14] from an algebraic point of view. However, when the system is subject to time delay, such analysis is more complicated; see the surveys [37, 40]. For linear time-delay systems, various aspects of observability have been studied in the literature, using different methods such as the functional analytic approach [9] or the algebraic approach [10, 17, 41].

The aim of this chapter is to first identify the time delay of nonlinear time-delay systems with unknown inputs and then study observability for the systems with identified delay. The work is based on the theory of non-commutative rings, which was first proposed in [30] for the disturbance decoupling problem of nonlinear time-delay system. Then this method was applied to study observability of nonlinear time-delay systems with known inputs in [44], to analyze identifiability of parameter for nonlinear time-delay systems in [47], and to study state elimination and delay identification of nonlinear time-delay systems with known inputs in [1]. The motivation to study nonlinear time-delay systems with unknown inputs is due to the fact that there exist some cases, such as observer design for time-delay systems, in which the inputs can be unknown [13, 25, 38, 46]. Moreover, some proposed unknown input observer design methods depend on the known delay, which should be identified in advance. Motivated by this requirement, this chapter investigates both the delay identification problem and the observability problem for nonlinear time-delay systems with unknown inputs.

This chapter is organized as follows. Section 2 recalls the algebraic framework proposed in [44]. Notations and preliminary result are given in Sect. 3. Necessary and sufficient conditions are discussed for identifying the delay in two different cases, namely, dependent outputs over non-commutative rings, and then independent ones. Section 5 deduces necessary and sufficient conditions for causal and non-causal observability for nonlinear time-delay systems with unknown inputs, and the proposed result is applied to analyze the identifiability and observability for an illustrative example in Sect. 6.

2 Algebraic Framework

It is assumed that the delays are constant and commensurate, that is, all of them are multiples of an elementary unknown delay τ. Under this assumption, the nonlinear time-delay system is described as follows:

$$\begin{cases} \dot{x} = f(x(t-i\tau)) + \sum_{j=0}^{s} g^j(x(t-i\tau))u(t-j\tau), \\ y = h(x(t-i\tau)) = [h_1(x(t-i\tau)), \ldots, h_p(x(t-i\tau))]^T, \\ x(t) = \psi(t),\ u(t) = \varphi(t),\ t \in [-s\tau, 0], \end{cases} \tag{1}$$

where the constant delays $i\tau$ are associated with the finite set of integers $i \in S_- = \{0, 1, \ldots, s\}$; $x \in W \subseteq R^n$ refers to the state variables; $u = [u_1, \ldots, u_m]^T \in R^m$ is the unknown input; $y \in R^p$ is the measurable output; f, g^j, and h are meromorphic functions (meaning their entries are quotients of convergent power series with real coefficients [12, 44]); and $f(x(t - i\tau)) = f(x(t), x(t - \tau), \ldots, x(t - s\tau))$; $\psi : [-s\tau, 0] \to R^n$, and $\varphi : [-s\tau, 0] \to R^m$ denote unknown continuous functions of initial conditions. Throughout this chapter, it is assumed that, for any initial conditions ψ and φ, system (1) admits a unique solution.

Based on the algebraic framework introduced in [44], consider the field $\mathscr{K}$ of meromorphic functions of a finite number of the variables from $\{x_j(t - i\tau), j \in [1, n], i \in S_-\}$. For the sake of simplicity, we introduce the delay operator δ, which means the following for all $i \in Z^+$:

$$\delta^i \xi(t) = \xi(t - i\tau),\ \xi(t) \in \mathscr{K}, \tag{2}$$

$$\begin{aligned} \delta^i\left(a(t)\xi(t)\right) &= \delta^i a(t)\delta^i \xi(t) \\ &= a(t - i\tau)\xi(t - i\tau). \end{aligned} \tag{3}$$

Let $\mathscr{K}(\delta]$ denote the set of all polynomials in δ over $\mathscr{K}$ of the form

$$a(\delta] = a_0(t) + a_1(t)\delta + \cdots + a_{r_a}(t)\delta^{r_a}, \tag{4}$$

where $a_i(t) \in \mathscr{K}$ and $r_a \in Z^+$. The addition in $\mathscr{K}(\delta]$ is defined as usual, but the multiplication is given as:

$$a(\delta]b(\delta] = \sum_{k=0}^{r_a+r_b} \sum_{i+j=k}^{i\le r_a, j\le r_b} a_i(t)b_j(t - i\tau)\delta^k. \tag{5}$$

Considering (1) without input, differentiation of an output component $h_j(x(t-i\tau))$ with respect to time t is defined as follows:

$$\dot{h}_j(x(t - i\tau)) = \sum_{i=0}^{s} \frac{\partial h_j}{\partial x(t - i\tau)}\delta^i f.$$

By the definition of $\mathscr{K}(\delta]$, the system (1) can be rewritten in this more compact form:

$$\begin{cases} \dot{x} = f(x, \delta) + G(x, \delta)u = f(x, \delta) + \sum_{i=1}^{m} G_i(x, \delta)u_i(t) \\ y = h(x, \delta) \\ x(t) = \psi(t),\ u(t) = \varphi(t),\ t \in [-s\tau, 0], \end{cases} \tag{6}$$

where $f(x, \delta) = f(x(t - i\tau))$ and $h(x, \delta) = h(x(t - i\tau))$ having all entries belonging to $\mathscr{K}$, $u = u(t)$, and $G(x, \delta) = [G_1, \cdots, G_m]$ with $G_i(x, \delta) = \sum_{l=0}^{s} g_i^l \delta^l$.

With the standard differential operator d, we let $\mathcal{M}$ denote the left module over $\mathcal{K}(\delta]$:

$$\mathcal{M} = span_{\mathcal{K}(\delta]}\{d\xi,\ \xi \in \mathcal{K}\} \tag{7}$$

where $\mathcal{K}(\delta]$ acts on $d\xi$ according to (2) and (3). Note that $\mathcal{K}(\delta]$ is a non-commutative ring, but it is proved that it is a left Ore ring [24, 44], which enables us to define the rank of a left module over $\mathcal{K}(\delta]$.

Define this vector space $\mathcal{E}$ over $\mathcal{K}$:

$$\mathcal{E} = span_{\mathcal{K}}\{d\xi : \xi \in \mathcal{K}\}$$

Then $\mathcal{E}$ is the set of linear combinations of a finite number of elements from $dx_j(t-i\tau)$ with row vector coefficients in $\mathcal{K}$. Since the delay operator δ and the standard differential operator are commutative, the one-form of $\omega \in \mathcal{M}$ can be written as $\omega = \sum_{j=1}^{n} a_j(\delta]dx_j$, where $a(\delta] \in \mathcal{K}(\delta]$. For a given vector field

$$\beta = \sum_{j=1}^{n} b_j(\delta]\frac{\partial}{\partial x_j}$$

with $b_j(\delta] \in \mathcal{K}(\delta]$, the inner product of ω and β is defined as follows:

$$\omega\beta = \sum_{j=1}^{n} a_j(\delta]b_j(\delta] \in \mathcal{K}(\delta].$$

3 Notation and Preliminary Result

Some efforts have been made to extend the Lie derivative [23] to nonlinear time-delay systems in the framework of commutative rings [11, 19–21, 32, 33]. In what follows, we define the derivative and Lie derivative for nonlinear time-delay systems from the non-commutative point of view.

For any j and s such that $0 \leq j \leq s$, let $f(x(t-j\tau))$ and $h(x(t-j\tau))$ be an n and p dimensional vector, respectively, with entries $f_r \in \mathcal{K}$ for $1 \leq r \leq n$ and $h_i \in \mathcal{K}$ for $1 \leq i \leq p$. Let

$$\frac{\partial h_i}{\partial x} = \left[\frac{\partial h_i}{\partial x_1}, \cdots, \frac{\partial h_i}{\partial x_n}\right] \in \mathcal{K}^{1\times n}(\delta], \tag{8}$$

where, for $1 \leq r \leq n$, we have

$$\frac{\partial h_i}{\partial x_r} = \sum_{j=0}^{s} \frac{\partial h_i}{\partial x_r(t-j\tau)}\delta^j \in \mathcal{K}(\delta].$$

Then, the Lie derivative for nonlinear systems without delays can be extended to nonlinear time-delay systems in the framework of [44] as follows:

$$L_f h_i = \frac{\partial h_i}{\partial x}(f) = \sum_{r=1}^{n}\sum_{j=0}^{s} \frac{\partial h_i}{\partial x_r(t-j\tau)} \delta^j (f_r) \tag{9}$$

and in the same way one can define $L_{G_i} h_i$. Based on the above notation, the relative degree can be defined in the following way:

Definition 1 (*Relative Degree*) System (6) has relative degree $(\nu_1, \cdots, \nu_p)$ in an open set $W \subseteq R^n$ provided the following conditions are satisfied for $1 \le i \le p$:

1. for all $x \in W$, we have $L_{G_j} L_f^r h_i(x) = 0$ for all $1 \le j \le m$ and $0 \le r < \nu_i - 1$;
2. there exists $x \in W$ such that there is a $j \in \{1, \cdots, m\}$ satisfying $L_{G_j} L_f^{\nu_i - 1} h_i(x) \neq 0$.

If the first condition is satisfied for all $r \ge 0$ and some $i \in \{1, \cdots, p\}$, then we set $\nu_i = \infty$.

Moreover, for the system (6), one can also define observability indices introduced in [27] over non-commutative rings. For any k and n such that $1 \le k \le n$, let $\mathscr{F}_k$ be the following left module over $\mathscr{K}(\delta]$:

$$\mathscr{F}_k := span_{\mathscr{K}(\delta]} \left\{ dh, dL_f h, \cdots, dL_f^{k-1} h \right\}.$$

It was shown that the filtration of the $\mathscr{K}(\delta]$-module satisfies $\mathscr{F}_1 \subseteq \mathscr{F}_2 \subseteq \cdots \subseteq \mathscr{F}_n$. We then define

$$d_1 = rank_{\mathscr{K}(\delta]} \mathscr{F}_1 \text{ and } d_k = rank_{\mathscr{K}(\delta]} \mathscr{F}_k - rank_{\mathscr{K}(\delta]} \mathscr{F}_{k-1}$$

when $2 \le k \le n$. Let $k_i = card\{d_k \ge i, 1 \le k \le n\}$. Then $(k_1, \cdots, k_p)$ are the observability indices. Reorder, if necessary, the output components of (6) so that

$$rank_{\mathscr{K}(\delta]} \{ \frac{\partial h_1}{\partial x}, \cdots, \frac{\partial L_f^{k_1 - 1} h_1}{\partial x}, \cdots, \frac{\partial h_p}{\partial x}, \cdots, \frac{\partial L_f^{k_p - 1} h_p}{\partial x} \} = k_1 + \cdots + k_p.$$

Based on the above definitions, let us define the following notation, which will be used in the sequel. For all i and p such that $1 \le i \le p$, we let k_i denote the observability indices, ν_i the relative degree for y_i of (6), and $\rho_i = \min\{\nu_i, k_i\}$.

Without loss of generality, suppose

$$\sum_{i=1}^{p} \rho_i = j,$$

so

$$\{dh_1, \cdots, dL_f^{\rho_1-1}h_1, \cdots, dh_p, \cdots, dL_f^{\rho_p-1}h_p\}$$

are j linearly independent vectors over $\mathscr{K}(\delta]$. Then note that

$$\Phi = \{dh_1, \cdots, dL_f^{\rho_1-1}h_1, \cdots, dh_p, \cdots, dL_f^{\rho_p-1}h_p\}, \tag{10}$$

and that

$$\pounds = span_{R[\delta]}\left\{h_1, \cdots, L_f^{\rho_1-1}h_1, \cdots, h_p, \cdots, L_f^{\rho_p-1}h_p\right\}, \tag{11}$$

where $R[\delta]$ is the commutative ring of polynomials in δ with coefficients belonging to the field R, and let $\pounds(\delta]$ be the set of all polynomials in δ with coefficients over £. The module spanned by element of Φ over $\pounds(\delta]$ is defined as follows:

$$\Omega = span_{\pounds(\delta]}\{\xi, \xi \in \Phi\}. \tag{12}$$

Define

$$\mathscr{G} = span_{R[\delta]}\{G_1, \ldots, G_m\},$$

where G_i is given in (6), and its *left* annihilator:

$$\mathscr{G}^{\perp} = span_{\pounds(\delta]}\{\omega \in \mathscr{M} \mid \omega\beta = 0, \forall\beta \in \mathscr{G}\}, \tag{13}$$

where $\mathscr{M}$ is defined in (7).

After having defined the relative degree and observability indices using the extended Lie derivative for nonlinear time-delay systems in the framework of non-commutative rings, an observable canonical form can now be derived.

Theorem 1 *Consider the system* (6) *with outputs* $(y_1, \cdots, y_p)$ *and the corresponding* $(\rho_1, \cdots, \rho_p)$ *with* $\rho_i = \min\{k_i, \nu_i\}$ *where* k_i *and* ν_i *are the observability indices and the relative degree indices, respectively. There exists a change of coordinates* $\varphi(x, \delta) \in \mathscr{K}^{n\times 1}$, *such that* (6) *is transformed into the following form:*

$$\dot{z}_{i,j} = z_{i,j+1} \tag{14}$$

$$\dot{z}_{i,\rho_i} = V_i(x, \delta) = L_f^{\rho_i}h_i(x, \delta) + \sum_{j=1}^{m} L_{G_j}L_f^{\rho_i-1}h_i(x, \delta)u_j \tag{15}$$

$$y_i = C_i z_i = z_{i,1} \tag{16}$$

$$\dot{\xi} = \alpha(z, \xi, \delta) + \beta(z, \xi, \delta)u \tag{17}$$

where

$$z_i = \left(z_{i,1}, \cdots, z_{i,\rho_i}\right)^T = \left(h_i, \cdots, L_f^{\rho_i-1}h_i\right)^T \in \mathscr{K}^{\rho_i\times 1},$$

$\alpha \in \mathcal{K}^{\mu\times 1}$, and $\beta \in \mathcal{K}^{\mu\times 1}(\delta]$, where $C_i = (1, 0, \cdots, 0) \in R^{1\times\rho_i}$ and

$$\mu = n - \sum_{j=1}^{p} \rho_j.$$

Moreover, if $k_i < \nu_i$, then one has $V_i(x, \delta) = L_f^{\rho_i} h_i = L_f^{k_i} h_i$. □

Proof See [48].

Based on Theorem 1, and noting that $\rho_i = \min\{\nu_i, k_i\}$ for $1 \le i \le p$ where the k_i's represent the observability indices and ν_i stands for the relative degree of y_i for (6), the following equality can be derived:

$$\mathcal{H}(x, \delta) = \Psi(x, \delta) + \Gamma(x, \delta)u, \tag{18}$$

where

$$\mathcal{H}(x, \delta) = \left(h_1^{(\rho_1)}, \cdots, h_p^{(\rho_p)}\right)^T, \quad \Psi(x, \delta) = \left(L_f^{\rho_1} h_1, \cdots, L_f^{\rho_p} h_p\right)^T,$$

and

$$\Gamma(x, \delta) = \begin{pmatrix} L_{G_1} L_f^{\rho_1 - 1} h_1 & \cdots & L_{G_m} L_f^{\rho_1 - 1} h_1 \\ \vdots & \ddots & \vdots \\ L_{G_1} L_f^{\rho_p - 1} h_p & \cdots & L_{G_m} L_f^{\rho_p - 1} h_p \end{pmatrix}, \tag{19}$$

and where $\mathcal{H}(x, \delta) \in \mathcal{K}^{p\times 1}$, $\Psi(x, \delta) \in \mathcal{K}^{p\times 1}$ and $\Gamma(x, \delta) \in \mathcal{K}^{p\times m}(\delta]$. Assume that

$$rank_{\mathcal{K}(\delta]} \Gamma = m. \tag{20}$$

Since $\Gamma \in \mathcal{K}^{p\times m}(\delta]$ with $m \le p$, it follows from Lemma 4 in [29] that there exists a matrix $\Xi \in \mathcal{K}^{p\times p}(\delta]$ such that:

$$\Xi \Gamma = \left[\bar{\Gamma}^T, \mathbf{0}\right]^T, \tag{21}$$

where $\bar{\Gamma} \in \mathcal{K}^{m\times m}(\delta]$ has full rank m. With the compact equation (18), identifiability and observability will be analyzed separately in Sects. 4 and 5.

4 Identifiability

To study the delay identifiability of (6), we first introduce the following definition of identifiability of time delay, which is an adaptation of Definition 2 in [1].

Definition 2 For system (6), an equation with delays containing only the output and a finite number of its derivatives, the equation

$$\alpha(h, \dot{h} \ldots, h^{(k)}, \delta) = 0, \;\; k \in Z^+ \tag{22}$$

is said to be an *output delay equation* (of order k). Moreover, this equation is said to be an *output delay-identifiable equation* for (6) if it cannot be written as $\alpha(h, \dot{h} \ldots, h^{(k)}, \delta) = a(\delta]\tilde{\alpha}(h, \dot{h} \ldots, h^{(k)})$ with $a(\delta] \in \mathscr{K}(\delta]$.

In [1], the equation (22) is stated to involve the delay in an essential way. As stated in [1], if there exists an output delay-identifiable equation for (6) (i.e., involving the delay in an essential way), then the delay can be identified for almost all y by numerically finding zeros of such an equation. By almost all, we mean that singularity of the delay identification exists for a countable set of y, and this case is excluded in this chapter. For this issue, the interested reader can refer to [3] and the references therein. Thus, delay identification for (6) reduces to research on such an output delay equation.

4.1 Dependent Outputs Over $\mathscr{K}(\delta]$

Let us first consider the most simple case for identifying the delay for (6), i.e., from only the outputs of (6), which is stated in the following result.

Theorem 2 *There exists an output delay-identifiable equation (of order* 0*)* $\alpha(h, \delta)$ *for* (6) *if and only if*

$$rank_{\mathscr{K}(\delta]} \frac{\partial h}{\partial x} < rank_{\mathscr{K}} \frac{\partial h}{\partial x}. \tag{23}$$

holds. □

Proof See [48].

Example 1 Consider the following dynamical system:

$$\begin{cases} \dot{x} = f(x, u, \delta), \\ y_1 = x_1, \\ y_2 = x_1 \delta x_1 + x_1^2. \end{cases} \tag{24}$$

It can be seen that

$$\frac{\partial h}{\partial x} = \begin{pmatrix} 1, 0 \\ \delta x_1 + 2x_1 + x_1 \delta, 0 \end{pmatrix}$$

which yields $\text{rank}_{\mathscr{K}(\delta]} \frac{\partial h}{\partial x} = 1$ and $\text{rank}_{\mathscr{K}} \frac{\partial h}{\partial x} = 2$. Thus, Theorem 2 applies, and the delay of system (24) can be identified.

In fact, a straightforward calculation gives

$$y_2 = y_1 \delta y_1 + y_1^2,$$

which makes it possible to identify the delay δ by applying an algorithm to detect zero-crossing when varying δ. □

Inequality (23) implies that the outputs of (6) are dependent over $\mathscr{K}(\delta]$. Theorem 2 can be seen as a special case of Theorem 2 in [1]. However, as will be shown in the next section, this condition is not necessary for the case where the output of (6) is independent over $\mathscr{K}(\delta]$.

4.2 Independent Outputs Over $\mathscr{K}(\delta]$

Theorem 2 analyzed the case where the outputs of (6) are dependent over $\mathscr{K}(\delta]$. In the contrary case (i.e., independence over $\mathscr{K}(\delta]$), the dynamics of system (6) must be involved in order to deduce some output delay equations, which might be used to identify the delay. In the following, we first give a sufficient condition for the existence of a delay output equation for system (6) when the output is independent over $\mathscr{K}(\delta]$. Then a necessary and sufficient condition will be provided. Based on the deduction of (18), we can state the following result:

Theorem 3 *There exists an output delay equation for* (6), *if there exists a non zero*

$$\omega = \sum_{c=1}^{n}\sum_{j=1}^{p} q_j \frac{\partial L_f^{\rho_j-1} h_j}{\partial x_c} dx_c$$

with $q_j \in \mathscr{K}(\delta]$ *for* $1 \leq j \leq p$, *such that* $\omega \in \mathscr{G}^{\perp} \cap \Omega$ *and* $\omega f \in £$, *where* $\mathscr{G}^{\perp}$ *is defined in* (13), Ω *in* (12), *and* £ *in* (11). □

Proof We use $Q = [q_1, \cdots, q_p]$ which is a $1 \times p$ vector with $q_j \in \mathscr{K}(\delta]$ for $1 \leq j \leq p$. Because of the associativity law over $\mathscr{K}(\delta]$, one has:

$$Q\Gamma = Q \begin{pmatrix} L_{G_1} L_f^{\rho_1-1} h_1 & \cdots & L_{G_m} L_f^{\rho_1-1} h_1 \\ \vdots & \ddots & \vdots \\ L_{G_1} L_f^{\rho_p-1} h_p & \cdots & L_{G_m} L_f^{\rho_p-1} h_p \end{pmatrix} = Q \begin{bmatrix} \frac{\partial L_f^{\rho_1-1} h_1}{\partial x} \\ \vdots \\ \frac{\partial L_f^{\rho_p-1} h_p}{\partial x} \end{bmatrix} [G_1, \cdots, G_m]$$

Then, according to the definition in (8), one gets:

$$\begin{aligned} Q\Gamma &= \omega\,[G_1, \cdots, G_m] = \omega G, \quad \text{where} \\ \omega &= \textstyle\sum_{c=1}^{n} \sum_{j=1}^{p} q_j \frac{\partial L_f^{\rho_j-1} h_j}{\partial x_c} dx_c. \end{aligned} \tag{25}$$

Moreover, one can check that:

$$\omega f = \left(Q \begin{bmatrix} \frac{\partial L_f^{\rho_1 - 1} h_1}{\partial x} \\ \vdots \\ \frac{\partial L_f^{\rho_p - 1} h_p}{\partial x} \end{bmatrix} \right) f = Q \left(\begin{bmatrix} \frac{\partial L_f^{\rho_1 - 1} h_1}{\partial x} \\ \vdots \\ \frac{\partial L_f^{\rho_p - 1} h_p}{\partial x} \end{bmatrix} f \right) = Q \begin{bmatrix} L_f^{\rho_1} h_1 \\ \vdots \\ L_f^{\rho_p} h_p \end{bmatrix} = Q\Psi.$$

According to (18), one has:

$$Q\mathscr{H} = Q(\Psi + \Gamma u) = \omega f + \omega G u, \tag{26}$$

where

$$\mathscr{H} = \left[y_1^{(\rho_1)}, \cdots, y_p^{(\rho_p)} \right]^T.$$

Thus, if $\omega \in \mathscr{G}^{\perp} \cap \Omega$ and $\omega f \in$ £, which implies there exists Q with entries belonging to £$[\delta]$, one has $Q\Gamma = \omega G = 0$ and $Q\mathscr{H} = \omega f \in$ £. Finally, one obtains the following relation:

$$Q(\mathscr{H} - \Psi) = 0, \tag{27}$$

which is exactly the output delay equation, since it contains only the output, its derivatives and delays. □

If, in addition, the output delay equation (27) we deduced is an output delay-identifiable equation, i.e., containing the delay δ in an essential way, then the delay of (6) can be identified (at least locally) by detecting a zero-crossing of (27). The following will give necessary and sufficient conditions guaranteeing the essential involvement of δ in (27). However, before this, let us define:

$$\mathscr{Y} = \left(h_1, \ldots, L_f^{\rho_1 - 1} h_1, \ldots, h_p, \ldots, L_f^{\rho_p - 1} h_p \right)^T,$$

and let $\mathscr{K}_0 \subseteq \mathscr{K}$ denote the field of meromorphic functions of x, which will be used in the following theorem, which also involves the Ψ defined in (18):

Theorem 4 *The output delay equation* (27) *is an output delay-identifiable equation if and only if one of the following is true: Either*

$$rank_{\mathscr{K}(\delta]} \frac{\partial \mathscr{Y}}{\partial x} < rank_{\mathscr{K}} \frac{\partial \{\mathscr{Y}, \Psi\}}{\partial x}, \tag{28}$$

or else for any element q_j of $Q \in \mathscr{K}^{1\times p}(\delta]$, there do not exist a $a(\delta] \in \mathscr{K}(\delta]$ and $\bar{q}_j \in \mathscr{K}_0$ such that

$$q_j = a(\delta]\bar{q}_j \text{ for } 1 \le j \le p \tag{29}$$

and

$$rank_{\mathscr{K}(\delta]} \frac{\partial \mathscr{Y}}{\partial x} = rank_{\mathscr{K}} \frac{\partial \{\mathscr{Y}, \Psi\}}{\partial x}. \tag{30}$$

holds. □

Proof See [48].

Remark 1 It is clear that Theorem 2 is a special case of Theorem 4, since the output delay-identifiable equation stated in Theorem 2 does not contain any derivative of the output. In [1], a condition similar to (28) of Theorem 4 is stated as a necessary and sufficient condition for delay identification for nonlinear systems with *known* inputs. However, as we proved above, in the case of *unknown* inputs, this condition is sufficient, but not necessary.

5 Observability

Similarly to the observability definitions given in [22] and [14] for nonlinear delay-free systems, the work [28] gives a definition of observability for nonlinear time-delay systems. The following gives a more generic definition of observability in the case of systems with unknown inputs:

Definition 3 System (6) is locally observable if the state $x(t)$ can be expressed as a function of the output and a finite number of its time derivatives with their backward and forward shifts. A locally observable system is locally causally observable if its state can be written as a function of only the output and its derivatives with their backward shifts. Otherwise, it is locally non-causally observable (and it also depends on the forward shifts).

In the same way, the following definition is given of systems with unknown inputs:

Definition 4 The unknown input $u(t)$ can be locally estimated if it can be written as a function of the output and a finite number of its time derivatives with backward and forward shifts. The input can be locally causally estimated if $u(t)$ can be expressed as a function of only the output and its time derivatives with backward shifts. Otherwise, it can be non causally estimated (and it also depends on the forward shifts).

Theorem 5 *Consider the system* (6) *with outputs* $(y_1, \cdots, y_p)$ *and their corresponding* $(\rho_1, \cdots, \rho_p)$ *with each* $\rho_i = \min\{k_i, \nu_i\}$ *where* k_i *and* ν_i *are the observability indices and the relative degree indices, respectively. Consider* Φ *and* $\bar{\Gamma}$ *defined in* (10) *and* (21)*, respectively. If* $rank_{\mathscr{K}(\delta]}\Phi = n$,, *then there exists a change of coordinates* $\varphi(x, \delta)$ *such that* (6) *can be transformed into* (14–17) *with dim* $\xi = 0$. *Moreover, if the change of coordinates is locally bicausal over* $\mathscr{K}$*, then the state* $x(t)$ *of* (6) *is locally causally observable; if, in addition,* $\bar{\Gamma} \in \mathscr{K}^{m\times m}(\delta]$ *is unimodular over* $\mathscr{K}(\delta]$*, then the unknown input* $u(t)$ *of* (6) *can be locally causally estimated.* □

Proof According to Theorem 1, the system (6) can be transformed into (14–17) by using the change of coordinates $(z, \xi) = \varphi(x, \delta)$. Hence, if $\text{rank}_{\mathcal{K}(\delta]}\Phi = n$, then one has

$$\sum_{j=1}^{p} \rho_j = n,$$

which implies that (6) can be transformed into (14–17) with dim $\xi = 0$ and the change of coordinates is given by $z = \varphi(x, \delta)$, where

$$z = (z_i^T, \cdots, z_p^T)^T \text{ and } z_i = (h_i, \cdots, L_f^{\rho_i - 1} h_i)^T.$$

Moreover, if $\varphi(x, \delta) \in \mathcal{K}^{n\times 1}$ is locally bicausal over $\mathcal{K}$, then one can write x as a function of y_i, its derivative, and backward shifts, which implies that the state x is locally causally observable.

Concerning the reconstruction of the unknown inputs, rewrite (18) as follows:

$$\Gamma u = H(x, \delta) - \Psi(x, \delta) = \Upsilon(x, \delta). \tag{31}$$

Since $\text{rank}_{\mathcal{K}(\delta]}\Phi = n$ and x is causally observable, we know that $\Upsilon(x, \delta)$ is a vector of known meromorphic functions belonging to $\mathcal{K}$. If $\bar{\Gamma} \in \mathcal{K}^{m\times m}(\delta]$ is unimodular over $\mathcal{K}(\delta]$, then there exists a matrix $\bar{\Gamma}^{-1} \in \mathcal{K}^{m\times m}(\delta]$ such that

$$\left[\bar{\Gamma}^{-1}\ \mathbf{0}\right] \Xi \Gamma = I_{m\times m} \text{ and } u = \left[\bar{\Gamma}^{-1}\ \mathbf{0}\right] \Xi \Upsilon.$$

Since $\bar{\Gamma}^{-1} \in \mathcal{K}^{m\times m}(\delta]$, $\Xi \in \mathcal{K}^{p\times p}$ and $\Upsilon \in \mathcal{K}^{p\times 1}$, it follows that u is also causally observable. □

For the case where the condition $\text{rank}_{\mathcal{K}(\delta]}\Phi = n$ in Theorem 5 is not satisfied, a constructive algorithm was proposed in [2] to solve this problem for nonlinear systems without delays. In the following, we extend this idea to treat the observation problem for time-delay systems with unknown inputs. The objective is to generate additional variables from the available measurement that are unaffected by the unknown input, such that an extended canonical form similar to (14)–(15) can be obtained for the estimation of the remaining state ξ.

Theorem 6 *Consider the system* (6) *with outputs* $y = (y_1, \cdots, y_p)^T$ *and the corresponding* $(\rho_1, \cdots, \rho_p)$ *with* $\rho_i = \min\{k_i, v_i\}$ *where* k_i *and* v_i *are the observability indices and the relative degree indices, respectively. Suppose* $\text{rank}_{\mathcal{K}(\delta]}\Phi < n$ *where* Φ *is defined in* (10)*. Then there exist* l *new independent outputs over* $\mathcal{K}$ *suitable to the causal estimation problem if and only if* $\text{rank}_{\mathcal{K}}\mathcal{L} = l$ *where*

$$\mathcal{L} = span_{R[\delta]}\{\omega \in \mathcal{G}^{\perp} \cap \Omega \mid \omega f \notin £\} \tag{32}$$

with f defined in (6), *£ defined in* (11), *Ω defined in* (12), *and $\mathscr{G}^{\perp}$ defined in* (13). *Moreover, the l additional outputs, denoted as $\bar{y}_i$, $1 \le i \le l$, are given by:*

$$\bar{y}_i = \omega_i f \ mod\, £$$

where $\omega_i \in \mathscr{L}$. □

Proof See [49].

Remark 2 Theorem 6 gives a constructive way to treat the case where $\text{rank}_{\mathscr{K}(\delta]}\Phi < n$. Once additional new outputs are deduced according to Theorem 6, it enables us to define a new Φ. If $\text{rank}_{\mathscr{K}(\delta]}\Phi = n$, then Theorem 5 can be applied. Otherwise, if $\text{rank}_{\mathscr{K}(\delta]}\Phi < n$ and if Theorem 6 is still valid, then one can still deduce new outputs for the studied system. Thus a "Check-Extend" procedure is iterated until $\text{rank}_{\mathscr{K}(\delta]}\Phi = n$ is obtained.

5.1 Non-causal Observability

The previous results can be extended to the case of non-causal observations of the state and the unknown inputs, which can be very useful in some applications. For instance, some proposed delay feedback control methods can be applied for stabilizing nonlinear time-delay systems [39]. Furthermore, other applications, such as cryptography based on chaotic systems, do not require real-time estimation, so non-causal observations can still play an important role in those applications.

To treat the non-causal case, we use the forward time-shift operator $\nabla f(t) = f(t+\tau)$, similarly to the backward time-shift operator δ from Sect. 2, and for all i and j in Z^+, we set

$$\nabla^i \delta^j f(t) = \delta^j \nabla^i f(t) = f(t-(j-i)\tau). \tag{33}$$

Following the principle of Sect. 2, let $\bar{\mathscr{K}}$ denote the field of all meromorphic functions of a finite number of variables from $\{x_j(t-i\tau), j \in [1,n], i \in S\}$ where $S = \{-s, \cdots, 0, \cdots, s\}$ is a finite set of relative integers. One has $\mathscr{K} \subseteq \bar{\mathscr{K}}$. Let $\bar{\mathscr{K}}(\delta, \nabla]$ denote the set of polynomials of the form:

$$a(\delta, \nabla] = \bar{a}_{r_{\bar{a}}}\nabla^{r_{\bar{a}}} + \cdots + \bar{a}_1\nabla + a_0(t) + a_1(t)\delta + \cdots + a_{r_a}(t)\delta^{r_a}, \tag{34}$$

with $a_i(t)$ and $\bar{a}_i(t)$ belonging to $\bar{\mathscr{K}}$. We maintain the usual definition of addition for $\bar{\mathscr{K}}(\delta, \nabla]$ and define the multiplication as follows:

$$\begin{aligned} a(\delta,\nabla]b(\delta,\nabla] = &\sum_{i=0}^{r_a}\sum_{j=0}^{r_b} a_i\delta^i b_j \delta^{i+j} + \sum_{i=0}^{r_a}\sum_{j=1}^{r_{\bar{b}}} a_i\delta^i \bar{b}_j \delta^i \nabla^j \\ &+ \sum_{i=1}^{r_{\bar{a}}}\sum_{j=0}^{r_b} \bar{a}_i\nabla^i b_j \nabla^i \delta^j + \sum_{i=1}^{r_{\bar{a}}}\sum_{j=1}^{r_{\bar{b}}} \bar{a}_i\nabla^i \bar{b}_j \nabla^{i+j}. \end{aligned} \tag{35}$$

It is clear that $\mathscr{K}(\delta] \subseteq \bar{\mathscr{K}}(\delta, \nabla]$ and that the ring $\bar{\mathscr{K}}(\delta, \nabla]$ possesses the same properties as $\mathscr{K}(\delta]$. Thus, a module

$$\bar{\mathscr{M}} = span_{\bar{\mathscr{K}}(\delta,\nabla]}\{d\xi, \xi \in \bar{\mathscr{K}}\}$$

can also be defined over $\bar{\mathscr{K}}(\delta, \nabla]$. Given the above definitions, Theorem 5 is now extended so to deal with non-causal observability for nonlinear time-delay systems, as follows:

Theorem 7 *Consider the system* (6) *with outputs* $(y_1, \cdots, y_p)$ *and the corresponding* $(\rho_1, \cdots, \rho_p)$ *with* $\rho_i = \min\{k_i, \nu_i\}$ *where* k_i *and* ν_i *are the observability indices and the relative degree indices, respectively. If* $\text{rank}_{\mathscr{K}(\delta]}\Phi = n$, *where* Φ *is defined in* (10), *then there exists a change of coordinates* $\varphi(x, \delta)$ *such that* (6) *can be transformed into* (14–17) *with* $\dim \xi = 0$.

Moreover, if the change of coordinates is locally bicausal over $\bar{\mathscr{K}}$, *then the state* $x(t)$ *of* (6) *is at least locally non-causally observable; if, in addition,* $\bar{\Gamma} \in \mathscr{K}^{m\times m}(\delta]$ *is unimodular over* $\bar{\mathscr{K}}(\delta, \nabla]$, *then the unknown input* $u(t)$ *of* (6) *can be at least locally non-causally estimated.* □

Proof See [49].

6 Illustrative Example

The following example highlights the proposed results in the case of delay identification and causal observability. Consider

$$\begin{cases} \dot{x}_1 = -\delta x_1^2 + \delta x_4 u_1, \quad \dot{x}_2 = -x_1^2 \delta x_3 + x_2 + x_1 \delta x_4 u_1, \\ \dot{x}_3 = x_4 - x_1^2 \delta x_4 u_1, \quad \dot{x}_4 = x_5 + \delta x_1, \quad \dot{x}_5 = \delta x_1 \delta x_3 + u_2, \\ y_1 = x_1, \quad y_2 = x_2, \quad y_3 = x_1 \delta x_1 + x_3. \end{cases} \tag{36}$$

One can check that $\nu_1 = k_1 = \nu_2 = k_2 = 1$, $\nu_3 = 1$, $k_3 = 3$, yielding $\rho_1 = \rho_2 = \rho_3 = 1$ and $\Phi = \{dx_1, dx_2, (\delta x_1 + x_1 \delta)dx_1 + dx_3\}$. One has $\text{rank}_{\mathscr{K}(\delta]}\Phi = 3 < n$.

Set $\mathscr{G} = span_{R[\delta]}\{G_1, \cdots, G_m\}$. Then

$$\mathscr{G}^\perp = span_{R[\delta]}\left\{x_1 dx_1 - dx_2, x_1^2 dx_1 + dx_3, dx_4\right\}.$$

Since $\text{rank}_{\mathscr{K}(\delta]}\Phi = 3$, we have $\pounds = span_{R[\delta]}\{x_1, x_2, x_1\delta x_1 + x_3\}$ and

$$\Omega = span_{\pounds(\delta]}\{dx_1, dx_2, dx_3\},$$

which yields:

$$\Omega \cap \mathscr{G}^\perp = span_{\pounds(\delta]}\left\{x_1 dx_1 - dx_2, x_1^2 dx_1 + dx_3\right\}.$$

In the following, identifiability and observability will be successively checked for (36).

Identifiability Analysis:

Following Theorem 1, one has

$$\mathscr{H} = \left[\dot{y}_1, \dot{y}_2, \dot{y}_3\right]^T, \quad \Psi = \left[-\delta x_1^2, -x_1^2\delta x_3 + x_2, x_4\right]^T,$$

and

$$\Gamma = \begin{bmatrix} \delta x_4, & 0 \\ x_1\delta x_4, & 0 \\ -x_1^2\delta x_4, & 0 \end{bmatrix}.$$

Thus, by choosing $Q = [x_1, -1, 0]$, a non zero one-form can be found, such as:

$$\omega = x_1 dx_1 - dx_2 \in \Omega \cap \mathscr{G}^\perp,$$

satisfying

$$\omega f = -x_1\delta x_1^2 + x_1^2\delta x_3 - x_2 \in \pounds.$$

According to Theorem 3, the following equation is an output delay equation:

$$Q\left(\mathscr{H} - \Psi\right) = 0, \tag{37}$$

since it contains only the output, its derivatives and delays.

Since $\mathscr{Y} = (x_1, x_2, x_1\delta x_1 + x_3)^T$, one has:

$$\frac{\partial \mathscr{Y}}{\partial x} = \begin{pmatrix} 1, & 0,\ 0,\ 0,\ 0 \\ 0, & 1,\ 0,\ 0,\ 0 \\ \delta x_1 + x_1\delta, & 0,\ 1,\ 0,\ 0 \end{pmatrix} \quad \text{and} \quad \frac{\partial \Psi}{\partial x} = \begin{pmatrix} -2\delta x_1\delta, & 0, & 0, & 0,\ 0 \\ -2x_1\delta x_3, & 1, & -x_1^2\delta, & 0,\ 0 \\ 0, & 0, & 0, & 1,\ 0 \end{pmatrix}.$$

Thus, one obtains:

$$rank_{\mathscr{K}(\delta]}\frac{\partial \mathscr{Y}}{\partial x} = 3 < rank_{\mathscr{K}}\frac{\partial\{\mathscr{Y}, \Psi\}}{\partial x} = 6.$$

Theorem 4 applies, and (37) involves δ in an essential way. A straightforward calculation gives $y_1\dot{y}_1 - \dot{y}_2 = -y_1\delta y_1^2 + y_1^2\delta y_3 - y_1^2\delta y_1\delta^2 y_1 - y_2$, which allows us to identify the delay.

Observability Analysis:

From the definition of $\mathscr{L}$ in (32), one can check that $\text{rank}_{\mathscr{K}}\mathscr{L} = 1$, which gives the one-form $\omega = x_1^2 dx_1 + dx_3$, satisfying $\omega \in \Omega \cap \mathscr{G}^\perp$ and $\omega f = -x_1^2\delta x_1^2 + x_4 \notin \pounds$. Thus, according to Theorem 6, a new output $\bar{y}_1 = h_4$ is given by:

$$\bar{y}_1 = h_4 = \omega f \bmod \pounds = x_4 = y_1^2\dot{y}_1 + \dot{y}_3 + y_1^2\delta y_1^2. \tag{38}$$

For the new output $\bar{y}_1$, one has $k_i = \nu_i = 1$ for $1 \leq i \leq 3$, $k_4 = \nu_4 = 2$, so $\rho_i = 1$ for $1 \leq i \leq 3$ and $\rho_4 = 2$. Finally, one obtains the new Φ as follows:

$$\Phi = \{dx_1, dx_2, (\delta x_1 + x_1\delta)dx_1 + dx_3, dx_4, \delta dx_1 + dx_5\}.$$

It can be checked that $\text{rank}_{\mathscr{K}(\delta]}\Phi = 5 = n$, and the new £ is:

$$\pounds = span_{R[\delta]}\{x_1, x_2, x_1\delta x_1 + x_3, x_4, x_5 + \delta x_1\}.$$

This gives the following change of coordinates:

$$z = \varphi(x, \delta) = (x_1, x_2, x_1\delta x_1 + x_3, x_4, x_5 + \delta x_1)^T.$$

It is easy to check that it is bicausal over $\mathscr{K}(\delta]$, since:

$$x = \varphi^{-1} = (z_1, z_2, z_3 - z_1\delta z_1, z_4, z_5 - \delta z_1)^T.$$

When $t \geq \tau$, one gets the following estimations of the states:

$$\begin{cases} x_1 = y_1, \; x_2 = y_2, \; x_3 = y_3 - y_1\delta y_1, \\ x_4 = \bar{y}_1, \; x_5 = -\delta y_1 + \dot{\bar{y}}_1, \end{cases}$$

where $\bar{y}_1$ defined in (38).

Moreover, the matrix Γ with the new output $\bar{y}_1$ can be obtained as follows:

$$\Gamma = \begin{pmatrix} \delta x_4, & 0 \\ x_1\delta x_4, & 0 \\ x_1^2\delta x_4, & 0 \\ 0, & 1 \end{pmatrix},$$

with $rank_{\mathscr{K}(\delta]}\Gamma = 2$. One can find matrices

$$\Xi = \begin{pmatrix} 1 & 0 & 0 & 0 \\ 0 & 0 & 0 & 1 \\ x_1 & -1 & 0 & 0 \\ x_1^2 & 0 & 1 & 0 \end{pmatrix}, \quad \bar{\Gamma} = \begin{pmatrix} \delta x_4 & 0 \\ 0 & 1 \end{pmatrix}, \quad \text{and} \quad \bar{\Gamma}^{-1} = \begin{pmatrix} \frac{1}{\delta x_4} & 0 \\ 0 & 1 \end{pmatrix}$$

such that

$$\left[\bar{\Gamma}^{-1} \; \mathbf{0}\right] \Xi \Gamma = I_{2\times 2}.$$

Consequently, according to Theorem 5, u_1 and u_2 can be causally estimated. When $t \geq 3\tau$, a straightforward computation yields the following estimates for the unknown inputs:

$$\begin{cases} u_1 = \frac{\dot{y}_1 + \delta y_1^2}{\delta \bar{y}_1}, \\ u_2 = \ddot{y}_1 - \delta \dot{y}_1 - \delta y_1 \delta y_3 + \delta y_1^2 \delta^2 y_1. \end{cases}$$

7 Conclusion

This chapter studied identifiability and observability for nonlinear time-delay systems with unknown inputs. Concerning the identification of the delay, dependent and independent outputs over the non-commutative rings have been analyzed. Concerning the observability, necessary and sufficient conditions have been deduced for both causal and non-causal cases. The causal and non-causal estimations of unknown inputs of the systems we studied have been analyzed as well.

References

1. Anguelova, M., Wennberg, B.: State elimination and identifiability of the delay parameter for nonlinear time-delay systems. Automatica **44**(5), 1373–1378 (2008)
2. Barbot, J.-P., Boutat, D., Floquet, T.: An observation algorithm for nonlinear systems with unknown inputs. Automatica **45**(8), 1970–1974 (2009)
3. Barbot, J-P., Zheng, G., Floquet, T., Boutat, D., Richard, J-P.: Delay estimation algorithm for nonlinear time-delay systems with unknown inputs. In: Proceedings of the IFAC Workshop on Time Delay Systems, pp. 237–241 (2012)
4. Belkoura, L.: Identifiability of systems described by convolution equations. Automatica **41**(3), 505–512 (2005)
5. Belkoura, L., Dambrine, M., Orlov, Y., Richard, J.-P.: Identifiability and identification of linear systems with delays. In: Niculescu, S.-I., Gu, K. (eds.) Advances in Time-Delay Systems. Lectures Notes in Computational Science and Engineering, vol. 38, pp. 123–136. Springer, New York (2004)
6. Belkoura, L., Richard, J-P., Fliess, M.: On-line identification of systems with delayed inputs. In: Proceedings of the 17th International Symposium on Mathematical Theory of Networks and Systems (2006)
7. Belkoura, L., Richard, J-P., Fliess, M.: Real time identification of delay systems. In: Proceedings of the IFAC Workshop on Time Delay Systems (2007)
8. Belkoura, L., Richard, J.-P., Fliess, M.: Parameters estimation of systems with delayed and structured entries. Automatica **45**(5), 1117–1125 (2009)
9. Bhat, K., Koivo, H.: Modal characterizations of controllability and observability in time delay systems. IEEE Trans. Autom. Control **21**(2), 292–293 (1976)
10. Brewer, J., Bunce, J., Van Vleck, F.: Linear Systems Over Commutative Rings. Marcel Dekker, New York (1986)
11. Califano, C., Marquez-Martinez, L., Moog, C.: Extended lie brackets for nonlinear time-delay systems. IEEE Trans. Autom. Control **56**(9), 2213–2218 (2011)
12. Conte, G., Moog, C., Perdon, A.: Nonlinear control systems: an algebraic setting. Series Lecture Notes in Control and Information Sciences, vol. 242. Springer, London (1999)

13. Darouach, M., Zasadzinski, M., Xu, S.: Full order observers for linear systems with unknown inputs. IEEE Trans. Autom. Control **39**(3), 606–609 (1994)
14. Diop, S., Fliess, M.: Nonlinear observability, identifiability and persistent trajectories. In: Proceedings of the 36th IEEE Conference on Decision and Control, pp. 714–719 (1991)
15. Drakunov, S., Perruquetti, W., Richard, J.-P., Belkoura, L.: Delay identification in time-delay systems using variable structure observers. Ann. Rev. Control **30**(2), 143–158 (2006)
16. Erneux, T.: Applied delay differential equations. Series Surveys and Tutorials in the Applied Mathematical Sciences, vol. 3. Springer, New York (2009)
17. Fliess, M., Mounier, H.: Controllability and observability of linear delay systems: an algebraic approach. ESAIM: Control. Optimisation Calc. Var. **3**, 301–314 (1998)
18. Fliess, M., Sira-Ramirez, H.: Reconstructeurs d'etat. C. R. Acad. Sci. **338**(1), 91–96 (2004)
19. Germani, A., Manes, C., Pepe, P.: Linearization of input-output mapping for nonlinear delay systems via static state feedback. In: Proceedings of the CESA IMACS Multiconference on Computational Engineering in Systems Application, pp. 599–602 (1996)
20. Germani, A., Manes, C., Pepe, P.: An asymptotic state observer for a class of nonlinear delay systems. Kybernetika **37**(4), 459–478 (2001)
21. Germani, A., Manes, C., Pepe, P.: A new approach to state observation of nonlinear systems with delayed output. IEEE Trans. Autom. Control **47**(1), 96–101 (2002)
22. Hermann, R., Krener, A.: Nonlinear controllability and observability. IEEE Trans. Autom. Control **22**(5), 728–740 (1977)
23. Isidori, A.: Nonlinear Control Systems, 3rd edn. Springer, London (1995)
24. Jezek, J.: Rings of skew polynomials in algebraical approach to control theory. Kybernetika **32**(1), 63–80 (1996)
25. Koenig, D., Bedjaoui, N., Litrico, X.: Unknown input observers design for time-delay systems application to an open-channel. In: Proceedings of the IEEE Conference on Decision and Control, pp. 5794–5799 (2005)
26. Kolmanovskii, V., Myshkis, A.: Introduction to the Theory and Application of Functional Differential Equations. Kluwer Academic Publishers, Dordrecht (1999)
27. Krener, A.: $(Ad_{f,g})$, $(ad_{f,g})$ and locally $(ad_{f,g})$ invariant and controllability distributions. SIAM J. Control Optim. **23**(4), 523–549 (1985)
28. Marquez-Martinez, L., Moog, C., Martin, V.: Observability and observers for nonlinear systems with time delays. Kybernetika **38**(4), 445–456 (2002)
29. Marquez-Martinez, L., Moog, C.: New insights on the analysis of nonlinear time-delay systems: Application to the triangular equivalence. Syst. Control Lett. **56**(2), 133–140 (2007)
30. Moog, C., Castro-Linares, R., Velasco-Villa, M., Marque-Martinez, L.: The disturbance decoupling problem for time-delay nonlinear systems. IEEE Trans. Autom. Control **45**(2), 305–309 (2000)
31. Niculescu, S-I.: Delay effects on stability: a robust control approach. Lecture Notes in Control and Information Sciences, vol. 269. Springer, New York (2001)
32. Oguchi, T., Richard, J.-P.: Sliding-mode control of retarded nonlinear systems via finite spectrum assignment approach. IEEE Trans. Autom. Control **51**(9), 1527–1531 (2006)
33. Oguchi, T., Watanabe, A., Nakamizo, T.: Input-output linearization of retarded non-linear systems by using an extension of Lie derivative. Int. J. Control **75**(8), 582–590 (2002)
34. Orlov, Y., Belkoura, L., Dambrine, M., Richard, J.-P.: On identifiability of linear time-delay systems. IEEE Trans. Autom. Control **47**(8), 1319–1324 (2002)
35. Orlov, Y., Belkoura, L., Richard, J.-P., Dambrine, M.: Adaptive identification of linear time-delay systems. Int. J. Robust Nonlinear Control **13**(9), 857–872 (2003)
36. Ren, X., Rad, A., Chan, P., Lo, W.: Online identification of continuous-time systems with unknown time delay. IEEE Trans. Autom. Control **50**(9), 1418–1422 (2005)
37. Richard, J.-P.: Time-delay systems: an overview of some recent advances and open problems. Automatica **39**(10), 1667–1694 (2003)
38. Sename, O.: Unknown input robust observers for time-delay systems. In Proceedings of the IEEE Conference on Decision and Control, pp. 1629–1630 (1994)

39. Sename, O.: Is a mixed design of observer-controllers for time-delay systems interesting? Asian J. Control **9**(2), 180–189 (2007)
40. Sename, O., Briat, C.: New trends in design of observers for time-delay systems. Kybernetica **37**(4), 427–458 (2001)
41. Sontag, E.: Linear systems over commutative rings: a survey. Ricerche di Automatica **7**(1), 1–34 (1976)
42. Sontag, E.: A concept of local observability. Syst. Control Lett. **5**(1), 41–47 (1984)
43. Tami, R., Boutat, D., Zheng, G.: Extended output depending normal form. Automatica **49**(7), 2192–2198 (2013)
44. Xia, X., Marquez, L., Zagalak, P., Moog, C.: Analysis of nonlinear time-delay systems using modules over non-commutative rings. Automatica **38**(9), 1549–1555 (2002)
45. Yadaiah, N., Deekshatulu, B.L., Sivakumar, L., Sree Hari Rao, V.: Neural network algorithm for parameter identification of dynamical systems involving time delays. Appl. Soft Comput. **7**(3), 1084–1091 (2007)
46. Yang, F.: Observers for linear systems with unknown inputs. IEEE Trans. Autom. Control **33**(7), 677–681 (1988)
47. Zhang, J., Xia, X., Moog, C.: Parameter identifiability of nonlinear systems with time-delay. IEEE Trans. Autom. Control **51**(2), 371–375 (2006)
48. Zheng, G., Barbot, J.-P., Boutat, D.: Identification of the delay parameter for nonlinear time-delay systems with unknown inputs. Automatica **49**(6), 1755–1760 (2013)
49. Zheng, G., Barbot, J.-P., Floquet, T., Boutat, D., Richard, J.-P.: On observability of nonlinear time-delay systems with unknown inputs. IEEE Trans. Autom. Control **56**(8), 1973–1978 (2011)

Erratum to: Recent Results on Nonlinear Delay Control Systems

Iasson Karafyllis, Michael Malisoff, Frederic Mazenc and Pierdomenico Pepe

Erratum to: M. Malisoff et al. (eds.), *Recent Results on Nonlinear Delay Control Systems*, Advances in Delays and Dynamics 4, DOI 10.1007/978-3-319-18072-4

The sequence of the editors name was incorrect. The correct sequence is given below:

Iasson Karafyllis, Michael Malisoff, Frederic Mazenc, Pierdomenico Pepe

The spelling of the book title was incorrect in the cover. The correct book title should read as "Recent Results on Nonlinear Delay Control Systems".

The online version of the original book can be found under
DOI 10.1007/978-3-319-18072-4

I. Karafyllis (✉)
Department of Mathematics, National Technical University of Athens, Athens, Greece
e-mail: iasonkar@central.ntua.gr

M. Malisoff
Department of Mathematics, Louisiana State University (LSU), Baton Rouge, LA, USA

F. Mazenc
EPI Inria DISCO, Laboratory of Signals and Systems (L2S, UMR CNRS 8506), CNRS-CentraleSupélec-Université Paris Sud, Gif-sur-Yvette, France

P. Pepe
Department of Information Engineering, Computer Science and Mathematics, University of L'Aquila, L'Aquila, Italy

I. Karafyllis et al. (eds.), *Recent Results on Nonlinear Delay Control Systems*, Advances in Delays and Dynamics 4, DOI 10.1007/978-3-319-18072-4_21

Index

I. Karafyllis et al. (eds.), *Recent Results on Nonlinear Delay Control Systems*,
Advances in Delays and Dynamics 4, DOI 10.1007/978-3-319-18072-4

MIX
Papier aus verantwortungsvollen Quellen
Paper from responsible sources
FSC® C105338

If you have any concerns about our products,
you can contact us on
ProductSafety@springernature.com

In case Publisher is established outside the EU,
the EU authorized representative is:
Springer Nature Customer Service Center GmbH
Europaplatz 3, 69115 Heidelberg, Germany

Printed by Libri Plureos GmbH
in Hamburg, Germany